VOLUME ONE HUNDRED AND THIRTY SIX

Current Topics in DEVELOPMENTAL BIOLOGY

Gastrulation: From Embryonic Pattern to Form

CURRENT TOPICS IN DEVELOPMENTAL BIOLOGY

"A meeting-ground for critical review and discussion of developmental processes"
A.A. Moscona and Alberto Monroy (Volume 1, 1966)

VOLUME ONE HUNDRED AND THIRTY SIX

CURRENT TOPICS IN DEVELOPMENTAL BIOLOGY

Gastrulation: From Embryonic Pattern to Form

Edited by

LILIANNA SOLNICA-KREZEL
Department of Developmental Biology,
Washington University in St. Louis, School of Medicine,
St. Louis, MO, United States

ELSEVIER

ACADEMIC PRESS
An imprint of Elsevier

Academic Press is an imprint of Elsevier
50 Hampshire Street, 5th Floor, Cambridge, MA 02139, United States
525 B Street, Suite 1650, San Diego, CA 92101, United States
The Boulevard, Langford Lane, Kidlington, Oxford OX5 1GB, United Kingdom
125 London Wall, London, EC2Y 5AS, United Kingdom

First edition 2020

ISBN: 978-0-12-812798-8
ISSN: 0070-2153

For information on all Academic Press publications
visit our website at https://www.elsevier.com/books-and-journals

Publisher: Zoe Kruze
Acquisition Editor: Zoe Kruze
Editorial Project Manager: Shellie Bryant
Production Project Manager: Denny Mansingh
Cover Designer: Alan Studholme

Typeset by SPi Global, India

Contents

Contributors

Ashley E.E. Bruce
Department of Cell & Systems Biology, University of Toronto, Toronto, ON, Canada

Anthony W. DeTomaso
Department of Molecular, Cellular and Developmental Biology, University of California, Santa Barbara, CA, United States

Anna-Katerina Hadjantonakis
Developmental Biology Program, Sloan Kettering Institute, Memorial Sloan Kettering Cancer Center, New York, NY, United States

Carl-Philipp Heisenberg
Institute of Science and Technology Austria, Klosterneuburg, Austria

Ray Keller
Department of Biology, University of Virginia, Charlottesville, VA, United States

Clint S. Ko
Department of Biology, Massachusetts Institute of Technology, Cambridge, MA, United States

Matthew J. Kourakis
Department of Molecular, Cellular and Developmental Biology, University of California, Santa Barbara, CA, United States

Florence L. Marlow
Icahn School of Medicine Mount Sinai Department of Cell, Developmental and Regenerative Biology, New York, NY, United States

Megan Martik
Division of Biology and Biological Engineering, California institute of Technology, Pasadena, CA, United States

Adam C. Martin
Department of Biology, Massachusetts Institute of Technology, Cambridge, MA, United States

David R. McClay
Department of Biology, Duke University, Durham, NC, United States

Esther Miranda
Department of Biology, Duke University, Durham, NC, United States

Matteo A. Molè
University of Cambridge, Cambridge, United Kingdom

Susan Newcomb
Division of Biology & Biological Engineering, California Institute of Technology, Pasadena, CA, United States

Sonja Nowotschin
Developmental Biology Program, Sloan Kettering Institute, Memorial Sloan Kettering Cancer Center, New York, NY, United States

Adam C. Paré
Department of Biological Sciences, University of Arkansas, Fayetteville, AR, United States

Diana Pinheiro
Institute of Science and Technology Austria, Klosterneuburg, Austria

Ana Raffaelli
Department of Cell & Developmental Biology, University College London, London, United Kingdom

Leslie Slota
Department of Biology, Duke University, Durham, NC, United States

William C. Smith
Department of Molecular, Cellular and Developmental Biology, University of California, Santa Barbara, CA, United States

Lilianna Solnica-Krezel
Department of Developmental Biology, Washington University in St. Louis, School of Medicine, St. Louis, MO, United States

Angelike Stathopoulos
Division of Biology & Biological Engineering, California Institute of Technology, Pasadena, CA, United States

Claudio D. Stern
Department of Cell & Developmental Biology, University College London, London, United Kingdom

Ann Sutherland
Department of Biology, University of Virginia, Charlottesville, VA, United States

Michael T. Veeman
Division of Biology, Kansas State University, Manhattan, KS, United States

Octavian Voiculescu
Department of Physiology, Development and Neuroscience, University of Cambridge, Cambridge, United Kingdom

Jacob Warner
Department of Biology, University of North Carolina, Wilmington, NC, United States

Antonia Weberling
University of Cambridge, Cambridge, United Kingdom

Margot L.K. Williams
Department of Developmental Biology, Washington University in St. Louis, School of Medicine, St. Louis, MO, United States

Rudolf Winklbauer
Department of Cell and Systems Biology, University of Toronto, Toronto, ON, Canada

Konner M. Winkley
Division of Biology, Kansas State University, Manhattan, KS, United States

Jennifer A. Zallen
Howard Hughes Medical Institute and Developmental Biology Program, Sloan Kettering Institute, New York, NY, United States

Magdalena Zernicka-Goetz
University of Cambridge, Cambridge, United Kingdom

Preface

How linear genetic information is translated into form and function is the key question that continues to fascinate and preoccupy biologists. During embryonic development, this question is posed in a repeated fashion: first, when a zygote starts to proliferate and then take the form of a blastula or blastocyst, then during gastrulation when germ layers are formed and shaped into a body plan, next when neural plate folds into a neural tube, and finally, during organogenesis when each organ rudiment has to specify a variety of cell types and assemble them into unique morphology and function. Naturally, as the Editor of the book titled "*Gastrulation: From embryonic pattern to form*," I find the quest for understanding how genomes instruct gastrulation the most fascinating one. After all, it is during gastrulation when embryonic rudiment, often in a form of a sheet of morphologically uniform and pluripotent cells, becomes diversified into germ layers, endoderm, mesoderm, and ectoderm. In the course of gastrulation, the germ layers are patterned with respect to anteroposterior and dorsoventral embryonic axes. Endoderm will take the most internal position and mesoderm will move between the endoderm and the ectoderm. These germ layers will undergo additional wholesale morphogenetic movements to form an embryonic body, elongated from head to tail and narrow from back to belly, and to position the organ rudiments. The following chapters review the exciting progress in our understanding of these fundamental processes in several research models. As our understanding of gastrulation in different animals deepens, similarities and differences of the cellular and molecular mechanisms start to emerge.

Part I of the book titled "*Setting up for gastrulation*" focuses on the key early stages of embryogenesis that precede gastrulation, during which the anteroposterior (AP) and dorsoventral (DV) embryonic axes and germ layers are established, as well as the position of the blastopore, the embryonic site through which mesodermal and endodermal precursors will begin their journey to set up the internal layers of the embryo.

In Chapter 1, Angelike Stathopoulos and Susan Newcomb describe these early embryonic processes in the well-studied *Drosophila* embryo, in which over 100 years of forward genetic screens, embryology, and molecular genetics yielded mechanistic insights. They lay out a beautiful example of mosaic development, where asymmetrically deposited maternal transcripts and proteins initiate specification of AP and DV embryonic polarities that are further elaborated by the zygotic genome. *Drosophila*'s blastopore, or ventral furrow, forms on the ventral side of the embryo, where mesoderm is specified at the highest levels of the Dorsal protein, and there is expression of transcription factors Snail and Twist. Interestingly, two endoderm anlagen form at the anterior and posterior poles of the embryo and invaginate independently to join by cell migration only later in embryogenesis.

In Chapter 2, Florence Marlow describes how zebrafish early embryo prepares for gastrulation. Similar to *Drosophila*, the maternal genome governs early cleavages, as well as specification of the AP and DV polarity of the zebrafish embryo, which, at the time of zygotic genome activation, features a mound of pluripotent blastomeres atop of a large syncytial yolk cell. The dorsal Nieuwkoop center in the nuclei of the syncytium and marginal blastomeres is marked by accumulation of β-catenin, which upon the zygotic genome activation will initiate a genetic cascade to specify the Spemann-Mangold organizer. Induction and patterning of mesendoderm in the marginal region of the blastoderm is accomplished by a concerted action of conserved signaling pathways, including Nodal, FGF, Wnt/β-catenin, and BMP. The endodermal and mesodermal precursors will internalize at the blastoderm margin, the zebrafish blastopore.

In Chapter 3, Ana Raffaelli and Claudio Stern review the signaling events that regulate embryonic polarity and formation of the blastopore, the primitive streak in the chick embryo. Like other amniote embryos, chick embryo elaborates an epiblast epithelial layer from which all germ layers and later embryonic tissues will arise, and extraembryonic structures that will both protect and sustain developing embryo, as well as secrete inductive signals that will break the radial symmetry of the epiblast. Also, like other amniotes, chick embryogenesis represents a regulative development, with fragments of embryonic disc being able to give rise to a complete embryo in isolation. But unlike small mammalian embryos, chick epiblast is composed of tens of thousands of cells when starting to establish embryonic axes, featuring the central area pellucida and a peripheral area opaca. Key organizing centers in the chick embryo are the marginal zone that forms in a ring of epiblast separating the area opaca and area pellucida. In particular, its posterior region was proposed to be an equivalent of the Nieuwkoop Center. Another key organizing center is Kohler's sickle, a cell population underlying the epiblast just anterior to the posterior marginal zone, but grafting experiments indicate that this cell population can contribute to primitive streak formation in addition to its inductive properties. The molecular actors involved in the induction of primitive streak that will bisect the epiblast from its prospective posterior end toward the anterior end of the embryo and will form the node, equivalent of the Spemann-Mangold organizer, include Nodal, Vg1, and Tgfβ ligands, as well as Wnt/β-catenin pathway ligand Wnt8c, which are expressed in the posterior marginal zone and posterior epiblast. Nodal activity is initially restrained by the pathway inhibitors, which are secreted by the underlying hypoblast. This inhibition is relieved as the hypoblast is displaced anteriorly by the endoblast tissue. Primitive streak morphogenesis in the chick's epiblast entails large-scale Polonaise cell movements. These swirling cell movements on both sides of the nascent primitive streak are driven by polarized cell intercalations instructed by the Planar Cell Polarity (PCP) signaling downstream of FGF, although a role of cell divisions associated with cell intercalations and chemotactic movements has been also proposed.

In Chapter 4 by Magdalena Zernicka-Goetz, Matteo Molè, and Antonia Weberling, we learn about early inductive and morphogenetic processes preceding gastrulation in mammalian embryos. They review the emerging similarities as well as unexpected differences between mouse, cynomolgus monkey, and human embryos obtained from analyses of both in vivo and in vitro cultured embryos. Contrasting mosaic development of *Drosophila* and zebrafish embryos instructed by maternally deposited gene products and the activation of zygotic genome that occurs only at ca. 1000 cell stage, mammalian embryos are highly regulative. This is paralleled by the early activation of the zygotic genome, which occurs at 1–2 cell stage in the mouse and at 4–8 cell stage in human embryos. The authors describe the first cleavages, followed by the formation of the blastocyst that involves segregation of extraembryonic lineages, trophoblast, and primitive endoderm from the inner cell mass. Whereas these processes and blastocyst morphology appear similar among mammalian embryos, some fascinating differences emerge. In the mouse blastocyst, the extraembryonic cell fates fully separate from the inner cell mass before implantation, but in human embryos some pluripotency markers continue to be expressed in extraembryonic lineages to be downregulated only in the late blastocyst around implantation. Contrasting mouse embryos, which implant via mural trophectoderm, distal to the inner cell mass of the blastocyst, human embryos use the polar trophectoderm, directly surrounding the inner cell mass. Finally, in preparation for the onset of gastrulation, morphogenesis of the inner cell mass leads to formation of a cup-shaped epiblast in the mouse, but to a flat epithelial disc in human. Braking symmetry and specification of mesendoderm and primitive is mediated in the mouse by a cascade of BMP, Wnt/β-catenin, Nodal, and FGF signaling that involves interactions between the embryonic epiblast and extraembryonic tissues.

Part II titled "*Many ways to gastrulate*" focuses on the business of gastrulation itself, with different chapters reviewing the current understanding of gastrulation movements of internalization, epiboly, convergence, and extension in a variety of species, including sea urchins, fruit fly, zebrafish, frog, chick, and mouse. The emerging picture of gastrulation is that one type of gastrulation movement can be mediated by different morphogenetic cell behaviors or their combinations. On the other hand, a specific morphogenetic cell behavior can contribute to different tissue transformations.

In Chapter 5 by Clint Ko and Adam Martin, we learn that in *Drosophila* the mesoderm is internalized as a coherent epithelial sheet of cells that maintain apical-basal polarity and intercellular adhesions. This process of mesoderm invagination is accompanied by cell shape changes, including apical constriction, which is driven by actomyosin contractility, followed by apical-basal elongation and subsequent basal relaxation. Myosin activity is enriched medioapically and its contractile pulses that are regulated by the Snail, Twist, and Rho/Rho kinase pathway, pull centripetally on the F-actin network associated with apical junctions. The mechanical linkage via adherens junctions allows forces that are generated by cells to be integrated across the epithelium, what ensures coordination of cellular behaviors during mesoderm invagination. Upon invagination, mesodermal cells undergo epithelial mesenchymal transition (EMT) and spread.

In Chapter 7, David McClay and colleagues review the cellular and molecular mechanisms of gastrulation in the sea urchin embryo, which prior to gastrulation forms a simple spherical monolayer. Mesoderm internalization occurs via two distinct cellular mechanisms. In the first, prospective mesodermal skeletogenic cells and later pigment cells in the vegetal epithelium undergo EMT and ingress into the blastocoel. In the second, nonskeletal mesoderm cells located at the center of the vegetal plate epithelium assume a shape of "bottle cells" to initiate inward bending followed by invagination of the archenteron.

In tunicates, such as *Ciona intestinalis*, as we learn from Chapter 8 by William Smith and colleagues, early embryogenic processes preceding gastrulation are controlled by asymmetrically deposited maternal gene products, including β-catenin that accumulates in the most vegetal cells to specify mesendodermal cell fates. Gastrulation entails a simple cup-shaped gastrula and is driven chiefly by invagination of the vegetal epithelium comprising endodermal progenitors. The latter undergo cell-shape changes, starting with apical constriction, followed by apical-basal cell body shortening. Driven by the RhoA/Rho kinase and myosin pathway, this process resembles the well-studied mesoderm invagination of mesoderm in *Drosophila*, reviewed in Chapter 5.

Mesoderm and endoderm internalization in *Xenopus* embryo are described in Chapter 9 by Rudolph Winklbauer. Contrasting monolayer-epithelium gastrulae described above, in *Xenopus* gastrulae, mesoderm precursors reside in the multilayered wall surrounding the blastocoel above larger vegetal endodermal precursors. Blastopore formation is initiated on the dorsal side by appearance of bottle cells in the superficial layer, followed by involution of the mesenchymal mesoderm population and vegetal rotation of the endodermal cell mass. The involution movements of mesoderm are driven in part by the concurrent epiboly and convergent extension movements.

In Chapter 12, Diana Pinheiro and Carl-Philipp Heisenberg review the process of emboly in zebrafish gastrula, in which mesenchymal mesendodermal progenitors are specified and reside at the margin of the multilayered blastoderm covering the animal hemisphere of the syncytial yolk cell. Mesendoderm internalization initiates and is more pronounced on the dorsal side, with endodermal cells internalizing ahead of mesodermal cells. Dissimilar to involution of mesodermal layer in *Xenopus*, in zebrafish mesenchymal mesendodermal progenitors ingress as individuals or in a coordinated manner, via a "synchronized ingression." Because these internalization movements are driven largely by individual cells, they do not depend on the concurrent movements of epiboly, convergence, and extension. Because of the mesenchymal nature of the mesendodermal progenitors, their internalization movements are not preceded by a classical EMT.

Internalization of mesendoderm by ingression of individual cells also predominates in amniote embryos. However, as reviewed by Octavian Voiculescu in Chapter 14 for chick embryos and by Sonja Nowotschin and Anna-Katerina Hadjantonakis in Chapter 15 for mouse embryos, as the mesendodermal progenitors arise in epithelial epiblast, their ingression via primitive streak is preceded by EMT. Hence, mesendoderm in different species forms via invagination, ingression, involution movements, or their combination. Most commonly, mesendoderm is specified in single-layer epithelium via a signaling cascade that activates Snail and Twist proteins, which in turn downregulate E-cadherin to drive EMT, dynamic actomyosin contractility, and internalization. Also, commonly, internalized mesendodermal progenitors migrate away from the blastopore (ventral furrow, blastoderm margin, and primitive streak).

Epiboly, the morphogenetic process that leads to thinning and spreading of tissues, is discussed for zebrafish by Ashley Bruce and Carl-Philipp Heisenberg in Chapter 11. Featuring the large syncytial yolk cell, epiboly in zebrafish entails spreading of the superficial enveloping layer, as well as all the germ layer precursors and nascent germ layers around the yolk cell toward the vegetal pole. The yolk syncytial layer also spreads vegetally. The authors review the recent studies demonstrating that the initial thinning of the blastoderm and associated doming of the yolk cell involve a strong decrease in cell viscosity in the central blastoderm region that occurs likely via changes in cell adhesion. Predominant cell behavior driving epiboly of the germ layers is radial cell intercalation, which is bidirectional with nearly equal movements toward superficial and deeper layers, arguing against this movement being instructed by adhesion or chemokine gradients. Interestingly, as discussed by Sonja Nowotschin and Anna-Katerina Hadjantonakis in Chapter 15, the intercalation of definitive endoderm progenitors emerging from the primitive streak into the underlying layer of primitive endoderm represents epiboly during mouse gastrulation.

Several chapters discuss the convergence and extension (C&E) gastrulation movements, which can engage various cell behaviors to narrow and elongate embryonic tissues. One type of C&E is convergent extension, in which polarized cell intercalations drive simultaneous tissue narrowing and lengthening. In Chapter 6, Adam Pare and Jennifer Zallen discuss cellular and molecular underpinnings of convergent extension in the *Drosophila* germband epithelium, in which polarized cell behaviors are controlled by asymmetrically distributed proteins regulating cytoskeletal tension and cell adhesion that drive polarized junction remodeling. It is in this system where the volume's title "*Gastrulation: From embryonic pattern to form*" is well illustrated, as recent advances have been made in our understanding of the links between the AP embryo patterning and polarized cell behaviors that drive germband extension. The authors discuss their discoveries demonstrating that the AP patterning system regulates striped expression of three receptors in the Toll receptor family that in turn control planar polarity of actomyosin and adherence junction components to guide intercalary cell behaviors. The Toll receptor system cooperates with the leucine-rich repeat receptor Tartan and the teneurin Teneurin-m, which regulate planar polarity of cytoskeletal elements at compartment boundaries.

Convergent extension movements elongate the invaginated archenteron in sea urchin gastrulae as reviewed by David McClay and colleagues in Chapter 7. This process is driven by a combination of cell behaviors, including directed cell movement in the direction of archenteron elongation and their convergence occurring as some migrating cells move in front of their neighbors. Oriented cell divisions, with division planes biased along the axis of the elongating archenteron, as well as repeated cycles of extension and contraction of filopodia elaborated by cells at the archenteron tip, also contribute to extension.

A highly conserved convergent extension movement driven by biased planar and radial intercalations of mesenchymal cells that preferentially separate anterior and posterior neighbors and are polarized by noncanonical Wnt/Planar Cell Polarity (Wnt/PCP) signaling shapes chordamesoderm tissues during gastrulation of tunicates (Chapter 8 by William Smith and colleagues), zebrafish (Chapter 13 by Margot Williams and Lilianna Solnica-Krezel), as well as *Xenopus* (Chapter 10 by Ray Keller and Anna Sutherland), or powers Polonaise movements that position primitive streak in the chick gastrula (Chapter 3 by Ana Raffaelli and Claudio Stern and Chapter 14 by Octavian Voiculescu). Like in *Drosophila* germband extension, convergent extension of chordamesoderm (and other tissues in some species) is linked to AP patterning. Moreover, the PCP pathway is thought to act as a molecular compass, with some of its components accumulating at the anterior and others at the posterior cell edges, to coordinate embryo patterning with morphogenetic cell behaviors, but the molecular links between AP embryonic patterning and PCP signaling in chordates remain to be discovered. The cellular mechanism of mediolateral intercalation underlying convergent extension is discussed in detail by Ray Keller and Ann Sutherland in Chapter 10 and involves mediolateral cell elongation and formation of mediolaterally polarized protrusions. The extension of these polarized protrusions, their adherence to medial and lateral neighbors, and subsequent contraction are thought to power C&E and are driven by mediolaterally oriented actomyosin cytoskeleton that forms a supracellular network. Contrasting this model of cell intercalation by crawling is the proposal that polarized "junction shrinking" of mediolaterally aligned cell-cell junctions in mesenchymal cells drives C&E in a manner similar to epithelial cell intercalation during *Drosophila* germband extension. However, cell intercalation does not account for all C&E processes during gastrulation. Indeed, directed cell migration contributes to this process in zebrafish, chick, and mouse.

As the chapters for this volume were being written, several new studies analyzing gastrulation by single-cell RNA sequencing (scRNA-seq) were published, ushering studies of gastrulation into a new era of analyzing cell behaviors and transcriptomes one cell at a time. The power of this approach is illustrated by Sonja Nowotschin and Anna-Katerina Hadjantonakis in Chapter 15, providing an orthogonal line of evidence that endoderm in the mouse gastrula is formed by intercalation of definitive endoderm cells into the primitive endoderm layer. The scRNA-seq data sets provide a comprehensive source of candidate genes to be studied, raw material for interspecies comparisons and for formulating hypotheses. It is intriguing how these new single-cell genomic and constantly improving imaging approaches will shape studies and knowledge of gastrulation in the next decade.

Acknowledgments

I would like to express my appreciation to all colleagues who contributed to the volume, Elsevier editors, Shellie Bryant and Zoe Kruze, for their professionalism, patience, and persistence to ensure that all the contributions were on time. I am also grateful to the members of my research group who share with me enthusiasm for gastrulation and for tolerating my more severe than usual shortage of time in the course of the last year. Work in my laboratory and efforts for this volume were supported in part by R35 Maximizing Investigators' Research Award from the National Institute of General Medical Sciences/ National Institutes of Health, Alan A. and Edith L. Wolff Distinguished Professorship in Developmental Biology and Washington University School of Medicine in St. Louis.

Lilianna Solnica-Krezel
December 2019

PART I

Setting up for gastrulation

CHAPTER ONE

Setting up for gastrulation: *D. melanogaster*

Angelike Stathopoulos*, Susan Newcomb
Division of Biology & Biological Engineering, California Institute of Technology, Pasadena, CA, United States
*Corresponding author: e-mail address: angelike@caltech.edu

Contents

Abstract

Drosophila melanogaster embryos develop initially as a syncytium of totipotent nuclei and subsequently, once cellularized, undergo morphogenetic movements associated with gastrulation to generate the three somatic germ layers of the embryo: mesoderm, ectoderm, and endoderm. In this chapter, we focus on the first phase of gastrulation in *Drosophila* involving patterning of early embryos when cells differentiate their gene expression programs. This patterning process requires coordination of multiple developmental processes including genome reprogramming at the maternal-to-zygotic transition, combinatorial action of transcription factors to support distinct gene expression, and dynamic feedback between this genetic patterning by transcription factors and changes in cell morphology. We discuss the gene regulatory programs acting during patterning to specify the three germ layers, which involve the regulation of spatiotemporal gene expression coupled to physical tissue morphogenesis.

1. Introduction

Gene regulatory programs drive developmental progression during early *Drosophila* embryo development (rev. in Briscoe & Small, 2015; Stathopoulos & Levine, 2005). Dynamic gene expression is initiated during

Current Topics in Developmental Biology, Volume 136
ISSN 0070-2153
https://doi.org/10.1016/bs.ctdb.2019.11.004

the early syncytial stage, when nuclei are totipotent, and continues to the cellularized blastoderm stage at which point cells are thought to exhibit a stepwise restriction of developmental potential (Beer, Technau, & Campos-Ortega, 1987; rev. in Technau, 1987). While many of the genes responsible for this patterning are conserved in other animals, the *Drosophila* embryo exhibits several atypical mechanisms that support early patterning.

First, while mammals rely on a symmetry-breaking event of sperm entry at fertilization to provide a polarity cue, embryonic polarity in *Drosophila* is established during oogenesis when the mother deposits mRNAs into the developing oocyte (Ajduk & Zernicka-Goetz, 2016; Moussian & Roth, 2005; Fig. 1A). These maternally-provided transcripts encode transcription factors or signaling pathway components that influence gene expression in the early embryo. A subset provide positional information by establishing gradients of maternal proteins in the embryo or within the extracellular space that impact spatial zygotic gene expression (rev. in Reeves & Stathopoulos, 2009; Wieschaus, 2016).

Second, the *Drosophila* embryo develops as a syncytium. The first 13 nuclear division cycles include a short DNA replication S phase and no G2 phase, and the nuclei are not enclosed in separate membrane compartments but instead present in a joint cytoplasm (Foe & Alberts, 1983). This streamlined division cycle likely relates to the speed at which *Drosophila* embryos develop as it permits a rapid increase in cell number before gastrulation. Indeed, short synchronous nuclear cycles are also employed by other fast-developing embryos such as amphibians (*Xenopus*) and teleost fish (zebrafish). Further, the *Drosophila* maternal-to-zygotic transition (MZT), the phase during which the zygotic genome is activated and maternal transcripts are degraded, occurs over the course of only a few hours, compared to days in pre-implantation mammalian embryos (Jukam, Shariati, & Skotheim, 2017). Once the length of the cell cycle increases, the *Drosophila* embryo switches from relying on maternal instructions to a program that is driven predominantly by the zygotic products (rev. in Blythe & Wieschaus, 2015a).

Lastly, it is only after the 14th nuclear division that the length of the cycle increases and cellularization occurs. After 14 nuclear cycles, the syncytial embryo is comprised of roughly 6000 nuclei within a common cytoplasm. During cellularization individual nuclei that had previously migrated to the embryo periphery become encased by individual membrane compartments. After this point, precise cell-to-cell signaling, which had been limited

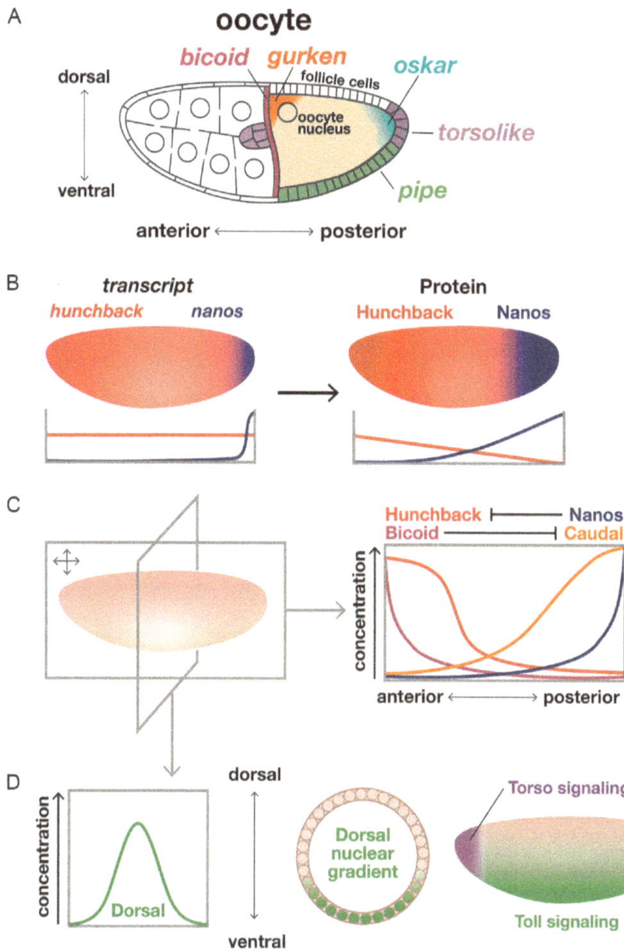

Fig. 1 Establishment of embryonic polarity. (A) Key maternal transcripts are specifically localized within the oocyte by RNA binding proteins establishing basic polarity that is transferred to the zygote. (B) Oskar-dependent posterior localization of *nanos* transcript establishes an anterior-to-posterior gradient of Hunchback protein through translational repression of uniformly-distributed *hunchback* mRNA. (C) Schematic of distribution of four key patterning factors along the A-P axis of the pre-blastoderm embryo. (D) D-V biased Toll signaling resulting from pipe localization in the ventral oocyte produces a nuclear gradient of the transcription factor Dorsal. *tsl* transcript deposited to the anterior and posterior of the oocyte ultimately activates Torso signaling in corresponding terminal domains in the early embryo. *Note*: All follicle and embryo diagrams are oriented with anterior left and dorsal up unless otherwise specified.

pre-cellularization, becomes widespread (rev. in Perrimon, Pitsouli, & Shilo, 2012; Stathopoulos & Levine, 2004).

Despite, but in many cases because of, these atypical properties of its early development, the *Drosophila* embryo offers an attractive system to study the mechanisms that govern embryonic patterning. Although early development is rapid, it also is tractable, occurring externally over the course of a few hours. Furthermore, *Drosophila* embryonic development has been well characterized over many decades, forming a strong basis for mechanistic studies; numerous genetic tools exist that have and will continue to catalyze progress. Lastly, precisely staged embryos can be obtained with relative ease to support a variety of methodological approaches including genomics, biochemistry, and *in vivo* imaging studies. In the following chapter, we discuss progress made toward understanding the gene regulatory program that establishes patterning of the embryo and prepares it to undergo cell movements associated with gastrulation.

2. Establishment of embryonic polarity occurs in oocytes

Gradients of maternal proteins are responsible for establishing polarity within the early embryo to specify its anterior-posterior (A-P) and dorsal-ventral (D-V) axes. A-P axis orientation relates to localization in the oocyte of transcripts, which provide the positional information necessary to specify anterior and posterior ends of the embryo. The localization of transcripts to the presumptive anterior and posterior poles within developing oocytes pre-fertilization depends on the RNA binding protein Staufen (Stau) (Johnston, St Johnston, Beuchle, & Nüsslein-Volhard, 1991). In particular, *bicoid* (*bcd*) and *oskar* (*osk*) mRNAs are localized to presumptive anterior and posterior poles, respectively, of the developing oocyte (Fig. 1A).

This localization of transcripts to the oocyte (i.e., *bcd*, *osk*) results in the production of at least five gradients of maternal factors in the early embryo that are important for patterning. The Bcd protein gradient is produced from a gradient of *bcd* mRNA in embryos, that stems from transcript localization to the anterior of oocytes (Ali-Murthy & Kornberg, 2016; Spirov et al., 2009). Bcd, which encodes a homeodomain protein, is pivotal for patterning along the A-P axis, but additional input also comes from maternal Hunchback (Hb) and Caudal (Cad) transcription factor gradients (Driever & Nüsslein-Volhard, 1988; Ochoa-Espinosa, Yu, Tsirigos, Struffi, & Small, 2009; Schulz & Tautz, 1995). The posterior maternal system is specified from local translation of *osk* transcripts, producing Osk protein which acts to localize another posterior maternal factor, Nanos (Nos), to the posterior. *nos* transcript localization as well as its translation is Osk-dependent (Ephrussi, Dickinson, & Lehmann, 1991; Wang & Lehmann, 1991). Furthermore, the posterior gradient of Nos acts to establish an anterior gradient of maternal Hb through translational repression of the initially uniformly-distributed *hb* mRNA in posterior regions of the embryo (Fig. 1B; Irish, Lehmann, & Akam, 1989; Struhl, 1989; Tautz, 1988). Both maternal and zygotic *hb* expression contribute to setting borders of the earliest zygotic genes expressed along the A-P axis, the so-called gap genes (Jaeger, 2011; Porcher et al., 2010). In addition to its role as a transcription factor in which it binds to DNA and regulates transcription (Niessing et al., 2000; Struhl, Struhl, & Macdonald, 1989), Bcd also mediates translational repression using its same homedomain to bind RNA and regulate translation (Dubnau & Struhl, 1996; Niessing et al., 2000; Rivera-Pomar, Niessing, Schmidt-Ott, Gehring, & Jäckle, 1996). In this role, Bcd forms an A-P-oriented concentration-gradient and acts to repress translation of *cad* mRNA in the anterior of the embryo, forming an inverse gradient of Cad protein (Macdonald & Struhl, 1986; Mlodzik & Gehring, 1987). Cad is important for specification of abdominal as well as posterior segments including the hindgut (Fig. 1C; Schulz & Tautz, 1995; Wu & Lengyel, 1998).

On the other hand, positional information that contributes to establishing the D-V axis and to patterning the embryonic anterior and posterior termini derives from signaling pathways active earlier in the oocyte. A transcriptional program in developing oocytes regulates the expression of two extracellular matrix proteins: Pipe and Torso-like (rev. in Johnson, Henstridge, & Warr, 2017; Moussian & Roth, 2005). *pipe* expression is influenced by the positioning of the nucleus within the oocyte. The oocyte nucleus expresses the ligand Gurken, which activates EGFR signaling in dorsal follicle cells and indirectly limits expression of *pipe* to the ventral-most follicle cells (Schüpbach, 1987; Sen, Goltz, Stevens, & Stein, 1998). *pipe* encodes a protein sharing homology with vertebrate heparan sulfate 2-*O*-sulfotransferases, but likely functions independent of heparan in the extracellular matrix on the ventral side of the egg and ultimately impacts signaling in the embryo (Zhu, Sen, Stevens, Goltz, & Stein, 2005). The *torso-like* (*tsl*) gene, encoding a secreted protein with homology to perforins, exhibits localized expression to a distinct subset of follicle cells at the anterior and posterior ends of the oocyte (Stevens, Beuchle, Jurcsak, Tong, & Stein, 2003). Tsl is incorporated into the vitelline membrane, the inner layer of the eggshell, and is subsequently transferred to the plasma membrane of the embryo thereby conveying spatial information from the follicle cells surrounding the oocyte to the developing embryo (Mineo, Furriols, & Casanova, 2015). Pipe and Tsl proteins serve as initial polarity cues for the embryo to define dorsal-ventral polarity and terminal ends, respectively.

Pipe and Tsl expression leads to spatial activation of signaling pathways within the early embryo. Ventrally-deposited Pipe protein leads to activation of Toll receptor signaling through the localized activation of a protease cascade in the fertilized egg (Cho, Stevens, & Stein, 2010; rev. in Moussian & Roth, 2005). Toll signals intracellularly through a conserved NF-kappaB pathway and establishes a ventral-to-dorsal nuclear gradient of the Rel transcription factor Dorsal (DL) (Fig. 1D; Roth, Stein, & Nüsslein-Volhard, 1989; Rushlow, Han, Manley, & Levine, 1989; Steward, 1989). The DL gradient directs D-V patterning and this transcription factor is thought to act as a morphogen, controlling gene expression along the D-V axis in a concentration-dependent fashion (rev. in Hong, Hendrix, Papatsenko, & Levine, 2008; Jiang & Levine, 1993; Reeves et al., 2012). On the other hand, extracellular Tsl facilitates spatially-localized activation of signaling through the Torso (Tor) receptor in the early embryo in cooperation with Torso's presumed ligand Trunk (Trk) (Amarnath, Stevens, & Stein, 2017; Mineo, Furriols, & Casanova, 2018). Tor is activated at the embryonic poles by Trk, a broadly-expressed secreted molecule, which relies on terminally-localized Tsl for its activation (Casanova, Furriols, McCormick, & Struhl, 1995; Casanova & Struhl, 1989; Sprenger & Nüsslein-Volhard, 1992). Activation of Tor signaling supports expression of particular genes at the embryo termini responsible for the formation of terminal embryonic structures, the anterior acron and posterior telson (Fig. 1D; Weigel, Jurgens, Klingler, & Jackle, 1990). Tor transduces signals mainly by means of the Ras–extracellular signal regulated kinase (ERK) signaling cassette (Duffy & Perrimon, 1994; Lu, Perkins, & Perrimon, 1993; Perrimon et al., 1995), but a subset of Tor signaling utilizes the *Drosophila* STAT (DStat92E) instead (Li, Xia, & Li, 2003; Li, Agaisse, Mathey-Prevot, & Perrimon, 2002). In particular, MAPK signaling through Tor has been shown to support degradation of a maternally-provided ubiquitous repressor, Capicua (Cic), limiting its levels at the embryonic termini (Astigarraga et al., 2007; de las Heras & Casanova, 2006). The Cic protein gradient is important for patterning of the trunk regions of embryos, contributing both to A-P and D-V patterning (Jiménez, Guichet, Ephrussi, & Casanova, 2000).

In this manner, despite lack of cell membranes while the embryo is a syncytium, broad signaling domains are established by molecules spatially deposited to the eggshell during oogenesis. These domains permit the spatial activation of signaling pathways downstream of Toll and Torso receptors in the early embryo to establish D-V axis and terminal end positional information, respectively. In the fertilized embryo, gene expression programs require input both from maternal transcription factors including those that provide polarity cues, as discussed above, as well as additional ubiquitous maternal transcription factors that serve as timing factors that help to manage the MZT, discussed in the following section.

3. Navigating the maternal-to-zygotic transition

Upon fertilization, the transition from dependence on maternal transcripts deposited into the egg to newly transcribed zygotic transcripts is carefully regulated to ensure proper development of early embryos. This maternal-to-zygotic transition, or MZT, requires reprogramming of the early embryonic genome by a network of maternal transcription factors, ultimately resulting in a hand off of the gene regulatory program to zygotic transcription factors. In *Drosophila*, this transition from dependence on maternal to zygotic products requires precise regulation of the division cycle, DNA replication, initiation of zygotic transcription, maternal RNA clearance and chromatin remodeling (rev. in Blythe & Wieschaus, 2015a; Hamm & Harrison, 2018; Vastenhouw, Cao, & Lipshitz, 2019; Yuan, Seller, Shermoen, & O'Farrell, 2016).

In the *Drosophila* embryo, the division cycles and DNA replication start off short and then increase in length (Fig. 2A). The first 2 hours of development comprise 13 rapid nuclear division cycles within a shared cytoplasm. Nuclei are located in the center within the yolk, but then migrate to the periphery during nuclear cycle 8–10. At this point, the yolk becomes localized in the center and cleavage occurs in the cortical area; but cleavage is superficial and does not bisect the yolk. The first 13 nuclear divisions occur rapidly, roughly every 8–10 min and without cytokinesis, within a shared cytoplasmic space (Hamm & Harrison, 2018). These mitoses lack gap phases and therefore involve repeating cycles of rate-limiting DNA synthesis (Foe, 1989; McCleland, Farrell, & O'Farrell, 2009; Shermoen, McCleland, & O'Farrell, 2010). However, after the 14th nuclear division, the cycle slows dramatically and nuclei become cellularized (Foe & Alberts, 1983; Shermoen et al., 2010). Cell membranes emerge to encapsulate nuclei, forming a single layered epithelium. In addition, at nuclear cycle 14,

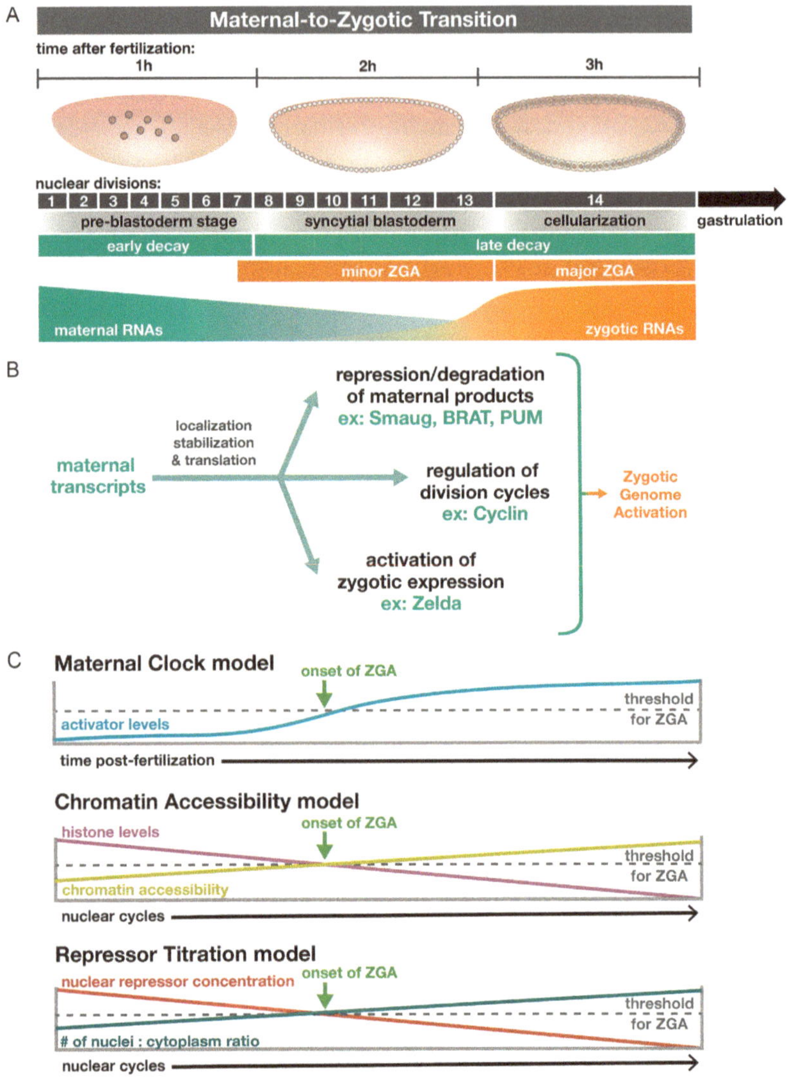

Fig. 2 Maternal to zygotic transition. (A) Early zygotic development entails 13 rapid nuclear divisions within a common cytoplasm to produce a syncytium. Nuclei migrate to the embryo periphery and then cellularize forming a blastoderm of ~6000 cells surrounding a central yolk. In order to transition from control of development by maternal RNA and proteins to regulation driven by zytotic products, maternal factors are systematically cleared from the embryo in two phases. (B) Maternal transcripts deposited into the zygote affect early development and the onset of zygotic genome activation (ZGA) by their precise localization within the embryo, stabilization and translation of RNA into factors that regulate key phenomena: the repression and degradation of maternal factors, regulation of nuclear divisions as early blastoderm development

developmental changes relating to DNA replication occur; namely a lengthened S phase and the introduction of G2 phase. This key milestone, the so-called midblastula transition (MBT) when embryonic programs of morphogenesis and differentiation initiate, is also associated with clearance of a subset of maternally-provided mRNAs, large-scale transcriptional activation of the zygotic genome, and an increase in cell cycle length (rev. in Blythe & Wieschaus, 2015a; Yuan et al., 2016).

The MZT, on the other hand, encompasses the entire period during which maternal products are cleared and zygotic genome activation occurs, roughly from fertilization to the onset of gastrulation (Fig. 2). Furthermore, although widespread zygotic transcription is delayed until the 14th nuclear cycle when cellularization occurs, some zygotic transcripts are produced earlier and likely contribute to the gene regulatory decisions that influence patterning and development in concert with maternal products. Zygotic expression of a small number of genes has been observed as early as nuclear cycle 7, even before nuclei migrate to the cortex at nuclear cycle 9 (Ali-Murthy & Kornberg, 2016; Ali-Murthy, Lott, Eisen, & Kornberg, 2013). It is possible that these early zygotic transcripts contribute to patterning, but not much is known about their roles. However, transcripts expressed early are on average shorter in size and contain small introns if any, possibly constrained in length by the short interphase; whereas transcripts present at nuclear cycle 14 on average are longer and contain larger introns (Chen et al., 2013; Kwasnieski, Orr-Weaver, & Bartel, 2019). A recent study found that truncated forms of long genes are expressed prior to nuclear cycle 14, and that these products act to regulate the timing of signal transduction either by generating dominant negative or constitutively active forms of signaling molecules (Sandler et al., 2018).

---

proceeds, and activation of zygotic gene expression. (C) Models for timing of ZGA. (Top) "Maternal clock" model: Time required for buildup of key activators determines onset of ZGA. (Middle) Chromatin accessibility model: As nuclei divide, the concentration of soluble histones declines. Combined with other mechanisms, including activity of pioneer factors, chromatin accessibility (particularly at enhancers) increases and zygotic genes begin to be expressed. (Bottom) Repressor titration: As the ratio of nuclei to cytoplasm increases during early divisions, concentration of maternal repressors is diluted within individual nuclei; once below repression threshold, zygotic targets of repression begin to be expressed. *Note*: Models shown are conceptual, exact timing of ZGA and levels of relevant factors likely differ depending on context. *Adapted from Hamm, D. C., & Harrison, M. M. (2018). Regulatory principles governing the maternal-to-zygotic transition: Insights from Drosophila melanogaster.* Open Biology, 8 *(12), 180183.*

The exact molecular mechanism that coordinates division cycle slowing and widespread activation of the zygotic genome at cycle 14 is not known, but several regulatory mechanisms have been proposed (Fig. 2C). One model suggests that a "maternal clock" acts to track time elapsed since fertilization (Blythe & Wieschaus, 2015b; Tadros & Lipshitz, 2009). A set amount of developmental time following fertilization may be necessary for levels of maternal factors that trigger zygotic genome activation to accumulate. Furthermore, the concentration of soluble histones in the nucleus drops within embryos as they approach to zygotic genome activation (Shindo & Amodeo, 2019). Decreasing histone levels advances zygotic transcription, cell cycle elongation, cell cycle checkpoint kinase Chk1 activation, and gastrulation; whereas increasing histone levels has the opposite effect, causing delays in transcription and the addition of an extra cell cycle (Chari, Wilky, Govindan, & Amodeo, 2019). A drop in the concentration of soluble histones in the nucleus likely provides an opportunity for the pioneer factors that herald zygotic gene expression to successfully compete for DNA access and activate transcription. For example, during this time, the ubiquitous, maternal transcription factor Zelda is a pivotal activator as it directly binds to and opens chromatin at enhancers, facilitates binding of other transcriptional activators to these DNA sequences, and allows initiation of zygotic gene expression (Harrison, Li, Kaplan, Botchan, & Eisen, 2011; Liang et al., 2008; Nien et al., 2011). Zelda binds nucleosomes and is considered a pioneer factor (McDaniel et al., 2019). Loss of Zelda therefore leads to a large scale, global decrease in zygotic gene expression as many enhancer regions remain inaccessible and thus non-functional (Schulz et al., 2015). As such, binding of Zelda to promoters and enhancers may be facilitated when histone levels decrease as chromatin may be more easily accessed.

Furthermore, it is likely not only the balance between histones and pioneering activators that is responsible for timing MZT, but also levels of repressors that can impact timing as well. For instance, an alternative model to explain MZT timing posits the need to titrate away an additional repressor in order to allow zygotic transcription to initiate. Titration may be accomplished by increasing the nuclear to cytoplasmic ratio, which would decrease the effective repressor concentration within individual nuclei. Some broadly-expressed repressors may fulfill this role and contribute to timing of the MZT. The transcription factors Tramtrack (Ttk), Suppressor of Hairless [Su(H)], and Runt are maternally loaded and have been shown to broadly repress the transcription of genes in the early embryo (Brown & Wu, 1993; Koromila & Stathopoulos, 2017; Read, Levine, &

Manley, 1992). Reducing the amount of these factors or the number of binding sites within particular enhancers of Ttk and Runt target genes *fushi-tarazu* and *short-gastrulation* (*sog*), respectively, results in earlier initiation of transcription, while increasing the amount of these proteins has the opposite effect—to delay transcription (Koromila & Stathopoulos, 2017; Pritchard & Schubiger, 1996). While the molecular mechanisms these repressors use to impact gene expression at MZT remain unknown, it has been proposed that they act as a counterbalance to Zelda to influence chromatin accessibility or other transcriptional activators (i.e. Bicoid or Dorsal) and consequently to influence levels of expression (Koromila & Stathopoulos, 2019; Ozdemir, Ma, White, & Stathopoulos, 2014).

Lastly, MZT also involves degradation of maternal transcripts and products once zygotic transcription has been initiated. At this time, the maternal instructions (i.e. mRNAs and proteins) are actively cleared from the embryo. mRNA degradation programs targeting maternal transcripts account for clearance of over half of all maternal mRNAs degraded during the MZT (Bashirullah et al., 1999; Lefebvre & Lécuyer, 2018; Tadros & Lipshitz, 2009). This process is regulated largely by RNA binding proteins (RBPs) including Smaug, Brain tumor (BRAT), and Pumilio (PUM) that bind to *cis*-acting elements within the open reading frames or 3′-untranslated regions of maternal transcripts (Laver et al., 2015; Tadros et al., 2007). In addition, microRNAs also support maternal transcript degradation. A major player in this pathway is the zygotically-expressed miR-309 cluster that contains eight microRNA genes (Bushati, Stark, Brennecke, & Cohen, 2008) and is required for the degradation of mRNAs encoded by approximately 400 maternally-expressed genes (Bushati et al., 2008). This clearance, through action of RBPs and microRNAs, allows the transition from dependence on the maternal program, which is erased, to dependence on the newly-transcribed zygotic program.

4. Gene expression patterns establish the prospective germ layers

By 3 hours of development, the MZT and cellularization of the blastoderm are complete and have resulted in approximately 6000 cells forming a single-layered, tall columnar epithelium surrounding the yolk, with the apical surface of cells facing outward. Germ cells are set aside earlier, and lie at the posterior pole of the embryo on the surface of the blastoderm epithelium (Fig. 3A). By this stage, the embryo is already divided into territories with specified fates, judged by gene expression patterns and transplantation experiments (Campos-Ortega & Hartenstein, 1997; Leptin in Stern, 2004). The prospective mesoderm occupies, ventrally, approximately 20% of the circumference and 70% of the length of the embryo; whereas the endoderm anlagen invaginate from two distinct blastoderm regions, one anterior to the ventral furrow and the other at the posterior pole (Figs. 3 and 4B). These two endoderm anlagen develop independently of each other but are joined by cell migration later in embryogenesis (Reuter, Grunewald, & Leptin, 1993). The ectodermal primordium is composed of several regions with different fates. Along the D-V axis and dorsal to

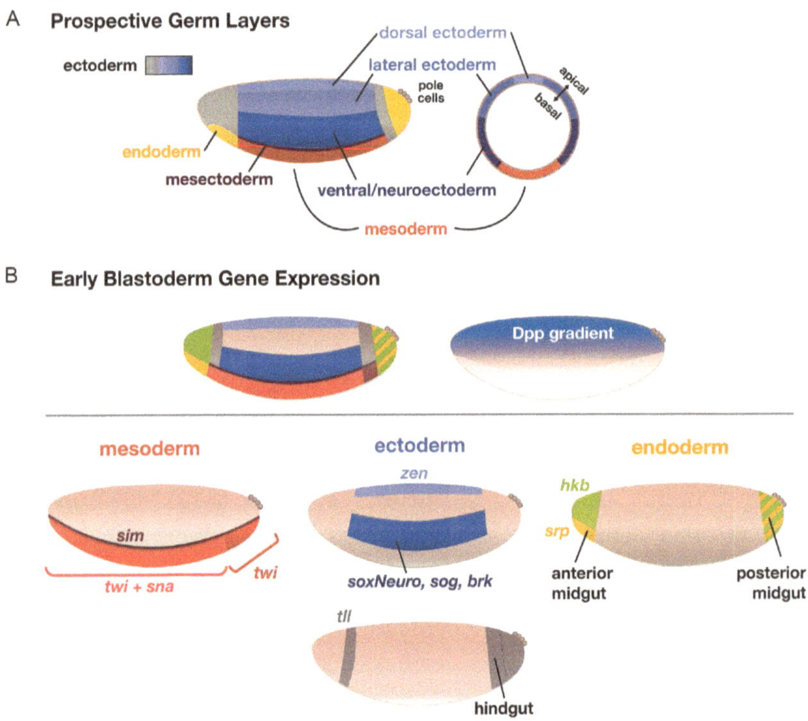

Fig. 3 Gene expression patterns establish prospective germ layers. (A) Broad patterns of gene expression subdivide the embryo into domains that prefigure the germ layers. (B) Ventral cells specified by expression of *twist* and *snail* and delineated by expression of *sim* will develop into the mesoderm. Domains along the trunk dorsal to *sim* comprise the future ectoderm. A dorsal gradient of Dpp signaling contributes to further subdivision of the ectodermal primordia into ventral/neuroectoderm and non-neuroectoderm (the lateral ectoderm and dorsal ectoderm/amnioserosa). Anterior and posterior ectodermal domains marked by *tll* will form the fore- and hindgut, respectively. Domains at the embryo termini marked by the expression of *hkb*, *tll*, and *srp* will form the endoderm.

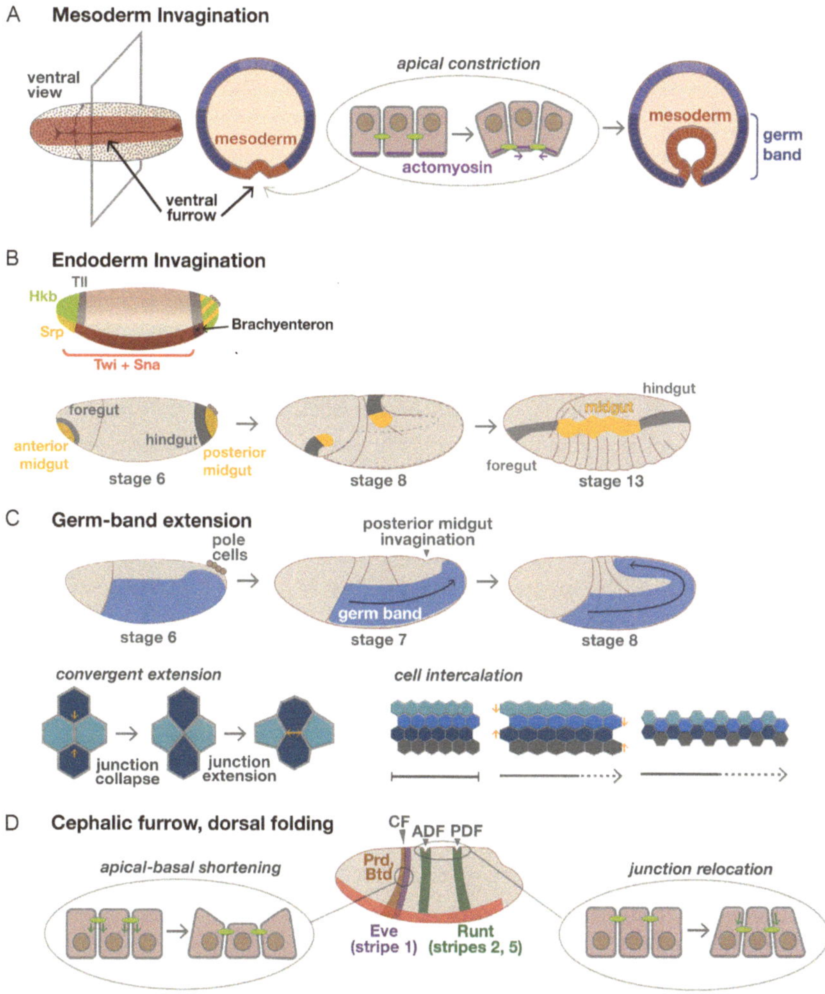

Fig. 4 Gene regulatory interactions prepare cells for diverse cell movements at gastrulation. (A) Mesoderm invagination: Expression of *twist* and *snail* lead to apical constriction which generates a ventral furrow. (B) Endoderm invagination: The anterior and posterior endoderms are specified by distinct mechanisms and invaginate separately. They eventually merge to form the gut. (C) Germ-band extension: GBE is largely driven by cell intercalation and cell shape changes that are a passive response to extrinsic tensile forces. Cells undergo convergent extension, shape changes and intercalation which leads to the overall lengthening of the ectoderm along the A-P axis, orthogonal to Toll gene expression (not shown). (D) Cephalic furrow formation: *eve* expression leads to myosin-independent cell shortening. Actomyosin-independent dorsal fold formation occurs at stripes of *runt* expression and results from basal junctional repositioning. *Note*: For simplicity, only the relevant stripes of *eve* and *runt* expression are shown. *Adapted from Gilmour, D., Rembold, M., & Leptin, M. (2017). From morphogen to morphogenesis and back.* Nature, 541, *311–320. https://doi.org/10.1038/nature21348.*

the mesoderm primordia on each side of the embryo lie the mesectoderm, large regions of neuroectoderm, and the lateral ectoderm (Fig. 3). Neuroblasts will delaminate from the neuroectoderm, moving toward the inside of the embryo (Hartenstein & Wodarz, 2013), whereas the lateral and dorsal ectoderm will form the majority of the larval epidermis and the tracheal system (Ray, Arora, Nüsslein-Volhard, & Gelbart, 1991). The dorsal side of the embryo will develop into amnioserosa. The proctodeal ring, which lies just anterior to the posterior endoderm and is composed of ectodermal tissue, will form the hindgut (Kispert, Herrmann, Leptin, & Reuter, 1994), while the anterior tip of the blastoderm lying above the anterior endodermal primordium will form the foregut (Figs. 3B and 4B; Rehorn, Thelen, Michelson, & Reuter, 1996). Borders of the future germ layers correlate with the expression patterns of several genes at the cellular blastoderm stage. The focus of this section is a discussion of how the gene expression patterns that define these prospective germ layers arise.

The borders of the future mesoderm germ layer can be visualized by the expression patterns in cells located most ventrally on the D-V axis, which express the important mesodermal determinant genes *snail* and *twist* (Alberga, Boulay, Kempe, Dennefeld, & Haenlin, 1991; Thisse, Perrin-Schmitt, Stoetzel, & Thisse, 1991). The presumptive mesodermal domain covers most, but not all, of the length of the embryo as it is excluded from both anterior and posterior ends. *twist* expression extends the length of the embryo including ends; whereas, *snail* expression extends the length of the embryo including the anterior end but is repressed at posterior end (Reuter & Leptin, 1994). The posterior border of *snail* expression coincides with the mesoderm primordium posterior border (Fig. 3).

The anterior and posterior terminal regions of the embryo contribute to endoderm as well as to ectodermal derivatives. Expression of the gap genes *tailless* (*tll*) and *huckebein* (*hkb*) mark the borders of germ layers along the A-P axis and are expressed both at the posterior and anterior caps of the embryo in response to Tor signaling (Klingler, Erdélyi, Szabad, & Nüsslein-Volhard, 1988; Strecker, Kongsuwan, Lengyel, & Merriam, 1986; Weigel et al., 1990). *tll* encodes a conserved, orphan nuclear receptor (Pignoni et al., 1990), whereas *hkb* encodes an Sp1/Egr-like Zn-finger protein (Brönner et al., 1994). The presumptive endoderm is also marked by *serpent* (*srp*), encoding a GATA transcription factor, which is expressed on the ventral side of the anterior-most part of the embryo and in a cap at the most posterior end of the embryo, specifying the anterior and posterior midgut primoridia, respectively (Fig. 3B; Rehorn et al., 1996).

The ectoderm ventral boundary is positioned by *snail*, encoding a transcription factor that predominantly acts as a repressor (Kosman, Ip, Levine, & Arora, 1991). Immediately adjacent to the mesoderm lies the mesectoderm, which is defined by expression of the gene *single-minded* (*sim*) and subsequently will form the midline glia (Fig. 3B; Crews, Thomas, & Goodman, 1988). Notch signaling is important for constraining *sim* expression to only a single row of mesectodermal cells along the D-V axis (Morel & Schweisguth, 2000). Expression of *zerknüllt* (*zen*) marks a band of cells on the dorsal side of the embryo that will ultimately encompass the amnioserosa (Rushlow, Frasch, Doyle, & Levine, 1987). The presumptive neuroectoderm lies between the *sim* and *zen* expression domains and expresses genes including *SoxNeuro*, encoding a Sox HMG DNA binding domain protein, *short-gastrulation* (*sog*), encoding a secreted inhibitor of BMP signaling, and *brinker* (*brk*) encoding a transcriptional repressor (Fig. 3B; Biehs, François, & Bier, 1996; Crémazy, Berta, & Girard, 2000). Sog functions to antagonize the activity of the ligand Decapentaplegic (Dpp), limiting activation of BMP signaling to dorsal regions and inhibiting its activity locally within the presumptive neurogenic region (Ashe & Levine, 1999; Biehs et al., 1996). Brk represses expression of Dpp target genes, excluding them from ventrolateral regions (Jaźwińska, Kirov, Wieschaus, Roth, & Rushlow, 1999; Zhang, 2001). Limiting Dpp activity in ventral regions is necessary to support formation of the neurogenic ectoderm, which gives rise to both the ventral epidermis and the central nervous system.

In the dorsal region lies the non-neurogenic ectoderm. Dpp signaling both suppresses neurogenesis and maintains expression of genes that promote dorsal cell fates (Ferguson & Anderson, 1992a, 1992b). Dpp is present in a gradient and acts as a morphogen, in that it specifies many different cell fates via intercellular signaling in a concentration-dependent fashion (Fig. 3B; Ferguson & Anderson, 1992a, 1992b; Ray et al., 1991). High levels of Dpp activity specify the amnioserosa, while progressively lower levels specify dorsal and lateral ectoderm (Rushlow & Roth, 1996). Lower Dpp levels in the dorsolateral region specify the non-neurogenic ectoderm harboring peripheral nervous system precursors. This potential for Dpp to specify cell fate is highly dosage sensitive (Wharton, Ray, & Gelbart, 1993). In the wild-type embryo, increasing the gene dosage of Dpp can shift cell fates along the D-V axis. Sog inhibitor and Brk repressor proteins work in concert to specify the dorsolateral limits of Dpp target genes (Ashe, Mannervik, & Levine, 2000; Jaźwińska, Rushlow, & Roth, 1999).

Thereby, the establishment of mesoderm and ectoderm relies predominantly on D-V axis positional information, whereas endoderm requires input from terminally-localized cell signaling along the A-P axis. It is unclear if the neurogenic ectoderm requires positive input to define this cell state or whether, rather, it represents a default state.

5. Gene regulatory interactions prepare cells for diverse cell movements at gastrulation

Germ layer formation begins in the cellularized blastoderm with two major cell movements (Campos-Ortega & Hartenstein, 1997). Ectoderm and mesoderm separate as a result of the ventral furrow formation and invagination of the presumptive mesoderm in ventral regions. The developing anterior and posterior endoderm anlagen grow toward the center where they combine to form the midgut, the only endodermal derivative of the fly (Skaer in Bate & Arias, 1993). The cells that remain at the surface of the gastrulating embryo develop into mainly epidermis and neural tissue (Campos-Ortega & Hartenstein, 1997). In addition to these mesodermal and endodermal invaginations, the morphogenetic movements of *Drosophila* gastrulation also include germ-band extension, cephalic furrow invagination, and formation of dorsal folds (Leptin & Grunewald, 1990; Spencer, Siddiqui, & Thomas, 2015; Sweeton, Parks, Costa, & Wieschaus, 1991). In this section and discussed in the subsequent chapters of this book (Chapters 6 and 7), we will focus on the genes and regulatory interactions known to support specification and differentiation of the cells that encompass germ layers, as well as changes in gene expression that occur concurrently with or just preceding major cell movements at gastrulation.

In the embryo, maternal polarity cues are used to establish two morphogen gradients of transcription factors that help to direct gene expression along the two axes. Bcd and DL maternal gradients regulate the expression of target genes along the A-P and D-V axes, respectively, in a concentration-dependent manner (rev. in Briscoe & Small, 2015; Hong et al., 2008; Porcher & Dostatni, 2010; Fig. 1). These threshold outputs are also supported by input from the ubiquitous, maternal activator Zelda and broadly-expressed repressors (Foo et al., 2014; Liberman & Stathopoulos, 2009; Löhr, Chung, Beller, & Jäckle, 2009; Ozdemir et al., 2014; Xu et al., 2014). In addition, spatially-localized repressors help define boundaries of gene expression patterns (Chen, Xu, Mei, Yu, & Small, 2012; e.g. Ip, Park, Kosman, Bier, & Levine, 1992). These broad domains established by the Bcd and DL morphogen gradients are subsequently refined by hierarchical zygotic gene regulatory networks to produce differentiated tissues. Most target genes receive both A-P and D-V axes positional information, and contribute to cell specific gene expression profiles and changes in morphology.

The mesoderm layer is formed through invagination of ventral cells during gastrulation (Ip, Park, Kosman, Yazdanbakhsh, & Levine, 1992; Kosman et al., 1991; Thisse et al., 1991). The nuclear DL gradient along the D-V axis of the embryo (Fig. 1D) results in ventral expression of the genes *twist* and *snail*, which encode transcription factors important for the positioning and initiation of cell shape changes associated with ventral furrow formation (Leptin & Grunewald, 1990). Both Twist and Snail regulate gene expression in the mesoderm as well as in neighboring regions, but each employs a different mechanism (Leptin, 1991). Snail, a characterized transcriptional repressor, is expressed in the presumptive mesoderm but excluded from the posterior endoderm and inhibits mesodermal expression of genes that are destined to be active only in more lateral or dorsal regions. Conversely, Twist is required for the activation of downstream mesodermal genes (Ip, Park, Kosman, Yazdanbakhsh, et al., 1992; Kosman et al., 1991; Simpson, 1983). Only in the absence of both Twist and Snail are mesodermal characteristics lost. Genes activated as a result of Twist and Snail expression are important for modulating the activity or subcellular localization of components of cell-cell junctions and the contractile actomyosin network; thereby providing a link between patterns of upstream gene expression established in the blastoderm and proteins that produce mechanical forces necessary for tissue-specific cell shape changes and movement.

In the mesoderm, Snail represses expression of *neuralized* (*neur*), which encodes an E3 ubiquitin ligase needed to disassemble subapical adherens junctions (Chanet & Schweisguth, 2012). Proteins of the Bearded (Brd) family interact with Neur, providing spatial regulation of internalization of the Notch ligand Delta along the D-V axis important for precise positioning of the mesectoderm (Bardin and Schweisguth, 2006). Additionally, Snail, together with Twist, supports expression of three genes encoding key modulators of the contractile actomyosin network: the G protein-coupled receptor, Mist, and its apically-secreted ligand, Folded gastrulation (Fog) (Costa, Wilson, & Wieschaus, 1994; Manning, Peters, Peifer, & Rogers, 2013) as well as the transmembrane protein T48. Live imaging studies have shown that the transcriptional dynamics of *fog* and *T48* relate to spatial activation of myosin, foreshadowing mesoderm invagination, and suggest that this morphogenetic process of invagniation is controlled through the regulation of spatiotemporal gene expression (Kölsch, Seher,

Fernandez-Ballester, Serrano, & Leptin, 2007; Lim, Levine, & Yamazaki, 2017). Together, Mist/Fog and T48 serve to recruit Ras-like GTP-binding protein (Rho) guanine nucleotide exchange factor (RhoGEF)2 to the apical cell cortex. Further, TNF-receptor-associated factor 4 (Traf4), also induced by Twist and Snail, is required for apical accumulation of adherens junctions (rev. in Gilmour, Rembold, & Leptin, 2017). Specifically, Snail induces stochastic fluctuation of mesoderm apical cell surface size through Fog and RhoGEF2 to promote medio-apical accumulation of Myosin-II, which, along with junctional repositioning, leads to apical constriction of the presumptive mesoderm cells (Fig. 4A; Martin, Kaschube, & Wieschaus, 2009; Mitrossilis et al., 2017; Pouille, Ahmadi, Brunet, & Farge, 2009). Mesoderm invagination is an instructive example of how broad domains established by morphogens (i.e. DL) permit subdivision of epithelial sheets into discrete domains of expression of tissue-specific transcription factors (i.e. Twist, Snail) that drive the regulation of cell machinery to modulate the activity of broadly-expressed mechanosensory components. Coordinated arrangement of these components, including actomyosin and adherens junctions, permit stereotyped cell shape changes and mechanical forces that are translated across interconnected epithelial sheets to produce tissue-scale morphological outputs.

Whereas the entire mesoderm undergoes coordinate invagination regulated by a single set of inputs, the endoderm invaginates as two distinct parts (Fig. 4B) in striking difference to formation of this germ layer in most other animals (Technau & Scholz, 2003). The anterior and posterior endoderm primordia are not only specified separately but also utilize distinct morphogenetic mechanisms to invaginate. At the posterior of the embryo, *hkb*'s anterior-ventral boundary abuts the *sna* posterior boundary. In this posterior region, *tll* overlaps the *hkb* expression domain, but also extends further anteriorly (Fig. 4B). *hkb* is necessary for the expression of *srp*, which is able to not only specify endoderm from ectoderm but is also required for the morphogenesis and function of the endoderm (Reuter, 1994). The region where *tll* overlaps with *sna*, and *hkb* is also absent, expresses *brachyenteron* (*byn*) and will invaginate together with the posterior midgut primordium to specify ectodermal hingut and anal pads, and caudal visceral mesoderm (Kispert et al., 1994; Kusch & Reuter, 1999; Singer, Harbecke, Kusch, Reuter, & Lengyel, 1996; Fig. 4C). At the opposite end of the embryo, the anterior midgut primordium is specified from a region in which *hkb*, *snail*, and *twist* are coexpressed, and which corresponds precisely to the anterior expression domain of *srp* (Rehorn et al., 1996). Endoderm morphogenesis emerges from a combination of local transcriptional initiation as well as mechanically-driven cell deformation resulting from either mesoderm invagination (posterior endoderm) or germ-band elongation (anterior endoderm), discussed below (Bailles et al., 2019; Mitrossilis et al., 2017). Notably, the posterior endoderm invaginates by apical constriction governed by a pathway highly similar to that employed during mesoderm invagination including recruitment of RhoGEF2 (Sweeton et al., 1991). Moreover, while Sna and Fog are required for apical accumulation of myosin to support invagination of the mesoderm, Fog, but not Snail, is expressed in the posterior endoderm. Stretching of the posterior endoderm cells, as a result of mesoderm invagination, activates Myosin-II in these cells about 80 seconds later (Mitrossilis et al., 2017). Alternatively, the anterior midgut is thought to invaginate using a different mechanism that depends instead on germ-band elongation-induced expression of *twist* (Farge, 2003).

A few minutes following mesoderm invagination, cell intercalation within the lateral epidermis initiates and leads to germ-band extension (GBE) (Fig. 4C). During this process following mesoderm invagination, the germ band from the left and right side of the embryo join at the ventral midline. GBE is largely driven by cell intercalation and cell shape changes that are a passive response to extrinsic tensile forces that deform the germ band (Bertet, Sulak, & Lecuit, 2004; Blankenship, Backovic, Sanny, Weitz, & Zallen, 2006; Irvine & Wieschaus, 1994). In short, cells converge in the D-V direction while extending in the A-P direction causing overall lengthening of the germ band along the sides of the embryo (Fig. 4C). Cell intercalation supports GBE and is regulated by the pair-rule transcription factors Even-skipped (Eve) and Runt (Zallen & Wieschaus, 2004). Eve and Runt direct the spatial expression of three Toll-family receptors in partially overlapping patterns, creating asymmetry along the A-P axis. Heterophilic interactions between Toll receptors are necessary for the polarized distribution of myosin to cell surfaces orthogonal to the plane of extension (Paré et al., 2014). Some evidence also supports a role for the classical Frizzled-based planar cell polarity pathways in cell polarity of the germ band (Warrington, Strutt, & Strutt, 2013), but whether Toll and Frizzled pathways cooperate in generating planar polarity is unclear (Tepass, 2014). In addition, tissue-level pulling forces generated by invaginations of the posterior midgut contribute to GBE (Collinet, Rauzi, Lenne, & Lecuit, 2015; Lye et al., 2015), but GBE is thought to be independent of mesoderm invagination (Irvine & Wieschaus, 1994).

Additional cell movements during gastrulation contribute to the formation of folds along the A-P axis; however, these events differ in the mechanism of fold formation from either mesoderm or endoderm invagination (Fig. 4D). For example, Bcd regulates expression of the gap genes, *paired* and *buttonhead*, which subsequently interact to specify the anterior-most expression of the pair-rule gene *eve* (Vincent, Blankenship, & Wieschaus, 1997). This particular domain of *eve* expression directs cell behavior at the cephalic furrow that involves actin- (but not myosin) dependent shortening of cells along their apical–basal axis (Spencer et al., 2015; Fig. 4). Further, dorsoventral differences in cephalic furrow depth depend on D-V axis positional information (e.g. Irish & Gelbart, 1987), providing an example of how cells at distinct positions along the A-P and D-V axes of the embryo come to acquire a specific fate and morphology determined by the complement of transcription factors they express.

The anterior and posterior dorsal transverse folds (ADF & PDF, respectively—Fig. 4D) are epithelial folds that form on the dorsal side of the gastrulating *Drosophila* embryo at stereotypical locations coincident with the second and fifth stripes of *runt* expression (Wang, Khan, Kaschube, & Wieschaus, 2012). These dorsal folds are formed via a myosin-independent mechanism (i.e. actin) to apically-constrict cells involving differential positioning of adherens junctions and controlled by modulation of epithelial apical–basal polarity. The dorsal folds do not appear to contribute to germ-band extension; it is conceivable that D-V convergence of the extending germ-band mechanically pulls on the amnioserosa to support cell flattening (Lye et al., 2015; Pope & Harris, 2008). Finally, the dorsal-most ectoderm, which does not participate in GBE, develops into the extra-embryonic amnioserosa and non-neural ectoderm. The amnioserosa is a transient structure that is only present for half of embryogenesis and does not contribute to the larva, but has been shown to support the process of germ-band retraction later in embryogenesis following gastrulation (Lynch et al., 2013). In this way, broad domains established in the blastoderm adopt distinct programs of downstream gene expression to activate effectors of cellular machinery that subsequently produce cell movements and tissue morphologies resulting in a gastrulated embryo.

6. Dynamic feedback between genetic patterning and physical tissue morphogenesis

In this chapter, we have provided a framework for understanding patterning of the embryo as it relates to germ layer formation, focusing on the gene regulatory interactions that control specification of distinct cell types.

However, studies of gene expression *in vivo* have provided more evidence that patterning is a dynamic process. Cell-fate decisions are likely not fixed, but evolving; able to refine as tissues physically take shape. Therefore, rather than patterning of the embryo being organized in a strict hierarchical relationship, in which transcription factors work in series to change expression patterns, it is likely that genetic patterning involves significant feedback. Intercellular signaling pathways that control cell-fate decisions and tissue patterning are emerging as central mediators of morphogenetic feedback control (rev. in Gilmour et al., 2017).

Any morphogenetic event that changes the relative position of cells that produce or receive signals has the potential to provide feedback on the patterning process. This includes, but is not limited to, alterations in the number or arrangement of cells through proliferation or convergent extension, or changes in cell adhesion or polarity, which can influence the trafficking and presentation of ligands to impact cell signaling. In addition, mechanical forces have been shown to contribute to driving tissue-shaping processes, and provide another important channel for feedback between cell form and fate. For example, a mechanosensitive pathway, involving the Wnt–cadherin effector β-catenin, has been shown to regulate *Drosophila* mesodermal identity *in vivo* and has been found to work similarly in zebrafish (Brunet et al., 2013). Further, cell movements during gastrulation have been shown to be affected by the mechanical properties of neighboring germ layers suggesting complex interdependencies between tissues that likely depend on crosstalk between the gene networks that regulate cell tension throughout the embryo (Perez-Mockus et al., 2017; Rauzi et al., 2015). Mechanistic investigations into how tissue shaping is encoded genetically and how, conversely, tissue shaping feeds back into gene expression is an important area of future study.

References

Ajduk, A., & Zernicka-Goetz, M. (2016). Polarity and cell division orientation in the cleavage embryo: From worm to human. *Molecular Human Reproduction, 22*(10), 691–703.

Alberga, A., Boulay, J. L., Kempe, E., Dennefeld, C., & Haenlin, M. (1991). The snail gene required for mesoderm formation in Drosophila is expressed dynamically in derivatives of all three germ layers. *Development, 111*(4), 983–992.

Ali-Murthy, Z., & Kornberg, T. B. (2016). Bicoid gradient formation and function in the Drosophila pre-syncytial blastoderm. *eLife, 5*, e13222. https://doi.org/10.7554/eLife.13222.

Ali-Murthy, Z., Lott, S. E., Eisen, M. B., & Kornberg, T. B. (2013). An essential role for zygotic expression in the pre-cellular Drosophila embryo. *PLoS Genetics, 9*(4), e1003428.

Amarnath, S., Stevens, L. M., & Stein, D. S. (2017). Reconstitution of Torso signaling in cultured cells suggests a role for both Trunk and Torso-like in receptor activation. *Development, 144*(4), 677–686.
Ashe, H. L., & Levine, M. (1999). Local inhibition and long-range enhancement of Dpp signal transduction by Sog. *Nature, 398*(6726), 427–431.
Ashe, H. L., Mannervik, M., & Levine, M. (2000). Dpp signaling thresholds in the dorsal ectoderm of the Drosophila embryo. *Development, 127*(15), 3305–3312.
Astigarraga, S., Grossman, R., Díaz-Delfín, J., Caelles, C., Paroush, Z. 'e., & Jiménez, G. (2007). A MAPK docking site is critical for downregulation of Capicua by Torso and EGFR RTK signaling. *The EMBO Journal, 26*(3), 668–677.
Bailles, A., Collinet, C., Philippe, J.-M., Lenne, P.-F., Munro, E., & Lecuit, T. (2019). Genetic induction and mechanochemical propagation of a morphogenetic wave. *Nature, 572*(7770), 467–473.
Bardin, A. J., & Schweisguth, F. (2006). Bearded family members inhibit neuralized-mediated endocytosis and signaling activity of delta in Drosophila. *Developmental Cell, 10*(2), 245–255.
Bashirullah, A., Halsell, S. R., Cooperstock, R. L., Kloc, M., Karaiskakis, A., Fisher, W. W., et al. (1999). Joint action of two RNA degradation pathways controls the timing of maternal transcript elimination at the midblastula transition in Drosophila melanogaster. *The EMBO Journal, 18*(9), 2610–2620.
Bate, M., & Arias, A. M. (1993). *The development of* Drosophila melanogaster. CSHL Press.
Beer, J., Technau, G. M., & Campos-Ortega, J. A. (1987). Lineage analysis of transplanted individual cells in embryos of Drosophila melanogaster: IV. Commitment and proliferative capabilities of mesodermal cells. *Roux's Archives of Developmental Biology: The Official Organ of the EDBO, 196*(4), 222–230.
Bertet, C., Sulak, L., & Lecuit, T. (2004). Myosin-dependent junction remodelling controls planar cell intercalation and axis elongation. *Nature, 429*(6992), 667–671.
Biehs, B., François, V., & Bier, E. (1996). The Drosophila short gastrulation gene prevents Dpp from autoactivating and suppressing neurogenesis in the neuroectoderm. *Genes & Development, 10*(22), 2922–2934.
Blankenship, J. T., Backovic, S. T., Sanny, J. S. P., Weitz, O., & Zallen, J. A. (2006). Multicellular rosette formation links planar cell polarity to tissue morphogenesis. *Developmental Cell, 11*(4), 459–470.
Blythe, S. A., & Wieschaus, E. F. (2015a). Coordinating cell cycle remodeling with transcriptional activation at the Drosophila MBT. *Current Topics in Developmental Biology, 113*, 113–148.
Blythe, S. A., & Wieschaus, E. F. (2015b). Zygotic genome activation triggers the DNA replication checkpoint at the midblastula transition. *Cell, 160*(6), 1169–1181.
Briscoe, J., & Small, S. (2015). Morphogen rules: Design principles of gradient-mediated embryo patterning. *Development, 142*, 3996–4009. https://doi.org/10.1242/dev.129452.
Brönner, G., Chu-LaGraff, Q., Doe, C. Q., Cohen, B., Weigel, D., Taubert, H., et al. (1994). Sp1/egr-like zinc-finger protein required for endoderm specification and germ-layer formation in Drosophila. *Nature, 369*, 664–668. https://doi.org/10.1038/369664a0.
Brown, J. L., & Wu, C. (1993). Repression of Drosophila pair-rule segmentation genes by ectopic expression of tramtrack. *Development, 117*(1), 45–58.
Brunet, T., Bouclet, A., Ahmadi, P., Mitrossilis, D., Driquez, B., Brunet, A.-C., et al. (2013). Evolutionary conservation of early mesoderm specification by mechanotransduction in bilateria. *Nature Communications, 4*, 2821.
Bushati, N., Stark, A., Brennecke, J., & Cohen, S. M. (2008). Temporal reciprocity of miRNAs and their targets during the maternal-to-zygotic transition in Drosophila. *Current Biology, 18*(7), 501–506.

Campos-Ortega, J. A., & Hartenstein, V. (1997). *The embryonic development of* Drosophila melanogaster. Springer-Verlag Berlin Heidelberg. https://doi.org/10.1007/978-3-662-22489-2.

Casanova, J., Furriols, M., McCormick, C. A., & Struhl, G. (1995). Similarities between trunk and spätzle, putative extracellular ligands specifying body pattern in Drosophila. *Genes & Development*, *9*(20), 2539–2544.

Casanova, J., & Struhl, G. (1989). Localized surface activity of torso, a receptor tyrosine kinase, specifies terminal body pattern in Drosophila. *Genes & Development*, *3*(12B), 2025–2038.

Chanet, S., & Schweisguth, F. (2012). Regulation of epithelial polarity by the E3 ubiquitin ligase neuralized and the bearded inhibitors in Drosophila. *Nature Cell Biology*, *14*(5), 467–476.

Chari, S., Wilky, H., Govindan, J., & Amodeo, A. A. (2019). Histone concentration regulates the cell cycle and transcription in early development. *Development*, *146*(19), dev177402. https://doi.org/10.1242/dev.177402.

Chen, K., Johnston, J., Shao, W., Meier, S., Staber, C., & Zeitlinger, J. (2013). A global change in RNA polymerase II pausing during the Drosophila midblastula transition. *eLife*, *2*, e00861.

Chen, H., Xu, Z., Mei, C., Yu, D., & Small, S. (2012). A system of repressor gradients spatially organizes the boundaries of bicoid-dependent target genes. *Cell*, *149*(3), 618–629.

Cho, Y. S., Stevens, L. M., & Stein, D. (2010). Pipe-dependent ventral processing of Easter by snake is the defining step in Drosophila embryo DV axis formation. *Current Biology*, *20*(12), 1133–1137.

Collinet, C., Rauzi, M., Lenne, P.-F., & Lecuit, T. (2015). Local and tissue-scale forces drive oriented junction growth during tissue extension. *Nature Cell Biology*, *17*(10), 1247–1258.

Costa, M., Wilson, E. T., & Wieschaus, E. (1994). A putative cell signal encoded by the folded gastrulation gene coordinates cell shape changes during Drosophila gastrulation. *Cell*, *76*(6), 1075–1089.

Crémazy, F., Berta, P., & Girard, F. (2000). Sox neuro, a new Drosophila Sox gene expressed in the developing central nervous system. *Mechanisms of Development*, *93*(1–2), 215–219.

Crews, S. T., Thomas, J. B., & Goodman, C. S. (1988). The Drosophila single-minded gene encodes a nuclear protein with sequence similarity to the per gene product. *Cell*, *52*(1), 143–151.

de las Heras, J. M., & Casanova, J. (2006). Spatially distinct downregulation of Capicua repression and tailless activation by the Torso RTK pathway in the Drosophila embryo. *Mechanisms of Development*, *123*(6), 481–486.

Driever, W., & Nüsslein-Volhard, C. (1988). The bicoid protein determines position in the Drosophila embryo in a concentration-dependent manner. *Cell*, *54*(1), 95–104.

Dubnau, J., & Struhl, G. (1996). RNA recognition and translational regulation by a homeodomain protein. *Nature*, *379*(6567), 694–699.

Duffy, J. B., & Perrimon, N. (1994). The torso pathway in Drosophila: Lessons on receptor tyrosine kinase signaling and pattern formation. *Developmental Biology*, *166*(2), 380–395.

Ephrussi, A., Dickinson, L. K., & Lehmann, R. (1991). Oskar organizes the germ plasm and directs localization of the posterior determinant nanos. *Cell*, *66*(1), 37–50.

Farge, E. (2003). Mechanical induction of Twist in the Drosophila foregut/stomodeal primordium. *Current Biology*, *13*(16), 1365–1377.

Ferguson, E. L., & Anderson, K. V. (1992a). Decapentaplegic acts as a morphogen to organize dorsal-ventral pattern in the Drosophila embryo. *Cell*, *71*, 451–461. https://doi.org/10.1016/0092-8674(92)90514-d.

Ferguson, E. L., & Anderson, K. V. (1992b). Localized enhancement and repression of the activity of the TGF-beta family member, decapentaplegic, is necessary for dorsal-ventral pattern formation in the Drosophila embryo. *Development*, *114*(3), 583–597.

Foe, V. E. (1989). Mitotic domains reveal early commitment of cells in Drosophila embryos. *Trends in Genetics*, *5*, 322. https://doi.org/10.1016/0168-9525(89)90120-0.

Foe, V. E., & Alberts, B. M. (1983). Studies of nuclear and cytoplasmic behaviour during the five mitotic cycles that precede gastrulation in Drosophila embryogenesis. *Journal of Cell Science*, *61*, 31–70.

Foo, S. M., Sun, Y., Lim, B., Ziukaite, R., O'Brien, K., Nien, C.-Y., et al. (2014). Zelda potentiates morphogen activity by increasing chromatin accessibility. *Current Biology*, *24*(12), 1341–1346.

Gilmour, D., Rembold, M., & Leptin, M. (2017). From morphogen to morphogenesis and back. *Nature*, *541*, 311–320. https://doi.org/10.1038/nature21348.

Hamm, D. C., & Harrison, M. M. (2018). Regulatory principles governing the maternal-to-zygotic transition: Insights from Drosophila melanogaster. *Open Biology*, *8*(12), 180183.

Harrison, M. M., Li, X.-Y., Kaplan, T., Botchan, M. R., & Eisen, M. B. (2011). Zelda binding in the early Drosophila melanogaster embryo marks regions subsequently activated at the maternal-to-zygotic transition. *PLoS Genetics*, 7(10), e1002266.

Hartenstein, V., & Wodarz, A. (2013). Initial neurogenesis in Drosophila. *Wiley Interdisciplinary Reviews: Developmental Biology*, *2*, 823. https://doi.org/10.1002/wdev.117.

Hong, J.-W., Hendrix, D. A., Papatsenko, D., & Levine, M. S. (2008). How the dorsal gradient works: Insights from postgenome technologies. *Proceedings of the National Academy of Sciences of the United States of America*, 105(51), 20072–20076.

Ip, Y. T., Park, R. E., Kosman, D., Bier, E., & Levine, M. (1992). The dorsal gradient morphogen regulates stripes of rhomboid expression in the presumptive neuroectoderm of the Drosophila embryo. *Genes & Development*, *6*(9), 1728–1739.

Ip, Y. T., Park, R. E., Kosman, D., Yazdanbakhsh, K., & Levine, M. (1992). Dorsal-twist interactions establish snail expression in the presumptive mesoderm of the Drosophila embryo. *Genes & Development*, *6*(8), 1518–1530.

Irish, V. F., & Gelbart, W. M. (1987). The decapentaplegic gene is required for dorsal-ventral patterning of the Drosophila embryo. *Genes & Development*, *1*(8), 868–879.

Irish, V., Lehmann, R., & Akam, M. (1989). The Drosophila posterior-group gene nanos functions by repressing hunchback activity. *Nature*, *338*(6217), 646–648.

Irvine, K. D., & Wieschaus, E. (1994). Cell intercalation during Drosophila germband extension and its regulation by pair-rule segmentation genes. *Development*, *120*(4), 827–841.

Jaeger, J. (2011). The gap gene network. *Cellular and Molecular Life Sciences*, *68*(2), 243–274.

Jaźwińska, A., Kirov, N., Wieschaus, E., Roth, S., & Rushlow, C. (1999). The Drosophila gene brinker reveals a novel mechanism of Dpp target gene regulation. *Cell*, *96*(4), 563–573.

Jaźwińska, A., Rushlow, C., & Roth, S. (1999). The role of brinker in mediating the graded response to Dpp in early Drosophila embryos. *Development*, *126*(15), 3323–3334.

Jiang, J., & Levine, M. (1993). Binding affinities and cooperative interactions with bHLH activators delimit threshold responses to the dorsal gradient morphogen. *Cell*, *72*(5), 741–752.

Jiménez, G., Guichet, A., Ephrussi, A., & Casanova, J. (2000). Relief of gene repression by torso RTK signaling: Role of capicua in Drosophila terminal and dorsoventral patterning. *Genes & Development*, *14*(2), 224–231.

Johnson, T. K., Henstridge, M. A., & Warr, C. G. (2017). MACPF/CDC proteins in development: Insights from Drosophila torso-like. *Seminars in Cell & Developmental Biology*, 72, 163–170.

Johnston, D. S., St Johnston, D., Beuchle, D., & Nüsslein-Volhard, C. (1991). Staufen, a gene required to localize maternal RNAs in the Drosophila egg. *Cell*, *66*, 51–63. https://doi.org/10.1016/0092-8674(91)90138-o.

Jukam, D., Shariati, S. A. M., & Skotheim, J. M. (2017). Zygotic genome activation in vertebrates. *Developmental Cell*, *42*(4), 316–332.

Kispert, A., Herrmann, B. G., Leptin, M., & Reuter, R. (1994). Homologs of the mouse Brachyury gene are involved in the specification of posterior terminal structures in Drosophila, Tribolium, and Locusta. *Genes & Development*, *8*, 2137–2150. https://doi.org/10.1101/gad.8.18.2137.

Klingler, M., Erdélyi, M., Szabad, J., & Nüsslein-Volhard, C. (1988). Function of torso in determining the terminal anlagen of the Drosophila embryo. *Nature*, *335*, 275–277. https://doi.org/10.1038/335275a0.

Kölsch, V., Seher, T., Fernandez-Ballester, G. J., Serrano, L., & Leptin, M. (2007). Control of Drosophila gastrulation by apical localization of adherens junctions and RhoGEF2. *Science*, *315*(5810), 384–386.

Koromila, T., & Stathopoulos, A. (2017). Broadly expressed repressors integrate patterning across orthogonal axes in embryos. *Proceedings of the National Academy of Sciences of the United States of America*, *114*(31), 8295–8300. https://doi.org/10.1073/pnas.1703001114.

Koromila, T., & Stathopoulos, A. (2019). Distinct roles of broadly expressed repressors support dynamic enhancer action and change in time. *Cell Reports*, *28*(4), 855–863.e5.

Kosman, D., Ip, Y. T., Levine, M., & Arora, K. (1991). Establishment of the mesoderm-neuroectoderm boundary in the Drosophila embryo. *Science*, *254*(5028), 118–122.

Kusch, T., & Reuter, R. (1999). Functions for Drosophila brachyenteron and forkhead in mesoderm specification and cell signalling. *Development*, *126*(18), 3991–4003.

Kwasnieski, J. C., Orr-Weaver, T. L., & Bartel, D. P. (2019). Early genome activation in is extensive with an initial tendency for aborted transcripts and retained introns. *Genome Research*, *29*(7), 1188–1197.

Laver, J. D., Li, X., Ray, D., Cook, K. B., Hahn, N. A., Nabeel-Shah, S., et al. (2015). Brain tumor is a sequence-specific RNA-binding protein that directs maternal mRNA clearance during the Drosophila maternal-to-zygotic transition. *Genome Biology*, *16*, 94.

Lefebvre, F. A., & Lécuyer, É. (2018). Flying the RNA nest: Drosophila reveals novel insights into the transcriptome dynamics of early development. *Journal of Developmental Biology*, *6*(1), 5. https://doi.org/10.3390/jdb6010005.

Leptin, M. (1991). Twist and snail as positive and negative regulators during Drosophila mesoderm development. *Genes & Development*, *5*(9), 1568–1576.

Leptin, M., & Grunewald, B. (1990). Cell shape changes during gastrulation in Drosophila. *Development*, *110*(1), 73–84.

Li, W. X., Agaisse, H., Mathey-Prevot, B., & Perrimon, N. (2002). Differential requirement for STAT by gain-of-function and wild-type receptor tyrosine kinase Torso in Drosophila. *Development*, *129*(18), 4241–4248.

Li, J., Xia, F., & Li, W. X. (2003). Coactivation of STAT and Ras is required for germ cell proliferation and invasive migration in Drosophila. *Developmental Cell*, *5*(5), 787–798.

Liang, H.-L., Nien, C.-Y., Liu, H.-Y., Metzstein, M. M., Kirov, N., & Rushlow, C. (2008). The zinc-finger protein Zelda is a key activator of the early zygotic genome in Drosophila. *Nature*, *456*(7220), 400–403.

Liberman, L. M., & Stathopoulos, A. (2009). Design flexibility in cis-regulatory control of gene expression: Synthetic and comparative evidence. *Developmental Biology*, *327*(2), 578–589.

Lim, B., Levine, M., & Yamazaki, Y. (2017). Transcriptional pre-patterning of Drosophila gastrulation. *Current Biology*, *27*(4), 610.

Löhr, U., Chung, H.-R., Beller, M., & Jäckle, H. (2009). Antagonistic action of Bicoid and the repressor Capicua determines the spatial limits of Drosophila head gene expression domains. *Proceedings of the National Academy of Sciences of the United States of America*, *106*(51), 21695–21700.

Lu, X., Perkins, L. A., & Perrimon, N. (1993). The torso pathway in Drosophila: A model system to study receptor tyrosine kinase signal transduction. *Development*, *119*(Supplement), 47–56.

Lye, C. M., Blanchard, G. B., Naylor, H. W., Muresan, L., Huisken, J., Adams, R. J., et al. (2015). Mechanical coupling between endoderm invagination and axis extension in Drosophila. *PLoS Biology, 13*(11), e1002292.

Lynch, H. E., Crews, S. M., Rosenthal, B., Kim, E., Gish, R., Echiverri, K., et al. (2013). Cellular mechanics of germ band retraction in Drosophila. *Developmental Biology, 384*(2), 205–213.

Macdonald, P. M., & Struhl, G. (1986). A molecular gradient in early Drosophila embryos and its role in specifying the body pattern. *Nature, 324*(6097), 537–545.

Manning, A. J., Peters, K. A., Peifer, M., & Rogers, S. L. (2013). Regulation of epithelial morphogenesis by the G protein-coupled receptor mist and its ligand Fog. *Science Signaling, 6*, ra98. https://doi.org/10.1126/scisignal.2004427.

Martin, A. C., Kaschube, M., & Wieschaus, E. F. (2009). Pulsed contractions of an actin-myosin network drive apical constriction. *Nature, 457*(7228), 495–499.

McCleland, M. L., Farrell, J. A., & O'Farrell, P. H. (2009). Influence of cyclin type and dose on mitotic entry and progression in the early Drosophila embryo. *The Journal of Cell Biology, 184*, 639–646. https://doi.org/10.1083/jcb.200810012.

McDaniel, S. L., Gibson, T. J., Schulz, K. N., Fernandez Garcia, M., Nevil, M., Jain, S. U., et al. (2019). Continued activity of the pioneer factor Zelda is required to drive zygotic genome activation. *Molecular Cell, 74*(1), 185–195.e4.

Mineo, A., Furriols, M., & Casanova, J. (2015). Accumulation of the Drosophila Torso-like protein at the blastoderm plasma membrane suggests that it translocates from the eggshell. *Development, 142*, 1299–1304. https://doi.org/10.1242/dev.117630.

Mineo, A., Furriols, M., & Casanova, J. (2018). The trigger (and the restriction) of Torso RTK activation. *Open Biology, 8*(12), 180180.

Mitrossilis, D., Röper, J.-C., Le Roy, D., Driquez, B., Michel, A., Ménager, C., et al. (2017). Mechanotransductive cascade of Myo-II-dependent mesoderm and endoderm invaginations in embryo gastrulation. *Nature Communications, 8*, 13883.

Mlodzik, M., & Gehring, W. J. (1987). Expression of the caudal gene in the germ line of Drosophila: Formation of an RNA and protein gradient during early embryogenesis. *Cell, 48*, 465–478. https://doi.org/10.1016/0092-8674(87)90197-8.

Morel, V., & Schweisguth, F. (2000). Repression by suppressor of hairless and activation by Notch are required to define a single row of single-minded expressing cells in the Drosophila embryo. *Genes & Development, 14*(3), 377–388.

Moussian, B., & Roth, S. (2005). Dorsoventral axis formation in the Drosophila embryo—Shaping and transducing a morphogen gradient. *Current Biology*, 15(21), R887–R899.

Nien, C.-Y., Liang, H.-L., Butcher, S., Sun, Y., Fu, S., Gocha, T., et al. (2011). Temporal coordination of gene networks by Zelda in the early Drosophila embryo. *PLoS Genetics*, 7(10), e1002339.

Niessing, D., Driever, W., Sprenger, F., Taubert, H., Jäckle, H., & Rivera-Pomar, R. (2000). Homeodomain position 54 specifies transcriptional versus translational control by bicoid. *Molecular Cell, 5*(2), 395–401.

Ochoa-Espinosa, A., Yu, D., Tsirigos, A., Struffi, P., & Small, S. (2009). Anterior-posterior positional information in the absence of a strong Bicoid gradient. *Proceedings of the National Academy of Sciences, 106*, 3823–3828. https://doi.org/10.1073/pnas.0807878105.

Ozdemir, A., Ma, L., White, K. P., & Stathopoulos, A. (2014). Su(H)-mediated repression positions gene boundaries along the dorsal-ventral axis of Drosophila embryos. *Developmental Cell, 31*(1), 100–113.

Paré, A. C., Vichas, A., Fincher, C. T., Mirman, Z., Farrell, D. L., Mainieri, A., et al. (2014). A positional Toll receptor code directs convergent extension in Drosophila. *Nature, 515*(7528), 523–527.

Perez-Mockus, G., Mazouni, K., Roca, V., Corradi, G., Conte, V., & Schweisguth, F. (2017). Spatial regulation of contractility by neuralized and bearded during furrow invagination in Drosophila. *Nature Communications*, *8*(1), 1594.

Perrimon, N., Lu, X., Hou, X. S., Hsu, J. C., Melnick, M. B., Chou, T. B., et al. (1995). Dissection of the Torso signal transduction pathway in Drosophila. *Molecular Reproduction and Development*, *42*(4), 515–522.

Perrimon, N., Pitsouli, C., & Shilo, B.-Z. (2012). Signaling mechanisms controlling cell fate and embryonic patterning. *Cold Spring Harbor Perspectives in Biology*, *4*(8), a005975.

Pignoni, F., Baldarelli, R. M., Steingrímsson, E., Diaz, R. J., Patapoutian, A., Merriam, J. R., et al. (1990). The Drosophila gene tailless is expressed at the embryonic termini and is a member of the steroid receptor superfamily. *Cell*, *62*(1), 151–163.

Pope, K. L., & Harris, T. J. C. (2008). Control of cell flattening and junctional remodeling during squamous epithelial morphogenesis in Drosophila. *Development*, *135*(13), 2227–2238.

Porcher, A., Abu-Arish, A., Huart, S., Roelens, B., Fradin, C., & Dostatni, N. (2010). The time to measure positional information: Maternal hunchback is required for the synchrony of the bicoid transcriptional response at the onset of zygotic transcription. *Development*, *137*, 2795–2804. https://doi.org/10.1242/dev.051300.

Porcher, A., & Dostatni, N. (2010). The bicoid morphogen system. *Current Biology*, *20*(5), R249–R254.

Pouille, P.-A., Ahmadi, P., Brunet, A.-C., & Farge, E. (2009). Mechanical signals trigger myosin II redistribution and mesoderm invagination in Drosophila embryos. *Science Signaling*, *2*(66) ra16.

Pritchard, D. K., & Schubiger, G. (1996). Activation of transcription in Drosophila embryos is a gradual process mediated by the nucleocytoplasmic ratio. *Genes & Development*, *10*(9), 1131–1142.

Rauzi, M., Krzic, U., Saunders, T. E., Krajnc, M., Ziherl, P., Hufnagel, L., et al. (2015). Embryo-scale tissue mechanics during Drosophila gastrulation movements. *Nature Communications*, *6*, 8677.

Ray, R. P., Arora, K., Nüsslein-Volhard, C., & Gelbart, W. M. (1991). The control of cell fate along the dorsal-ventral axis of the Drosophila embryo. *Development*, *113*(1), 35–54.

Read, D., Levine, M., & Manley, J. L. (1992). Ectopic expression of the Drosophila tramtrack gene results in multiple embryonic defects, including repression of even-skipped and fushi tarazu. *Mechanisms of Development*, *38*(3), 183–195.

Reeves, G. T., & Stathopoulos, A. (2009). Graded dorsal and differential gene regulation in the Drosophila embryo. *Cold Spring Harbor Perspectives in Biology*, *1*(4), a000836.

Reeves, G. T., Trisnadi, N., Truong, T. V., Nahmad, M., Katz, S., & Stathopoulos, A. (2012). Dorsal-ventral gene expression in the Drosophila embryo reflects the dynamics and precision of the dorsal nuclear gradient. *Developmental Cell*, *22*(3), 544–557.

Rehorn, K. P., Thelen, H., Michelson, A. M., & Reuter, R. (1996). A molecular aspect of hematopoiesis and endoderm development common to vertebrates and Drosophila. *Development*, *122*(12), 4023–4031.

Reuter, R. (1994). The gene serpent has homeotic properties and specifies endoderm versus ectoderm within the Drosophila gut. *Development*, *120*(5), 1123–1135.

Reuter, R., Grunewald, B., & Leptin, M. (1993). A role for the mesoderm in endodermal migration and morphogenesis in Drosophila. *Development*, *119*(4), 1135–1145.

Reuter, R., & Leptin, M. (1994). Interacting functions of snail, twist and huckebein during the early development of germ layers in Drosophila. *Development*, *120*(5), 1137–1150.

Rivera-Pomar, R., Niessing, D., Schmidt-Ott, U., Gehring, W. J., & Jäckle, H. (1996). RNA binding and translational suppression by bicoid. *Nature*, *379*(6567), 746–749.

Roth, S., Stein, D., & Nüsslein-Volhard, C. (1989). A gradient of nuclear localization of the dorsal protein determines dorsoventral pattern in the Drosophila embryo. *Cell*, *59*, 1189–1202. https://doi.org/10.1016/0092-8674(89)90774-5.

Rushlow, C., Frasch, M., Doyle, H., & Levine, M. (1987). Maternal regulation of zerknüllt: A homoeobox gene controlling differentiation of dorsal tissues in Drosophila. *Nature*, *330*(6148), 583–586.

Rushlow, C. A., Han, K., Manley, J. L., & Levine, M. (1989). The graded distribution of the dorsal morphogen is initiated by selective nuclear transport in Drosophila. *Cell*, *59*(6), 1165–1177.

Rushlow, C., & Roth, S. (1996). The role of the dpp-group genes in dorsoventral patterning of the Drosophila embryo. *Advances in Developmental Biology (1992)*, *4*, 27–82. https://doi.org/10.1016/s1566-3116(08)60016-x.

Sandler, J. E., Irizarry, J., Stepanik, V., Dunipace, L., Amrhein, H., & Stathopoulos, A. (2018). A developmental program truncates long transcripts to temporally regulate cell signaling. *Developmental Cell*, *47*(6), 773–784.e6.

Schulz, K. N., Bondra, E. R., Moshe, A., Villalta, J. E., Lieb, J. D., Kaplan, T., et al. (2015). Zelda is differentially required for chromatin accessibility, transcription factor binding, and gene expression in the early Drosophila embryo. *Genome Research*, *25*(11), 1715–1726.

Schulz, C., & Tautz, D. (1995). Zygotic caudal regulation by hunchback and its role in abdominal segment formation of the Drosophila embryo. *Development*, *121*(4), 1023–1028.

Schüpbach, T. (1987). Germ line and soma cooperate during oogenesis to establish the dorsoventral pattern of egg shell and embryo in Drosophila melanogaster. *Cell*, *49*, 699–707. https://doi.org/10.1016/0092-8674(87)90546-0.

Sen, J., Goltz, J. S., Stevens, L., & Stein, D. (1998). Spatially restricted expression of pipe in the Drosophila egg chamber defines embryonic dorsal–ventral polarity. *Cell*, *95*, 471–481. https://doi.org/10.1016/s0092-8674(00)81615-3.

Shermoen, A. W., McCleland, M. L., & O'Farrell, P. H. (2010). Developmental control of late replication and S phase length. *Current Biology*, *20*, 2067–2077. https://doi.org/10.1016/j.cub.2010.10.021.

Shindo, Y., & Amodeo, A. A. (2019). Dynamics of free and chromatin-bound histone H3 during early embryogenesis. *Current Biology*, *29*(2), 359–366.e4.

Simpson, P. (1983). Maternal-zygotic gene interactions during formation of the dorsoventral pattern in Drosophila embryos. *Genetics*, *105*(3), 615–632.

Singer, J. B., Harbecke, R., Kusch, T., Reuter, R., & Lengyel, J. A. (1996). Drosophila brachyenteron regulates gene activity and morphogenesis in the gut. *Development*, *122*(12), 3707–3718.

Spencer, A. K., Siddiqui, B. A., & Thomas, J. H. (2015). Cell shape change and invagination of the cephalic furrow involves reorganization of F-actin. *Developmental Biology*, *402*(2), 192–207.

Spirov, A., Fahmy, K., Schneider, M., Frei, E., Noll, M., & Baumgartner, S. (2009). Formation of the bicoid morphogen gradient: An mRNA gradient dictates the protein gradient. *Development*, *136*(4), 605–614.

Sprenger, F., & Nüsslein-Volhard, C. (1992). Torso receptor activity is regulated by a diffusible ligand produced at the extracellular terminal regions of the Drosophila egg. *Cell*, 71(6), 987–1001.

Stathopoulos, A., & Levine, M. (2004). Whole-genome analysis of Drosophila gastrulation. *Current Opinion in Genetics & Development*, *14*(5), 477–484.

Stathopoulos, A., & Levine, M. (2005). Genomic regulatory networks and animal development. *Developmental Cell*, *9*(4), 449–462.

Stern, C. D. (2004). *Gastrulation: From cells to embryo*. CSHL Press.

Stevens, L. M., Beuchle, D., Jurcsak, J., Tong, X., & Stein, D. (2003). The Drosophila embryonic patterning determinant torsolike is a component of the eggshell. *Current Biology*, *13*(12), 1058–1063.
Steward, R. (1989). Relocalization of the dorsal protein from the cytoplasm to the nucleus correlates with its function. *Cell*, *59*(6), 1179–1188.
Strecker, T. R., Kongsuwan, K., Lengyel, J. A., & Merriam, J. R. (1986). The zygotic mutant tailless affects the anterior and posterior ectodermal regions of the Drosophila embryo. *Developmental Biology*, *113*(1), 64–76.
Struhl, G. (1989). Differing strategies for organizing anterior and posterior body pattern in Drosophila embryos. *Nature*, *338*(6218), 741–744.
Struhl, G., Struhl, K., & Macdonald, P. M. (1989). The gradient morphogen bicoid is a concentration-dependent transcriptional activator. *Cell*, *57*, 1259–1273. https://doi.org/10.1016/0092-8674(89)90062-7.
Sweeton, D., Parks, S., Costa, M., & Wieschaus, E. (1991). Gastrulation in Drosophila: The formation of the ventral furrow and posterior midgut invaginations. *Development*, *112*(3), 775–789.
Tadros, W., Goldman, A. L., Babak, T., Menzies, F., Vardy, L., Orr-Weaver, T., et al. (2007). SMAUG is a major regulator of maternal mRNA destabilization in Drosophila and its translation is activated by the PAN GU kinase. *Developmental Cell*, *12*(1), 143–155.
Tadros, W., & Lipshitz, H. D. (2009). The maternal-to-zygotic transition: A play in two acts. *Development*, 136(18), 3033–3042.
Tautz, D. (1988). Regulation of the Drosophila segmentation gene hunchback by two maternal morphogenetic centres. *Nature*, *332*(6161), 281–284.
Technau, G. M. (1987). A single cell approach to problems of cell lineage and commitment during embryogenesis of Drosophila melanogaster. *Development*, *100*(1), 1–12.
Technau, U., & Scholz, C. B. (2003). Origin and evolution of endoderm and mesoderm. *The International Journal of Developmental Biology*, *47*(7-8), 531–539.
Tepass, U. (2014). Review of developmental biology: Polarize to elongate. *Nature*, *515*(7528), 499–501.
Thisse, C., Perrin-Schmitt, F., Stoetzel, C., & Thisse, B. (1991). Sequence-specific trans-activation of the Drosophila twist gene by the dorsal gene product. *Cell*, *65*, 1191–1201. https://doi.org/10.1016/0092-8674(91)90014-p.
Vastenhouw, N. L., Cao, W. X., & Lipshitz, H. D. (2019). The maternal-to-zygotic transition revisited. *Development*, *146*(11). https://doi.org/10.1242/dev.161471.
Vincent, A., Blankenship, J. T., & Wieschaus, E. (1997). Integration of the head and trunk segmentation systems controls cephalic furrow formation in Drosophila. *Development*, *124*(19), 3747–3754.
Wang, Y.-C., Khan, Z., Kaschube, M., & Wieschaus, E. F. (2012). Differential positioning of adherens junctions is associated with initiation of epithelial folding. *Nature*, *484*(7394), 390–393.
Wang, C., & Lehmann, R. (1991). Nanos is the localized posterior determinant in Drosophila. *Cell*, *66*(4), 637–647.
Warrington, S. J., Strutt, H., & Strutt, D. (2013). The frizzled-dependent planar polarity pathway locally promotes E-cadherin turnover via recruitment of RhoGEF2. *Development*, *140*(5), 1045–1054.
Weigel, D., Jurgens, G., Klingler, M., & Jackle, H. (1990). Two gap genes mediate maternal terminal pattern information in Drosophila. *Science*, *248*, 495–498. https://doi.org/10.1126/science.2158673.
Wharton, K. A., Ray, R. P., & Gelbart, W. M. (1993). An activity gradient of decapentaplegic is necessary for the specification of dorsal pattern elements in the Drosophila embryo. *Development*, *117*(2), 807–822.

Wieschaus, E. (2016). Positional information and cell fate determination in the early Drosophila embryo. *Current Topics in Developmental Biology*, *117*, 567–579. https://doi.org/10.1016/bs.ctdb.2015.11.020.

Wu, L. H., & Lengyel, J. A. (1998). Role of caudal in hindgut specification and gastrulation suggests homology between Drosophila amnioproctodeal invagination and vertebrate blastopore. *Development*, *125*(13), 2433–2442.

Xu, Z., Chen, H., Ling, J., Yu, D., Struffi, P., & Small, S. (2014). Impacts of the ubiquitous factor Zelda on bicoid-dependent DNA binding and transcription in Drosophila. *Genes & Development*, *28*(6), 608–621.

Yuan, K., Seller, C. A., Shermoen, A. W., & O'Farrell, P. H. (2016). Timing the Drosophila mid-blastula transition: A cell cycle-centered view. *Trends in Genetics*, *32*(8), 496–507.

Zallen, J. A., & Wieschaus, E. (2004). Patterned gene expression directs bipolar planar polarity in Drosophila. *Developmental Cell*, *6*(3), 343–355.

Zhang, H. (2001). Brinker is a sequence-specific transcriptional repressor in the Drosophila embryo. *Genes & Development*, *15*, 261–266. https://doi.org/10.1101/gad.861201.

Zhu, X., Sen, J., Stevens, L., Goltz, J. S., & Stein, D. (2005). Drosophila pipe protein activity in the ovary and the embryonic salivary gland does not require heparan sulfate glycosaminoglycans. *Development*, *132*(17), 3813–3822.

CHAPTER TWO

Setting up for gastrulation in zebrafish

Florence L. Marlow*
Icahn School of Medicine Mount Sinai Department of Cell, Developmental and Regenerative Biology, New York, NY, United States
*Corresponding author: e-mail address: florence.marlow@mssm.edu

Contents

Abstract

Soon after fertilization the zebrafish embryo generates the pool of cells that will give rise to the germline and the three somatic germ layers of the embryo (ectoderm, mesoderm and endoderm). As the basic body plan of the vertebrate embryo emerges, evolutionarily conserved developmental signaling pathways, including Bmp, Nodal, Wnt, and Fgf, direct the nearly totipotent cells of the early embryo to adopt gene expression profiles and patterns of cell behavior specific to their eventual fates. Several decades of molecular genetics research in zebrafish has yielded significant insight into the maternal and zygotic contributions and mechanisms that pattern this vertebrate embryo. This new understanding is the product of advances in genetic manipulations and imaging technologies that have allowed the field to probe the cellular, molecular and biophysical aspects underlying early patterning. The current state of the field indicates that patterning is governed by the integration of key signaling pathways and physical interactions between cells, rather than a patterning system in which distinct pathways are deployed

Current Topics in Developmental Biology, Volume 136
ISSN 0070-2153
https://doi.org/10.1016/bs.ctdb.2019.08.002

to specify a particular cell fate. This chapter focuses on recent advances in our understanding of the genetic and molecular control of the events that impart cell identity and initiate the patterning of tissues that are prerequisites for or concurrent with movements of gastrulation.

1. Introduction

Well before the movements of gastrulation begin in fish, maternal pathways ensure that there are a sufficient number of cells, and that the embryo is poised to turn on the gene expression programs that are necessary to specify each germ layer and to drive cell behavior and morphogenesis according to or in parallel to cell fate. Not only is the fertilized egg poised molecularly, but it is also structurally prepared as the initial framework is in place to orient the cells as they navigate to the proper regions of the embryo to ensure that the head, the tail and all the organs in between form in the correct place and at the correct time. The events that prepare the embryo for gastrulation begin in the oocyte, with specification of the oocyte axes, completion of oocyte maturation and meiosis to prepare the egg for activation and fertilization. Soon after fertilization of the zebrafish egg, the blastoderm elevates at the animal pole as the cytoplasm and yolky content of the egg separate. During this process known as cytoplasmic segregation the cytoplasm moves in an animal-ward direction and the yolk remains at the vegetal pole (Fernandez, Valladares, Fuentes, & Ubilla, 2006; Fuentes & Fernandez, 2010; Shamipour et al., 2019) and reviewed in Marlow (2010). As this occurs, the fertilized egg completes the final meiotic division as the maternal chromosomes unite with the paternally contributed chromosomes to restore ploidy, and the centrosome is rebuilt from the paternally contributed centrosome to prepare the spindle that will mediate the early mitotic divisions, called cleavages. These synchronous and fast cell divisions divide the very large blastoderm, which sits atop the yolky mass of the egg, into progressively smaller cells until the time of midblastula transition or zygotic genome activation (reviewed in Marlow, 2010). These cleavage divisions occur without compensating cell growth through a truncated cell cycle until the nucleocytoplasmic ratio reaches a stable size after about 13 cell cycles. At this time, bulk activation of the genome occurs, and the cell cycle lengthens and becomes asynchronous (reviewed in Lee, Bonneau, & Giraldez, 2014; Marlow, 2010). These early cleavages ensure that the embryo has enough cells to initiate morphogenesis and patterning of the germ cells and is a prerequisite for gastrulation because mutants or chemical/pharmacological treatments that disrupt these early cell divisions delay the onset of genome activation and gastrulation proper (reviewed in Lee et al., 2014). These events occur prior to activation of the zygotic genome and are fully under the control of maternally supplied RNAs and proteins. In recent years, some of the genes and mechanisms that mediate the unique cell biology of these early cleavages and the activation of the zygotic genome, including roles for chromatin remodeling have been identified (Joseph et al., 2017; Meier et al., 2018; Veil, Yampolsky, Gruning, & Onichtchouk, 2019) and reviewed in Jukam, Shariati, and Skotheim, 2017, Lee et al. (2014), Marlow (2010). The important advances in these areas are not covered here due to space limitations but have been reviewed elsewhere (Jukam et al., 2017; Lee et al., 2014; Marlow, 2010). As mentioned, the key product of the cleavage divisions is the pool of cells that will give rise to the germ layers of the embryo, ectoderm, mesoderm, endoderm. In response to spatial and temporal cues, this troupe of nearly totipotent cells will adopt specific fates, initiate cell-type specific gene expression profiles, and undergo carefully orchestrated cell-type specific movements to form the basic body plan of the embryo. These cells also give rise to extraembryonic tissues when just before gastrulation, the marginal blastomeres collapse and donate their nuclei to form the yolk syncytial layer (YSL), which as you will see plays a key role in patterning and coordinates morphogenetic movements described in the following chapters. After several decades of molecular genetics research in zebrafish significant insight into the maternal and zygotic factors that pattern the early embryo has been attained. Excitingly, despite this wealth of knowledge, new genetic tools and technical advances have allowed the field to delve even deeper into the cellular and molecular aspects of early patterning. Emerging models indicate that patterning relies on the integration of key signaling pathways, including BMP, Nodal, Wnts, and Fgf, rather than employing distinct pathways with dedicated roles to specify a particular cell fate. This chapter focuses on recent advances in our understanding of the genetic and molecular control of acquisition of cell identity and embryo pattern that are prerequisites for or concurrent with morphogenetic movements of gastrulation.

2. Maternal factors and dorsal-ventral patterning

2.1 Defining and delimiting dorsal: Maternal β-catenin with or without Wnt

Although the early zebrafish embryo is polarized along the animal-vegetal axis, with the blastoderm positioned at the animal pole and the extraembryonic yolk at the vegetal pole, the blastoderm remains radially symmetric until the Nieuwkoop center induces the Spemann-Mangold organizer in the overlying blastoderm. Once specified, the organizer region confers dorsal identity and the ability to differentiate as prechordal plate and axial mesoderm in the milieu of global Bmp, a potent ventralizing signal. The first postfertilization steps toward patterning the dorsal-ventral (DV) axis of the embryo involve asymmetric animal-ward movement of the dorsal determinants (DDs) (Fig. 1). Surgical manipulation of early embryos

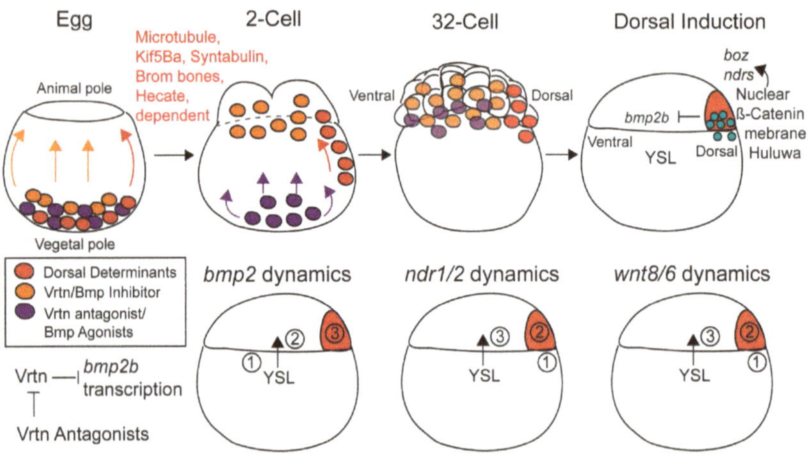

Fig. 1 Maternal factors in dorsal-ventral axis determination and dynamics of the potent ventralizer, *bmp2b* and dorsal organizer genes. Maternally deposited dorsal determinants, ventral agonists and inhibitors reside at the vegetal pole of the unactivated zebrafish egg. Immediately upon activation of the egg, the dorsal determinants and the Bmp regulator Vrtn move toward the animal pole. Translocation of the dorsal determinants requires a vegetal microtubule array and several maternal factors including Kinesin motors and their adaptor proteins (genes listed in red). Once the dorsal determinants reach the future dorsal side, the Nieuwkoop center and later the Spemann-Mangold organizer, or shield, is induced and can be seen in accumulation of Huluwa protein at the membrane and β-Catenin in the nuclei on the prospective dorsal side, which leads to activation of *bozozok* and *nodals*. With a roughly 45-min delay, antagonists of Vrtn translocate toward the animal pole. The early translocation of a *bmp2b* transcriptional repressor and the dorsal determinants are thought to help shape the expression of the organizer gene and *bmp2b* expression domains. Initially uniform *bmp2b* is confined to the blastoderm by the action of Vrtn, and zygotic domains are thought to be initiated from the YSL (1) to the blastoderm margin (2) by the later arriving Bmp agonists from the vegetal pole. Once induced Bmp positively feeds back to promote its own expression. In the gastrula, a new domain is induced within the organizer (3) that helps to shape the Bmp signaling gradient. New *nodal* ligand expression is also initiated from the YSL, by dorsally accumulating maternal *squint/ndr1* to induce organizer expression of *ndr* ligands (2) followed by expression around the margin (3). The dynamics of *wnt8a* and *wnt6* are similar to those of *squint/ndr1*.

indicates that factors important for dorsal axis formation reside at the vegetal pole of fertilized eggs and are moved to the prospective dorsal side very early, because bisecting embryos into animal and vegetal halves before the 2-cell stage produces ventralized animal halves, presumably by preventing relocation of these determinants (Mizuno, Yamaha, Kuroiwa, & Takeda, 1999). There is accumulating evidence that translocation of DDs requires an organized microtubule array and maternally supplied proteins that bind to microtubules. Loss of certain maternally deposited molecular motors and adaptor proteins the egg causes maternal-effect ventralization phenotypes due to deficits in nuclear accumulation of maternal Wnt/β-Catenin signaling upstream of the Nieuwkoop Center (Campbell, Heim, Smith, & Marlow, 2015; Dougan, Warga, Kane, Schier, & Talbot, 2003; Ge et al., 2014; Gore et al., 2005; Jesuthasan & Stahle, 1997; Lyman Gingerich, Westfall, Slusarski, & Pelegri, 2005; Mei, Lee, Marlow, Miller, & Mullins, 2009; Nojima et al., 2010; Schneider, Steinbesser, Warga, & Hausen, 1996; Tran et al., 2012). Consistent with this notion, ventralization of maternal-effect *hecate* and *syntabulin* mutants can be suppressed by overexpressing Wnt/β-Catenin pathway components (Ge et al., 2014; Lyman Gingerich et al., 2005; Mei et al., 2009; Nojima et al., 2010), thus activating the pathway in embryonic regions acts somewhat redundant to or bypasses the signaling from the extraembryonic Nieuwkoop Center that normally activates maternally supplied β-Catenin to induce organizer genes (Fig. 1) (Bellipanni et al., 2006; Fekany et al., 1999; Fekany-Lee, Gonzalez, Miller-Bertoglio, & Solnica-Krezel, 2000; Gonzalez et al., 2000; Kelly, Chin, Leatherman, Kozlowski, & Weinberg, 2000; Varga, Maegawa, Bellipanni, & Weinberg, 2007).

Canonical Wnt signaling acts through the effector protein Dsh and culminates in activation and nuclear translocation of β-Catenin to activate gene expression (Fig. 2) (reviewed in Nusse & Clevers, 2017). As mentioned, it is clear that maternal β-Catenin is required for organizer specification in zebrafish, but thus far no essential maternal Wnt ligand has been identified (Fig. 2) (Bellipanni et al., 2006; Varga et al., 2007). Wnt8a is a compelling candidate, because it can induce organizer genes when over expressed, and expression analyses show that *wnt8a* transcripts are initially localized at the vegetal pole, where the DDs reside, and then moved toward the prospective dorsal side in a microtubule-dependent manner (Fig. 1) (Ge et al., 2014; Lu, Thisse, & Thisse, 2011; Tran et al., 2012). However, maternal and zygotic Wnt8a are separately dispensable for maternal β-Catenin signaling (Hino et al., 2018). Instead, the more severe patterning phenotypes associated with

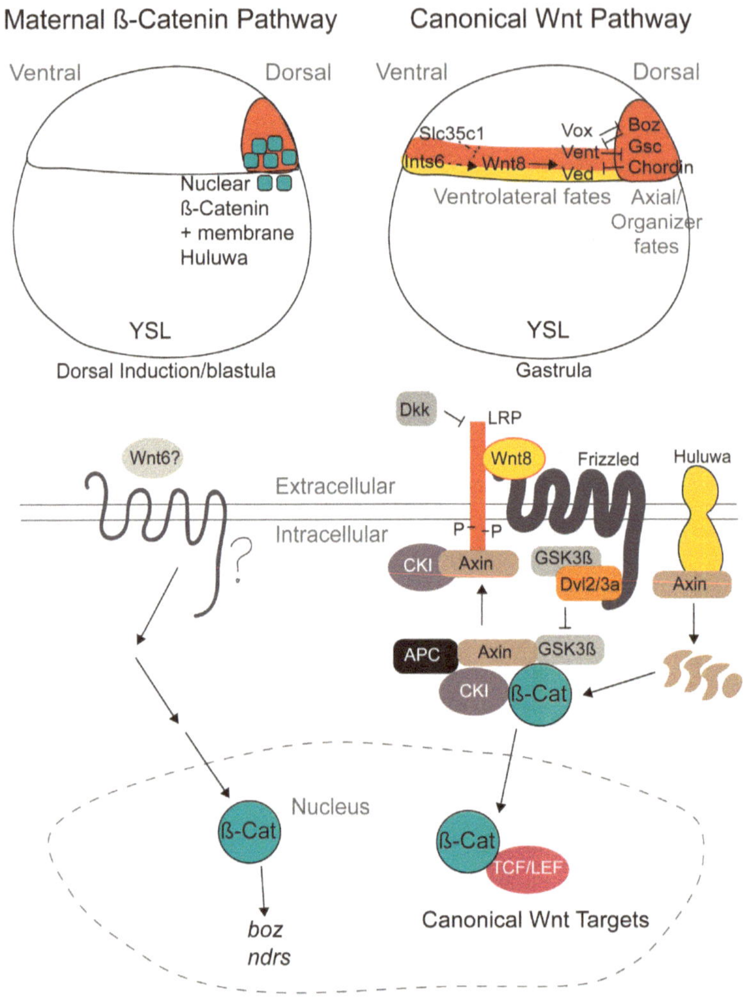

Fig. 2 Comparison of Maternal β-Catenin and Canonical Wnt Pathways. Activation of maternal β-Catenin pathway can be seen in the nuclear accumulation of β-Catenin in the cells on the dorsal side on the early blastula with membrane-localized Huluwa protein (top left). Although both pathways depend on β-Catenin, no ligand or receptor necessary for maternal β-Catenin signaling has been identified, and Disheveled (Dvl) does not seem to be an essential effector for this pathway (bottom left). Instead, the transmembrane maternal protein Huluwa binds to Axin and promotes its degradation, thereby promoting accumulation and signaling via β-Catenin (bottom left). The transcriptional targets of the maternal β-Catenin pathway include early organizer genes such as *bozozok* and *nodals*. The first sign of a required Wnt ligand for canonical Wnt signaling is detectable in the early gastrula (top right). At this time Wnt signaling requires zygotic and maternal Wnt8a, which initiates in the shield and then signals through Frizzled receptors and LRP co-receptors to promote association between Disheveled (Dvl) and GSK3β to promote liberation of β-Catenin from the destruction complex (Axin, APC). β-Catenin then translocates into the nucleus where it interacts with transcriptional cofactors, TCF/LEF to promote the expression of Wnt targets and lateral mesodermal fates.

the combined loss of maternal and zygotic *wnt8*, indicate that maternal Wnt8a cooperates with zygotic Wnt8a to limit the organizer and promote ventrolateral patterning (Fig. 2) (Hino et al., 2018). Thus, the elusive maternal Wnt remains to be identified; *wnt6a* which can induce immediate organizer genes and resides at the vegetal pole (Hino et al., 2018) seems to be the most promising of the remaining candidates. If a maternal Wnt is required for maternal β-Catenin signaling, then it likely utilizes a novel effector, as the five zebrafish *disheveled* (*dvl*) genes (Gray et al., 2009) have been excluded based on a lack of maternal expression or genetic data indicating that early dorsal fate specification is intact in single and double mutants lacking maternal and zygotic Dvl function (Xing et al., 2018). Moreover, unlike loss of or interference with maternal and zygotic Wnt8 or overexpression of negative regulators of canonical Wnt signaling, which causes expansion of organizer fates at the expense of lateral fates and anterior-posterior patterning defects (Feng, Jiang, Wu, & Marlow, 2014; Hino et al., 2018; Lekven, Thorpe, Waxman, & Moon, 2001; Mo et al., 2010; Xing et al., 2018; Yao et al., 2010), loss of Dvl does not, indicating that these Wnt8 patterning functions are Dvl independent (Fig. 2) (Xing et al., 2018).

The notion of a Wnt-ligand and Dvl-independent mechanism of axis specification in zebrafish has garnered support from the recent discovery of a ventralized maternal-effect mutant, *huluwa,* that disrupts a previously uncharacterized transmembrane protein (Yan et al., 2018). Elevated Axin expression in *huluwa* mutants and rescue experiments placed Huluwa upstream of β-Catenin (Yan et al., 2018). Consistent with a role in promoting β-Catenin mediated axis formation, Huluwa protein localizes to the dorsal blastomeres in which nuclear β-Catenin accumulates (Yan et al., 2018) (Fig. 1). Structure-function studies indicate that membrane-localized Huluwa protein promotes β-catenin and axis formation by directly binding to Axin to promote degradation of this key negative regulator of β-Catenin (Yan et al., 2018) (Fig. 2). Currently it is not known how or if Huluwa interacts with Wnt or other signaling pathways acting in the early embryo, but because overexpression of Huluwa protein lacking the extracellular domain restores patterning in *huluwa* mutants it is not thought to act as a receptor for a putative Wnt or other ligand.

bozozok (*boz*) and genes encoding *nodal-related signals* (*ndrs*) to be discussed later are among the earliest zygotic targets of the maternal β-Catenin (Dougan et al., 2003; Gore et al., 2005; Kelly et al., 2000; Koos & Ho, 1998, 1999; Leung, Soll, Arnold, Kemler, & Driever, 2003; Maegawa, Varga, & Weinberg, 2006; Nojima et al., 2004; Ryu et al., 2001; Shimizu et al., 2000;

Solnica-Krezel & Driever, 2001). The transcriptional repressor Sox3, counteracts β-Catenin by repressing organizer genes including *noggin*, *bozozok*, *chordin*, *ndrs* and *fgfs* (Kuo, Lam, Hewitt, & Scotting, 2013; Shih et al., 2010). Bozozok is a direct repressor of *bmp2b* and the zygotic homeodomain transcription factors, *vox*, *vent*, and *ved*, that are initially dependent on Wnt8a, maternal Pou5f3/Spg, maternal Runx2bt2, but eventually rely on zygotic expression of Bmp2b to limit the organizer domain and axial mesoderm to allow for nonaxial mesoderm and endodermal fates in lateral regions (Fig. 3A) (Belting et al., 2011; Burgess, Reim, Chen, Hopkins, & Brand, 2002; Fekany et al., 1999; Fekany-Lee et al., 2000; Flores, Lam, Crosier, & Crosier, 2008; Imai & Talbot, 2001; Kawahara, Wilm, Solnica-Krezel, & Dawid, 2000a, 2000b; Kramer et al., 2002; Koos & Ho, 1998, 1999; Leung, Bischof, et al., 2003; Leung, Soll, et al., 2003; Lunde, Belting, & Driever, 2004; Melby, Beach, Mullins, & Kimelman, 2000; Nguyen et al., 1998; Onichtchouk, Geier, Messerschmidt, et al., 2010; Onichtchouk, Geier, Polok, et al., 2010; Ramel & Lekven, 2004; Reim & Brand, 2006; Reim, Mizoguchi, Stainier, Kikuchi, & Brand, 2004; Schier et al., 1996; Shimizu et al., 2002, 2000; Solnica-Krezel & Driever, 2001; Yamanaka et al., 1998). In *bozozok* mutants, axial mesoderm or organizer derivatives do not form due to excess Wnt8b, as suppression of Wnt8b can restore axial fates in the absence of *boz* (Fekany et al., 1999; Fekany-Lee et al., 2000). *Bozozok* promotes expression of *goosecoid* by repressing *vox*, *vent* and *ved* (Imai & Talbot, 2001; Shimizu et al., 2002). Conversely, the absence of the maternally supplied E3 Ubiquitin ligase, Lnx2b, that destabilizes Boz and inhibits Nodal, or of direct repressors of *chordin* and *goosecoid* (*vox*, *vent* and *ved*) in ventrolateral regions of the blastoderm margin, causes ectopic Chordin and Goosecoid and consequently axial mesoderm and organizer fates expand at the expense of ventrolateral fates (Fekany et al., 1999; Fekany-Lee et al., 2000; Kawahara & Dawid, 2000; Kawahara et al., 2000b; Ro & Dawid, 2009, 2010). Initial induction of Fgf signaling components by Nodal activity subsequently promotes expression of *goosecoid* and the BMP antagonist *chordin* while repressing *bmps* (Feldman et al., 1998; Furthauer, Van Celst, Thisse, & Thisse, 2004; Gritsman, Talbot, & Schier, 2000; Joore et al., 1996; Kuo et al., 2013; Maegawa et al., 2006; Varga et al., 2007). Fgf limits the organizer domain and its own activity by directly inducing Fgf repressors *sprouty2/4* and *sefs* (Furthauer, Reifers, Brand, Thisse, & Thisse, 2001; Furthauer et al., 2004; Kovalenko et al., 2006) and indirectly inducing *vox, vent and ved* (Imai & Talbot, 2001; Kawahara et al., 2000a, 2000b; Yamanaka et al.,

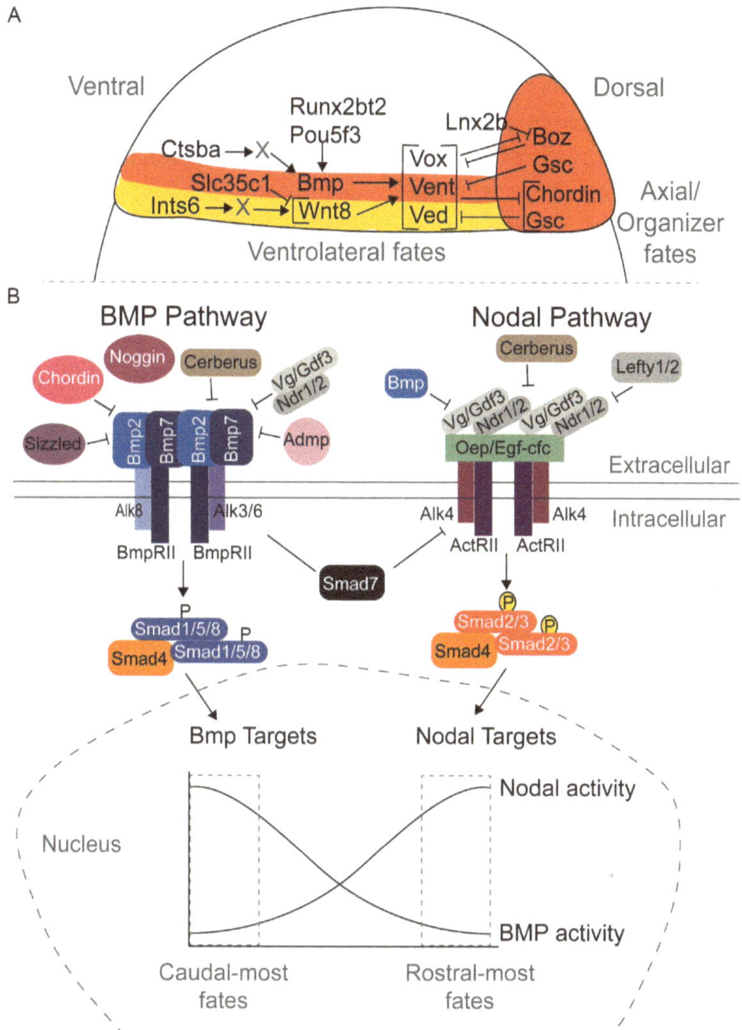

Fig. 3 TGFβ and Wnt signaling in axis formation and patterning. (A) Schematic of early gastrula showing antagonistic and cooperative transcriptional and repressive activities of Bmp, Nodal targets, and Wnt signaling pathways. (B) Bmp signaling, which is highest on the ventral side of the early embryo signals as a dimer of heterodimeric ligands (Bmp2/7) through a heterodimer of type I and type II receptors. The key type II receptor remains to be identified. A number of extracellular antagonists and intracellular repressive Smad7 act to block pathway activity. Once activated, the Bmp receptors activate phosphorylation of Smad1/5/8, which competes with Nodal Smads (2/3) for the common Smad4. Upon association with Smad4 the complex enters the nucleus and promotes transcription of Bmp targets. Nodal signaling is also activated heterodimers that complex with dimers of type II (ActRII) and type I, Alk4 receptors to promote phosphorylation of Smad2/3, which when associated with Smad4 enters the nucleus to activate Nodal target genes. A consequence of these mutually repressive interactions is high Bmp activity on the ventral side and high Nodal activity closest to the margin and on dorsal. Cell fates along the rostral caudal axis are thought to be determined by integrating the levels of both Bmp and Nodal activity such that high Nodal and low Bmp promotes rostral fates and high Bmp and low Nodal promotes caudal fates.

1998). Pou5f3 promotes *bmp2b* expression in part by repressing *fgf8a* by a mechanism that may involve *sprouty4* (Belting et al., 2011; Onichtchouk, Geier, Messerschmidt, et al., 2010; Onichtchouk, Geier, Polok, et al., 2010; Reim & Brand, 2006). The Wnt8 inhibitor, *dickkopf* (*dkk*) is induced in the organizer by maternal β-Catenin, as are the Wnt inhibitors, *sfrp1* and *frzb*, along with negative regulators of Nodal signaling, *dapper1* and *dapper2*, which are indirectly induced via Nodal; thus, feedback inhibition is an important factor in defining the organizer domain (Hashimoto et al., 2000; Lu et al., 2011; Nojima et al., 2004; Peng & Westerfield, 2006; Pezeron et al., 2006; Seiliez, Thisse, & Thisse, 2006; Shinya, Eschbach, Clark, Lehrach, & Furutani-Seiki, 2000; Tendeng & Houart, 2006; Waxman, 2005; Zhang et al., 2004). Expansion of the organizer at the expense of ventrolateral fates also occurs when maternal *ints6*, a component of the RNA processing complex that generates snRNAs or spliceosomal RNAs, is absent (Baillat et al., 2005; Ezzeddine et al., 2011; Kapp, Abrams, Marlow, & Mullins, 2013). Like, Wnt8, Ints6 does not mediate initial organizer formation as judged by the expression of maternal *wnts* and nuclear β-Catenin, but subsequently prevents expansion of markers of axial fates in mid-gastrulae (Kapp et al., 2013). Expansion of axial fates in *ints6* mutants is thought to be due to a primary function in promoting BMP and Wnt8a, which regulate expression of the genes encoding Chordin repressors, *vox*, *vent* and *ved* (Kapp et al., 2013); however, the mechanism of *ints6* action and its direct targets remain to be determined.

2.2 Determinants at the vegetal pole

Surgical manipulation of early embryos provided evidence that factors essential for dorsal axis formation such as maternal *wnt* RNAs discussed above reside at the vegetal pole initially but reach the blastoderm margin by the 2-cell stage (Mizuno et al., 1999). Consistent with this notion, embryo bisection prior to 2-cell stage caused ventralization, but bisection after the 2-cell stage had the surprising effect of dorsalizing rather than ventralizing the animal halves (Mizuno et al., 1999). Recently, Shao and colleagues sought to determine the mechanism by which this dorsalization occurs. They postulated that factors required for ventral fates or that limit dorsal fates might also reside at the vegetal pole. As discussed above, the main dorsal determinant is maternal β-Catenin, which activates canonical Wnt signaling on the future dorsal side; so, it was assumed that the factor missing in embryos bisected after 2-cell stage would be a Wnt inhibitor. However, the observation that expansion of dorsal at the expense of ventral occurred without an increase in maternal Wnt activity hinted at another mechanism (Shao et al., 2017). Instead, dorsalization was associated with decreased expression of Bmp target genes and decreased p-Smad, indicating impaired BMP signaling (Shao et al., 2017).

Bmp ligands are produced in the embryo zygotically, but their expression is regulated by maternal factors, including Gdf6/Radar, Pou5f3, Alk6/8, MGA, Max, Runx2bt, and Smad4 (Flores et al., 2008; Goutel, Kishimoto, Schulte-Merker, & Rosa, 2000; Reim & Brand, 2006; Reim et al., 2004; Sidi, Goutel, Peyrieras, & Rosa, 2003; Sun, Tseng, Fan, Ball, & Dougan, 2014; Wilm & Solnica-Krezel, 2003). Maternal MGA, Max and Smad4 in the YSL promote *bmp2b* expression and activate BMP signaling within the overlying blastoderm of gastrula (Fig. 1) (Sun et al., 2014). Initially, the dorsal factors like *bozozok* repress *bmp* expression on the dorsal side, while the antagonists like Chordin block its activity (Fisher, Amacher, & Halpern, 1997; Hammerschmidt, Pelegri, et al., 1996; Leung, Bischof, et al., 2003; Miller-Bertoglio et al., 1999; Schulte-Merker, Lee, McMahon, & Hammerschmidt, 1997). However, despite the concentration of BMP antagonists on dorsal, *bmp* ligands, including *bmp2b* and *bmp7* are expressed within the dorsal organizer and contribute to formation of the BMP-activity gradient (Figs. 1 and 3) (Bauer, Lele, Rauch, Geisler, & Hammerschmidt, 2001; Dick et al., 2000; Hammerschmidt, Serbedzija, & McMahon, 1996; Hild et al., 1999; Kishimoto, Lee, Zon, Hammerschmidt, & Schulte-Merker, 1997; Kramer et al., 2002; Little & Mullins, 2006, 2009; Nguyen et al., 1998; Schmid et al., 2000; Stickney, Imai, Draper, Moens, & Talbot, 2007; Tucker, Mintzer, & Mullins, 2008; Xue et al., 2014). Diminished *bmp2b* expression in gastrula stage embryos bisected at 2-cell and the observation that overexpression of *bmp2b* can suppress dorsalization of 2-cell bisected embryos led to models wherein removal of a vegetal pole Bmp agonist from 2-cell bisected embryos leaves repressors already positioned near the blastoderm to block BMP activation (Shao et al., 2017). Accordingly, if both activators and repressors of dorsal fates reside at the vegetal pole initially, then trapping of activators in the vegetal piece would account for the ventralization associated with early bisection (Fig. 1) (Kishimoto et al., 1997; Shao et al., 2017). Using this line of reasoning, *vertebrate development associated transcription repressor* (*vrtn*) was identified in an RNA sequencing based screen for differentially expressed animal and vegetal pole RNAs from fertilized eggs (Shao et al., 2017). The dynamic expression pattern of *vrtn*, initially at the vegetal pole and subsequently at the margin of intact embryos before the 2-cell stage along with its dorsalizing activity in overexpression (OE) assays provides support for *vrtn* as a compelling candidate regulator (Shao et al., 2017). Moreover, like dorsalization observed in surgically manipulated embryos, Vrtn induces dorsalization by reducing *bmp2b* expression without affecting maternal Wnt target expression (Shao et al., 2017). Reciprocally, maternal *vrtn* is required for dorsalization of 2-cell bisected and intact embryos, by a mechanism that does not involve blocking Bmp signal transduction (Shao et al., 2017). A combination of embryo, cell culture, ChIP, and reporter assays suggest that Vrtn functions in the nucleus as a transcriptional repressor of the *bmp2b* gene (Shao et al., 2017). The current model, based on the observation that isolated factors from vegetal pole pieces can antagonize Vrtn-associated dorsalization of animal pole pieces, involves yet to be identified repressors of Vrtn that also reside at the vegetal pole, but that move animal-ward with slower kinetics than *vrtn*. The current data suggest that maternally encoded products, including Vrtn and others still to be defined, are asymmetrically positioned in the oocyte and selectively moved from the extraembryonic yolk cell to sequentially signal to the overlying blastoderm, thereby triggering the exquisite spatial and temporal control of patterning. Recombining fluorescent tags into these maternal loci is now possible (Albadri, Del Bene, & Revenu, 2017; Auer & Del Bene, 2014; Auer, Duroure, Concordet, & Del Bene, 2014; Auer, Duroure, De Cian, Concordet, & Del Bene, 2014; Shin, Chen, & Solnica-Krezel, 2014) and will be required to analyze these early movements and rigorously explore this exciting hypothesis.

2.3 Turning BMP on by shutting down the organizer

Gdf6a/Radar and Pou5f3 positively regulate expression of *bmp* genes at the margin (Reim & Brand, 2006; Reim et al., 2004; Sidi et al., 2003; Wilm & Solnica-Krezel, 2003), whereas Vrtn acts also at the margin to repress them (Shao et al., 2017). Unlike *vrtn,* the other genes known to affect BMP signaling, including maternal *runx2bt2*, *ints6*, and *split top,* do not regulate initial induction of *bmp* or promote its expression at the margin; instead, these genes act to repress axial fates and to prevent loss of BMP in lateral regions (Flores et al., 2008; Kapp et al., 2013; Langdon et al., 2016). For example, *split top*, which encodes the lysosomal endopeptidase Cathepsin B (Ctsba) was identified in a maternal-effect screen and is required for patterning the DV axis upstream of Bmp signaling (Langdon et al., 2016). Consistent with Ctsba function in promoting Bmp gradient formation, *split top* mutants have disruptions in epiboly and the Bmp-regulated cell behaviors that will be discussed in detail in the chapter "Mechanisms of zebrafish epiboly: A current view" by Bruce and Heisenberg. Aspects of both the *split top* and *ints6* phenotypes (including ventral fate deficits and convergence extension defects of maternal *split top* mutants and ventralization of *ints6* mutants) can be suppressed by overexpression of Bmp signaling agonists (Langdon et al., 2016). Nevertheless, Bmp signaling is not directly regulated by these factors because knocking down Bmp antagonists is sufficient to restore DV patterning in these mutants (Kapp et al., 2013; Langdon et al., 2016). How Ctsba and Ints6 regulate DV patterning remains to be determined, but, like other maternal factors that regulate DV axis formation (e.g., Pou5f3, which directly induces repressors of organizer genes), and unlike zygotic *bmp* mutants, Ctsba and Ints6 are required to limit axial mesoderm; thus, loss of Bmp signaling is secondary to failure to repress the dorsal organizer genes discussed in the previous section (Belting et al., 2011; Kapp et al., 2013; Khan et al., 2012; Langdon et al., 2016; Nguyen et al., 1998; Reim & Brand, 2006; Schmid et al., 2000; Schulte-Merker et al., 1997).

Ventralizing BMP ligands belonging to the TGFβ superfamily establish and signal as heterodimers, primarily BMP 2/7 for early dorsoventral patterning, with contributions from BMP 4/7 for later caudal patterning. These BMP ligands signal through heterodimeric receptors that phosphorylate and activate a gradient of nuclear Smad1/5/8 that promotes expression of BMP target genes (Fig. 3). Among these target genes are those encoding the BMP ligands, particularly in ventrolateral regions where BMP signaling activity is the highest (Bauer et al., 2001; Dick et al., 2000; Hammerschmidt, Pelegri, et al., 1996; Hild et al., 1999; Kishimoto et al., 1997; Kramer et al., 2002; Little & Mullins, 2006, 2009; Nguyen et al., 1998; Schmid et al., 2000; Stickney et al., 2007; Tucker et al., 2008). Of the six known type II receptors in zebrafish, it remains unclear which are relevant for dorsal-ventral patterning (Monteiro et al., 2008; Yadin, Knaus, & Mueller, 2016). Of the type I receptors, two nonredundant classes, Alk8, *lost a fin,* and Alk3/6, are maternally and zygotically expressed and are required for ventral fates in zebrafish (Little & Mullins, 2009; Mintzer et al., 2001; Mullins et al., 1996; Wagner & Mullins, 2002). Several factors that antagonize or modulate BMP activity during dorsal ventral patterning of the zebrafish embryo have been identified, including Chordin, Noggin1, and Follistatin2 (Fig. 3B). For early DV patterning, Chordin seems to be the most significant of these dorsally localized BMP antagonists because the others are either not expressed at this stage or, like *noggin1* and *follistatin2*, are not required for DV patterning on their own but enhance loss of Chordin (Fisher et al., 1997; Hammerschmidt,

Pelegri, et al., 1996; Miller-Bertoglio et al., 1999; Schulte-Merker et al., 1997). Opposing the action of Chordin, BMP1 and Tolloid metalloproteases promote BMP signaling and ventral fates by cleaving Chordin to degrade it and liberate BMP ligand (Blader, Rastegar, Fischer, & Strahle, 1997; Connors, Trout, Ekker, & Mullins, 1999; Connors, Tucker, & Mullins, 2006; Jasuja et al., 2006). The activity of these metalloproteases is competitively blocked by Sizzled/Ogon, a secreted frizzled related protein (Sfrp) that is expressed ventrally and positively regulated by BMP but promotes dorsal fates by inhibiting Tolloid-dependent degradation of Chordin (Miller-Bertoglio et al., 1999; Yabe et al., 2003). In zebrafish, the BMP modulator Twisted gastrulation (Tsg) is a Chordin dependent antagonist of BMP ligands (Little & Mullins, 2004) unless Tolloid is also present, in which case, Tsg promotes Bmp by potentiating degradation of Chordin (Scott et al., 2001; Xie & Fisher, 2005). Crossveinless, also known as Bmper, is an extracellular modulator of BMP that is converted from an extracellular matrix (ECM) associated antagonistic form that is proteolytically cleaved to a proBmp form that competes with Chordin for BMP association (Rentzsch et al., 2006; Zhang et al., 2010). That both reduced and excess Bmper cause dorsalization of zebrafish embryos seems surprising at first glance but can likely be explained by the ability of both uncleaved and cleaved Bmper to directly bind to BMP (Rentzsch et al., 2006). Accordingly, one could imagine that in the overexpression scenario BMP associated with excess ECM-bound Bmper would be sequestered from receptors and thus inactive, leading to dorsalization, and in the loss of function scenario BMP would instead be associated with Chordin and thus also inactive. Determining the extent of redundancy between the known and putative BMP antagonists remains to be investigated by analysis of compound mutants.

Several models have been proposed to explain how the ventral to dorsal BMP gradient is established. These models factor in diffusivity and stability of Bmp ligands and key antagonists, including Chordin (Ben-Zvi, Fainsod, Shilo, & Barkai, 2014; Francois, Vonica, Brivanlou, & Siggia, 2009; Inomata, Shibata, Haraguchi, & Sasai, 2013; Pomreinke et al., 2017; Ramel & Hill, 2013; Zinski et al., 2017). The main differences in these models relates to the extent to which Bmp ligands diffuse and the stability of antagonists (Ben-Zvi et al., 2014; Francois et al., 2009; Inomata et al., 2013; Pomreinke et al., 2017; Ramel & Hill, 2013; Zinski et al., 2017). None of the models to date fully integrate all of the potential inputs from antagonists and proteolytic regulators, but application of new imaging and labeling tools and computational modeling based on biologically derived rules, including expression and activity patterns from zebrafish studies have made headway. In particular, key experimental data have come from transplantation studies, measurements of diffusivity and stability of tagged ligands and antagonists in wild-type and mutant backgrounds, and readouts of pathway activity based on phospho-Smad1/5/9. Collectively these results support models that depend on diffusion of ligands and antagonists such as the "graded source-sink model," wherein BMP diffuses from a ventral source and its activity is antagonized by interaction with a diffusible inhibitor (e.g., Chordin) that is expressed in a reciprocal gradient (Miller-Bertoglio et al., 1999; Pomreinke et al., 2017; Xu, Houssin, Ferri-Lagneau, Thisse, & Thisse, 2014; Zinski et al., 2017). These measurements as well as mathematical models indicate that gradient formation in zebrafish is not likely to involve shuttling by Chordin or Chordin-mediated attainment of peak levels of BMP activity, because co-expression of Chordin with BMP does not appreciably alter diffusivity or stability of BMP, and because BMP ligands still reach peak levels on ventral in *chordin* mutants (Pomreinke et al., 2017; Zinski et al., 2017). Although the BMP gradient has been visualized, there is currently no evidence distinguishing whether differences exist in target responsiveness to different doses or differing length of exposure to BMP signals (Hashiguchi & Mullins, 2013; Ramel & Hill, 2013; Schumacher, Hashiguchi, Nguyen, & Mullins, 2011; Tucker et al., 2008; Xue et al., 2014). Moreover, how the BMP gradient is translated into distinct ventral and dorsal identities is not fully understood, but cells are thought to periodically read the level of BMP signal in the dorsoventral dimension relative to Nodal, Fgf and Wnt cues along the rostral caudal axis and adopt their fate accordingly (Fig. 3) (Hashiguchi & Mullins, 2013; Ramel & Hill, 2013; Schumacher et al., 2011; Tucker et al., 2008; Xue et al., 2014). The ability of a global source of BMP2b or Chordin to rescue the respective mutant phenotypes suggests that establishing a difference is probably more important than equivalent opposing gradients (Fisher & Halpern, 1999; Kishimoto et al., 1997). Moreover, overexpression and transplantation studies indicate that juxtaposition of pathway activities and levels, particularly BMP and Nodal, is key to generating distinct fates and complete embryonic axes (Agathon, Thisse, & Thisse, 2003; Fauny, Thisse, & Thisse, 2009; Thisse & Thisse, 2015; Toyama, O'Connell, Wright, Kuehn, & Dawid, 1995; Xue et al., 2014). How cross talk and integration of pathway activities is mediated is not fully understood, but it is clear that both cooperative regulation and mutual antagonism contribute.

As mentioned above, the expression domains of *bmp* ligand genes are diverse and dynamic (Fig. 1). Although *bmps* are predominantly expressed in ventrolateral regions during gastrulation, a few ligands, including BMP2b, are expressed in the organizer, where expression of BMP antagonists is the highest. These dorsal antagonists directly bind BMP protein or directly repress *bmp* transcription (Furthauer et al., 2004; Hammerschmidt, Pelegri, et al., 1996; Hammerschmidt, Serbedzija, & McMahon, 1996; Kishimoto et al., 1997; Leung, Soll, et al., 2003; Ramel & Hill, 2013; Schmid et al., 2000; Schulte-Merker et al., 1997; Sidi et al., 2003; Wilm & Solnica-Krezel, 2003; Xue et al., 2014), raising the interesting question of what role if any BMP2b plays in the organizer. Evidence that BMP2b may function in the organizer includes detection of Bmp2b protein within the organizer using BMP2b antibodies, and expansion of *chordin* RNA and protein domains when BMP2b activity is specifically blocked within the organizer by expressing a mutant version of Alk3 receptor (tBR) under the control of the *goosecoid* promoter (Xue et al., 2014). Chordin is a dorsally expressed extracellular BMP antagonist that binds to Bmps and prevents their association with the Bmp receptors (Branam, Hoffman, Pelegri, & Greenspan, 2010; Miller-Bertoglio et al., 1999; Miller-Bertoglio, Fisher, Sanchez, Mullins, & Halpern, 1997; Ramel & Hill, 2013; Schulte-Merker et al., 1997; Shimizu et al., 2000; Troilo et al., 2014; Xue et al., 2014). Concomitant with lateral expansion of Chordin, BMP2b protein and phospho-Smad1/5/8 gradients decline at the onset of gastrulation but not before, likely due to feedback antagonism of BMP by the expanded Chordin (Xue et al., 2014). Although it had been postulated from work in other systems that interactions with antagonists might limit one another's diffusion or stability, the mechanism in the case of organizer BMP2b seems to involve direct BMP2b:Smad complex-dependent transcriptional repression of *chordin* (Xue et al., 2014). In support of this model, embryos treated with BMP inhibitors show reduced Smad1 association, including at the *chordin* locus, while embryos lacking maternal and zygotic function of One-eyed pinhead, an essential Nodal cofactor show increased Smad1 association, raising the possibility that Smad2/3 (Nodal) and Smad1/5/8 (Bmp) complexes compete for regulation of genes in the organizer, like *chordin* (Fig. 3) (Gritsman et al., 1999; Xue et al., 2014). Thus, BMP2b in the organizer is thought to limit expression of organizer genes and limit organizer fate and activity possibly by providing buffering in dorsolateral regions to shape and stabilize the reciprocal BMP and Chordin gradients.

In addition to BMP2b, the BMP-like molecule, Antidorsalizing morphogenetic protein (Admp) is induced by Nodal and Wnt signaling within the organizer where it is thought to antagonize extracellular Chordin

(Lele, Nowak, & Hammerschmidt, 2001; Willot et al., 2002). However, Admp cannot account for full repression of Chordin because its knock down only causes mild dorsalization and does not change Chordin protein distribution even though *chordin* transcripts expand laterally (Lele et al., 2001; Willot et al., 2002; Xue et al., 2014). Lateral expansion of *chordin* transcripts but not protein indicates that Admp is not likely to mediate transport and degradation of Chordin protein as proposed in other contexts (Dubrulle et al., 2015; Lele et al., 2001; Willot et al., 2002). Instead it may act through its interactions with Nodal ligands and in collaboration with BMP to confine *chordin*, *goosecoid* and other Nodal targets, such as and *noggin*, within the organizer domain (Dubrulle et al., 2015; Lele et al., 2001; Willot et al., 2002). This hypothesis and mechanism remain to be tested.

3. Transforming blastula tissues to form the germ layers

3.1 It takes two to tango: TGFβ heterodimers specify mesendoderm

In zebrafish as in other vertebrates, mesoderm and endoderm are patterned along the animal-vegetal axis of the embryo in a Nodal-dependent manner. During mesendodermal patterning, Nodal ligands, members of the Activin-like subgroup of the TGFβ superfamily, activate type I and type II activin receptor complex (Alk4 and ActRIIA/B), inducing phosphorylation and nuclear translocation of Smad complex comprised of Nodal-specific Smad2/3 and the effector Smad4, which also mediates Bmp signaling (Fig. 3B). This Smad complex activates expression of Nodal target genes, which include the *nodal* ligands themselves and *lefty* inhibitors of Nodal (reviewed in Schier, 2003; Shen, 2007). Zebrafish Nodals (also known as Squint/Ndr1 and Cyclops/Ndr2) and Vg1 (also known as Gdf3/Dvr1) are members of the TGFβ superfamily of cell signaling factors. Along with the maternal *wnts* discussed above, *Vg1 and sqt/ndr1* are expressed asymmetrically along the animal vegetal axis of the oocyte. In the embryo; however, *vg1* is ubiquitously expressed. Asymmetric *sqt/ndr1* has been observed in the early embryo, and *cyc/ndr2* and a third nodal gene *southpaw* are zygotically expressed and spatially restricted (Bally-Cuif, Schatz, & Ho, 1998; Feldman et al., 1998; Gore et al., 2005; Gore & Sampath, 2002; Helde & Grunwald, 1993; Long, Ahmad, & Rebagliati, 2003; Rebagliati, Toyama, Haffter, & Dawid, 1998; Sampath, Cheng, Frisch, & Wright, 1997). Thus, much like *bmp2b* discussed above, *sqt* and *cyc* transcripts are expressed in a dynamic manner in extraembryonic (YSL) and embryonic tissues of the blastula. Establishing this dynamic pattern of ligand transcripts, including new domains in extraembryonic regions and the overlying prospective mesendodermal cells requires active signaling, feedback regulation loops, and the *mix/bon* transcription factor Mxtx2 (Erter, Solnica-Krezel, & Wright, 1998; Fan et al., 2007; Feldman et al., 1998; Hirata et al., 2000; Hong, Jang, Brown, McBride, & Feldman, 2011; Rebagliati, Toyama, Fricke, Haffter, & Dawid, 1998; Rebagliati, Toyama, Haffter, et al., 1998; Sampath et al., 1998).

Despite their localization in oocytes and evidence that overexpression of *sqt* RNA leads to dorsalization, neither *sqt/ndr1, vg1/gdf3*, or Nodal cofactors appear to be essential for oocyte or maternal dorsal ventral patterning because mutant mothers for *sqt/ndr1* alleles that disrupt Sqt protein or alleles disrupting both protein and RNA produce eggs and embryos with normal animal vegetal and dorsal ventral axes (Feldman et al., 1998; Golling et al., 2002; Goudarzi, Berg, Pieper, & Schier, 2019; Hatta, Kimmel, Ho, & Walker, 1991; Heisenberg & Nusslein-Volhard, 1997; Lim et al., 2012; Pei, Williams, Clark, Stemple, & Feldman, 2007). Likewise, these embryonic axes are also normal in maternal-zygotic (MZ) mutants disrupting the cofactor *one-eyed pinhead* (*oep;* known as Cripto or Tgdf1 in mammals), *MZsmad2* mutants, and *MZvg1/gdf* mutants (Bisgrove, Su, & Yost, 2017; Dubrulle et al., 2015; Gritsman et al., 2000, 1999; Montague & Schier, 2017; Pelliccia, Jindal, & Burdine, 2017). Instead, these TGFβ family members appear to be mutually dependent cofactors that act zygotically to specify mesoderm and endoderm, mesendoderm (Bisgrove et al., 2017; Montague & Schier, 2017; Pelliccia et al., 2017). Loss of both maternal (M) and zygotic (Z) Nodal signaling as occurs in compound homozygotes for both ligands (*sqt; cyc* double mutants), *MZtdgf1/oep*, MZ*foxh1*, *MZsmad2*, and *Mvg1/gdf3* or *MZvg1/gdf3,* or overexpression of Lefty1 or Lefty2, Nodal antagonists, all result in severe mesendoderm deficits (Bisgrove, Essner, & Yost, 1999; Bisgrove et al., 2017; Dougan et al., 2003; Dubrulle et al., 2015; Erter et al., 1998; Gritsman et al., 1999; Montague & Schier, 2017; Pelliccia et al., 2017; Slagle, Aoki, & Burdine, 2011; Thisse & Thisse, 1999). Conversely, overexpression of Nodal ligands, Ndr1 or Ndr2, in the early embryo causes expansion of Nodal target genes in early gastrula and leads to dorsalization (Erter et al., 1998; Feldman et al., 1998). The RNA binding protein, Y box Protein 1 (Ybx1), binds to the *squint-3′utr*, and is thought to post-transcriptionally regulate *sqt* transcripts (Kumari et al., 2013). Mutations disrupting maternal *ybx1* cause premature YSL formation with excess nuclei (YSN), gastrulation defects, and embryonic lethality

(Kumari et al., 2013). That the observed YSN and gastrulation defects of *Mybx1* mutants resemble phenotypes reported for double morphants or mutants disrupting the Nodal antagonists *lefty1/lefty2* (Branford & Yost, 2002; Feldman, Dougan, Schier, & Talbot, 2000; Rogers et al., 2017) and can be suppressed by conditions that block Nodal signaling activity, indicates that Ybx1 is a negative regulator of Nodal signaling that likely acts by repressing *sqt* translation (Kumari et al., 2013).

Newly translated Nodals are produced in an immature form that must be processed for secretion and signaling (reviewed in Schier, 2003; Shen, 2007). Although examination of tagged Sqt/Ndr1 and Cyc/Ndr2 revealed that processing of these Nodals is Vg1/Gdf3 independent (Montague & Schier, 2017), Cyc/Ndr2 cannot induce dorsalization in overexpression assays in *Mvg1/gdf3* mutants, and Sqt/Ndr1 can do so only when expressed at high levels (Bisgrove et al., 2017; Montague & Schier, 2017; Pelliccia et al., 2017). Thus, full Nodal target activation in response to Nodal ligands requires Vg1/Gdf3 (Montague & Schier, 2017). In contrast, exogenous tagged Vg1/Gdf3-superfolderGFP can only be cleaved to the mature form and secreted when co-expressed with tagged Ndr1(Sqt) or Ndr2 (Cyc), indicating that Vg1 processing requires Ndr ligands (Montague & Schier, 2017).

The observation that some *Mvg1/gdf3* mutants are refractory to the effects of Nodal antagonist (Lefty) overexpression, but respond to overexpression of Activin or a constitutively active Alk4 receptor (CA-Alk4) in the same manner as wild-type embryos, further indicates that *Mvg1/gdf3* regulates Nodal activity (Bisgrove et al., 2017; Montague & Schier, 2017; Pelliccia et al., 2017). These observations demonstrate that the shared downstream components of the pathway are intact in Mvg1 mutants and indicate that Vg1/Gdf3, like Nodal, acts upstream of the Alk4 receptor. Although Vg1/Gdf3 is globally expressed, Nodal signaling is only activated at the blastoderm margin and in the midline (Bisgrove et al., 2017). More precisely, Vg1 acts within Nodal co-expressing cells since only transplanted cells that co-express both Nodal and Vg1/Gdf3 induce Nodal targets (Montague & Schier, 2017). Consistent with mutual dependence of Vg1/Gdf3 and Nodal, clusters of Nodal expressing cells can induce Nodal target gene expression in wild-type but not in *vg1/gdf3*$^{-/-}$ cells (Bisgrove et al., 2017). This is possible because global Vg1 is available to cooperate in Nodal signaling and facilitate activation of target genes when processed by Nodal that is expressed outside of its normal domain. Biochemical studies indicate that Nodal signaling is mediated by an obligate Nodal-Vg1/Gdf3 heterodimer, a discovery that echoes the requirement for Nodal-Vg1/Gdf heterodimers to pattern the germ layers and left right axis in mice (Andersson, Reissmann, Jornvall, & Ibanez, 2006; Tanaka, Sakuma, Nakamura, Hamada, & Saijoh, 2007) and for Bmp2-Bmp7 heterodimers in dorsal-ventral patterning of the zebrafish embryo (Little & Mullins, 2009).

3.2 Exposure matters, inhibition needed, no feedback required when nodal organizes

The marginal blastoderm cells closest to the source of Nodal signals in the YSL express distinct Nodal targets, high-threshold targets, from cells up to 7–10 tiers further away, which express low-threshold targets (Fig. 4). Lineage tracing indicates that cells expressing high- and low-threshold targets adopt distinct mesendodermal fates as indicated by gene expression (Bennett et al., 2007; Dubrulle et al., 2015; Harvey & Smith, 2009; Varga et al., 2007). Feedback regulation, both positive and negative, has been identified as a mechanistic feature utilized by developmental patterning mechanisms, including those that form Nodal gradients. A key question has been how does Nodal induction of positive and negative regulators of pathway activity lead to productive signaling and generate spatially restricted domains of expression/pattern? Models to explain how this might work are rooted in the classic autoactivation and feedback inhibition models proposed by Meinhard and Gierer wherein pattern is generated by localized sources of distinct and opposing morphogens with different "effective concentrations" lead to activation proximal to the source and inhibition over a long range (Gierer & Meinhardt, 1972). These include delays or differences in activation or transcription kinetics of positive factors versus inhibitors, region-specific differences in duration of exposure to signaling, as well as differential diffusion of the ligands and inhibitors through the tissue (Almuedo-Castillo et al., 2018; Chen & Schier, 2001; Dubrulle et al., 2015; Feldman et al., 1998; Hagos & Dougan, 2007; Hatta et al., 1991; Heisenberg & Nusslein-Volhard, 1997; Muller et al., 2012; Muller, Rogers, Yu, Brand, & Schier, 2013; Schier, 2009). Accordingly, earlier-induced or high-mobility ligands might diffuse further than low-mobility ligands but slower than the inhibitors with higher mobility. In Turing-type reaction diffusion mechanism, the signals induce inhibitors that move more quickly from source cells than the signals, thereby establishing a gradient of activity and consequently activate distinct target gene and different locations (Muller et al., 2012). For some time, it was difficult to test the predictions of these models owing to a lack of high-quality antibodies and the dynamic nature of the process. Advances in fluorescent tagging strategies and live-imaging

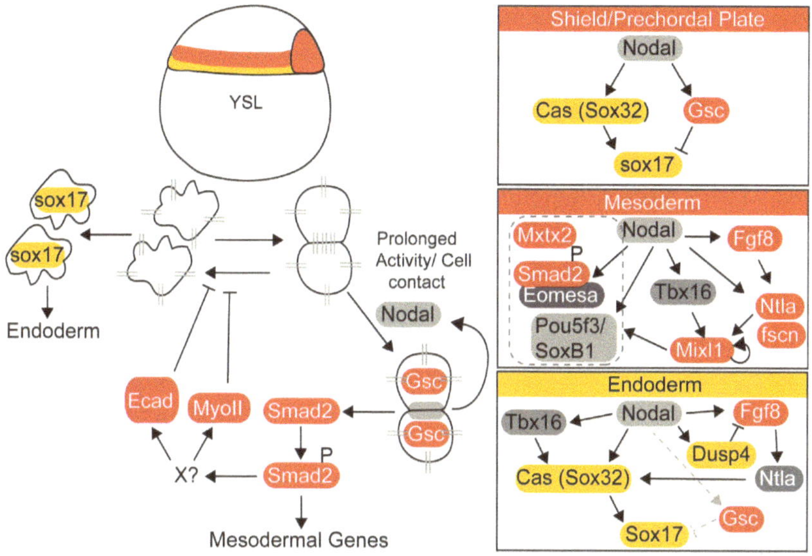

Fig. 4 Nodal signaling in mesendoderm specification. In the gastrula Nodal promotes mesoderm and endodermal fates around the embryo margin. Axial fates, such as prechordal plate require "high threshold" Nodal targets, such as goosecoid (Gsc). Lateral mesodermal fates require "intermediate threshold" Nodal targets, including Fgf and T-box transcription factors that promote mesoderm gene expression programs. Endoderm is induced by "low threshold" Nodal targets. In recent years, it has become clear that levels alone do not account for differential activation of Nodal targets, and that duration of Nodal signaling and timing are key factors. Within the early embryo the cells on dorsal and at the margin are more crowded than elsewhere in the embryo (Schematic on the left represents a lateral view of shield stage embryo with mesoderm in red and endoderm in yellow). Cell-cell contact, in particular duration of cell contact has emerged as a way to reinforce high levels of Nodal signaling to promote cell fate decisions. Accordingly, cells with the most sustained contacts have the highest level of Nodal signaling. These cells express "high-threshold targets" which are activated in part by Nodal induced Fgf to promote "long range" mesodermal fates. Whereas, cells with less sustained contacts, have less Nodal activity and adopt endodermal fates. In these cells, Nodal has been proposed to simultaneously activate Fgf and its inhibitor Dusp4 to promote endodermal fate.

technologies have made it possible to visualize diffusion of ligands and inhibitors, gradient formation, and pathway activation in vivo (Dubrulle et al., 2015; Harvey & Smith, 2009; Muller et al., 2012, 2013; Sako et al., 2016; van Boxtel, Economou, Heliot, & Hill, 2018).

Overexpression assays (OE) using tagged ligands and inhibitors generated support for involvement of Turing-type reaction diffusion mechanisms in Nodal signaling (Muller et al., 2012, 2013), but more recent studies indicate that duration of Nodal activity is also or primarily responsible for the diverse fates generated in response to Nodal signaling, giving way to revised models including a "cumulative dose model" acknowledging the contribution of dose and time of exposure, and a "temporal window model" in which a short-range Nodal domain is set by the period of *lefty* translational repression (Almuedo-Castillo et al., 2018; Hagos & Dougan, 2007; Hatta et al., 1991; Rogers & Schier, 2011; Sako et al., 2016; Schier, 2009; van Boxtel et al., 2018). Using a transgenic reporter for Smad2/Smad4/Foxh1 activity (Germain, Howell, Esslemont, & Hill, 2000), phospho-Smad2 antibodies and probes for two "low threshold or long range" Nodal targets found that activation of two of these targets (*ntla* and *fscn1*) outside of the Nodal producing cells is dependent on Fgf (van Boxtel et al., 2018). In that work no evidence was found for activation of Nodal targets beyond the active signaling cells, suggesting that long-range signaling is not necessary for mesoderm specification, consistent with earlier genetics observations that mesoderm is patterned normally, albeit with a lag in *MZsqt* mutants (Feldman et al., 1998; Heisenberg & Nusslein-Volhard, 1997; Lim et al., 2012).

The combination of tools for visualization with new gene interference technologies to disrupt or mutate pathway components supports the importance of both Nodal induced activators and inhibitors to mesendoderm patterning. Still the question of precisely how important feedback, negative feedback in particular, is for generating spatially and temporally restricted domains of expression, and how might this be achieved if cells are exposed to both ligand and inhibitor remained unclear. Earlier work established miR-430 as a negative regulator of *ndr1;sqt* and *lefty* RNAs (Bassett et al., 2014; Choi, Giraldez, & Schier, 2007), and more recently, it has been postulated that miR-430 repression can provide a delay in *lefty* translation that might account for the shape of the Nodal signaling domain (van Boxtel et al., 2018). According to this model, maternal Ndr1/Sqt from the YSL provides a head start over the antagonists, with Nodal signaling first detectable, using a transcriptional sensor transgene, and p-Smad2 antibodies, on the dorsal side where *sqt* RNA accumulates, and then spreading around the margin within the first tier of cells nearest the margin (Erter et al., 1998; Feldman et al., 1998; van Boxtel et al., 2018). In these marginal cells, signaling is postulated to reach higher levels due to a combination of a lag in Lefty production and continuous signaling from internalized receptors that are immune to Lefty inhibition; thus, allowing propagation of the signal to adjacent blastomeres 5–6 tiers from the margin (van Boxtel et al., 2018). It is unclear what factors and mechanisms alleviate miR-430 repression to allow for production of Lefty protein. It could be that saturation of available miR-430 frees *lefty* and other targets for translation, or more likely, that a RNA binding protein is produced that can compete with miR-430 for binding to *lefty* RNA to alleviate inhibition, but these models and the proposed mechanism remain to be tested. CRISPR/Cas9 and pharmacological tools provide insight into the contribution of negative feedback and the consequences to patterning when Nodal cannot activate its inhibitors, Lefty1 and Lefty2 (Rogers et al., 2017). As expected, removing Lefty inhibition leads to excess Nodal signaling and expansion of mesendoderm (Rogers et al., 2017). Somewhat surprisingly, broad and uniform application of Nodal inhibitors or supplying *lefty* to the mutant embryos outside of its normal expression domain, both of which should break the feedback inhibition loop in the Nodal source cells restores patterning and supports development of *lefty* mutants to adulthood (Rogers et al., 2017). Thus, although limiting Nodal or its activity is essential for normal mesendoderm patterning, negative feedback need not be tightly coupled to Nodal signaling as long as Nodal is constrained and the embryo experiences no fluctuation of Nodal activity during development. Similarly, negative feedback from Lefty becomes critical when embryonic size is perturbed (Almuedo-Castillo et al., 2018).

In addition to inducing expression of *lefty* antagonists, other extracellular antagonists and transcriptional repressors are among known Nodal target genes (Bennett et al., 2007; Dickmeis et al., 2001; Dixon Fox & Bruce, 2009; Latinkic et al., 1997). Insight into how activation of repressors contributes to specification of distinct fates within the organizer region has come from a recent study using transgenic reporter systems to examine Nodal target activation (Sako et al., 2016). One reporter of a direct Nodal target, *mezzo*, revealed similar levels of activation in cells within the organizer, where Nodal signaling is first induced and where the highest threshold targets are expressed, and cells within the blastoderm margin, further indicating that levels alone cannot account for fate differences between Nodal responsive cells (Poulain & Lepage, 2002; Sako et al., 2016). In support of the significance of duration to target gene expression, the use of a photoactivatable Nodal receptor in which the light-oxygen-voltage (LOV) domain is used to dimerize the ICD of Acvr1b and Acvr2b and anchor them to the plasma membrane upon exposure to blue light can induce distinct Nodal responsive genes with different lengths of exposure, but also revealed reduced expression of some targets after prolonged exposure (Sako et al., 2016; Toyooka, Hisatomi, Takahashi, Kataoka, & Terazima, 2011). Analysis of FACs sorted cells from a reporter line that simultaneously detects activation of two Nodal targets, *gsc* and *sox17*, as a read-out of prechordal plate specification (Schulte-Merker et al., 1994) and endoderm specification (Alexander, Rothenberg, Henry, & Stainier, 1999) revealed that the level of *gsc* inversely correlates with that of *sox17* (Sako et al., 2016). Seeding experiments with FACS sorted prechordal plate precursor cells, light inducible Nodal signaling, and the dual reporter system indicate that Nodal promotes cell contact via upregulation of *E*-Cadherin and MyosinII (Barone et al., 2017). Thus, internalized Nodal is expected to accumulate in endosomes at sites of sustained cell-cell contact, which in turn reinforces Nodal signaling (Barone et al., 2017; Blanchet et al., 2008; Jullien & Gurdon, 2005) (Fig. 4). Accordingly, prechordal plate precursor cells in contact the longest have the highest levels of *gsc* and are more likely to become mesoderm rather than endoderm (Fig. 4) (Barone et al., 2017). Moreover, analysis of reporters in conditions of extended Nodal signaling provide compelling evidence for a mechanism wherein sustained Nodal signaling, as occurs in the shield, limits upregulation of *sox17* by inducing *gsc*, which can in turn directly repress *sox17* expression and non-prechordal plate mesoderm fates (Barone et al., 2017; Dixon Fox & Bruce, 2009; Sako et al., 2016) (Fig. 4).

Mounting evidence favors a role for short-range Nodal signaling in patterning. A recent model posits that sustained cell-cell contact and enrichment of endocytosed Nodal allows cells to measure both the duration of contact and concentration of Nodal, thereby establishing a "kinetic proofreading" mechanism that allows cells to distinguish sustained and specific interactions from random transient exposures (Barone et al., 2017; Hopfield, 1974; McKeithan, 1995) (Fig. 4). Although elevated *gsc* can suppress endoderm formation, analysis of *gsc* mutants has not uncovered a requirement for *gsc* in mesendoderm specification in zebrafish (Dixon Fox & Bruce, 2009; Seiliez et al., 2006). This has been taken as evidence of potential for redundancy with other transcriptional repressors activated in the organizer that have yet to be discovered. Moreover, because *gsc* is not induced in the germ ring, it cannot account for differences in mesendodermal fates in lateral regions. It is possible that induction of other region-specific repressors in response to sustained Nodal could contribute to generating distinct mesodermal and endodermal fates. For example, the observation that Nodal simultaneously induces long-range Fgf signaling and the Fgf inhibitor Dusp4 within the first two tiers of cells from the margin has been proposed to generate pattern such that the margin proximal cells with high Nodal and low Fgf become endoderm (van Boxtel et al., 2018).

3.3 Cohabiting or dwelling alone, location matters in mesendoderm patterning

The maternally supplied T-box transcription factor, Eomesodermin, can induce excess endodermal and axial mesodermal marker expression in a manner that depends on maternal and zygotic Oep (Bjornson et al., 2005; Bruce et al., 2003; Xu et al., 2014). Thus, maternal *eomesodermin* (*eomesa*) was hypothesized to act upstream of Nodal or to have earlier roles in oocyte or embryonic development. In support of Eomesodermin function upstream or parallel to Nodal signaling, loss of maternal and zygotic *eomesa* disrupts endodermal marker expression (Du, Draper, Mione, Moens, & Bruce, 2012). The previously known interaction between Smad2 and Eomesa (Arnold, Hofmann, Bikoff, & Robertson, 2008; Picozzi, Wang, Cronk, & Ryan, 2009; Slagle et al., 2011; Teo et al., 2011) and the observation that Smad2 and Eomesa bind to common regions proximal to key mesoderm and endoderm genes established Eomesa as a component of Smad2 complexes in zebrafish (Nelson et al., 2014). Moreover, integration of ChIP, microarray, RNAseq, and in situ data from embryos overexpressing Ndr1 and mutants lacking Nodal signaling indicate that Eomesa and the Nodal effector Smad2 share common direct targets (Nelson et al., 2014). Despite multiple lines of evidence indicating that the most highly responsive Nodal targets are co-occupied by Smad2 and Eomesa, these targets are only transiently reduced in *MZeomesa* mutants (Nelson et al., 2014). Similarly, independent analyses of *MZ eomesa* mutants showed that induction of *mxtx2*, a direct Eomesodermin target, and other targets lag in *MZ eomesa* mutants (Bruce, Howley, Dixon Fox, & Ho, 2005; Du et al., 2012; Xu et al., 2014). However, despite the initial deficits in endodermal markers, including *sox32*, *bon*, and *og9*, Eomesodermin is not essential for endoderm formation because mutants lacking maternal and zygotic *eomesa* eventually express endodermal markers and form a gut (Du et al., 2012), raising the possibility that additional transcription factors might redundantly or cooperatively regulate Nodal mesodermal and endodermal targets.

This role could be fulfilled by any of the several transcription factors that bind to Smad2, including FoxH1, also known as Fast1 and Schmalspur (Pogoda, Solnica-Krezel, Driever, & Meyer, 2000; Sirotkin, Gates, Kelly, Schier, & Talbot, 2000). FoxH1 is required for axial mesoderm/notochord development but is dispensable for nonaxial mesoderm and endoderm formation presumably because other Smad2-associated transcription factors mediate Nodal signaling in those regions (Pogoda et al., 2000; Sirotkin et al., 2000; Slagle et al., 2011). Consistent with the notion that some Nodal activity remains intact, more severe deficits in mesendoderm formation arise upon morpholino knock down of the Nodal ligands, Cyc and Sqt, in *foxH1* mutants (Pogoda et al., 2000; Sirotkin et al., 2000; Slagle et al., 2011). Given its association with Smad2 and its known role in endoderm formation the Mix homeobox transcription factor Mixl1, also known as Bon became a compelling candidate; however, loss of Bon in *foxH1* mutants did not phenocopy *MZoep* phenotypes, and thus Bon could not account for the residual Nodal activity in *foxH1* mutants (Kikuchi et al., 2000; Slagle et al., 2011). Two independent studies support combinatorial activity of Eomesa and FoxH1 in regulation of mesendodermal targets of Nodal (Nelson et al., 2014; Slagle et al., 2011). In tests for functional compensation by FoxH1 for loss of *eomesa*, select Nodal target expression can be partially restored by OE of FoxH1 (Nelson et al., 2014). In reciprocal studies, expressing active Eomesa (Eomesa-VP16) restores notochord development in *MZ* mutants for the stronger *midway allele disrupting foxH1* even though Eomesa is not normally expressed in the notochord and is not required for notochord fate (Slagle et al., 2011). Moreover, the mesendoderm deficits caused by expressing a dominant negative Eomesa (DN-Eomesa), including loss of *bon* expression without disrupting the Nodal ligands (Slagle et al., 2011) suggests that Eomesa may account for the Nodal signaling remaining in the absence of FoxH1. However, because the DN-Eomesa phenotypes are more severe than the *MZeomesa* mutants, it is also possible that the DN strategy affects other T-box genes with highly similar DNA binding elements (Gentsch et al., 2013) or that genetic compensation ameliorates the phenotypes of mutants with the "null" allele (El-Brolosy et al., 2019; El-Brolosy & Stainier, 2017). Nonetheless, the cumulative evidence favors combinatorial regulation of Nodal signaling, a possibility that can only be clarified by analysis of double mutants.

Whether distinct FoxH1:Smad2 and Eomesa:Smad2 complexes act in parallel downstream of Nodal, or if all three form a complex that promotes mesendoderm gene expression is not clear. Attempts to map the FoxH1 binding sites have been hampered by the lack of available antibodies. However, phenotypic differences between mutants generating FoxH1 protein predicted not to interact with Smad2 and those disrupting its DNA binding domain indicate that FoxH1 can bind to genomic sites directly via its DNA binding domain or indirectly via association with Smad2 (Slagle et al., 2011). That some genes with elevated expression in *MZeomesa* mutants map to genomic sites occupied by Smad2 but not by Eomesa, including factors involved in chromatin assembly, implicates Eomesa in repression of these targets (Nelson et al., 2014). Given that chromatin assembly genes are among targets of TGFβ-dependent Smad repression (Gokhman, Livyatan, Sailaja, Melcer, & Meshorer, 2013), Eomesa could facilitate Smad mediated repression of Smad2 bound sites. Smad mediated repression of factors governing chromatin accessibility opens the possibility that differences in chromatin accessibility could underlie or contribute to differential dosage responsiveness of Nodal targets. In light of the recently appreciated role that signaling exposure duration plays in activation of distinct Nodal targets, it seems plausible that some targets might simply be less accessible and thus would require more time or additional factors to become transcriptionally available.

3.4 Pattering along, toddler loses endoderm

Toddler is a secreted peptide that signals through a G-protein coupled receptor pathway mediated by ApelinA/B receptors to mediate proper mesendoderm morphogenesis (Chng, Ho, Tian, & Reversade, 2013; Deshwar, Chng, Ho, Reversade, & Scott, 2016; Norris et al., 2017; Pauli et al., 2014). Endoderm deficiencies due to loss of Toddler function in morphants or mutants indicated that it may contribute to Nodal-mediated endoderm patterning or maintenance (Chng et al., 2013; Deshwar et al., 2016; Norris et al., 2017; Pauli et al., 2014). Based on the observations that initial patterning of mesendoderm is normal (Pauli et al., 2014), but later endodermal cells numbers decline in *toddler* mutants (Chng et al., 2013; Pauli et al., 2014), two models have been proposed to account for the endodermal deficits and for cell migration deficits (to be discussed in more detail in the chapters "Zebrafish gastrulation: Putting fate in motion" by Pinheiro and Heisenberg and "Cellular and molecular mechanisms of convergence and extension in zebrafish" by Williams and Solnica-Krezel) of *toddler* mutants. In one model Toddler is postulated to enhance Nodal signaling to promote endodermal fates. Here, the endoderm specification deficit in *toddler* mutants is posited to secondarily disrupt cell migration (Chng et al., 2013; Deshwar et al., 2016). In the second model Toddler's primary function is proposed to mediate mesendodermal cell migration downstream of Nodal (Pauli et al., 2014; Tucker et al., 2007). In this model the deficit in endodermal cells is then secondary to the migration deficit. Recent work strengthens the latter model by demonstrating first that increased cell death in *toddler* mutants affects also endodermal cells. Second, the Apelin receptors are expressed on mesodermal cells rather than endodermal cells, indicating that the pathway functions in mesoderm rather than endoderm, and finally that even when endodermal cell numbers are increased by removing the Nodal inhibitor *lefty* in *toddler* mutants, the cell migration deficits persist (Norris et al., 2017). This work strongly suggests that although Nodal induces expression of the *apelin* receptors that are responsive to Toddler, the primary function of these Nodal targets is to mediate cell migration rather than patterning of mesendoderm (Norris et al., 2017; Pauli et al., 2014; Tucker et al., 2007).

Endoderm formation downstream of Nodal requires a maternally supplied DEAH-box RNA helicase, Mission Impossible (Mis) or Dhx16 (Putiri & Pelegri, 2011). Zygotic mutants with null alleles disrupting *dhx16* exhibit normal patterning of the endoderm but are not compatible with development beyond 3–4 days postfertilization (Golling et al., 2002). The discovery of a hypomorphic allele of *dhx16*, that supports development to adulthood revealed the maternal requirement for this RNA helicase in patterning the endoderm (Putiri & Pelegri, 2011). Normal expression of *cyc* and *squint* and mesendodermal targets of Nodal signaling, and the lack of expression of endodermal targets such as *foxA2* and *sox17* show that maternal *dhx16* acts downstream of Nodal ligands (Putiri & Pelegri, 2011). Interestingly, among Nodal targets, only those involved in endoderm signaling are impaired in *mis* mutants, including Nodal-dependent but not initial expression of *Casanova* (*cas*), a Sox family transcription factor that is essential for endoderm induction (Dickmeis et al., 2001; Kikuchi et al., 2001; Putiri & Pelegri, 2011). The observation that early events of mesendoderm specification are intact indicate that maternal Dhx16 is required late in the process or is specifically required for endoderm formation. Molecular epistasis analyses indicate that overexpression of Casanova, but not Nodal ligands restores endoderm marker expression in *mis* mutants, placing *mis* at the level of or downstream of *cas* and upstream of *sox17* (Putiri & Pelegri, 2011) (Fig. 4). Transplantation, lineage tracing, and explant assays indicate that *dhx16/mis* acts cell autonomously within blastoderm cells to promote endoderm specification, but the underlying mechanisms are not clear. As mentioned, initial expression of *cas* in the zebrafish YSL is Nodal independent, but subsequent expression of *cas* in the endoderm requires Nodal (Kikuchi et al., 2001). In other contexts, Dhx16 regulates pre-mRNA splicing (Gencheva, Kato, Newo, & Lin, 2010; Gencheva, Lin, et al., 2010). Because *cas* is required for its own expression (Dickmeis et al., 2001), and given that Dhx16 functions in regulating splicing, it is tempting to speculate that *mis/dhx16* regulates splicing of *cas*, and thereby promotes Cas-dependent expression of *cas*. This model remains to be tested but is consistent with the initial induction of *cas* and subsequent diminished expression of the Nodal-dependent endodermal domain of *cas* and other endodermal genes in *mis* mutants. The targets of Dhx are largely not known, but like Dhx RNA interacting proteins are expected to play substantial maternal roles in regulating gene expression and setting up embryonic pattern, particularly at stages prior to activation of the zygotic genome.

3.5 Mom's got skin in the game, patterning the ectoderm and the enveloping layer

The zebrafish epidermis is composed of an outer epithelial layer known as the enveloping layer or EVL and an inner basal layer of distinct embryonic origins. The p63 + basal cell layer, which is derived from ventral ectoderm, forms later and subsequently migrates beneath the EVL to form the embryonic epidermis (Bakkers, Hild, Kramer, Furutani-Seiki, & Hammerschmidt, 2002; Lee & Kimelman, 2002). The outer layer EVL cells are specified in the early embryo and, as evidenced by explant assays, arise from the superficial but not deep blastomeres of the blastula (Kimmel, Warga, & Schilling, 1990; Sagerstrom, Gammill, Veale, & Sive, 2005). Lineage restriction of EVL cells occurs prior to and independent of zygotic genome activation as the first cellular changes, namely, asymmetric calcium signaling, formation of the tight junctions and spreading of the cells, occur even in the absence of transcription (Kiener, Selptsova-Friedrich, & Hunziker, 2008; Ma, Webb, Chan, Zhang, & Miller, 2009). Differentiating EVL cells have a dynamic cell cycle, with cells undergoing rapid proliferation in the early epiblast, slowing in midblastula and then speeding up again after epiboly (Kane, Warga, & Kimmel, 1992). Subsequent development, including a characteristic lengthening of the cell cycle of prospective EVL cells during midblastula stages and de novo expression of *keratins* require activation of the zygotic genome (Ho, 1992; Imboden, Goblet, Korn, & Vriz, 1997; Kane & Kimmel, 1993; Kimmel et al., 1990; Sagerstrom et al., 2005). The EVL, once viewed as a transient extraembryonic structure that would be lost upon differentiation of the skin (Bouvet, 1976; Kimmel et al., 1990) has more recently been shown through lineage tracing and clonal analysis to persist into larval stages and form the outer layer of the epidermis (Fukazawa et al., 2010; Slanchev et al., 2009; Sonawane, Martin-Maischein, Schwarz, & Nusslein-Volhard, 2009). Formation of both cell layers depends on the broadly expressed transcription factor Pou5f1 (Belting et al., 2001, 2011; Kotkamp, Mossner, Allen, Onichtchouk, & Driever, 2014; Lachnit, Kur, & Driever, 2008;

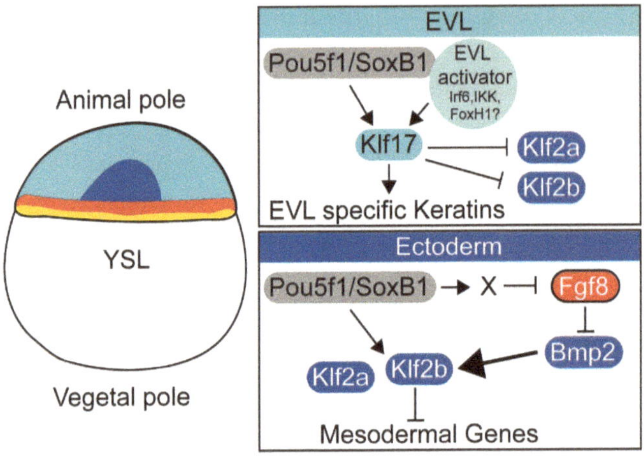

Fig. 5 Ectoderm and EVL specification. Schematic of blastula stage zebrafish embryo. Superficial cells of the blastula develop as EVL in response to a region-specific Pou5f3 dependent transcriptional program that converges on expression of *krupple-like* (*klf*) transcription factors. Klf17 represses the ectodermal *klf2s* and promotes the expression of *keratin* genes and EVL fate. Near the margin (red), prospective ectodermal cells (blue) are induced in response to a Pou5f3 dependent transcriptional program mediated by an unidentified regional cofactor that represses Fgf signaling, thus cooperatively promoting Bmp activation of Klf2b, which is a potent repressor of mesodermal genes. Yellow indicates endodermal tissue.

Onichtchouk, Geier, Messerschmidt, et al., 2010; Onichtchouk, Geier, Polok, et al., 2010; Reim et al., 2004; Reim & Brand, 2006) (Fig. 5). Loss of function and overexpression studies together with microarray analysis indicate that Pou5f1 promotes disparate fates through its association with distinct activators that are spatially restricted along the embryonic axes (Kotkamp et al., 2014; Onichtchouk, Geier, Messerschmidt, et al., 2010; Onichtchouk, Geier, Polok, et al., 2010).

EVL specification is thought to involve an incoherent feed-forward loop wherein Pou5f1 directly activates *kruppel-like zinc finger* transcription factors, *klf2a and klf2b*, within the EVL and simultaneously represses *klf2a and klf2b* via activation of *klf17* (Fig. 5). Consistent with this model, *klf17* expression is diminished in *MZspg/pou5f1* mutants, but *klf2b* is increased (Onichtchouk, Geier, Messerschmidt, et al., 2010; Onichtchouk, Geier, Polok, et al., 2010). Moreover, *klf17* overexpression and morpholino studies indicate that Klf17 promotes expression of EVL development genes but antagonizes that of *klf2a* and *klf2b* (Lachnit et al., 2008). Thus, elevated expression of *klf2b* in *MZspg* early gastrulae is attributed to lack of repression by Pou5f1-dependent EVL pathway specific repressors, such as Klf17. Further search for direct Klf17 targets identified several genes involved in EVL development, including several *keratins* that are known to be EVL specific (Kotkamp et al., 2014). The hypothesized regional cofactor for Pou5f1 in EVL differentiation has not been experimentally confirmed. However, because dominant negative disruption of Interferon regulatory factor 6 (Irf6) results in failure of superficial blastomeres to exit or lengthen the cell cycle, and failure to activate keratin expression, Irf6 has been postulated to be the region-specific cofactor for Pou5f1 in differentiation and formation of the EVL (Kotkamp et al., 2014; Sabel et al., 2009).

Another factor involved in EVL differentiation identified in a maternal-effect screen is inhibitor of NF-κB kinase (IKK), also known as Poky and Chuck (Fukazawa et al., 2010; Wagner, Dosch, Mintzer, Wiemelt, & Mullins, 2004). Like dominant negative Irf6 OE, blastomeres lacking maternal *poky/ikk* do not exit or lengthen their cell cycles, and do not initiate keratin expression (Fukazawa et al., 2010). However, loss of maternal *poky/ikk* does not disrupt initial EVL formation as early cell morphological changes and barrier function typical of EVL are initially intact in *poky/ikk* mutants treated with alpha-amanitin (Fukazawa et al., 2010). Although best known for its role in NF-κB signaling, Poky/Ikk is not thought to regulate EVL development through the NF-κB pathway because inhibition of NF-κB activity with pharmaceuticals or expressing a version of Ikk lacking the NEMO domain, which is thought to antagonize NF-κB activity, can rescue EVL development (Fukazawa et al., 2010). Instead, *poky* is thought to phosphorylate unknown downstream targets that are essential for EVL development, perhaps known Ikk targets that regulate proliferation, because mutant versions of Ikk lacking kinase activity are unable to rescue the EVL deficits of *poky* mutants (Descargues, Sil, & Karin, 2008; Fukazawa et al., 2010; Liu et al., 2008; Zhu et al., 2007). Underscoring the important contribution of intermediate filaments to development and integrity of the EVL, knockdown of *keratins*, dominant negative *irf6*, mutants lacking maternal *poky/ikk* or maternal and zygotic (MZ) *pou5f1/oct4*, *epcam*, *penner/lethal giant larvae lgl2*, *hai1/psoriasis*, and knock-down of *foxh1* or *claudin* all lead to deficits in *keratin* expression and cause blastoderm lysis due to EVL and epiboly deficits to be discussed further in the chapter "Mechanisms of zebrafish epiboly: A current view" by Bruce and Heisenberg (Carney et al., 2007; Fukazawa et al., 2010; Lachnit et al., 2008; Pei et al., 2007; Sabel et al., 2009; Siddiqui, Sheikh, Tran, & Bruce, 2010; Slanchev et al., 2009; Webb, Driever, & Kimelman, 2008).

As mentioned above the basal cell layer of the epidermis is derived from ventral ectoderm, which is induced and maintained by Pou5f1. Unlike *klf17 expression*, which is not asymmetric with respect to the dorsal-ventral axis, transcripts of the Pou5f1 targets *klf2a* and *klf2b* are restricted to the prospective ventral ectoderm (Kotkamp et al., 2014). Morpholino knockdown of *klf2a* or *klf2b* leads to elevated expression of mesodermal genes in *klf2/* ectodermal domains, indicating that a key role of *klf2b* is to repress non-ectodermal fate (Kotkamp et al., 2014). Given their ventral expression domains and diminished BMP2b in *MZspg* mutants due to elevated Fgf8 (Belting et al., 2001; Reim et al., 2004; Reim & Brand, 2006; Schier et al., 1996), make BMP2b a compelling cofactor for Pou5f1 in ventral ectoderm (Kotkamp et al., 2014). Indeed, *klf2s* dependence on Bmp signaling, can be seen in the strong (*klf2b*) and moderate (*klf2a*) induction of *klf2s* by BMP2b OE or Fgf8 inhibition, and reciprocally inhibition of *klf2b* when Bmp is blocked by Fgf8 OE (Furthauer et al., 2004) or Noggin OE (Furthauer, Thisse, & Thisse, 1999; Kotkamp et al., 2014). Although Bmp is important for *klf2b* expression in ventral ectoderm, it is not sufficient in the absence of Pou5f1, as restoring *bmp2b* expression in *MZspg* mutants by inhibiting Fgf8 fails to restore *klf2b* in ventral ectoderm (Kotkamp et al., 2014). Thus, Pou5f1 does not simply regulate *klf2b* by promoting *bmp2b* expression, but instead both factors are required for normal expression of *klf2b* in ventral ectoderm.

Eomesa, already mentioned for its role in Nodal signaling, is, like Pou5f1, not spatially restricted in the early embryo. Thus, Eomesa specificity in regulating gene expression is thought to be imparted by its association with other proteins, like Smad2. In addition to the genomic sites occupied by both Smad2 and Eomesa, or just Smad2, sites uniquely occupied by Eomesa have been defined (Nelson et al., 2014). Based on RNAseq experiments of *MZeomesa* mutants and Eomesa OE it appears that many of the genes elevated in *MZeomesa* mutants and those diminished in Eomesa OE encode factors involved in ectodermal/neuroectodermal fate specification (Nelson et al., 2014). The observation that these genes map to genomic sites occupied by Eomesa but not Smad2 in ChIP analysis indicates that Eomesa functions as a repressor of ectodermal genes in a Nodal-independent manner. How it does so and the region-specific cofactor, if any, for Eomesa in this context has yet to be identified.

4. Conclusion

The embryo domains defined in this chapter are not only important for establishing regional and cell-type specific identity, but also to provide starting material and orchestrate the cell movements and behaviors underlying gastrulation that will be discussed in the chapters that follow. As you can and will see patterning is continually modulated throughout morphogenesis and continues to rely on some of the same major pathways and cues that set the stage for cell identity even before movement begins. New tools and technological advances have and will continue to shape our view of patterning. The dynamic repertoire of tools for genome editing that allow reporters or specific mutations to be introduced into a target-loci to tag endogenous genes or for mutagenesis to disrupt a candidate regulator or pathway will allow the field to test predictions arising from the rapidly accumulating, increasingly detailed high-resolution view of gene and lineage expression profiles coming from single cell data (Farrell et al., 2018). Coupled with refined in vivo labeling and the increasingly sophisticated view of the cells afforded by today's imaging, quantitative and modeling approaches have provided insights into decades old questions while opening up new lines of investigation and unexplored territory. Genetics will undoubtedly continue to play a pivotal role and, in combination with biochemical, biophysical, and in vivo cell biology in the fish approaches, promises to further advance our understanding of how the interplay between and integration of inputs from the key developmental pathways discussed in this chapter and those yet to be discovered pattern the early zebrafish embryo as it embarks on the morphogenetic movements of gastrulation.

Acknowledgments

Research on oocyte polarity and the germ line in the Marlow lab is supported by NIHR01GM089979 and NIHR21HD091456 to F.L.M.

References

Agathon, A., Thisse, C., & Thisse, B. (2003). The molecular nature of the zebrafish tail organizer. *Nature*, *424*(6947), 448–452. https://doi.org/10.1038/nature01822.

Albadri, S., Del Bene, F., & Revenu, C. (2017). Genome editing using CRISPR/Cas9-based knock-in approaches in zebrafish. *Methods*, *121–122*, 77–85. https://doi.org/10.1016/j.ymeth.2017.03.005.

Alexander, J., Rothenberg, M., Henry, G. L., & Stainier, D. Y. (1999). Casanova plays an early and essential role in endoderm formation in zebrafish. *Developmental Biology*, *215*(2), 343–357. https://doi.org/10.1006/dbio.1999.9441.

Almuedo-Castillo, M., Blassle, A., Morsdorf, D., Marcon, L., Soh, G. H., Rogers, K. W., et al. (2018). Scale-invariant patterning by size-dependent inhibition of nodal signalling. *Nature Cell Biology*, *20*(9), 1032–1042. https://doi.org/10.1038/s41556-018-0155-7.

Andersson, O., Reissmann, E., Jornvall, H., & Ibanez, C. F. (2006). Synergistic interaction between Gdf1 and Nodal during anterior axis development. *Developmental Biology*, *293*(2), 370–381. Retrieved from <Go to ISI>://MEDLINE:16564040.

Arnold, S. J., Hofmann, U. K., Bikoff, E. K., & Robertson, E. J. (2008). Pivotal roles for eomesodermin during axis formation, epithelium-to-mesenchyme transition and endoderm specification in the mouse. *Development, 135*(3), 501–511. https://doi.org/10.1242/dev.014357.
Auer, T. O., & Del Bene, F. (2014). CRISPR/Cas9 and TALEN-mediated knock-in approaches in zebrafish. *Methods, 69*(2), 142–150. https://doi.org/10.1016/j.ymeth.2014.03.027.
Auer, T. O., Duroure, K., Concordet, J. P., & Del Bene, F. (2014). CRISPR/Cas9-mediated conversion of eGFP- into Gal4-transgenic lines in zebrafish. *Nature Protocols, 9*(12), 2823–2840. https://doi.org/10.1038/nprot.2014.187.
Auer, T. O., Duroure, K., De Cian, A., Concordet, J. P., & Del Bene, F. (2014). Highly efficient CRISPR/Cas9-mediated knock-in in zebrafish by homology-independent DNA repair. *Genome Research, 24*(1), 142–153. https://doi.org/10.1101/gr.161638.113.
Baillat, D., Hakimi, M. A., Naar, A. M., Shilatifard, A., Cooch, N., & Shiekhattar, R. (2005). Integrator, a multiprotein mediator of small nuclear RNA processing, associates with the C-terminal repeat of RNA polymerase II. *Cell, 123*(2), 265–276. https://doi.org/10.1016/j.cell.2005.08.019.
Bakkers, J., Hild, M., Kramer, C., Furutani-Seiki, M., & Hammerschmidt, M. (2002). Zebrafish DeltaNp63 is a direct target of Bmp signaling and encodes a transcriptional repressor blocking neural specification in the ventral ectoderm. *Developmental Cell, 2*(5), 617–627. Retrieved from https://www.ncbi.nlm.nih.gov/pubmed/12015969.
Bally-Cuif, L., Schatz, W. J., & Ho, R. K. (1998). Characterization of the zebrafish Orb/CPEB-related RNA binding protein and localization of maternal components in the zebrafish oocyte. *Mechanisms of Development*, 77(1), 31–47. Retrieved from http://www.ncbi.nlm.nih.gov/entrez/query.fcgi?cmd=Retrieve&db=PubMed&dopt=Citation&list_uids=9784598.
Barone, V., Lang, M., Krens, S. F. G., Pradhan, S. J., Shamipour, S., Sako, K., et al. (2017). An effective feedback loop between cell-cell contact duration and morphogen signaling determines cell fate. *Developmental Cell, 43*(2), 198–211. e112 https://doi.org/10.1016/j.devcel.2017.09.014.
Bassett, A. R., Azzam, G., Wheatley, L., Tibbit, C., Rajakumar, T., McGowan, S., et al. (2014). Understanding functional miRNA-target interactions in vivo by site-specific genome engineering. *Nature Communications, 5*, 4640. https://doi.org/10.1038/ncomms5640.
Bauer, H., Lele, Z., Rauch, G. J., Geisler, R., & Hammerschmidt, M. (2001). The type I serine/threonine kinase receptor Alk8/Lost-a-fin is required for Bmp2b/7 signal transduction during dorsoventral patterning of the zebrafish embryo. *Development, 128*(6), 849–858. Retrieved from http://www.ncbi.nlm.nih.gov/entrez/query.fcgi?cmd=Retrieve&db=PubMed&dopt=Citation&list_uids=11222140.
Bellipanni, G., Varga, M., Maegawa, S., Imai, Y., Kelly, C., Myers, A. P., et al. (2006). Essential and opposing roles of zebrafish {beta}-catenins in the formation of dorsal axial structures and neurectoderm. *Development, 133*(7), 1299–1309. Retrieved from http://www.ncbi.nlm.nih.gov/entrez/query.fcgi?cmd=Retrieve&db=PubMed&dopt=Citation&list_uids=16510506.
Belting, H. G., Hauptmann, G., Meyer, D., Abdelilah-Seyfried, S., Chitnis, A., Eschbach, C., et al. (2001). Spiel ohne grenzen/pou2 is required during establishment of the zebrafish midbrain-hindbrain boundary organizer. *Development, 128*(21), 4165–4176. Retrieved from https://www.ncbi.nlm.nih.gov/pubmed/11684654.
Belting, H. G., Wendik, B., Lunde, K., Leichsenring, M., Mossner, R., Driever, W., et al. (2011). Pou5f1 contributes to dorsoventral patterning by positive regulation of vox and modulation of fgf8a expression. *Developmental Biology, 356*(2), 323–336. https://doi.org/10.1016/j.ydbio.2011.05.660.

Bennett, J. T., Joubin, K., Cheng, S., Aanstad, P., Herwig, R., Clark, M., et al. (2007). Nodal signaling activates differentiation genes during zebrafish gastrulation. *Developmental Biology*, *304*(2), 525–540. https://doi.org/10.1016/j.ydbio.2007.01.012.

Ben-Zvi, D., Fainsod, A., Shilo, B. Z., & Barkai, N. (2014). Scaling of dorsal-ventral patterning in the Xenopus laevis embryo. *BioEssays*, *36*(2), 151–156. https://doi.org/10.1002/bies.201300136.

Bisgrove, B. W., Essner, J. J., & Yost, H. J. (1999). Regulation of midline development by antagonism of lefty and nodal signaling. *Development*, *126*(14), 3253–3262. Retrieved from http://www.ncbi.nlm.nih.gov/entrez/query.fcgi?cmd=Retrieve&db=PubMed&dopt=Citation&list_uids=10375514.

Bisgrove, B. W., Su, Y. C., & Yost, H. J. (2017). Maternal Gdf3 is an obligatory cofactor in nodal signaling for embryonic axis formation in zebrafish. *Elife*, *6*, e28534, https://doi.org/10.7554/eLife.28534.

Bjornson, C. R., Griffin, K. J., Farr, G. H., 3rd, Terashima, A., Himeda, C., Kikuchi, Y., et al. (2005). Eomesodermin is a localized maternal determinant required for endoderm induction in zebrafish. *Developmental Cell*, *9*(4), 523–533. https://doi.org/10.1016/j.devcel.2005.08.010.

Blader, P., Rastegar, S., Fischer, N., & Strahle, U. (1997). Cleavage of the BMP-4 Antagonist Chordin by Zebrafish Tolloid. *Science*, *278*, 1937–1939.

Blanchet, M. H., Le Good, J. A., Oorschot, V., Baflast, S., Minchiotti, G., Klumperman, J., et al. (2008). Cripto localizes Nodal at the limiting membrane of early endosomes. *Science Signaling*, *1*(45), ra13. https://doi.org/10.1126/scisignal.1165027.

Bouvet, J. (1976). Enveloping layer and periderm of the trout embryo (Salmo trutta fario L.) 20U. *Cell and Tissue Research*, *170*(3), 367–382. Retrieved from https://www.ncbi.nlm.nih.gov/pubmed/954062.

Branam, A. M., Hoffman, G. G., Pelegri, F., & Greenspan, D. S. (2010). Zebrafish chordin-like and chordin are functionally redundant in regulating patterning of the dorsoventral axis. *Developmental Biology*, *341*(2), 444–458. https://doi.org/10.1016/j.ydbio.2010.03.001.

Branford, W. W., & Yost, H. J. (2002). Lefty-dependent inhibition of nodal- and Wnt-responsive organizer gene expression is essential for normal gastrulation. *Current Biology*, *12*(24), 2136–2141. https://doi.org/S096098220201360X [pii].

Bruce, A. E., Howley, C., Dixon Fox, M., & Ho, R. K. (2005). T-box gene eomesodermin and the homeobox-containing Mix/Bix gene mtx2 regulate epiboly movements in the zebrafish. *Developmental Dynamics*, *233*(1), 105–114. https://doi.org/10.1002/dvdy.20305.

Bruce, A. E., Howley, C., Zhou, Y., Vickers, S. L., Silver, L. M., King, M. L., et al. (2003). The maternally expressed zebrafish T-box gene eomesodermin regulates organizer formation. *Development*, *130*(22), 5503–5517. https://doi.org/10.1242/dev.00763.130/22/5503 [pii].

Burgess, S., Reim, G., Chen, W., Hopkins, N., & Brand, M. (2002). The zebrafish spiel-ohne-grenzen (spg) gene encodes the POU domain protein Pou2 related to mammalian Oct4 and is essential for formation of the midbrain and hindbrain, and for pre-gastrula morphogenesis. *Development*, *129*(4), 905–916. Retrieved from http://www.ncbi.nlm.nih.gov/entrez/query.fcgi?cmd=Retrieve&db=PubMed&dopt=Citation&list_uids=11861474.

Campbell, P. D., Heim, A. E., Smith, M. Z., & Marlow, F. L. (2015). Kinesin-1 interacts with Bucky ball to form germ cells and is required to pattern the zebrafish body axis. *Development*, *142*(17), 2996–3008. https://doi.org/10.1242/dev.124586.

Carney, T. J., von der Hardt, S., Sonntag, C., Amsterdam, A., Topczewski, J., Hopkins, N., et al. (2007). Inactivation of serine protease Matriptase1a by its inhibitor Hai1 is required for epithelial integrity of the zebrafish epidermis. *Development*, *134*(19), 3461–3471. https://doi.org/10.1242/dev.004556.

Chen, Y., & Schier, A. F. (2001). The zebrafish Nodal signal squint functions as a morphogen. *Nature, 411*(6837), 607–610. https://doi.org/10.1038/35079121. 35079121 [pii].

Chng, S. C., Ho, L., Tian, J., & Reversade, B. (2013). ELABELA: A hormone essential for heart development signals via the apelin receptor. *Developmental Cell, 27*(6), 672–680. https://doi.org/10.1016/j.devcel.2013.11.002.

Choi, W. Y., Giraldez, A. J., & Schier, A. F. (2007). Target protectors reveal dampening and balancing of Nodal agonist and antagonist by miR-430. *Science, 318*(5848), 271–274. https://doi.org/10.1126/science.1147535.

Connors, S. A., Trout, J., Ekker, M., & Mullins, M. C. (1999). The role of tolloid/mini fin in dorsoventral pattern formation of the zebrafish embryo. *Development, 126*(14), 3119–3130.

Connors, S. A., Tucker, J. A., & Mullins, M. C. (2006). Temporal and spatial action of tolloid (mini fin) and chordin to pattern tail tissues. *Developmental Biology, 293*(1), 191–202. https://doi.org/10.1016/j.ydbio.2006.01.029.

Descargues, P., Sil, A. K., & Karin, M. (2008). IKKalpha, a critical regulator of epidermal differentiation and a suppressor of skin cancer. *The EMBO Journal, 27*(20), 2639–2647. https://doi.org/10.1038/emboj.2008.196.

Deshwar, A. R., Chng, S. C., Ho, L., Reversade, B., & Scott, I. C. (2016). The apelin receptor enhances nodal/TGFbeta signaling to ensure proper cardiac development. *eLife, 5*, e13758. https://doi.org/10.7554/eLife.13758.

Dick, A., Hild, M., Bauer, H., Imai, Y., Maifeld, H., Schier, A. F., et al. (2000). Essential role of bmp7 (snailhouse) and its prodomain in dorsoventral patterning of the zebrafish embryo [In process citation]. *Development, 127*(2), 343–354. Retrieved from http://www.ncbi.nlm.nih.gov/cgi-bin/Entrez/referer?http://www.biologists.com/Development/127/02/dev2495.html.

Dickmeis, T., Mourrain, P., Saint-Etienne, L., Fischer, N., Aanstad, P., Clark, M., et al. (2001). A crucial component of the endoderm formation pathway, CASANOVA, is encoded by a novel sox-related gene. *Genes & Development, 15*(12), 1487–1492. https://doi.org/10.1101/gad.196901.

Dixon Fox, M., & Bruce, A. E. (2009). Short- and long-range functions of Goosecoid in zebrafish axis formation are independent of chordin, noggin 1 and follistatin-like 1b. *Development, 136*(10), 1675–1685. https://doi.org/10.1242/dev.031161.

Dougan, S. T., Warga, R. M., Kane, D. A., Schier, A. F., & Talbot, W. S. (2003). The role of the zebrafish nodal-related genes squint and cyclops in patterning of mesendoderm. *Development, 130*(9), 1837–1851. Retrieved from http://www.ncbi.nlm.nih.gov/entrez/query.fcgi?cmd=Retrieve&db=PubMed&dopt=Citation&list_uids=12642489.

Du, S., Draper, B. W., Mione, M., Moens, C. B., & Bruce, A. (2012). Differential regulation of epiboly initiation and progression by zebrafish Eomesodermin A. *Developmental Biology, 362*(1), 11–23. https://doi.org/10.1016/j.ydbio.2011.10.036.

Dubrulle, J., Jordan, B. M., Akhmetova, L., Farrell, J. A., Kim, S.-H., Solnica-Krezel, L., et al. (2015). Response to Nodal morphogen gradient is determined by the kinetics of target gene induction. *Elife, 4*, https://doi.org/10.7554/eLife.05042.

El-Brolosy, M. A., Kontarakis, Z., Rossi, A., Kuenne, C., Gunther, S., Fukuda, N., et al. (2019). Genetic compensation triggered by mutant mRNA degradation. *Nature, 568*(7751), 193–197. https://doi.org/10.1038/s41586-019-1064-z.

El-Brolosy, M. A., & Stainier, D. Y. R. (2017). Genetic compensation: A phenomenon in search of mechanisms. *PLoS Genetics, 13*(7). e1006780https://doi.org/10.1371/journal.pgen.1006780.

Erter, C. E., Solnica-Krezel, L., & Wright, C. V. (1998). Zebrafish nodal-related 2 encodes an early mesendodermal inducer signaling from the extraembryonic yolk syncytial layer. *Developmental Biology, 204*(2), 361–372. S0012-1606(98)99097-2 [pii] https://doi.org/10.1006/dbio.1998.9097.

Ezzeddine, N., Chen, J., Waltenspiel, B., Burch, B., Albrecht, T., Zhuo, M., et al. (2011). A subset of Drosophila integrator proteins is essential for efficient U7 snRNA and spliceosomal snRNA 3'-end formation. *Molecular and Cellular Biology*, *31*(2), 328–341. https://doi.org/10.1128/MCB.00943-10.

Fan, X., Hagos, E. G., Xu, B., Sias, C., Kawakami, K., Burdine, R. D., et al. (2007). Nodal signals mediate interactions between the extra-embryonic and embryonic tissues in zebrafish. *Developmental Biology*, *310*(2), 363–378. https://doi.org/10.1016/j.ydbio.2007.08.008.

Farrell, J. A., Wang, Y., Riesenfeld, S. J., Shekhar, K., Regev, A., & Schier, A. F. (2018). Single-cell reconstruction of developmental trajectories during zebrafish embryogenesis. *Science*, *360*(6392). https://doi.org/10.1126/science.aar3131.

Fauny, J. D., Thisse, B., & Thisse, C. (2009). The entire zebrafish blastula-gastrula margin acts as an organizer dependent on the ratio of Nodal to BMP activity. *Development*, *136*(22), 3811–3819. https://doi.org/10.1242/dev.039693.

Fekany, K., Yamanaka, Y., Leung, T., Sirotkin, H. I., Topczewski, J., Gates, M. A., et al. (1999). The zebrafish *bozozok* locus encodes Dharma, a homeodomain protein essential for induction of gastrula organizer and dorsoanterior embryonic structures. *Development*, *126*(7), 1427–1438. Retrieved from http://www.ncbi.nlm.nih.gov/cgi-bin/Entrez/referer?http://www.biologists.com/Development/126/07/dev3955.html.

Fekany-Lee, K., Gonzalez, E., Miller-Bertoglio, V., & Solnica-Krezel, L. (2000). The homeobox gene *bozozok* promotes anterior neuroectoderm formation in zebrafish through negative regulation of BMP2/4 and Wnt pathways. *Development*, *127*(11), 2333–2345.

Feldman, B., Dougan, S. T., Schier, A. F., & Talbot, W. S. (2000). Nodal-related signals establish mesendodermal fate and trunk neural identity in zebrafish. *Current Biology*, *10*(9), 531–534. Retrieved from http://www.ncbi.nlm.nih.gov/entrez/query.fcgi?cmd=Retrieve&db=PubMed&dopt=Citation&list_uids=10801442.

Feldman, B., Gates, M. A., Egan, E. S., Dougan, S. T., Rennebeck, G., Sirotkin, H. I., et al. (1998). Zebrafish organizer development and germ-layer formation require nodal-related signals. *Nature*, *395*, 181–185.

Feng, L., Jiang, H., Wu, P., & Marlow, F. L. (2014). Negative feedback regulation of Wnt signaling via N-linked fucosylation in zebrafish. *Developmental Biology*, *395*(2), 268–286. https://doi.org/10.1016/j.ydbio.2014.09.010.

Fernandez, J., Valladares, M., Fuentes, R., & Ubilla, A. (2006). Reorganization of cytoplasm in the zebrafish oocyte and egg during early steps of ooplasmic segregation. *Developmental Dynamics*, *235*(3), 656–671. Retrieved from http://www.ncbi.nlm.nih.gov/entrez/query.fcgi?cmd=Retrieve&db=PubMed&dopt=Citation&list_uids=16425221.

Fisher, S., Amacher, S. L., & Halpern, M. E. (1997). Loss of cerebum function ventralizes the zebrafish embryo. *Development*, *124*, 1301–1311.

Fisher, S., & Halpern, M. E. (1999). Patterning the zebrafish axial skeleton requires early chordin function. *Nature Genetics*, *23*(4), 442–446. https://doi.org/10.1038/70557.

Flores, M. V., Lam, E. Y., Crosier, K. E., & Crosier, P. S. (2008). Osteogenic transcription factor Runx2 is a maternal determinant of dorsoventral patterning in zebrafish. *Nature Cell Biology*, *10*(3), 346–352. doi:ncb1697 [pii] https://doi.org/10.1038/ncb1697.

Francois, P., Vonica, A., Brivanlou, A. H., & Siggia, E. D. (2009). Scaling of BMP gradients in Xenopus embryos. *Nature*, *461*(7260), E1. discussion E2. Retrieved from. https://www.ncbi.nlm.nih.gov/pubmed/19736667.

Fuentes, R., & Fernandez, J. (2010). Ooplasmic segregation in the zebrafish zygote and early embryo: Pattern of ooplasmic movements and transport pathways. *Developmental Dynamics*, *239*(8), 2172–2189. https://doi.org/10.1002/dvdy.22349.

Fukazawa, C., Santiago, C., Park, K. M.,Deery, W. J., de la Gomez, T. C. S., Holterhoff, C. K., et al. (2010). poky/chuk/ikk1 is required for differentiation of the zebrafish embryonic epidermis. *Developmental Biology*, *346*(2), 272–283. https://doi.org/10.1016/j.ydbio.2010.07.037. PubMed PMID: 20692251; PubMed Central PMCID: PMC2956273.

Furthauer, M., Reifers, F., Brand, M., Thisse, B., & Thisse, C. (2001). sprouty4 acts in vivo as a feedback-induced antagonist of FGF signaling in zebrafish. *Development*, *128*(12), 2175–2186. Retrieved from https://www.ncbi.nlm.nih.gov/pubmed/11493538.

Furthauer, M., Thisse, B., & Thisse, C. (1999). Three different noggin genes antagonize the activity of bone morphogenetic proteins in the zebrafish embryo. *Developmental Biology*, *214*(1), 181–196. Retrieved from http://www.ncbi.nlm.nih.gov/cgi-bin/Entrez/referer?http://www.idealibrary.com/links/citation/0012-1606/214/181.

Furthauer, M., Van Celst, J., Thisse, C., & Thisse, B. (2004). Fgf signalling controls the dorsoventral patterning of the zebrafish embryo. *Development*, *131*(12), 2853–2864. https://doi.org/10.1242/dev.01156.

Ge, X., Grotjahn, D., Welch, E., Lyman-Gingerich, J., Holguin, C., Dimitrova, E., et al. (2014). Hecate/Grip2a acts to reorganize the cytoskeleton in the symmetry-breaking event of embryonic axis induction. *PLoS Genetics*, *10*(6). e1004422https://doi.org/10.1371/journal.pgen.1004422.

Gencheva, M., Kato, M., Newo, A. N., & Lin, R. J. (2010). Contribution of DEAH-box protein DHX16 in human pre-mRNA splicing. *The Biochemical Journal*, *429*(1), 25–32. https://doi.org/10.1042/BJ20100266.

Gencheva, M., Lin, T. Y., Wu, X., Yang, L., Richard, C., Jones, M., et al. (2010). Nuclear retention of unspliced pre-mRNAs by mutant DHX16/hPRP2, a spliceosomal DEAH-box protein. *The Journal of Biological Chemistry*, *285*(46), 35624–35632. https://doi.org/10.1074/jbc.M110.122309.

Gentsch, G. E., Owens, N. D., Martin, S. R., Piccinelli, P., Faial, T., Trotter, M. W., et al. (2013). In vivo T-box transcription factor profiling reveals joint regulation of embryonic neuromesodermal bipotency. *Cell Reports*, *4*(6), 1185–1196. https://doi.org/10.1016/j.celrep.2013.08.012.

Germain, S., Howell, M., Esslemont, G. M., & Hill, C. S. (2000). Homeodomain and winged-helix transcription factors recruit activated Smads to distinct promoter elements via a common Smad interaction motif. *Genes and Development*, *14*(4), 435–451. Retrieved from https://www.ncbi.nlm.nih.gov/pubmed/10691736.

Gierer, A., & Meinhardt, H. (1972). A theory of biological pattern formation. *Kybernetik*, *12*(1), 30–39. Retrieved from https://www.ncbi.nlm.nih.gov/pubmed/4663624.

Gokhman, D., Livyatan, I., Sailaja, B. S., Melcer, S., & Meshorer, E. (2013). Multilayered chromatin analysis reveals E2f, Smad and Zfx as transcriptional regulators of histones. *Nature Structural & Molecular Biology*, *20*(1), 119–126. https://doi.org/10.1038/nsmb.2448.

Golling, G., Amsterdam, A., Sun, Z., Antonelli, M., Maldonado, E., Chen, W., et al. (2002). Insertional mutagenesis in zebrafish rapidly identifies genes essential for early vertebrate development. *Nature Genetics*, *31*(2), 135–140.

Gonzalez, E., Fekany-Lee, K., Carmany-Rampey, A., Erter, C., Topczewski, J., Wright, C. V., et al. (2000). Head and trunk in zebrafish arise via coinhibition of BMP signaling by bozozok and chordino. *Genes and Develeopment*, *14*(24), 3087–3092. https://doi.org/10.1101/gad.852400. PubMed PMID: 11124801; PubMed Central PMCID: PMC317122.

Gore, A. V., Maegawa, S., Cheong, A., Gilligan, P. C., Weinberg, E. S., & Sampath, K. (2005). The zebrafish dorsal axis is apparent at the four-cell stage. *Nature*, *438*(7070), 1030–1035. Retrieved from http://www.ncbi.nlm.nih.gov/entrez/query.fcgi?cmd=Retrieve&db=PubMed&dopt=Citation&list_uids=16355228.

Gore, A. V., & Sampath, K. (2002). Localization of transcripts of the zebrafish morphogen Squint is dependent on egg activation and the microtubule cytoskeleton. *Mechanisms of Development*, *112*(1–2), 153–156. doi:S0925477301006220 [pii].

Goudarzi, M., Berg, K., Pieper, L. M., & Schier, A. F. (2019). Individual long non-coding RNAs have no overt functions in zebrafish embryogenesis, viability and fertility. *eLife*, *8*, e40815. https://doi.org/10.7554/eLife.40815.

Goutel, C., Kishimoto, Y., Schulte-Merker, S., & Rosa, F. (2000). The ventralizing activity of Radar, a maternally expressed bone morphogenetic protein, reveals complex bone morphogenetic protein interactions controlling dorso-ventral patterning in zebrafish. *Mechanisms of Development*, *99*(1–2), 15–27. Retrieved from http://www.ncbi.nlm.nih.gov/entrez/query.fcgi?cmd=Retrieve&db=PubMed&dopt=Citation&list_uids=11091070.

Gray, R. S., Bayly, R. D., Green, S. A., Agarwala, S., Lowe, C. J., & Wallingford, J. B. (2009). Diversification of the expression patterns and developmental functions of the dishevelled gene family during chordate evolution. *Developmental Dynamics*, *238*(8), 2044–2057. https://doi.org/10.1002/dvdy.22028.

Gritsman, K., Talbot, W. S., & Schier, A. F. (2000). Nodal signaling patterns the organizer. *Development*, *127*(5), 921–932. Retrieved from http://www.ncbi.nlm.nih.gov/cgi-bin/Entrez/referer?http://www.biologists.com/Development/127/05/dev3140.html.

Gritsman, K., Zhang, J., Cheng, S., Heckscher, E., Talbot, W. S., & Schier, A. F. (1999). The EGF-CFC protein one-eyed pinhead is essential for nodal signaling. *Cell*, *97*(1), 121–132. Retrieved from http://www.ncbi.nlm.nih.gov/cgi-bin/Entrez/referer?http://www.cell.com/cgi/content/full/97/1/121.

Hagos, E. G., & Dougan, S. T. (2007). Time-dependent patterning of the mesoderm and endoderm by Nodal signals in zebrafish. *BMC Developmental Biology*, *7*, 22. https://doi.org/10.1186/1471-213X-7-22.

Hammerschmidt, M., Pelegri, F., Mullins, M. C., Kane, D. A., van Eeden, F. J. M., Granato, M., et al. (1996). *dino* and *mercedes*, two genes regulating dorsal development in the zebrafish embryo. *Development*, *123*, 95–102.

Hammerschmidt, M., Serbedzija, G. N., & McMahon, A. (1996). Genetic analysis of dorsoventral pattern formation in the zebrafish: Requirement of BMP-like ventralizing activity and its dorsal repressor. *Genes and Development*, *10*, 2452–2461.

Harvey, S. A., & Smith, J. C. (2009). Visualisation and quantification of morphogen gradient formation in the zebrafish. *PLoS Biology*, 7(5) e1000101. https://doi.org/10.1371/journal.pbio.1000101.

Hashiguchi, M., & Mullins, M. C. (2013). Anteroposterior and dorsoventral patterning are coordinated by an identical patterning clock. *Development*, *140*(9), 1970–1980. https://doi.org/10.1242/dev.088104.

Hashimoto, H., Itoh, M., Yamanaka, Y., Yamashita, S., Shimizu, T., Solnica-Krezel, L., et al. (2000). Zebrafish Dkk1 functions in forebrain specification and axial mesendoderm formation. *Developmental Biology*, *217*(1), 138–152. Retrieved from http://www.ncbi.nlm.nih.gov/cgi-bin/Entrez/referer?http://www.idealibrary.com/links/citation/0012-1606/217/138.

Hatta, K., Kimmel, C. B., Ho, R. K., & Walker, C. (1991). The cyclops mutation blocks specification of the floor plate of the zebrafish central nervous system. *Nature*, *350*(6316), 339–341.

Heisenberg, C. P., & Nusslein-Volhard, C. (1997). The function of *silberblick* in the positioning of the eye anlage in the zebrafish embryo. *Developmental Biology*, *184*(1), 85–94.

Helde, K. A., & Grunwald, D. J. (1993). The DVR-1 (Vg1) transcript of zebrafish is maternally supplied and distributed throughout the embryo. *Developmental Biology*, *159*(2), 418–426.

Hild, M., Dick, A., Rauch, G. J., Meier, A., Bouwmeester, T., Haffter, P., et al. (1999). The smad5 mutation somitabun blocks Bmp2b signaling during early dorsoventral patterning of the zebrafish embryo. *Development, 126*(10), 2149–2159. Retrieved from http://www.ncbi.nlm.nih.gov/cgi-bin/Entrez/referer?http://www.biologists.com/Development/126/10/dev6387.html.

Hino, H., Nakanishi, A., Seki, R., Aoki, T., Yamaha, E., Kawahara, A., et al. (2018). Roles of maternal wnt8a transcripts in axis formation in zebrafish. *Developmental Biology, 434*(1), 96–107. https://doi.org/10.1016/j.ydbio.2017.11.016.

Hirata, T., Yamanaka, Y., Ryu, S. L., Shimizu, T., Yabe, T., Hibi, M., et al. (2000). Novel mix-family homeobox genes in zebrafish and their differential regulation. *Biochemical and Biophysical Research Communications, 271*(3), 603–609. https://doi.org/10.1006/bbrc.2000.2672.

Ho, R. K. (1992). Cell movements and cell fate during zebrafish gastrulation. *Development Supplement*, 65–73, PMID:1299369.

Hong, S. K., Jang, M. K., Brown, J. L., McBride, A. A., & Feldman, B. (2011). Embryonic mesoderm and endoderm induction requires the actions of non-embryonic Nodal-related ligands and Mxtx2. *Development, 138*(4), 787–795. https://doi.org/10.1242/dev.058974.

Hopfield, J. J. (1974). Kinetic proofreading: a new mechanism for reducing errors in biosynthetic processes requiring high specificity. *Proceedings of the National Academy of Sciences of the United States of America, 71*(10), 4135–4139. https://doi.org/10.1073/pnas.71.10.4135.

Imai, Y., & Talbot, W. S. (2001). Morpholino phenocopies of the bmp2b/swirl and bmp7/snailhouse mutations. *Genesis, 30*(3), 160–163. Retrieved from http://www.ncbi.nlm.nih.gov/pubmed/11477698.

Imboden, M., Goblet, C., Korn, H., & Vriz, S. (1997). Cytokeratin 8 is a suitable epidermal marker during zebrafish development. *Comptes Rendus de l'Académie des Sciences Série III, 320*(9), 689–700. Retrieved from https://www.ncbi.nlm.nih.gov/pubmed/9377174.

Inomata, H., Shibata, T., Haraguchi, T., & Sasai, Y. (2013). Scaling of dorsal-ventral patterning by embryo size-dependent degradation of Spemann's organizer signals. *Cell, 153*(6), 1296–1311. https://doi.org/10.1016/j.cell.2013.05.004.

Jasuja, R., Voss, N., Ge, G., Hoffman, G. G., Lyman-Gingerich, J., Pelegri, F., et al. (2006). bmp1 and mini fin are functionally redundant in regulating formation of the zebrafish dorsoventral axis. *Mechanisms of Development, 123*(7), 548–558. https://doi.org/10.1016/j.mod.2006.05.004.

Jesuthasan, S., & Stahle, U. (1997). Dynamic microtubules and specification of the zebrafish embryonic axis. *Current Biology*, 7(1), 31–42. Retrieved from http://www.ncbi.nlm.nih.gov/pubmed/9024620.

Joore, J., Fasciana, C., Speksnijder, J. E., Kruijer, W., Destree, O. H., van den Eijnden-van Raaij, A. J., et al. (1996). Regulation of the zebrafish goosecoid promoter by mesoderm inducing factors and Xwnt1. *Mechanisms of Development, 55*(1), 3–18. Retrieved from https://www.ncbi.nlm.nih.gov/pubmed/8734495.

Joseph, S. R., Palfy, M., Hilbert, L., Kumar, M., Karschau, J., Zaburdaev, V., et al. (2017). Competition between histone and transcription factor binding regulates the onset of transcription in zebrafish embryos. *eLife, 6,* e23326. https://doi.org/10.7554/eLife.23326.

Jukam, D., Shariati, S. A. M., & Skotheim, J. M. (2017). Zygotic genome activation in vertebrates. *Developmental Cell, 42*(4), 316–332. https://doi.org/10.1016/j.devcel.2017.07.026.

Jullien, J., & Gurdon, J. (2005). Morphogen gradient interpretation by a regulated trafficking step during ligand-receptor transduction. *Genes & Development, 19*(22), 2682–2694. https://doi.org/10.1101/gad.341605.

Kane, D. A., & Kimmel, C. B. (1993). The zebrafish midblastula transition. *Development, 119*(2), 447–456. Retrieved from http://www.ncbi.nlm.nih.gov/entrez/query.fcgi?cmd=Retrieve&db=PubMed&dopt=Citation&list_uids=8287796.

Kane, D. A., Warga, R. M., & Kimmel, C. B. (1992). Mitotic domains in the early embryo of the zebrafish. *Nature, 360*(6406), 735–737. https://doi.org/10.1038/360735a0.

Kapp, L. D., Abrams, E. W., Marlow, F. L., & Mullins, M. C. (2013). The integrator complex subunit 6 (ints6) confines the dorsal organizer in vertebrate embryogenesis. *PLoS Genetics, 9*(10). e1003822https://doi.org/10.1371/journal.pgen.1003822.

Kawahara, A., & Dawid, I. B. (2000). Expression of the Kruppel-like zinc finger gene biklf during zebrafish development. *Mechanisms of Development, 97*(1–2), 173–176. Retrieved from https://www.ncbi.nlm.nih.gov/pubmed/11025220.

Kawahara, A., Wilm, T., Solnica-Krezel, L., & Dawid, I. B. (2000a). Antagonistic role of vega1 and bozozok/dharma homeobox genes in organizer formation. *Proceedings of the National Academy of Sciences of the United States of America, 97*(22), 12121–12126. https://doi.org/10.1073/pnas.97.22.12121.

Kawahara, A., Wilm, T., Solnica-Krezel, L., & Dawid, I. B. (2000b). Functional interaction of vega2 and goosecoid homeobox genes in zebrafish. *Genesis, 28*(2), 58–67. Retrieved from https://www.ncbi.nlm.nih.gov/pubmed/11064422.

Kelly, C., Chin, A. J., Leatherman, J. L., Kozlowski, D. J., & Weinberg, E. S. (2000). Maternally controlled (beta)-catenin-mediated signaling is required for organizer formation in the zebrafish. *Development, 127*(18), 3899–3911. Retrieved from http://www.ncbi.nlm.nih.gov/entrez/query.fcgi?cmd=Retrieve&db=PubMed&dopt=Citation&list_uids=10952888.

Khan, A., Nakamoto, A., Tai, M., Saito, S., Nakayama, Y., Kawamura, A., et al. (2012). Mesendoderm specification depends on the function of Pou2, the class V POU-type transcription factor, during zebrafish embryogenesis. *Development, Growth & Differentiation, 54*(7), 686–701. https://doi.org/10.1111/j.1440-169X.2012.01369.x.

Kiener, T. K., Selptsova-Friedrich, I., & Hunziker, W. (2008). Tjp3/zo-3 is critical for epidermal barrier function in zebrafish embryos. *Developmental Biology, 316*(1), 36–49. https://doi.org/10.1016/j.ydbio.2007.12.047.

Kikuchi, Y., Agathon, A., Alexander, J., Thisse, C., Waldron, S., Yelon, D., et al. (2001). casanova encodes a novel Sox-related protein necessary and sufficient for early endoderm formation in zebrafish. *Genes & Development, 15*(12), 1493–1505. https://doi.org/10.1101/gad.892301.

Kikuchi, Y., Trinh, L. A., Reiter, J. F., Alexander, J., Yelon, D., & Stainier, D. Y. (2000). The zebrafish bonnie and clyde gene encodes a Mix family homeodomain protein that regulates the generation of endodermal precursors. *Genes & Development, 14*(10), 1279–1289. Retrieved from https://www.ncbi.nlm.nih.gov/pubmed/10817762.

Kimmel, C. B., Warga, R. M., & Schilling, T. F. (1990). Origin and organization of the zebrafish fate map. *Development, 108*(4), 581–594.

Kishimoto, Y., Lee, K. H., Zon, L., Hammerschmidt, M., & Schulte-Merker, S. (1997). The molecular nature of zebrafish swirl: BMP2 function is essential during early dorsoventral patterning. *Development, 124*(22), 4457–4466. Retrieved from http://www.ncbi.nlm.nih.gov/entrez/query.fcgi?cmd=Retrieve&db=PubMed&dopt=Citation&list_uids=9409664.

Koos, D. S., & Ho, R. K. (1998). The *nieuwkoid* gene characterizes and mediates a Nieuwkoop-center-like activity in zebrafish. *Current Biology, 8*, 1199–1206.

Koos, D. S., & Ho, R. K. (1999). The nieuwkoid/dharma homeobox gene is essential for bmp2b repression in the zebrafish pregastrula. *Developmental Biology, 215*(2), 190–207. Retrieved from http://www.ncbi.nlm.nih.gov/cgi-bin/Entrez/referer?http://www.idealibrary.com/links/citation/0012-1606/215/190.

Kotkamp, K., Mossner, R., Allen, A., Onichtchouk, D., & Driever, W. (2014). A Pou5f1/Oct4 dependent Klf2a, Klf2b, and Klf17 regulatory sub-network contributes to EVL and ectoderm development during zebrafish embryogenesis. *Developmental Biology*, *385*(2), 433–447. https://doi.org/10.1016/j.ydbio.2013.10.025.

Kovalenko, D., Yang, X., Chen, P. Y., Nadeau, R. J., Zubanova, O., Pigeon, K., et al. (2006). A role for extracellular and transmembrane domains of Sef in Sef-mediated inhibition of FGF signaling. *Cellular Signalling*, *18*(11), 1958–1966. https://doi.org/10.1016/j.cellsig.2006.03.001.

Kramer, C., Mayr, T., Nowak, M., Schumacher, J., Runke, G., Bauer, H., et al. (2002). Maternally supplied Smad5 is required for ventral specification in zebrafish embryos prior to zygotic Bmp signaling. *Developmental Biology*, *250*(2), 263–279. Retrieved from http://www.ncbi.nlm.nih.gov/entrez/query.fcgi?cmd=Retrieve&db=PubMed&dopt=Citation&list_uids=12376102.

Kumari, P., Gilligan, P. C., Lim, S., Tran, L. D., Winkler, S., Philp, R., et al. (2013). An essential role for maternal control of Nodal signaling. *eLife2.* , e00683https://doi.org/10.7554/eLife.00683.

Kuo, C. L., Lam, C. M., Hewitt, J. E., & Scotting, P. J. (2013). Formation of the embryonic organizer is restricted by the competitive influences of Fgf signaling and the SoxB1 transcription factors. *PLoS One*, *8*(2). e57698. https://doi.org/10.1371/journal.pone.0057698.

Lachnit, M., Kur, E., & Driever, W. (2008). Alterations of the cytoskeleton in all three embryonic lineages contribute to the epiboly defect of Pou5f1/Oct4 deficient MZspg zebrafish embryos. *Developmental Biology*, *315*(1), 1–17. doi:S0012-1606(07)01432-7 [pii] https://doi.org/10.1016/j.ydbio.2007.10.008.

Langdon, Y. G., Fuentes, R., Zhang, H., Abrams, E. W., Marlow, F. L., & Mullins, M. C. (2016). Split top: A maternal cathepsin B that regulates dorsoventral patterning and morphogenesis. *Development*, *143*(6), 1016–1028. https://doi.org/10.1242/dev.128900.

Latinkic, B. V., Umbhauer, M., Neal, K. A., Lerchner, W., Smith, J. C., & Cunliffe, V. (1997). The Xenopus Brachyury promoter is activated by FGF and low concentrations of activin and suppressed by high concentrations of activin and by paired-type homeodomain proteins. *Genes & Development*, *11*(23), 3265–3276. Retrieved from https://www.ncbi.nlm.nih.gov/pubmed/9389657.

Lee, M. T., Bonneau, A. R., & Giraldez, A. J. (2014). Zygotic genome activation during the maternal-to-zygotic transition. *Annual Review of Cell and Developmental Biology*, *30*, 581–613. https://doi.org/10.1146/annurev-cellbio-100913-013027.

Lee, H., & Kimelman, D. (2002). A dominant-negative form of p63 is required for epidermal proliferation in zebrafish. *Developmental Cell*, *2*(5), 607–616. Retrieved from https://www.ncbi.nlm.nih.gov/pubmed/12015968.

Lekven, A. C., Thorpe, C. J., Waxman, J. S., & Moon, R. T. (2001). Zebrafish wnt8 encodes two wnt8 proteins on a bicistronic transcript and is required for mesoderm and neurectoderm patterning. *Developmental Cell*, *1*(1), 103–114. Retrieved from https://www.ncbi.nlm.nih.gov/pubmed/11703928.

Lele, Z., Nowak, M., & Hammerschmidt, M. (2001). Zebrafish admp is required to restrict the size of the organizer and to promote posterior and ventral development. *Developmental Dynamics*, *222*(4), 681–687. https://doi.org/10.1002/dvdy.1222.

Leung, T., Bischof, J., Soll, I., Niessing, D., Zhang, D., Ma, J., et al. (2003). bozozok directly represses bmp2b transcription and mediates the earliest dorsoventral asymmetry of bmp2b expression in zebrafish. *Development*, *130*(16), 3639–3649. Retrieved from https://www.ncbi.nlm.nih.gov/pubmed/12835381.

Leung, T., Soll, I., Arnold, S. J., Kemler, R., & Driever, W. (2003). Direct binding of Lef1 to sites in the boz promoter may mediate pre-midblastula-transition activation of boz expression. *Developmental Dynamics*, *228*(3), 424–432. Retrieved from http://www.ncbi.nlm.nih.gov/entrez/query.fcgi?cmd=Retrieve&db=PubMed&dopt=Citation&list_uids=14579381.

Lim, S., Kumari, P., Gilligan, P., Quach, H. N., Mathavan, S., & Sampath, K. (2012). Dorsal activity of maternal squint is mediated by a non-coding function of the RNA. *Development, 139*(16), 2903–2915. https://doi.org/10.1242/dev.077081.

Little, S. C., & Mullins, M. C. (2004). Twisted gastrulation promotes BMP signaling in zebrafish dorsal-ventral axial patterning. *Development, 131*(23), 5825–5835. https://doi.org/10.1242/dev.01464.

Little, S. C., & Mullins, M. C. (2006). Extracellular modulation of BMP activity in patterning the dorsoventral axis. *Birth Defects Research. Part C, Embryo Today, 78*(3), 224–242. https://doi.org/10.1002/bdrc.20079.

Little, S. C., & Mullins, M. C. (2009). Bone morphogenetic protein heterodimers assemble heteromeric type I receptor complexes to pattern the dorsoventral axis. *Nature Cell Biology, 11*(5), 637–643. doi:ncb1870 [pii] https://doi.org/10.1038/ncb1870.

Liu, B., Xia, X., Zhu, F., Park, E., Carbajal, S., Kiguchi, K., et al. (2008). IKKalpha is required to maintain skin homeostasis and prevent skin cancer. *Cancer Cell, 14*(3), 212–225. https://doi.org/10.1016/j.ccr.2008.07.017.

Long, S., Ahmad, N., & Rebagliati, M. (2003). The zebrafish nodal-related gene southpaw is required for visceral and diencephalic left-right asymmetry. *Development, 130*(11), 2303–2316. Retrieved from http://www.ncbi.nlm.nih.gov/entrez/query.fcgi?cmd=Retrieve&db=PubMed&dopt=Citation&list_uids=12702646.

Lu, F. I., Thisse, C., & Thisse, B. (2011). Identification and mechanism of regulation of the zebrafish dorsal determinant. *Proceedings of the National Academy of Sciences of the United States of America, 108*(38), 15876–15880. https://doi.org/10.1073/pnas.1106801108.

Lunde, K., Belting, H. G., & Driever, W. (2004). Zebrafish pou5f1/pou2, homolog of mammalian Oct4, functions in the endoderm specification cascade. *Current Biology, 14*(1), 48–55. Retrieved from http://www.ncbi.nlm.nih.gov/entrez/query.fcgi?cmd=Retrieve&db=PubMed&dopt=Citation&list_uids=14711414.

Lyman Gingerich, J., Westfall, T. A., Slusarski, D. C., & Pelegri, F. (2005). hecate, a zebrafish maternal effect gene, affects dorsal organizer induction and intracellular calcium transient frequency. *Developmental Biology, 286*(2), 427–439. Retrieved from http://www.ncbi.nlm.nih.gov/entrez/query.fcgi?cmd=Retrieve&db=PubMed&dopt=Citation&list_uids=16154557.

Ma, L. H., Webb, S. E., Chan, C. M., Zhang, J., & Miller, A. L. (2009). Establishment of a transitory dorsal-biased window of localized Ca2+ signaling in the superficial epithelium following the mid-blastula transition in zebrafish embryos. *Developmental Biology, 327*(1), 143–157. https://doi.org/10.1016/j.ydbio.2008.12.015.

Maegawa, S., Varga, M., & Weinberg, E. S. (2006). FGF signaling is required for {beta}-catenin-mediated induction of the zebrafish organizer. *Development, 133*(16), 3265–3276. https://doi.org/10.1242/dev.02483.

Marlow, F. L. (2010). *Maternal Control of Development in Vertebrates: My Mother Made Me Do It!*. San Rafael (CA): Morgan & Claypool Life Sciences.

McKeithan, T. W. (1995). Kinetic proofreading in T-cell receptor signal transduction. *Proceedings of the National Academy of Sciences of the United States of America, 92*(11), 5042–5046. https://doi.org/10.1073/pnas.92.11.5042.

Mei, W., Lee, K. W., Marlow, F. L., Miller, A. L., & Mullins, M. C. (2009). hnRNP I is required to generate the Ca2+ signal that causes egg activation in zebrafish. *Development, 136*(17), 3007–3017. 136/17/3007 [pii] https://doi.org/10.1242/dev.037879.

Meier, M., Grant, J., Dowdle, A., Thomas, A., Gerton, J., Collas, P., et al. (2018). Cohesin facilitates zygotic genome activation in zebrafish. *Development, 145*(1). https://doi.org/10.1242/dev.156521.

Melby, A. E., Beach, C., Mullins, M., & Kimelman, D. (2000). Patterning the early zebrafish by the opposing actions of bozozok and vox/vent. *Developmental Biology, 224*(2), 275–285. https://doi.org/10.1006/dbio.2000.9780.

Miller-Bertoglio, V., Carmany-Rampey, A., Furthauer, M., Gonzalez, E. M., Thisse, C., Thisse, B., et al. (1999). Maternal and zygotic activity of the zebrafish *ogon* locus antagonizes BMP signaling. *Developmental Biology, 214*(1), 72–86. Retrieved from http://www.ncbi.nlm.nih.gov/cgi-bin/Entrez/referer?http://www.idealibrary.com/links/citation/0012-1606/214/72.
Miller-Bertoglio, V. E., Fisher, S., Sanchez, A., Mullins, M. C., & Halpern, M. E. (1997). Differential regulation of chordin expression domains in mutant zebrafish. *Developmental Biology, 192*(2), 537–550.
Mintzer, K. A., Lee, M. A., Runke, G., Trout, J., Whitman, M., & Mullins, M. C. (2001). Lost-a-fin encodes a type I BMP receptor, Alk8, acting maternally and zygotically in dorsoventral pattern formation. *Development, 128*(6), 859–869. Retrieved from http://www.ncbi.nlm.nih.gov/entrez/query.fcgi?cmd=Retrieve&db=PubMed&dopt=Citation&list_uids=11222141.
Mizuno, T., Yamaha, E., Kuroiwa, A., & Takeda, H. (1999). Removal of vegetal yolk causes dorsal deficencies and impairs dorsal-inducing ability of the yolk cell in zebrafish. *Mechanisms of Development, 81*(1-2), 35–47. Retrieved from http://www.ncbi.nlm.nih.gov/cgi-bin/Entrez/referer?http://www.elsevier.com:80/cgi-bin/cas/tree/store/mod/cas_sub/browse/browse.cgi%3fyear=1999&volume=81&issue=1-2&aid=1039.
Mo, S., Wang, L., Li, Q., Li, J., Li, Y., Thannickal, V. J., et al. (2010). Caveolin-1 regulates dorsoventral patterning through direct interaction with beta-catenin in zebrafish. *Developmental Biology, 344*(1), 210–223. https://doi.org/10.1016/j.ydbio.2010.04.033.
Montague, T. G., & Schier, A. F. (2017). Vg1-Nodal heterodimers are the endogenous inducers of mesendoderm. *eLife, 6*, e28183. https://doi.org/10.7554/eLife.28183.
Monteiro, R., van Dinther, M., Bakkers, J., Wilkinson, R., Patient, R., ten Dijke, P., et al. (2008). Two novel type II receptors mediate BMP signalling and are required to establish left-right asymmetry in zebrafish. *Developmental Biology, 315*(1), 55–71. https://doi.org/10.1016/j.ydbio.2007.11.038.
Muller, P., Rogers, K. W., Jordan, B. M., Lee, J. S., Robson, D., Ramanathan, S., et al. (2012). Differential diffusivity of Nodal and Lefty underlies a reaction-diffusion patterning system. *Science, 336*(6082), 721–724. https://doi.org/10.1126/science.1221920.
Muller, P., Rogers, K. W., Yu, S. R., Brand, M., & Schier, A. F. (2013). Morphogen transport. *Development, 140*(8), 1621–1638. https://doi.org/10.1242/dev.083519.
Mullins, M. C., Hammerschmidt, M., Kane, D. A., Odenthal, J., Brand, M., van Eeden, F. J., et al. (1996). Genes establishing dorsoventral pattern formation in the zebrafish embryo: The ventral specifying genes. *Development, 123*, 81–93. Retrieved from http://www.ncbi.nlm.nih.gov/entrez/query.fcgi?cmd=Retrieve&db=PubMed&dopt=Citation&list_uids=9007231.
Nelson, A. C., Cutty, S. J., Niini, M., Stemple, D. L., Flicek, P., Houart, C., et al. (2014). Global identification of Smad2 and Eomesodermin targets in zebrafish identifies a conserved transcriptional network in mesendoderm and a novel role for Eomesodermin in repression of ectodermal gene expression. *BMC Biology, 12*, 81. https://doi.org/10.1186/s12915-014-0081-5.
Nguyen, V. H., Schmid, B., Trout, J., Connors, S. A., Ekker, M., & Mullins, M. C. (1998). Ventral and lateral regions of the zebrafish gastrula, including the neural crest progenitors, are established by a *bmp2b/swirl* pathway of genes. *Developmental Biology, 199*, 93–110.
Nojima, H., Rothhamel, S., Shimizu, T., Kim, C. H., Yonemura, S., Marlow, F. L., et al. (2010). Syntabulin, a motor protein linker, controls dorsal determination. *Development, 137*(6), 923–933. dev.046425 [pii]. https://doi.org/10.1242/dev.046425.
Nojima, H., Shimizu, T., Kim, C. H., Yabe, T., Bae, Y. K., Muraoka, O., et al. (2004). Genetic evidence for involvement of maternally derived Wnt canonical signaling in dorsal determination in zebrafish. *Mechanisms of Development, 121*(4), 371–386. Retrieved from http://www.ncbi.nlm.nih.gov/entrez/query.fcgi?cmd=Retrieve&db=PubMed&dopt=Citation&list_uids=15110047.

Norris, M. L., Pauli, A., Gagnon, J. A., Lord, N. D., Rogers, K. W., Mosimann, C., et al. (2017). Toddler signaling regulates mesodermal cell migration downstream of nodal signaling. *eLife, 6,* e22626. https://doi.org/10.7554/eLife.22626.

Nusse, R., & Clevers, H. (2017). Wnt/beta-catenin signaling, disease, and emerging therapeutic modalities. *Cell, 169*(6), 985–999. https://doi.org/10.1016/j.cell.2017.05.016.

Onichtchouk, D., Geier, F., Messerschmidt, D. M., Mossner, R., Taylor, V., Timmer, J., et al. (2010). Oct4/Pou5f1 controls tissue-specific repressors in early zebrafish embryo. *Journal of Stem Cells & Regenerative Medicine, 6*(2), 82. Retrieved from https://www.ncbi.nlm.nih.gov/pubmed/24693100.

Onichtchouk, D., Geier, F., Polok, B., Messerschmidt, D. M., Mossner, R., Wendik, B., et al. (2010). Zebrafish Pou5f1-dependent transcriptional networks in temporal control of early development. *Molecular Systems Biology, 6,* 354. https://doi.org/10.1038/msb.2010.9.

Pauli, A., Norris, M. L., Valen, E., Chew, G. L., Gagnon, J. A., Zimmerman, S., et al. (2014). Toddler: An embryonic signal that promotes cell movement via apelin receptors. *Science, 343*(6172), 1248636. https://doi.org/10.1126/science.1248636.

Pei, W., Noushmehr, H., Costa, J., Ouspenskaia, M. V., Elkahloun, A. G., & Feldman, B. (2007). An early requirement for maternal FoxH1 during zebrafish gastrulation. *Developmental Biology, 310*(1), 10–22. S0012-1606(07)01186-4 [pii]. https://doi.org/10.1016/j.ydbio.2007.07.011.

Pei, W., Williams, P. H., Clark, M. D., Stemple, D. L., & Feldman, B. (2007). Environmental and genetic modifiers of squint penetrance during zebrafish embryogenesis. *Developmental Biology, 308*(2), 368–378. S0012-1606(07)01083-4 [pii]. https://doi.org/10.1016/j.ydbio.2007.05.026.

Pelliccia, J. L., Jindal, G. A., & Burdine, R. D. (2017). Gdf3 is required for robust nodal signaling during germ layer formation and left-right patterning. *Elife, 6,* e28635. https://doi.org/10.7554/eLife.28635.

Peng, G., & Westerfield, M. (2006). Lhx5 promotes forebrain development and activates transcription of secreted Wnt antagonists. *Development, 133*(16), 3191–3200. https://doi.org/10.1242/dev.02485.

Pezeron, G., Anselme, I., Laplante, M., Ellingsen, S., Becker, T. S., Rosa, F. M., et al. (2006). Duplicate sfrp1 genes in zebrafish: sfrp1a is dynamically expressed in the developing central nervous system, gut and lateral line. *Gene Expression Patterns, 6*(8), 835–842. https://doi.org/10.1016/j.modgep.2006.02.002.

Picozzi, P., Wang, F., Cronk, K., & Ryan, K. (2009). Eomesodermin requires transforming growth factor-beta/activin signaling and binds Smad2 to activate mesodermal genes. *The Journal of Biological Chemistry, 284*(4), 2397–2408. https://doi.org/10.1074/jbc.M808704200.

Pogoda, H., Solnica-Krezel, L., Driever, W., & Meyer, D. (2000). The zebrafish forkhead transcription factor FoxH1/Fast1 is a modulator of nodal signaling required for organizer formation [In process citation]. *Current Biology, 10*(17), 1041–1049.

Pomreinke, A. P., Soh, G. H., Rogers, K. W., Bergmann, J. K., Blassle, A. J., & Muller, P. (2017). Dynamics of BMP signaling and distribution during zebrafish dorsal-ventral patterning. *eLife, 6,* e25861. https://doi.org/10.7554/eLife.25861.

Poulain, M., & Lepage, T. (2002). Mezzo, a paired-like homeobox protein is an immediate target of Nodal signalling and regulates endoderm specification in zebrafish. *Development, 129*(21), 4901–4914. Retrieved from https://www.ncbi.nlm.nih.gov/pubmed/12397099.

Putiri, E., & Pelegri, F. (2011). The zebrafish maternal-effect gene mission impossible encodes the DEAH-box helicase Dhx16 and is essential for the expression of downstream endodermal genes. *Developmental Biology, 353*(2), 275–289. https://doi.org/10.1016/j.ydbio.2011.03.001.

Ramel, M. C., & Hill, C. S. (2013). The ventral to dorsal BMP activity gradient in the early zebrafish embryo is determined by graded expression of BMP ligands. *Developmental Biology, 378*(2), 170–182. https://doi.org/10.1016/j.ydbio.2013.03.003.

Ramel, M. C., & Lekven, A. C. (2004). Repression of the vertebrate organizer by Wnt8 is mediated by Vent and Vox. *Development, 131*(16), 3991–4000. https://doi.org/10.1242/dev.01277.

Rebagliati, M. R., Toyama, R., Fricke, C., Haffter, P., & Dawid, I. B. (1998). Zebrafish nodal-related genes are implicated in axial patterning and establishing left-right asymmetry. *Developmental Biology, 199*(2), 261–272. https://doi.org/10.1006/dbio.1998.8935. S0012-1606(98)98935-7 [pii].

Rebagliati, M. R., Toyama, R., Haffter, P., & Dawid, I. B. (1998). *cyclops* encodes a nodal-related factor involved in midline signaling. *Proceedings of the National Academy of Sciences of the United States of America, 95*, 9932–9937.

Reim, G., & Brand, M. (2006). Maternal control of vertebrate dorsoventral axis formation and epiboly by the POU domain protein Spg/Pou2/Oct4. *Development, 133*(14), 2757–2770. doi:dev.02391 [pii] https://doi.org/10.1242/dev.02391.

Reim, G., Mizoguchi, T., Stainier, D. Y., Kikuchi, Y., & Brand, M. (2004). The POU domain protein spg (pou2/Oct4) is essential for endoderm formation in cooperation with the HMG domain protein casanova. *Developmental Cell, 6*(1), 91–101. Retrieved from https://www.ncbi.nlm.nih.gov/pubmed/14723850.

Rentzsch, F., Anton, R., Saina, M., Hammerschmidt, M., Holstein, T. W., & Technau, U. (2006). Asymmetric expression of the BMP antagonists chordin and gremlin in the sea anemone Nematostella vectensis: implications for the evolution of axial patterning. *Developmental Biology, 296*(2), 375–387. https://doi.org/10.1016/j.ydbio.2006.06.003.

Ro, H., & Dawid, I. B. (2009). Organizer restriction through modulation of Bozozok stability by the E3 ubiquitin ligase Lnx-like. *Nature Cell Biology, 11*(9), 1121–1127. https://doi.org/10.1038/ncb1926.

Ro, H., & Dawid, I. B. (2010). Lnx-2b restricts gsc expression to the dorsal mesoderm by limiting Nodal and Bozozok activity. *Biochemical and Biophysical Research Communications, 402*(4), 626–630. https://doi.org/10.1016/j.bbrc.2010.10.070.

Rogers, K. W., Lord, N. D., Gagnon, J. A., Pauli, A., Zimmerman, S., Aksel, D. C., et al. (2017). Nodal patterning without Lefty inhibitory feedback is functional but fragile. *eLife, 6*, e28785. https://doi.org/10.7554/eLife.28785.

Rogers, K. W., & Schier, A. F. (2011). Morphogen gradients: from generation to interpretation. *Annual Review of Cell and Developmental Biology, 27*, 377–407. https://doi.org/10.1146/annurev-cellbio-092910-154148.

Ryu, S. L., Fujii, R., Yamanaka, Y., Shimizu, T., Yabe, T., Hirata, T., et al. (2001). Regulation of dharma/bozozok by the Wnt pathway. *Developmental Biology, 231*(2), 397–409. https://doi.org/10.1006/dbio.2000.0150.

Sabel, J. L., d'Alencon, C., O'Brien, E. K., Van Otterloo, E., Lutz, K., Cuykendall, T. N., et al. (2009). Maternal interferon regulatory factor 6 is required for the differentiation of primary superficial epithelia in Danio and Xenopus embryos. *Developmental Biology, 325*(1), 249–262. S0012-1606(08)01295-5 [pii]. https://doi.org/10.1016/j.ydbio.2008.10.031.

Sagerstrom, C. G., Gammill, L. S., Veale, R., & Sive, H. (2005). Specification of the enveloping layer and lack of autoneuralization in zebrafish embryonic explants. *Developmental Dynamics, 232*(1), 85–97. https://doi.org/10.1002/dvdy.20198.

Sako, K., Pradhan, S. J., Barone, V., Ingles-Prieto, A., Muller, P., Ruprecht, V., et al. (2016). Optogenetic control of nodal signaling reveals a temporal pattern of nodal signaling regulating cell fate specification during gastrulation. *Cell Reports, 16*(3), 866–877. https://doi.org/10.1016/j.celrep.2016.06.036.

Sampath, K., Cheng, A. M., Frisch, A., & Wright, C. V. (1997). Functional differences among Xenopus nodal-related genes in left-right axis determination. *Development, 124*(17), 3293–3302. Retrieved from http://www.ncbi.nlm.nih.gov/entrez/query.fcgi?cmd=Retrieve&db=PubMed&dopt=Citation&list_uids=9310324.

Sampath, K., Rubinstein, A. L., Cheng, A. M., Liang, J. O., Fekany, K., Solnica-Krezel, L., et al. (1998). Induction of the zebrafish ventral brain and floorplate requires cyclops/nodal signalling. *Nature, 395*(6698), 185–189. https://doi.org/10.1038/26020.

Schier, A. F. (2003). Nodal signaling in vertebrate development. *Annual Review of Cell and Developmental Biology, 19*, 589–621. https://doi.org/10.1146/annurev.cellbio.19.041603.094522.

Schier, A. F. (2009). Nodal morphogens. *Cold Spring Harbor Perspectives in Biology, 1*(5), a003459. https://doi.org/10.1101/cshperspect.a003459.

Schier, A. F., Neuhauss, S. C., Harvey, M., Malicki, J., Solnica-Krezel, L., Stainier, D. Y., et al. (1996). Mutations affecting the development of the embryonic zebrafish brain. *Development, 123*, 165–178. Retrieved from https://www.ncbi.nlm.nih.gov/pubmed/9007238.

Schmid, B., Furthauer, M., Connors, S. A., Trout, J., Thisse, B., Thisse, C., et al. (2000). Equivalent genetic roles for bmp7/snailhouse and bmp2b/swirl in dorsoventral pattern formation. *Development, 127*(5), 957–967. Retrieved from http://www.ncbi.nlm.nih.gov/cgi-bin/Entrez/referer?http://www.biologists.com/Development/127/05/dev2506.html.

Schneider, S., Steinbesser, H., Warga, R. M., & Hausen, P. (1996). β-Catenin translocation into nuclei demarcates the dorsalizing centers in frog and fish embryos. *Mechanisms of Development, 57*, 191–198.

Schulte-Merker, S., Hammerschmidt, M., Beuchle, D., Cho, K. W., Derobertis, E. M., & Nusslein-Volhard, C. (1994). Expression of zebrafish goosecoid and no tail gene products in wild-type and mutant no tail embryos. *Development, 120*(4), 843–852.

Schulte-Merker, S., Lee, K. J., McMahon, A. P., & Hammerschmidt, M. (1997). The zebrafish organizer requires *chordino*. *Nature, 387*, 862–863. 26 June 1997.

Schumacher, J. A., Hashiguchi, M., Nguyen, V. H., & Mullins, M. C. (2011). An intermediate level of BMP signaling directly specifies cranial neural crest progenitor cells in zebrafish. *PLoS One, 6*(11). e27403https://doi.org/10.1371/journal.pone.0027403.

Scott, I. C., Blitz, I. L., Pappano, W. N., Maas, S. A., Cho, K. W., & Greenspan, D. S. (2001). Homologues of twisted gastrulation are extracellular cofactors in antagonism of BMP signalling. *Nature, 410*(6827), 475–478. https://doi.org/10.1038/35068572.

Seiliez, I., Thisse, B., & Thisse, C. (2006). FoxA3 and goosecoid promote anterior neural fate through inhibition of Wnt8a activity before the onset of gastrulation. *Developmental Biology, 290*(1), 152–163. https://doi.org/10.1016/j.ydbio.2005.11.021.

Shamipour, S., Kardos, R., Xue, S. L., Hof, B., Hannezo, E., & Heisenberg, C. P. (2019). Bulk actin dynamics drive phase segregation in zebrafish oocytes. *Cell, 177*(6), 1463–1479. e1418 https://doi.org/10.1016/j.cell.2019.04.030.

Shao, M., Wang, M., Liu, Y. Y., Ge, Y. W., Zhang, Y. J., & Shi, D. L. (2017). Vegetally localised Vrtn functions as a novel repressor to modulate bmp2b transcription during dorsoventral patterning in zebrafish. *Development, 144*(18), 3361–3374. https://doi.org/10.1242/dev.152553.

Shen, M. M. (2007). Nodal signaling: Developmental roles and regulation. *Development, 134*(6), 1023–1034. https://doi.org/10.1242/dev.000166.

Shih, Y. H., Kuo, C. L., Hirst, C. S., Dee, C. T., Liu, Y. R., Laghari, Z. A., et al. (2010). SoxB1 transcription factors restrict organizer gene expression by repressing multiple events downstream of Wnt signalling. *Development, 137*(16), 2671–2681. https://doi.org/10.1242/dev.054130.

Shimizu, T., Yamanaka, Y., Nojima, H., Yabe, T., Hibi, M., & Hirano, T. (2002). A novel repressor-type homeobox gene, ved, is involved in dharma/bozozok-mediated dorsal organizer formation in zebrafish. *Mechanisms of Development*, *118*(1–2), 125–138. Retrieved from https://www.ncbi.nlm.nih.gov/pubmed/12351176.

Shimizu, T., Yamanaka, Y., Ryu, S., Hashimoto, H., Yabe, T., Hirata, T., et al. (2000). Cooperative roles of Bozozok/Dharma and nodal-related proteins in the formation of the dorsal organizer in zebrafish. *Mechanisms of Development*, *91*(1–2), 293–303. Retrieved from http://www.ncbi.nlm.nih.gov/cgi-bin/Entrez/referer?http://www.elsevier.com:80/cgi-bin/cas/tree/store/mod/cas_sub/browse/browse.cgi%3fyear=2000&volume=91&issue=1-2&aid=1379.

Shin, J., Chen, J., & Solnica-Krezel, L. (2014). Efficient homologous recombination-mediated genome engineering in zebrafish using TALE nucleases. *Development*, *141*(19), 3807–3818. https://doi.org/10.1242/dev.108019.

Shinya, M., Eschbach, C., Clark, M., Lehrach, H., & Furutani-Seiki, M. (2000). Zebrafish Dkk1, induced by the pre-MBT Wnt signaling, is secreted from the prechordal plate and patterns the anterior neural plate. *Mechanisms of Development*, *98*(1–2), 3–17. Retrieved from https://www.ncbi.nlm.nih.gov/pubmed/11044603.

Siddiqui, M., Sheikh, H., Tran, C., & Bruce, A. E. (2010). The tight junction component claudin E is required for zebrafish epiboly. *Developmental Dynamics*, *239*(2), 715–722. https://doi.org/10.1002/dvdy.22172.

Sidi, S., Goutel, C., Peyrieras, N., & Rosa, F. M. (2003). Maternal induction of ventral fate by zebrafish radar. *Proceedings of the National Academy of Sciences of the United States of America*, *100*(6), 3315–3320. Retrieved from http://www.ncbi.nlm.nih.gov/entrez/query.fcgi?cmd=Retrieve&db=PubMed&dopt=Citation&list_uids=12601179.

Sirotkin, H. I., Gates, M. A., Kelly, P. D., Schier, A. F., & Talbot, W. S. (2000). Fast1 is required for the development of dorsal axial structures in zebrafish. *Current Biology*, *10*(17), 1051–1054. Retrieved from http://www.ncbi.nlm.nih.gov/entrez/query.fcgi?cmd=Retrieve&db=PubMed&dopt=Citation&list_uids=10996072.

Slagle, C. E., Aoki, T., & Burdine, R. D. (2011). Nodal-dependent mesendoderm specification requires the combinatorial activities of FoxH1 and Eomesodermin. *PLoS Genetics*, 7(5) e1002072. Retrieved from <Go to ISI>://MEDLINE:21637786.

Slanchev, K., Carney, T. J., Stemmler, M. P., Koschorz, B., Amsterdam, A., Schwarz, H., et al. (2009). The epithelial cell adhesion molecule EpCAM is required for epithelial morphogenesis and integrity during zebrafish epiboly and skin development. *PLoS Genetics*, *5*(7). e1000563https://doi.org/10.1371/journal.pgen.1000563.

Solnica-Krezel, L., & Driever, W. (2001). The role of the homeodomain protein Bozozok in zebrafish axis formation. *International Journal of Developmental Biology*, *45*, 299–310. 1 Spec No.

Sonawane, M., Martin-Maischein, H., Schwarz, H., & Nusslein-Volhard, C. (2009). Lgl2 and E-cadherin act antagonistically to regulate hemidesmosome formation during epidermal development in zebrafish. *Development*, *136*(8), 1231–1240. https://doi.org/10.1242/dev.032508.

Stickney, H. L., Imai, Y., Draper, B., Moens, C., & Talbot, W. S. (2007). Zebrafish bmp4 functions during late gastrulation to specify ventroposterior cell fates. *Developmental Biology*, *310*(1), 71–84. https://doi.org/10.1016/j.ydbio.2007.07.027.

Sun, Y., Tseng, W. C., Fan, X., Ball, R., & Dougan, S. T. (2014). Extraembryonic signals under the control of MGA, Max, and Smad4 are required for dorsoventral patterning. *Developmental Cell*, *28*(3), 322–334. https://doi.org/10.1016/j.devcel.2014.01.003.

Tanaka, C., Sakuma, R., Nakamura, T., Hamada, H., & Saijoh, Y. (2007). Long-range action of Nodal requires interaction with GDF1. *Genes & Development*, *21*(24), 3272–3282. Retrieved from <Go to ISI>://MEDLINE:18079174.

Tendeng, C., & Houart, C. (2006). Cloning and embryonic expression of five distinct sfrp genes in the zebrafish Danio rerio. *Gene Expression Patterns*, *6*(8), 761–771. https://doi.org/10.1016/j.modgep.2006.01.006.

Teo, A. K., Arnold, S. J., Trotter, M. W., Brown, S., Ang, L. T., Chng, Z., et al. (2011). Pluripotency factors regulate definitive endoderm specification through eomesodermin. *Genes & Development*, *25*(3), 238–250. https://doi.org/10.1101/gad.607311.

Thisse, C., & Thisse, B. (1999). Antivin, a novel and divergent member of the TGFbeta superfamily, negatively regulates mesoderm induction. *Development*, *126*(2), 229–240.

Thisse, B., & Thisse, C. (2015). Formation of the vertebrate embryo: Moving beyond the Spemann organizer. *Seminars in Cell & Developmental Biology*, *42*, 94–102. https://doi.org/10.1016/j.semcdb.2015.05.007.

Toyama, R., O'Connell, M. L., Wright, C. V., Kuehn, M. R., & Dawid, I. B. (1995). Nodal induces ectopic goosecoid and lim1 expression and axis duplication in zebrafish. *Development*, *121*(2), 383–391. Retrieved from https://www.ncbi.nlm.nih.gov/pubmed/7768180.

Toyooka, T., Hisatomi, O., Takahashi, F., Kataoka, H., & Terazima, M. (2011). Photoreactions of aureochrome-1. *Biophysical Journal*, *100*(11), 2801–2809. https://doi.org/10.1016/j.bpj.2011.02.043.

Tran, L. D., Hino, H., Quach, H., Lim, S., Shindo, A., Mimori-Kiyosue, Y., et al. (2012). Dynamic microtubules at the vegetal cortex predict the embryonic axis in zebrafish. *Development*, *139*(19), 3644–3652. https://doi.org/10.1242/dev.082362.

Troilo, H., Zuk, A. V., Tunnicliffe, R. B., Wohl, A. P., Berry, R., Collins, R. F., et al. (2014). Nanoscale structure of the BMP antagonist chordin supports cooperative BMP binding. *Proceedings of the National Academy of Sciences of the United States of America*, *111*(36), 13063–13068. https://doi.org/10.1073/pnas.1404166111.

Tucker, B., Hepperle, C., Kortschak, D., Rainbird, B., Wells, S., Oates, A. C., et al. (2007). Zebrafish Angiotensin II Receptor-like 1a (agtrl1a) is expressed in migrating hypoblast, vasculature, and in multiple embryonic epithelia. *Gene Expression Patterns*, 7(3), 258–265. https://doi.org/10.1016/j.modgep.2006.09.006.

Tucker, J. A., Mintzer, K. A., & Mullins, M. C. (2008). The BMP signaling gradient patterns dorsoventral tissues in a temporally progressive manner along the anteroposterior axis. *Developmental Cell*, *14*(1), 108–119. https://doi.org/10.1016/j.devcel.2007.11.004.

van Boxtel, A. L., Economou, A. D., Heliot, C., & Hill, C. S. (2018). Long-range signaling activation and local inhibition separate the mesoderm and endoderm lineages. *Developmental Cell*, *44*(2), 179-191.e175. https://doi.org/10.1016/j.devcel.2017.11.021.

Varga, M., Maegawa, S., Bellipanni, G., & Weinberg, E. S. (2007). Chordin expression, mediated by Nodal and FGF signaling, is restricted by redundant function of two beta-catenins in the zebrafish embryo. *Mechanisms of Development*, *124*(9–10), 775–791. https://doi.org/10.1016/j.mod.2007.05.005.

Veil, M., Yampolsky, L. Y., Gruning, B., & Onichtchouk, D. (2019). Pou5f3, SoxB1, and Nanog remodel chromatin on high nucleosome affinity regions at zygotic genome activation. *Genome Research*, *29*(3), 383–395. https://doi.org/10.1101/gr.240572.118.

Wagner, D. S., Dosch, R., Mintzer, K. A., Wiemelt, A. P., & Mullins, M. C. (2004). Maternal control of development at the midblastula transition and beyond: mutants from the zebrafish II. *Developmental Cell*, *6*(6), 781–790. Retrieved from http://www.ncbi.nlm.nih.gov/entrez/query.fcgi?cmd=Retrieve&db=PubMed&dopt=Citation&list_uids=15177027.

Wagner, D. S., & Mullins, M. C. (2002). Modulation of BMP activity in dorsal-ventral pattern formation by the chordin and ogon antagonists. *Developmental Biology*, *245*(1), 109–123. Retrieved from http://www.ncbi.nlm.nih.gov/entrez/query.fcgi?cmd=Retrieve&db=PubMed&dopt=Citation&list_uids=11969259.

Waxman, J. S. (2005). Regulation of the early expression patterns of the zebrafish Dishevelled-interacting proteins Dapper1 and Dapper2. *Developmental Dynamics*, *233*(1), 194–200. https://doi.org/10.1002/dvdy.20301.

Webb, A. E., Driever, W., & Kimelman, D. (2008). psoriasis regulates epidermal development in zebrafish. *Developmental Dynamics*, *237*(4), 1153–1164. https://doi.org/10.1002/dvdy.21509.

Willot, V., Mathieu, J., Lu, Y., Schmid, B., Sidi, S., Yan, Y. L., et al. (2002). Cooperative action of ADMP- and BMP-mediated pathways in regulating cell fates in the zebrafish gastrula. *Developmental Biology*, *241*(1), 59–78. https://doi.org/10.1006/dbio.2001.0494.

Wilm, T. P., & Solnica-Krezel, L. (2003). Radar breaks the fog: Insights into dorsoventral patterning in zebrafish. *Proceedings of the National Academy of Sciences of the United States of America*, *100*(8), 4363–4365. https://doi.org/10.1073/pnas.0931010100.

Xie, J., & Fisher, S. (2005). Twisted gastrulation enhances BMP signaling through chordin dependent and independent mechanisms. *Development*, *132*(2), 383–391. https://doi.org/10.1242/dev.01577.

Xing, Y. Y., Cheng, X. N., Li, Y. L., Zhang, C., Saquet, A., Liu, Y. Y., et al. (2018). Mutational analysis of dishevelled genes in zebrafish reveals distinct functions in embryonic patterning and gastrulation cell movements. *PLoS Genetics*, *14*(8). e1007551https://doi.org/10.1371/journal.pgen.1007551.

Xu, P. F., Houssin, N., Ferri-Lagneau, K. F., Thisse, B., & Thisse, C. (2014). Construction of a vertebrate embryo from two opposing morphogen gradients. *Science*, *344*(6179), 87–89. https://doi.org/10.1126/science.1248252.

Xue, Y., Zheng, X., Huang, L., Xu, P., Ma, Y., Min, Z., et al. (2014). Organizer-derived Bmp2 is required for the formation of a correct Bmp activity gradient during embryonic development. *Nature Communications*, *5*, 3766. https://doi.org/10.1038/ncomms4766.

Yabe, T., Shimizu, T., Muraoka, O., Bae, Y. K., Hirata, T., Nojima, H., et al. (2003). Ogon/Secreted frizzled functions as a negative feedback regulator of Bmp signaling. *Development*, *130*(12), 2705–2716. Retrieved from http://www.ncbi.nlm.nih.gov/entrez/query.fcgi?cmd=Retrieve&db=PubMed&dopt=Citation&list_uids= 12736214.

Yadin, D., Knaus, P., & Mueller, T. D. (2016). Structural insights into BMP receptors: Specificity, activation and inhibition. *Cytokine & Growth Factor Reviews*, *27*, 13–34. https://doi.org/10.1016/j.cytogfr.2015.11.005.

Yamanaka, Y., Mizuna, T., Sasai, Y., Khishi, M., Takeda, H., Kim, C.-H., et al. (1998). A novel homeobox gene, dharma, can induce the organizer in a non-cell-autonomous manner. *Genes and Development*, *12*(15), 2345–2353.

Yan, L., Chen, J., Zhu, X., Sun, J., Wu, X., Shen, W., et al. (2018). Maternal Huluwa dictates the embryonic body axis through beta-catenin in vertebrates. *Science*, *362*(6417). https://doi.org/10.1126/science.aat1045.

Yao, S., Qian, M., Deng, S., Xie, L., Yang, H., Xiao, C., et al. (2010). Kzp controls canonical Wnt8 signaling to modulate dorsoventral patterning during zebrafish gastrulation. *The Journal of Biological Chemistry*, *285*(53), 42086–42096. https://doi.org/10.1074/jbc.M110.161554.

Zhang, C., Basta, T., Hernandez-Lagunas, L., Simpson, P., Stemple, D. L., Artinger, K. B., et al. (2004). Repression of nodal expression by maternal B1-type SOXs regulates germ layer formation in Xenopus and zebrafish. *Developmental Biology*, *273*(1), 23–37. https://doi.org/10.1016/j.ydbio.2004.05.019. S001216060400363X [pii].

Zhang, J. L., Patterson, L. J., Qiu, L. Y., Graziussi, D., Sebald, W., & Hammerschmidt, M. (2010). Binding between crossveinless-2 and chordin von willebrand factor type

C domains promotes BMP signaling by blocking Chordin activity. *PLoS One*, *5*(9). e12846https://doi.org/10.1371/journal.pone.0012846.

Zhu, F., Xia, X., Liu, B., Shen, J., Hu, Y., Person, M., et al. (2007). IKKalpha shields 14-3-3sigma, a G(2)/M cell cycle checkpoint gene, from hypermethylation, preventing its silencing. *Molecular Cell*, *27*(2), 214–227. https://doi.org/10.1016/j.molcel.2007.05.042.

Zinski, J., Bu, Y., Wang, X., Dou, W., Umulis, D., & Mullins, M. C. (2017). Systems biology derived source-sink mechanism of BMP gradient formation. *Elife*, *6*, e22199. https://doi.org/10.7554/eLife.22199.

CHAPTER THREE

Signaling events regulating embryonic polarity and formation of the primitive streak in the chick embryo

Ana Raffaelli, Claudio D. Stern*

Department of Cell & Developmental Biology, University College London, London, United Kingdom
*Corresponding author: e-mail address: c.stern@ucl.ac.uk

Contents

Abstract

The avian embryo is a key experimental model system for early development of amniotes. One key difference with invertebrates and "lower" vertebrates like fish and amphibians is that amniotes do not rely so heavily on maternal messages because the zygotic genome is activated very early. Early development also involves considerable growth in volume and mass of the embryo, with cell cycles that include G1 and G2 phases from very early cleavage. The very early maternal to zygotic transition also allows the embryo to establish its own polarity without relying heavily on maternal determinants. In many amniotes including avians and non-rodent mammals, this enables an ability of the embryo to "regulate": a single multicellular embryo can give rise to more

Current Topics in Developmental Biology, Volume 136
ISSN 0070-2153
https://doi.org/10.1016/bs.ctdb.2019.10.001

than one individual—monozygotic twins. Here we discuss the embryological, cellular, molecular and evolutionary underpinnings of gastrulation in avian embryos as a model amniote embryo. Many of these properties are shared by human embryos.

1. Introduction

Like human embryos and most mammals, avian embryos are particularly useful for studying twinning and axis formation because the early embryo has a flat disc shape (Pasteels, 1940). Research over the last two decades has uncovered signaling pathways and molecules involved in initiation of chick primitive streak formation. One particularly remarkable feature of many amniote embryos, including avian embryos, is the ability of fragments of the embryonic disc (right until the time at which primitive streak formation begins) to form a complete embryo when isolated from the rest of the embryo. This property is called "embryonic regulation." During normal development only one embryo forms, but regulative ability has provided unique tools to help us understand the processes that initiate normal primitive streak development, to analyze how cells communicate across the embryo, and revealed some of the mechanisms that ensure that only a single embryo normally develops from each blastoderm. This review provides an overview of our current knowledge.

Upon cleavage, which occurs in the uterus (before egg laying, where the egg spends about 20 h), a semi-transparent sheet of cells appears at the posterior half, subsequently spreading in an anterior direction and eventually forming the center of the embryo. This central region of the epiblast (ectoderm) is known as the area pellucida (meaning "clear area") (Fig. 1A) and all embryonic tissues will eventually arise from it. The peripheral part of the embryo is known as the area opaca (opaque area) and only gives rise to extraembryonic tissue (Eyal-Giladi & Kochav, 1976; Pasteels, 1940; Spratt, 1942; Vakaet, 1970). Separating the area opaca and area pellucida is a ring of epiblast known as the marginal zone (MZ) (Azar & Eyal-Giladi, 1979; Bachvarova, Skromne, & Stern, 1998; Eyal-Giladi & Khaner, 1989; Khaner & Eyal-Giladi, 1986, 1989; Khaner, Mitrani, & Eyal-Giladi, 1985; Stern, 1990) (Fig. 1). The MZ, in contrast to the rest of the epiblast, has a lucent appearance and its cells have a cuboidal shape, while the rest are cylindrical. The cells of the MZ do not contribute to the embryo proper—they remain extraembryonic (Bachvarova et al., 1998).

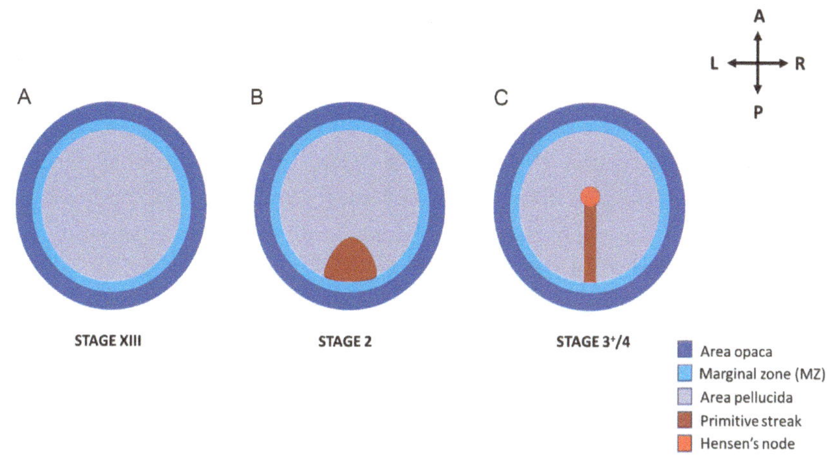

Fig. 1 The chick epiblast and primitive streak development. (A) Epiblast at EGK stage XIII, before primitive streak formation. (B) At HH stage 2, the primitive streak becomes visible in the posterior area pellucida as a wide thickening. (C) At stages 3^{+}–4, the primitive streak is fully elongated, extending approximately two-thirds of the length of the area pellucida.

In the chick embryo, the most visible manifestation of the process of gastrulation is the formation of the primitive streak, which appears as a triangular thickening at the posterior border of the area pellucida. Pre-primitive-streak stages are labeled by the system of Eyal-Giladi and Kochav (EGK) using Roman numerals I–XIV (Eyal-Giladi & Kochav, 1976). Subsequent stages are classified according to Hamburger and Hamilton, starting at stage 2 when the primitive streak first becomes visible (Hamburger & Hamilton, 1951) (Fig. 1). As the primitive streak elongates, the whole embryo elongates posteriorly, acquiring a pear shape (Spratt, 1946). The first pair of somites comes from approximately the point of primitive streak initiation. After gastrulation, the point at which primitive streak formation started becomes far removed from the posterior edge of the area pellucida (Spratt, 1947).

1.1 Embryonic regulation

In many invertebrates and in anamniote vertebrates, each of the two cells of a two-cell-stage embryo has the potential to form a complete embryo (Driesch, 1892; Herbst, 1900; Spemann, 1919). This property is called "embryonic regulation." In anamniotes, this property is lost after the second or third cleavage, the main reason being that the cells of the early embryo acquire their identities largely by localization of maternal components. In contrast, avian and most mammalian embryos retain the ability for embryonic regulation for a long time, even to a stage of development when the embryo contains many thousands of cells. This has been most clearly shown in avian embryos (Lutz, 1949; Spratt & Haas, 1960a; Ulshafer & Clavert, 1979), where cutting an embryo into several fragments can generate a complete embryo from each fragment. Chick embryos can do this until just before the formation of the primitive streak.

Spratt and Haas carried out many experiments in which they cut the chick embryo into fragments that were cultured separately; they observed that as little as 1/8 of the area pellucida (with the corresponding MZ) is enough to form a complete embryo (Spratt & Haas, 1960a). This experiment begs the questions: what prevents the normal blastoderm from forming more than one embryo, and what mechanisms are involved in initiating the formation of a whole embryo from an isolated blastoderm fragment? In their experiments, Spratt and Haas cut the embryos in different ways, symmetrically and asymmetrically, in different positions and orientations.

When the embryo was cut in half, with subsequent separation of anterior and posterior parts, each half formed a complete embryo. The posterior halves always retained their polarity of axis formation, while the anterior halves initiated primitive streak formation at the left or right posterior-most part, close to the MZ, 90° relative to the prospective axis (Spratt & Haas, 1960a). These experiments were also the first to show the significance of the MZ for axis development, as axis formation can be initiated at any point along the MZ ring but not from the middle of the area pellucida. Due to the higher frequency of embryo formation from the posterior MZ (PMZ), Spratt and Haas proposed that a gradient in "embryo-forming potential" exists in the MZ, being highest posteriorly, adjacent to where the streak and the subsequent axis will arise (Spratt & Haas, 1960a).

2. Role of the posterior marginal zone in initiation of primitive streak formation

The importance of the MZ, and particularly the PMZ, for axis development was further emphasized through a number of different studies (Azar & Eyal-Giladi, 1979; Eyal-Giladi & Khaner, 1989; Khaner & Eyal-Giladi, 1986, 1989; Mitrani, Shimoni, & Eyal-Giladi, 1983). The MZ first becomes visible when Koller's sickle appears at stage X (Khaner & Eyal-Giladi, 1989) (Fig. 2). The sickle is a thickened ridge attached to the ventral

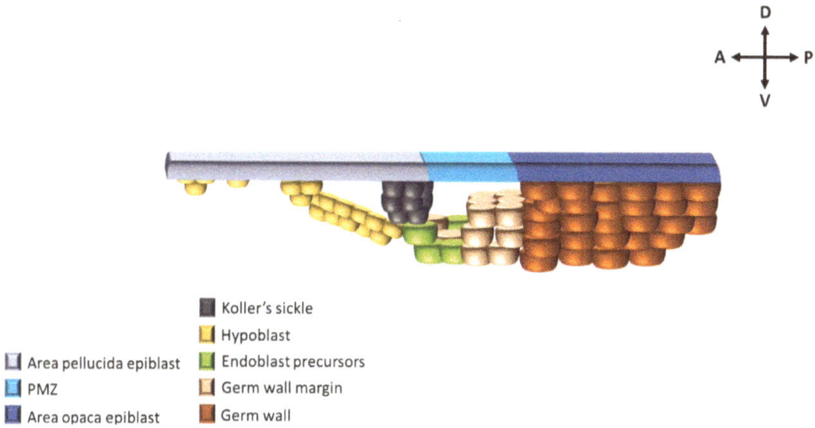

Fig. 2 Sagittal section of the posterior region of the chick embryo at stage XI. In this sagittal section of a stage XI embryo, the arrangement of lower layers of cells underlying the area pellucida, PMZ and area opaca are shown. Posterior to the right, the center of the embryo to the left. The hypoblast (yellow) forms by the fusion of hypoblast islands from posterior to the anterior. The cells shown in green (part of the germ wall margin, cream color) lying ventrally from Koller's sickle (gray) will give rise to the endoblast starting from stage XIII/XIV. The brown cells are the germ wall: yolky cells of the peripheral area opaca.

side of the epiblast, situated posteriorly, over about 60° arc. Its position separates the inner area pellucida from the PMZ (Bachvarova et al., 1998; Callebaut & Van Nueten, 1994; Kochav, Ginsburg, & Eyal-Giladi, 1980; Koller, 1882).

Further experiments designed to shed light on the role of the PMZ in axis formation were done by variations of PMZ excision and replacement with lateral MZ at different stages (X, XI, XII). A series of experiments involving homo- and heterotopic as well as heterochronic transplantations of the PMZ (Khaner & Eyal-Giladi, 1989), and others exploring the ability of various sizes of PMZ transplant to induce a primitive streak (Eyal-Giladi & Khaner, 1989) all suggested that the inductive ability of the PMZ changes from stage X to stage XII, being highest at stage X. A later study (Bachvarova et al., 1998) argued that the above PMZ transplantation experiments probably included some future organizer cells as well as part or all of Koller's sickle (see Fig. 2), because fate mapping experiments showed that while the PMZ itself does not contribute to the primitive streak, the sickle and adjacent inner epiblast do (Bachvarova et al., 1998; Izpisúa-Belmonte, De Robertis, Storey, & Stern, 1993; Streit, Berliner, Papanayotou, Sirulnik, & Stern, 2000). The boundary for cells contributing to the primitive streak is the posterior edge of the sickle at stage XI–XIII: cells posterior to the sickle are not precursors for the primitive streak and do not give rise to embryonic tissue. Furthermore, grafting X–XII quail PMZ to the anterior side of the anterior halves of the chick MZ without Koller's sickle did not result in a contribution to the streak, node or epiblast, unlike grafts containing the sickle, which contributed to the node and the primitive streak and their derivatives (Bachvarova et al., 1998).

The same grafting experiments, along with labeling of anterior cells of the anterior half, showed that formation of the ectopic primitive streak in the anterior halves involves a change of cell fate, through induction. This means that anteriorly-located streaks do not recruit cells from the posterior-most part of the anterior half for their formation, but the PMZ induces the nearby epiblast cells to give rise to the primitive streak even though this is not their usual fate. It was therefore proposed that the PMZ is the avian equivalent of the amphibian Nieuwkoop center (Bachvarova et al., 1998; Skromne & Stern, 2001) as it can induce adjacent cells to form the primitive streak and "primary" (Spemann/Mangold) organizer (Hensen's node), without contributing cells to these resulting structures.

3. Molecular basis of primitive streak induction by the posterior marginal zone

3.1 cVg1 (GDF1)

With the experimental results showing the importance of the MZ for induction of the primitive streak and, thus, axis formation, a search began for a possible molecular basis for that induction. Activin was a good candidate, as it had been reported to elicit formation of axial structures in isolated epiblasts lacking the MZ (Mitrani & Shimoni, 1990). In addition, it can also induce mesoderm in the area opaca (Stern et al., 1995), even though it was previously thought (Khaner et al., 1985) that the area opaca epiblast is not able to develop axial structures of any kind. Nevertheless, its expression is not regionally localized in the chick (Connolly, Patel, Seleiro, Wilkinson, & Cooke, 1995), which is why it was ruled out as a candidate for axis induction. On the other hand, Thomsen and Melton showed that the activin-related molecule Vg1 can initiate axis development in *Xenopus* (Thomsen & Melton, 1993). Shortly after, chick *Vg1* (*cVg1*) encoding a functional ligand, with similar activity to *Xenopus* Vg1 and Activin, was discovered (Seleiro, Connolly, & Cooke, 1996).

The earliest expression of *cVg1* (annotated as GDF3 in the current release of the chicken genome, Gal6, but more closely homologous to mammalian GDF1) is seen at stage X, at low levels, in the posterior blastoderm (mainly in the MZ, then increasingly in the adjacent posterior area pellucida) (Fig. 3). This expression becomes stronger at stages XI–XII. When the primitive streak starts forming, cVg1 expression is seen in the streak itself, but by stage 4 it becomes downregulated in most of the streak. cVg1 protein can induce dorsal mesoderm in *Xenopus* animal caps, as revealed by expression of *Brachyury* (TBXT), a mesodermal marker, as well as the organizer markers *Goosecoid* and *Chordin* (Shah et al., 1997). The expression pattern of cVg1 along with its ability to induce dorsal mesoderm in *Xenopus* animal caps were compelling reasons to investigate whether cVg1 is the PMZ-derived factor responsible for induction of the primitive streak. When mammalian COS cells transfected with a *cVg1* expression construct were grafted to the MZ at stages X–XII, 180° from the putative posterior site, an ectopic primitive streak arose in more than 50% of the cases at the point of grafting. When the same was repeated at stages HH 2–4, no induction took place (Shah et al., 1997). The ectopic primitive streak expressed *Brachyury* and *Goosecoid*, primitive streak markers, and subsequently two normal axes formed in those

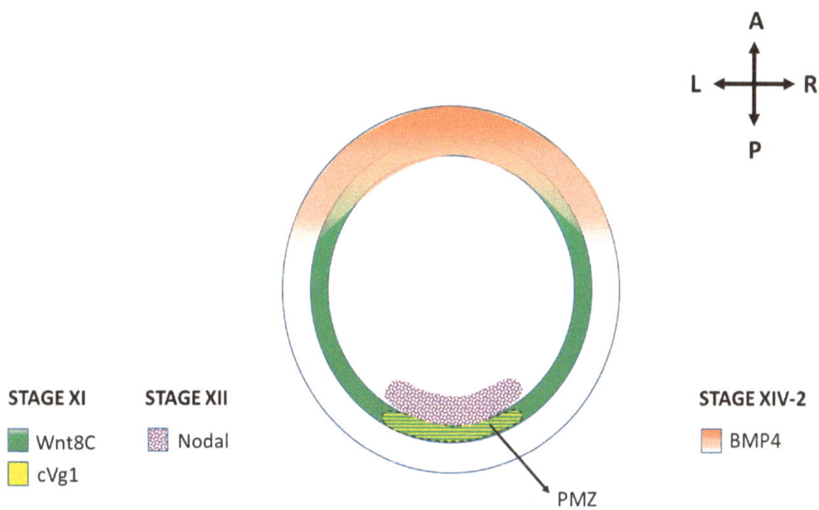

Fig. 3 Expression domains of some key genes implicated in axis formation. At stage XI, *Wnt8C* is expressed throughout the MZ in a posterior-anterior gradient. In the PMZ, *Wnt8C* expression overlaps with *cVg1* expression. Together, they induce *Nodal* expression in the area pellucida, from stage XII to XIII. An antagonistic signal is provided by BMP4, expressed as a gradient in the area opaca and MZ, strongest anteriorly.

embryos (Shah et al., 1997). Also, 8–9 h after being cut, the anterior half of the embryo starts to express *cVg1* either at the left or right side and forms the primitive streak (Bertocchini, Skromne, Wolpert, & Stern, 2004). Moreover, COS cells transfected with *cVg1* are able to induce expression of primitive streak and organizer markers, like *Brachyury* and *Chordin*, when grafted to the area opaca or area pellucida but are not able to initiate primitive streak formation in that location (Shah et al., 1997). This result suggests that there should be other factors in extraembryonic regions that are necessary for primitive streak initiation but are not expressed in central regions of the embryo.

3.2 Wnt8C

Since the PMZ is comparable to the Nieuwkoop center in amphibians (Bachvarova et al., 1998; Skromne & Stern, 2001) and since in *Xenopus* Vg1 cooperates with the Wnt signaling pathway to induce organizer genes (Sokol & Melton, 1992), Wnt seemed like a likely candidate factor for primitive streak induction. Indeed, *Wnt8C* is expressed throughout the MZ (Hume & Dodd, 1993) in a posterior-to-anterior gradient (Fig. 3). A graft of cVg1-secreting COS cells together with fibroblasts secreting Wnt1 (a member of the same functional subclass of Wnt proteins as Wnt8C) to the anterior third of the area pellucida induces *cVg1* and *Nodal* (ENSGALG00000003209, Chromosome 22) followed by an ectopic primitive streak with *Brachyury* and *Chordin* expression. None of these genes can be induced in the area pellucida by either cVg1 or Wnt1 alone (Skromne & Stern, 2001).

The combination of Wnt1 and cVg1 in the anterior area pellucida also induces expression of *Lef1*, which is usually only expressed in the PMZ at stages XI-2 (Skromne & Stern, 2001). As the PMZ is the only region where *cVg1* expression (Shah et al., 1997) overlaps with the ubiquitous expression of *Wnt8C* in the MZ (Fig. 3), this is the most likely place where the two proteins interact to regulate downstream genes in the Wnt pathway such as *Lef1* and *β-catenin* (Skromne & Stern, 2001). In other model organisms, intertwining of the TGFβ and Wnt signaling pathways has been reported through the interaction between Smad4 and β-catenin—TCF/Lef1 complex, downstream components of the TGFβ (Massague & Chen, 2000) and Wnt (Behrens et al., 1996; Cadigan & Nusse, 1997) signaling pathways, respectively. In *Xenopus*, their interaction leads to activation of *twin* (*CNOT6*) transcription (Nishita et al., 2000).

Wnt inhibitors Dkk1 and Crescent grafted together with cVg1 secreting cells in the anterior MZ interfere with formation of the (ectopic) primitive streak. The same result was obtained when Wnt signaling was inhibited by a truncated form of the Frizzled-8 receptor in the PMZ. This suggests that Wnt signaling is necessary for primitive streak formation and that Wnt signaling in the MZ may account for why *cVg1*-transfected cells can induce a streak if placed in the MZ or in the area opaca, but not when it is grafted to the area pellucida (Skromne & Stern, 2001).

3.3 Pitx2

Since cVg1 and Wnt signaling cooperate for the initiation of axis formation and *Wnt8C* is expressed throughout the MZ, a key question is what positions *cVg1* expression within the PMZ. Using differential transcriptome analysis of the early AMZ and PMZ, as well as of the *cVg1*-expressing and non-expressing sides of isolated anterior halves, the transcription factor Pitx2 was uncovered as a candidate regulator (Torlopp et al., 2014). *Pitx2* expression is almost 10-fold higher in the PMZ than in the AMZ (Torlopp et al., 2014).

Normally, *cVg1* expression is observed from around stage XI. However, *Pitx2* expression is already observed in the PMZ at stage X where it remains until just after primitive streak formation. These observations made Pitx2 a good candidate regulator of *cVg1* expression. When *Pitx2* morpholino is electroporated in one corner of an isolated anterior half, primitive streak formation predominantly occurs on the opposite side. However, knocking down Pitx2 in normal embryos only delays primitive streak formation. It was found that this is caused by upregulation of expression of *Pitx1* after *Pitx2*-morpholino electroporation. When *Pitx1* and *Pitx2* morpholinos are introduced together this causes loss of cVg1 and little or no recovery. These findings strongly suggest that Pitx2/Pitx1 is required *cVg1* expression in both normal embryos and in isolated fragments (Torlopp et al., 2014).

A few genes are present in the neighborhood of *cVg1/GDF3* on chromosome-28, including *CERS1/LASS1* and *HOMER3*. Four conserved domains (named E1, E3, E5 and E6) that contain the consensus binding sequence for Pitx2 were found in the introns of *CERS1/LASS1* and *HOMER3*. Pitx2 binding to these was confirmed by chromatin immunoprecipitation with Pitx2 antibody. Each of these putative enhancers was then used in constructs with a minimal promoter (TK) and a reporter (EGFP or RFP), which were then electroporated into early embryos. E3 and E5 were found to drive expression precisely in the PMZ and in the *cVg1*-expressing side of the isolated anterior half. Furthermore, mutation of Pitx2- and Pitx1-binding sites caused loss of activity of these enhancers in the PMZ and in isolated anterior halves (Torlopp et al., 2014). These results make Pitx2 a key regulator of *cVg1* expression in the early embryo and therefore the earliest known gene involved in axis formation and positioning and in embryonic regulation.

4. The hypoblast inhibits primitive streak formation

In addition to the MZ, the hypoblast also plays a role in primitive streak development. The hypoblast is a layer of cells situated under the epiblast, which does not contribute to embryo itself (Azar & Eyal-Giladi, 1979; Vakaet, 1970) (Fig. 4). It arises by fusion of 30–200 "islands," each containing 5–20 cells distributed over the ventral (basal) surface of the area pellucida epiblast at stage X (Fig. 4), which themselves are thought to have

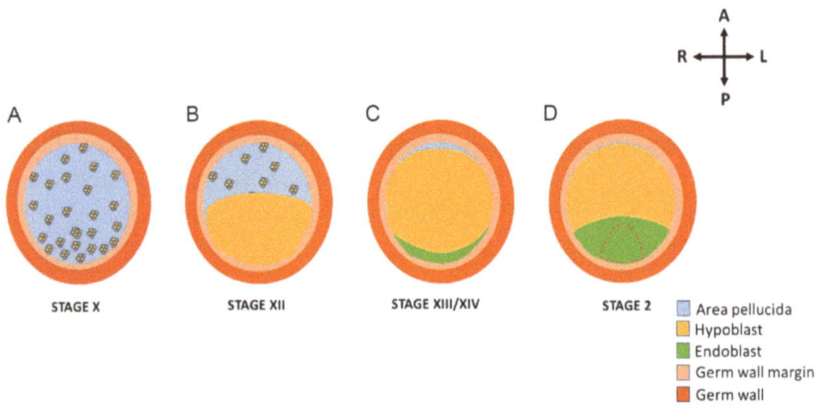

Fig. 4. Cellular composition of the lower layer of the chick embryo. The figure shows the embryo viewed from its ventral side. The primitive streak is drawn in red dashed line for perspective. (A) At stage X, precursor cells of the hypoblast (yellow) are present as "islands" of a few cells, with a slight posterior-to-anterior gradient. Thereafter (from stage XI), the islands begin to fuse together in posterior-to-anterior direction to form a sheet. (B) At stage XII, the continuous layer has expanded to cover about half of the ventral side of the area pellucida. (C) At stage XIII/XIV, the hypoblast is complete, underlying nearly the whole area pellucida. The endoblast (green) starts forming from the cells of the posterior germ wall margin, displacing the hypoblast in an anterior direction. (D) At stage 2, the endoblast has expanded further, removing the hypoblast from the primitive-streak-forming area; the primitive streak formation now begins from the overlying epiblast.

arisen by polyingression (Peter, 1938). Fusion of the islands occurs in a posterior-to-anterior direction, which results in generation of a continuous sheet (Fig. 4). This tissue, the hypoblast, is subsequently joined by a new cell layer (endoblast) derived from a subset of yolky cells underlying the PMZ ("posterior germ wall margin") (see Figs. 2 and 4) (Bertocchini & Stern, 2002, 2008; Stern, 1990; Stern & Ireland, 1981; Vakaet, 1970). Both endoblast and hypoblast have an extraembryonic fate and are not thought to contribute significantly to the embryonic gut. Formation of the hypoblast is complete at stage XIII (Khaner et al., 1985; Stern, 1990; Vakaet, 1970).

The significance of the hypoblast in axis development was first shown through rotation experiments, which showed that the orientation of the primitive streak is re-aligned after hypoblast rotation (Azar & Eyal-Giladi, 1981; Vakaet, 1967; Waddington, 1932, 1933). The ability of the hypoblast to shift the primitive streak declines gradually and ends at stage 3. These experiments were first interpreted as implying that the hypoblast causes induction of the primitive streak (Azar & Eyal-Giladi, 1981; Waddington, 1933). Later experiments, in which the cells that will give rise to the organizer were labeled with DiI/DiO and their movements upon hypoblast rotation monitored, showed that the rotation of the lower layer changes only the movement of the cells but not their fate (Foley, Skromne, & Stern, 2000; Voiculescu, Bertocchini, Wolpert, Keller, & Stern, 2007).

Rather than an inducing function, it now appears that the hypoblast actually inhibits primitive streak formation. The first finding which suggested a possible inhibitory role was that generation of a naked area by purposely placing the hypoblast away from the center of the area pellucida is sufficient to form twins (Azar & Eyal-Giladi, 1981), implying that hypoblast removal allows formation of another axis. A subsequent study showed that when the hypoblast was completely removed from stage XII–XIII embryos, one or more ectopic primitive streaks expressing *Brachyury* appeared. However, these streaks would later regress and only one axis would develop further. Moreover, *Nodal* and *Chordin* gene expression was found ectopically. These genes are found downstream of Vg1 + Wnt, which means that the hypoblast acts in antagonistic manner downstream of Vg1 + Wnt (Bertocchini & Stern, 2002). Grafting of *Nodal*-transfected COS cells in the presence of the hypoblast had no effect; however, when the same was done in the absence of the hypoblast, an ectopic primitive streak expressing *Brachyury* arose adjacent to the location of *Nodal* misexpression. In addition, grafts of COS cells expressing the Nodal antagonists

Cerberus or CerS at stage XIII to the posterior of area pellucida either prevented formation of the primitive streak or caused its displacement (Bertocchini & Stern, 2002). Since the hypoblast, but not the endoblast, expresses Nodal and Wnt antagonists (such as *Cerberus*, *Dkk* and *Crescent*), and since during normal development the primitive streak appears posteriorly immediately after the endoblast starts to displace the hypoblast at this site, it was suggested that hypoblast functions to delay primitive streak formation. Together, this led to the model that the cascade is initiated in the MZ by cVg1 and Wnt8C, Nodal signaling is then induced in the posterior area pellucida (Skromne & Stern, 2002), but its action is inhibited by the underlying hypoblast which expresses the Nodal antagonist Cerberus. Only when this is displaced away by the endoblast is Nodal freed to act (perhaps as a dimer with Vg1, as reported in zebrafish; see below), inducing primitive streak formation (Bertocchini & Stern, 2002) (Fig. 5). A similar role has been proposed for the anterior visceral endoderm, the mouse equivalent of the hypoblast (Perea-Gómez et al., 2002; Stern & Downs, 2012).

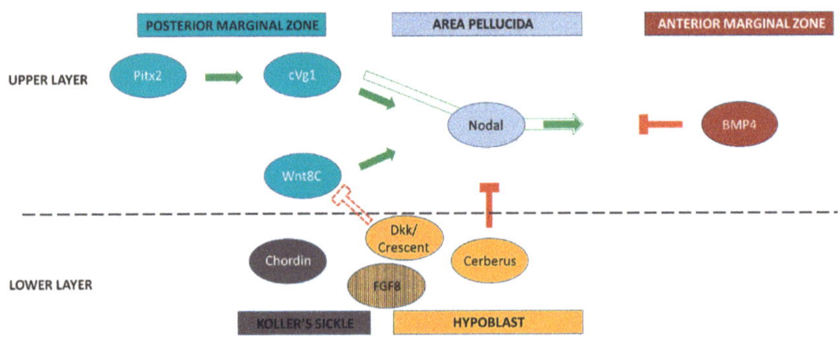

Fig. 5 Signaling network regulating the site of primitive streak formation in the chick embryo. In the PMZ, Pitx2, the earliest known regulator of streak formation, promotes transcription of *cVg1*. cVg1 together with Wnt8C, the latter present all around the MZ, induces *Nodal* expression, required for primitive streak formation, in the posterior area pellucida. However, Nodal is initially inhibited by Cerberus, an antagonist secreted from the underlying hypoblast. The hypoblast in addition expresses Wnt antagonists *Dkk* and *Crescent* which may inhibit Wnt8C. Once the hypoblast is displaced by the endoblast, primitive streak formation is allowed, induced by Nodal (perhaps acting as a heterodimer with cVg1). BMP (expressed as a gradient, strongest anteriorly) provides an antagonistic signal. Details of BMP antagonism by Chordin (expressed posteriorly in Koller's sickle) and of the poorly understood primitive-streak-promoting effect of FGF8 (expressed in both the hypoblast and Koller's sickle) are not shown.

5. Hypoblast, endoblast and definitive endoderm

As mentioned above, the initial hypoblast ("entophyll" in the older literature) is displaced anteriorly by the incoming endoblast ("junctional endoblast" in older papers) (Figs. 2 and 4), which spreads anteriorly from the posterior edge of the deep part of the marginal zone (Bertocchini & Stern, 2002; Foley et al., 2000; Spratt & Haas, 1960b; Stern, 1990; Stern & Ireland, 1981; Vakaet, 1970). The endoblast starts appearing at stage XIV, just before the primitive streak arises (Fig. 4), and releasing that area from molecular inhibitors of Nodal, ultimately allowing the primitive streak and, thus, axis formation from the posterior part of the embryo (Foley et al., 2000).

The endoblast, as previously mentioned, arises from the posterior germ wall margin (the lip of yolky cells underlying the MZ). Precursors of the endoblast in the germ wall margin label with HNK-1 antibody (Canning & Stern, 1988; Stern, 1990). More localized fate mapping experiments later confirmed that the part of the germ wall margin that lies ventral to Koller's sickle gives rise to the endoblast (Fig. 2) (Bachvarova et al., 1998). A differential screen of stage 3 hypoblast and endoblast cells was conducted to search for an endoblast marker (Bertocchini & Stern, 2008). Apolipoprotein A1, a member of a protein family implicated in the transport of lipids and lipid-linked morphogens (Srivastava & Srivastava, 2000; Willnow, Hammes, & Eaton, 2007), was found to be expressed in both hypoblast and endoblast, however, with a higher level in the former (Bertocchini & Stern, 2008). Unlike the hypoblast, whose markers include antagonists of Nodal and Wnt (Cerberus, Dickkopf1 and Crescent), markers specific for the endoblast have not yet been found.

After appearance of the primitive streak and the onset of its elongation, at stages HH 3–4, a population of epiblast cells ingressing through the most anterior tip of the streak insert into the lower layer (probably around the border between hypoblast and endoblast, which more or less tracks the tip of the elongating primitive streak). These are the precursors of the definitive endoderm (entoderm in the older literature) and will later form the lining of the gut tube and contribute to many internal organs associated with it (pancreas, liver, lung, etc.) (Bellairs, 1953a, 1953b, 1955, 1957; Kimura, Yasugi, Stern, & Fukuda, 2006; Kirby et al., 2003; Lawson & Schoenwolf, 2003). It is possible, as shown recently in the mouse (Kwon, Viotti, & Hadjantonakis, 2008), that some hypoblast and/or endoblast cells are retained among definitive endoderm and contribute to the embryonic gut, but the bulk of hypoblast and endoblast cells only give rise to extraembryonic tissues, such as the yolk sac stalk. The combination of endoblast and definitive endoderm cause the remnants of the hypoblast to become located at the most anterior part of the area pellucida, forming the "germinal crescent," so named because this movement is also responsible for concentrating germ cell precursors, which are interspersed with hypoblast cells, to this location from where they will later migrate to the circulatory system and from there to the gonads (Eyal-Giladi, Kochav, & Menashi, 1976; Ginsburg & Eyal-Giladi, 1986; Ginsburg, Hochman, & Eyal-Giladi, 1989; Karagenc, Cinnamon, Ginsburg, & Petitte, 1996).

6. Formation and shaping of the primitive streak

At pre-primitive streak stages, cells of the epiblast move bilaterally toward the posterior and then through the midline anteriorly. These specific movements are known as "Polonaise movements" (Gräper, 1929; Vakaet, 1970). Different models have been invoked to explain those movements, including differential cell division (Firmino, Rocancourt, Saadaoui, Moreau, & Gros, 2016; Wei & Mikawa, 2000) and chemotactic signals (Cui, Yang, Chuai, Glazier, & Weijer, 2005). However, the mechanism that provides the best-supported explanation involves cell intercalation.

Just prior to primitive streak formation (stages XII–XIII), posterior area pellucida epiblast cells (the domain expressing *Nodal*) intercalate preferentially at right angles to the radius of the blastoderm, causing convergence translated into extension (Voiculescu et al., 2007). The non-canonical WNT planar cell polarity (PCP) pathway has been implicated in later convergence-extension movements of the mesoderm in *Xenopus* and zebrafish (Heisenberg et al., 2000). Although the PCP ligand *Wnt11* is expressed ubiquitously in early chick embryos, components of the PCP pathway (*PRICKLE1*, *FMI1/CELSR1* and *VANGL2*) are expressed in the same domain of the posterior epiblast where the cells intercalate (Voiculescu et al., 2007). *FGF8*, which is expressed in the hypoblast, was found to induce expression of *CELSR1* and *PRICKLE1*. This suggests that FGF secretion by the hypoblast can induce PCP genes and thus drive cell intercalation and extension of the *Nodal*-domain (prospective primitive streak) along the midline of the embryo (Voiculescu et al., 2007). A dominant-negative version of Disheveled (Dsh-ΔDEP), which specifically inhibits the Wnt-PCP pathway but does not affect the canonical (β-catenin-dependent) Wnt signaling, does not interfere with mesoderm specification but abolishes primitive streak elongation. A similar result was achieved by electroporation of a mixture of morpholinos against *PRICKLE1, CELSR1* and *VANGL2* (Voiculescu et al., 2007). Intercalation also involves myosin-II downstream of the Wnt-PCP pathway (Rozbicki et al., 2015).

In addition to being necessary for primitive streak formation, intercalation results in movement of the cells from the anterior and lateral epiblast to the posterior medial area to compensate for the cells that have moved away. At the same time, the primitive streak undergoes anteriorly-directed movement along the midline, driven by convergence and extension. Hence, in addition to explaining formation of the primitive streak shape, cell intercalation can explain Polonaise movements without the need to invoke long-range gradients (Rozbicki et al., 2015; Voiculescu, Bodenstein, Lau, & Stern, 2014; Voiculescu et al., 2007).

Apart from intercalation, an important process for primitive streak formation is cell ingression, by which epiblast cells leave that layer and accumulate below it and become mesendoderm precursors. Primitive streak cells express the HNK-1 epitope. However, prior to primitive streak formation, a scattered population of epiblast cells (salt-and-pepper pattern) also expresses the epitope (Canning & Stern, 1988). These cells ingress prior to gastrulation and are implicated in primitive streak formation, eventually giving rise to mesoderm and endoderm (Stern & Canning, 1990). Ablation of the HNK-1 positive cells at pre-primitive streak stages caused loss of the primitive streak and mesodermal structures since the HNK-1-negative cells are not able to compensate for that loss (Stern & Canning, 1990). It is still not known how these early-ingressing mesendoderm precursor cells are specified. Upon ingression, the mesendodermal cells are responsible for triggering EMT in the nearby cells, thus establishing the community effect, which maintains the form of the primitive streak (Stern & Canning, 1990; Voiculescu et al., 2014). This positive feedback is most likely mediated through Nodal signaling. It was proposed that the fact that Nodal is inhibited by antagonists (Cerberus) from the hypoblast, EMT only takes place at a low rate prior to hypoblast displacement. Displacement of the hypoblast by the endoblast causes a more emphasized EMT, further amplified by positive feedback by early-ingressing cells, increased ingression resulting in primitive streak formation (Voiculescu et al., 2014).

Taken together, these findings suggest that the hypoblast has a dual role in primitive streak formation. On one hand, it secretes FGF which controls the Wnt-PCP pathway that drives elongation of the primitive streak forming domain, placing it at the midline (Voiculescu et al., 2007), and on the other hand, it secretes the Nodal antagonist Cerberus, which delays primitive streak formation until the hypoblast has been displaced by the incoming endoblast (which does not express *Cerberus*) (Bertocchini & Stern, 2002; Voiculescu et al., 2014). The disappearance of Cerberus releases the posterior epiblast from inhibition, now allowing Nodal to act and induce cells to ingress in this posterior domain, initiating primitive streak formation. This posterior domain co-expresses Vg1/GDF1 along with Nodal, and by analogy to other vertebrates, it is possible that these two TGFβ-related factors act as a heterodimer (Bisgrove, Su, & Yost, 2017; Montague, Gagnon, & Schier, 2018; Montague & Schier, 2017; Pelliccia, Jindal, & Burdine, 2017). As cells are lost through ingression, this creates a force for lateral cells to be pulled in. By entering the primitive streak, those cells also become exposed to the Wnt-PCP system, driving further intercalation and extension along the midline (Voiculescu et al., 2014).

7. Mechanisms ensuring that gastrulation is initiated only in one place

7.1 Inhibitors

The above-mentioned molecular factors that are implicated in axis formation positively regulate primitive streak formation. Since normal embryo develops only a single axis, yet has the ability to form up to eight complete embryos when divided into parts (Spratt & Haas, 1960a), there must be mechanisms that prevent the embryo from generating multiple axes (Bertocchini et al., 2004). BMP4 (Streit et al., 1998; Streit & Stern, 1999) and its target GATA2 (Bertocchini & Stern, 2012; Sheng & Stern, 1999) have a negative impact on primitive streak formation. Both are expressed anteriorly in the chick, at least as early as *cVg1* (Bertocchini & Stern, 2012) and appear to form a crescent-shaped gradient (decreasing posteriorly).

GATA2 is expressed from very early stages (VIII–IX) but ubiquitously. By stage X, an obvious anterior bias in expression is seen. This is concomitant with the first expression of *cVg1* posterior MZ around stage X. Introducing a morpholino against *GATA2* in the anterior part of the embryo caused formation of either two primitive streaks or one ectopically positioned streak, as well as ectopic expression of both *cVg1* and *Wnt8C* 6 h after electroporation. This shows that GATA2 influences embryonic polarity and can act upstream of *cVg1* and *Wnt8C*, possibly restricting expression of these factors to their respective domains (Bertocchini & Stern, 2012). However, although it is a transcription factor, the effects of this manipulation are not local to the targeted cells suggesting that GATA2 is part of a signaling loop that acts over long distances.

GATA2 knockdown with a morpholino placed anteriorly in an anterior half causes ectopic primitive streak formation, paradoxically in the posterior-most edge of the fragment, implying that GATA2 is not an absolute determinant of the polarity (Bertocchini & Stern, 2012). Since misexpression of *cVg1* does not affect *GATA2* expression, it was suggested that *cVg1* and *GATA2* are controlled independently of each other, which means that the embryo does not only need to correctly localize the expression of one of these genes, but both expressions must be regulated separately. Bertocchini and Stern named this the "global positioning system." This is an important finding because previously it was thought that only the posterior part of the embryo needs to be biased by molecular cues.

GATA seems to interact with BMPs both upstream (being down-regulated by Smad2-dependent Nodal or Activin activity) (Read et al., 1998; Walmsley et al., 1994) and downstream (upregulating their expression) (Linker et al., 2009; Sykes, Rodaway, Walmsley, & Patient, 1998). BMP, on the other hand, has been shown to be lost from the posterior area pellucida as the primitive streak forms, which suggested that BMP4 extinction is required for primitive streak formation. Indeed, grafting BMP4-transfected COS cells in the posterior part of the area pellucida causes loss of primitive streak formation and *Brachyury* expression (Streit et al., 1998). It has been shown in other organisms that BMPs can positively regulate their own expression (Jones, Lyons, Lapan, Wright, & Hogan, 1992; Schmidt, von Dassow, & Kimelman, 1996); however, BMP activity is also regulated by inhibitors like short gastrulation (sog) in *Drosophila* (Biehs, Francois, & Bier, 1996) and its vertebrate homologue, Chordin (Holley et al., 1995; Schulte-Merker, Lee, McMahon, & Hammerschmidt, 1997). The chick homologue of the secreted protein Chordin was found to be initially expressed in the epiblast, in the region adjacent to where the primitive streak will form, slightly anterior to Koller's sickle (Hatada & Stern, 1994; Streit et al., 1998; Streit & Stern, 1999).

When misexpressed in the anterior area pellucida, *Chordin* induces formation of a short primitive streak-like structure that expresses *Brachyury*. In contrast to cVg1 (Shah et al., 1997), Chordin can induce streak formation even at stage 2, when the original primitive streak had already started forming (Streit et al., 1998). This suggests that BMP4 normally inhibits primitive streak formation, and that Chordin endogenously inhibits this to mediate primitive streak formation (Streit et al., 1998). Interestingly, however, cells secreting another BMP antagonist, Noggin, do not induce an ectopic axis when grafted to the anterior area pellucida of either XII–XIII-stage or stage 2–3 embryos (Streit & Stern, 1999), suggesting ligand specificity and distinct functions of these two inhibitors.

7.2 Communication

In normal development, the posterior (primitive streak initiating) region of the embryo prevents the same process from occurring elsewhere in the embryo. Likewise, an isolated anterior half only initiates streak formation from one posterior edge, not both. Moreover, *cVg1* misexpression in the MZ at stage 2 (just after the endogenous primitive streak starts to form) is unable to induce an ectopic primitive streak (Skromne & Stern, 2001), even though this is possible at stages X–XII (Shah et al., 1997). These observations imply that the embryo has the ability to transmit information so that once one axis starts to form, formation of another axis in the embryo is inhibited.

To test the idea that some early event during primitive streak initiation propagates across the embryo and to measure the speed of this propagation, an experiment involving sequential *cVg1* misexpression was performed. Grafting two pellets with Vg1 in the left and right lateral MZ at the same time always induced two primitive streaks; however, when one was grafted 6 h after the other, only the first pellet caused streak formation. Since the diameter of the embryo at this stage is about 3 mm, it was suggested that the inhibitor must travel across the embryo at a speed of at least 500 μm/h, perhaps much faster depending on how many steps intervene between cells receiving the Vg1 signal and induction/deployment of the signal conveying the inhibitory information (Bertocchini et al., 2004). The experiments tested Cerberus as the possible inhibitor, as well as whether the inhibitor acts on Wnt signaling; however, neither was found to be the case. Grafting of a second cVg1 bead together with Nodal and Chordin, 6 h after the first one, on the other hand, allowed for the secondary axis formation, which suggested that the inhibition takes place upstream of Nodal and Chordin but downstream of cVg1 (Bertocchini et al., 2004). Furthermore, FGF was able to induce formation of an ectopic streak when a FGF8-soaked bead was grafted 6 h after cVg1. When Chordin or Nodal were grafted with FGF 6 h after the first cVg1 pellet, the axis developed more frequently than it did when any one of the three was grafted alone. These findings suggest that FGF interacts with Nodal and Chordin in primitive streak initiation (Bertocchini et al., 2004).

These findings, along with the observation that although a stage-2 (early primitive streak stage) embryo cannot initiate a second primitive streak in response to *cVg1* misexpression anteriorly (Shah et al., 1997) whereas an isolated stage-2 anterior half can suggested that the primitive streak itself could emit an inhibitor. As mentioned briefly above, in the mouse embryo, the anterior visceral endoderm (hypoblast equivalent) inhibits primitive streak formation through its expression of two Nodal antagonists: *Cerberus* and *Leftb (Lefty1)*. In *Cerberus/Leftb* double-mutant embryos multiple primitive streaks arise (Perea-Gómez et al., 2002). A further *Lefty*-related gene is expressed a little later, in the primitive streak. However, chick embryos possess only one *Lefty* gene, expressed in the primitive streak (like mouse *Lefty-2*). Chick *Lefty* could therefore be responsible for preventing primitive streak formation close to the endogenous one (Bertocchini et al., 2004; Bertocchini & Stern, 2002).

8. Comparison to other model organisms

Some of the above-mentioned processes of axis initiation, as well as the molecular players involved, seem to be at least partly conserved, not only among the amniotes, but even in *Drosophila*, *Xenopus* and zebrafish. However, differences between species can also be informative. The most significant evolutionary change seen in amniotes compared to anamniotes is the significant decrease in the importance of maternal components in axis development (Stern & Downs, 2012). Other differences also exist among amniotes, including the inability of rodents to give rise to monozygotic twins (beyond initiation of a supernumerary primitive streak in certain experimental situations) (Merrill et al., 2004; Perea-Gómez et al., 2002). This is in contrast with other mammals which can form monozygotic twins that develop to term. An extreme case of this is the armadillo, which generates obligate quadruplets by two sequential splitting events (Eakin & Behringer, 2004; Enders, 2002), but identical twins can also arise spontaneously in humans and other placental mammals. One reason why rodents may be unable to do this is the unusual cup shape of the embryo, perhaps because of the enforced proximity of the most extreme anterior and posterior ends of the epiblast, which may require stronger inhibitory mechanisms (Hoodless et al., 2001).

The layer underlying the epiblast of the mouse, known as the visceral endoderm (VE), forms at the embryonic day E4.5–5.5, when the embryo elongates and takes up the form of a cylinder. The VE plays the key role in the axial development of the embryo (Beddington & Robertson, 1999; Rhinn et al., 1998; Varlet, Collignon, & Robertson, 1997). One of the first changes in the VE that affects the axis formation happens at the distal tip of the embryo. A morphologically distinct domain known as the distal visceral endoderm (DVE) forms and migrates anteriorly until it reaches the embryonic/extra embryonic junction. When it has reached that point (E6.0), the DVE becomes known as the anterior visceral endoderm (AVE) (Thomas & Beddington, 1996).

The DVE/AVE, strikingly, undergoes the same movement as the hypoblast in the chick and it expresses similar inhibitory components including *Cerberus* and *Dickkopf1*, (Foley et al., 2000; Stern & Downs, 2012). Thus, the AVE has the ability to inhibit the expression of primitive streak markers in the anterior epiblast (Kimura et al., 2000), as does the hypoblast (Bertocchini & Stern, 2002). Furthermore, double mutants of asymmetrically expressed antagonists of Nodal, *Cerl* and *Lefty-1* (Yamamoto et al., 2004), form ectopic primitive streaks (Perea-Gómez et al., 2002). Lefty1 has additionally been found to be implicated in the DVE specification by regulating the number of DVE cells (Takaoka, Nishimura, & Hamada, 2017). The chick hypoblast, on the other hand, only expresses *Cerberus* to suppress Nodal activity. *Lefty1* does not have an orthologue in the chick. Based on its expression, the gene known as *Lefty1* in the chick is more similar to mouse *Lefty2*, which is only expressed in the primitive streak itself rather than in the early embryo (Stern & Downs, 2012). Interestingly, patterns of expression of *Nodal*, *Cerberus* and *Lefty1* in the rabbit suggest that the mechanism by which hypoblast/AVE prevent premature primitive streak formation is broadly conserved among amniotes (Stern & Downs, 2012).

At the posterior end of the mouse embryo, there is another population of cells that seem to be distinct from the rest of the VE—posterior visceral endoderm (PVE). The PVE is associated with the primitive streak and is not displaced until the early neural plate stage (E8.0) and expresses *Wnt3a* before migration of the DVE/AVE (Rivera-Perez & Magnuson, 2005). Loss of *Wnt3a* expression prevents primitive streak formation (Liu et al., 1999). The PVE is similar to the endoblast in the chick (see Figs. 2 and 4), which is not displaced from the lower layer until stage 4; however, *Wnt3a* expression has not been seen in the endoblast of the chick (Stern & Downs, 2012).

Unfortunately, no specific endoblast or PVE marker has yet been found (Bertocchini & Stern, 2008; Gonçalves et al., 2011).

9. Summary and conclusions

Embryonic regulation, the ability of the embryo to compensate for the loss of its parts, is the phenomenon that underlies monozygotic twinning in amniotes. The existence of this property also offers a unique opportunity to address experimentally the mechanisms that position the embryonic axis and to uncover cell interactions that prevent ectopic axis formation in normal embryos to ensure that only a single embryo develops. Some of the relatively recent discoveries presented in this review have started to reveal the network of molecular interactions underlying primitive streak initiation; a range of proteins is involved, from transcription factors to signaling proteins and possible morphogens, their receptors and inhibitors. Nevertheless, many pieces of the puzzle are still missing, not least the identity of the critical inhibitor(s), as well as the mechanisms that initiate the cascade that positions Pitx2 and cVg1 in a posterior location.

A recent study has shown that placing four BMP4 beads around the MZ at 90° to each other caused formation of multiple primitive streaks in the space between the beads (Arias, Herrero, Stern, & Bertocchini, 2017). It was suggested that communication happens within the MZ and that an active, dynamic process of competition occurs between remote regions of the MZ. The mechanism of such a competition remains to be discovered, but the speed at which it occurs suggests that non-genetic mechanisms, perhaps mediated by small molecules such as ions, or mechanical events, might be involved. Moreover, the observation that an isolated anterior half is able to initiate primitive streak formation even when it is cut at stage 2 suggest multiple inhibitory mechanisms that operate at various stages of development. Much remains to be understood.

References

Arias, C. F., Herrero, M. A., Stern, C. D., & Bertocchini, F. (2017). A molecular mechanism of symmetry breaking in the early chick embryo. *Scientific Reports*, *7*, 15776. https://doi.org/10.1038/s41598-017-15883-8.

Azar, Y., & Eyal-Giladi, H. (1979). Marginal zone cells—The primitive streak-inducing component of the primary hypoblast in the chick. *Journal of Embryology and Experimental Morphology*, *52*, 79–88.

Azar, Y., & Eyal-Giladi, H. (1981). Interaction of epiblast and hypoblast in the formation of the primitive streak and the embryonic axis in chick, as revealed by hypoblast-rotation experiments. *Journal of Embryology and Experimental Morphology*, *61*, 133–144.

Bachvarova, R. F., Skromne, I., & Stern, C. D. (1998). Induction of primitive streak and Hensen's node by the posterior marginal zone in the early chick embryo. *Development, 125*, 3521–3534.

Beddington, R. S., & Robertson, E. J. (1999). Axis development and early asymmetry in mammals. *Cell, 96*, 195–209.

Behrens, J., von Kries, J. P., Kuhl, M., Bruhn, L., Wedlich, D., Grosschedl, R., et al. (1996). Functional interaction of beta-catenin with the transcription factor LEF-1. *Nature, 382*, 638–642. https://doi.org/10.1038/382638a0.

Bellairs, R. (1953a). Studies on the development of the foregut in the chick blastoderm. 1. The presumptive foregut area. *Journal of Embryology and Experimental Morphology, 1*, 115–124.

Bellairs, R. (1953b). Studies on the development of the foregut in the chick blastoderm. 2. The morphogenetic movements. *Journal of Embryology and Experimental Morphology, 1*, 369–385.

Bellairs, R. (1955). Studies on the development of the foregut in the chick embryo. 3. The role of mitosis. *Journal of Embryology and Experimental Morphology, 3*, 242–250.

Bellairs, R. (1957). Studies on the development of the foregut in the chick embryo. 4. Mesodermal induction and mitosis. *Journal of Embryology and Experimental Morphology, 5*, 340–350.

Bertocchini, F., Skromne, I., Wolpert, L., & Stern, C. D. (2004). Determination of embryonic polarity in a regulative system: Evidence for endogenous inhibitors acting sequentially during primitive streak formation in the chick embryo. *Development, 131*, 3381–3390.

Bertocchini, F., & Stern, C. D. (2002). The hypoblast of the chick embryo positions the primitive streak by antagonizing nodal signaling. *Developmental Cell, 3*, 735–744.

Bertocchini, F., & Stern, C. D. (2008). A differential screen for genes expressed in the extraembryonic endodermal layer of pre-primitive streak stage chick embryos reveals expression of Apolipoprotein A1 in hypoblast, endoblast and endoderm. *Gene Expression Patterns, 8*, 477–480.

Bertocchini, F., & Stern, C. D. (2012). Gata2 provides an early anterior bias and uncovers a global positioning system for polarity in the amniote embryo. *Development, 139*, 4232–4238. https://doi.org/10.1242/dev.081901.

Biehs, B., Francois, V., & Bier, E. (1996). The Drosophila short gastrulation gene prevents Dpp from autoactivating and suppressing neurogenesis in the neuroectoderm. *Genes & Development, 10*, 2922–2934.

Bisgrove, B. W., Su, Y. C., & Yost, H. J. (2017). Maternal Gdf3 is an obligatory cofactor in Nodal signaling for embryonic axis formation in zebrafish. *eLife, 6*. https://doi.org/10.7554/eLife.28534.

Cadigan, K. M., & Nusse, R. (1997). Wnt signaling: A common theme in animal development. *Genes & Development, 11*, 3286–3305.

Callebaut, M., & Van Nueten, E. (1994). Rauber's (Koller's) sickle: The early gastrulation organizer of the avian blastoderm. *European Journal of Morphology, 32*, 35–48.

Canning, D. R., & Stern, C. D. (1988). Changes in the expression of the carbohydrate epitope HNK-1 associated with mesoderm induction in the chick embryo. *Development, 104*, 643–655.

Connolly, D. J., Patel, K., Seleiro, E. A., Wilkinson, D. G., & Cooke, J. (1995). Cloning, sequencing, and expressional analysis of the chick homologue of follistatin. *Developmental Genetics, 17*, 65–77.

Cui, C., Yang, X., Chuai, M., Glazier, J. A., & Weijer, C. J. (2005). Analysis of tissue flow patterns during primitive streak formation in the chick embryo. *Developmental Biology, 284*, 37–47.

Driesch, H. (1892). Entwicklungsmechanische Studien: I. Der Werth der beiden ersten Furschungszellen in der Echinodermenentwicklung. Experimentelle Erzeugen von Teil und Doppelbildung. *Zeitschrift für Wissenschaftliche Zoologie, 53*, 160–178.

Eakin, G. S., & Behringer, R. R. (2004). Gastrulation in other mammals and humans. In C. D. Stern (Ed.), *Gastrulation: From cells to embryo* (pp. 275–287). New York: Cold Spring Harbor Press.

Enders, A. C. (2002). Implantation in the nine-banded armadillo: How does a single blastocyst form four embryos? *Placenta, 23*, 71–85.

Eyal-Giladi, H., & Khaner, O. (1989). The chick's marginal zone and primitive streak formation. II. Quantification of the marginal zone's potencies—Temporal and spatial aspects. *Developmental Biology, 134*, 215–221.

Eyal-Giladi, H., & Kochav, S. (1976). From cleavage to primitive streak formation: A complementary normal table and a new look at the first stages of the development of the chick. I. General morphology. *Developmental Biology, 49*, 321–337.

Eyal-Giladi, H., Kochav, S., & Menashi, M. K. (1976). On the origin of primordial germ cells in the chick embryo. *Differentiation, 6*, 13–16.

Firmino, J., Rocancourt, D., Saadaoui, M., Moreau, C., & Gros, J. (2016). Cell division drives epithelial cell rearrangements during gastrulation in chick. *Developmental Cell, 36*, 249–261. https://doi.org/10.1016/j.devcel.2016.01.007.

Foley, A. C., Skromne, I., & Stern, C. D. (2000). Reconciling different models of forebrain induction and patterning: A dual role for the hypoblast. *Development, 127*, 3839–3854.

Ginsburg, M., & Eyal-Giladi, H. (1986). Temporal and spatial aspects of the gradual migration of primordial germ cells from the epiblast into the germinal crescent in the avian embryo. *Journal of Embryology and Experimental Morphology, 95*, 53–71.

Ginsburg, M., Hochman, J., & Eyal-Giladi, H. (1989). Immunohistochemical analysis of the segregation process of the quail germ cell lineage. *The International Journal of Developmental Biology, 33*, 389–395.

Gonçalves, L., Filipe, M., Marques, S., Salgueiro, A. M., Becker, J. D., & Belo, J. A. (2011). Identification and functional analysis of novel genes expressed in the anterior visceral endoderm. *The International Journal of Developmental Biology, 55*, 281–295. https://doi.org/10.1387/ijdb.103273lg.

Gräper, L. (1929). Die Primitiventwicklung des Hühnchens nach stereokinematographischen Untersuchungen, kontrolliert durch vitale Farbmarkierung und verglichen mit der Entwicklung anderer Wirbeltiere. *Wilhelm Roux' Archiv für Entwicklungsmechanik der Organismen, 116*, 382–429.

Hamburger, V., & Hamilton, H. L. (1951). A series of normal stages in the development of the chick embryo. *Journal of Morphology, 88*, 49–92.

Hatada, Y., & Stern, C. D. (1994). A fate map of the epiblast of the early chick embryo. *Development, 120*, 2879–2889.

Heisenberg, C. P., Tada, M., Saude, L., Concha, M., Rauch, J., Geisler, R., et al. (2000). Silberblick/Wnt11 activity in paraxial tissue mediates convergent extension movements during zebrafish gastrulation. *Nature, 405*, 76–81.

Herbst, C. (1900). Über das Auseinandergehen von Furschungs- und Gewebszellen im kalkfreien Medium. *Archiv für Entwicklungsmechanik, 9*, 424–463.

Holley, S. A., Jackson, P. D., Sasai, Y., Lu, B., De Robertis, E. M., Hoffmann, F. M., et al. (1995). A conserved system for dorsal-ventral patterning in insects and vertebrates involving sog and chordin. *Nature, 376*, 249–253.

Hoodless, P. A., Pye, M., Chazaud, C., Labbe, E., Attisano, L., Rossant, J., et al. (2001). FoxH1 (fast) functions to specify the anterior primitive streak in the mouse. *Genes & Development, 15*, 1257–1271.

Hume, C. R., & Dodd, J. (1993). Cwnt-8C: A novel Wnt gene with a potential role in primitive streak formation and hindbrain organization. *Development, 119*, 1147–1160.

Izpisúa-Belmonte, J. C., De Robertis, E. M., Storey, K. G., & Stern, C. D. (1993). The homeobox gene goosecoid and the origin of organizer cells in the early chick blastoderm. *Cell, 74*, 645–659.

Jones, C. M., Lyons, K. M., Lapan, P. M., Wright, C. V. E., & Hogan, B. M. (1992). DVR-4 (bone morphogentic protein-4) as a posterior-ventralizing factor in Xenopus mesoderm induction. *Development, 115*, 639–647.

Karagenc, L., Cinnamon, Y., Ginsburg, M., & Petitte, J. N. (1996). Origin of primordial germ cells in the prestreak chick embryo. *Developmental Genetics, 19*, 290–301.

Khaner, O., & Eyal-Giladi, H. (1986). The embryo-forming potency of the posterior marginal zone in stages X through XII of the chick. *Developmental Biology, 115*, 275–281.

Khaner, O., & Eyal-Giladi, H. (1989). The chick's marginal zone and primitive streak formation. I. Coordinative effect of induction and inhibition. *Developmental Biology, 134*, 206–214.

Khaner, O., Mitrani, E., & Eyal-Giladi, H. (1985). Developmental potencies of area opaca and marginal zone areas of early chick blastoderms. *Journal of Embryology and Experimental Morphology, 89*, 235–241.

Kimura, W., Yasugi, S., Stern, C. D., & Fukuda, K. (2006). Fate and plasticity of the endoderm in the early chick embryo. *Developmental Biology, 289*, 283–295.

Kimura, C., Yoshinaga, K., Tian, E., Suzuki, M., Aizawa, S., & Matsuo, I. (2000). Visceral endoderm mediates forebrain development by suppressing posteriorizing signals. *Developmental Biology, 225*, 304–321.

Kirby, M. L., Lawson, A., Stadt, H. A., Kumiski, D. H., Wallis, K. T., McCraney, E., et al. (2003). Hensen's node gives rise to the ventral midline of the foregut: Implications for organizing head and heart development. *Developmental Biology, 253*, 175–188.

Kochav, S., Ginsburg, M., & Eyal-Giladi, H. (1980). From cleavage to primitive streak formation: A complementary normal table and a new look at the first stages of the development of the chick. II. Microscopic anatomy and cell population dynamics. *Developmental Biology, 79*, 296–308.

Koller, C. (1882). Untersuchungen über die Blätterbildung im Hühnerkeim. *Archiv für Mikroskopische Anatomie, 20*, 174–211.

Kwon, G. S., Viotti, M., & Hadjantonakis, A. K. (2008). The endoderm of the mouse embryo arises by dynamic widespread intercalation of embryonic and extraembryonic lineages. *Developmental Cell, 15*, 509–520.

Lawson, A., & Schoenwolf, G. C. (2003). Epiblast and primitive-streak origins of the endoderm in the gastrulating chick embryo. *Development, 130*, 3491–3501.

Linker, C., De Almeida, I., Papanayotou, C., Stower, M., Sabado, V., Ghorani, E., et al. (2009). Cell communication with the neural plate is required for induction of neural markers by BMP inhibition: Evidence for homeogenetic induction and implications for Xenopus animal cap and chick explant assays. *Developmental Biology, 327*, 478–486.

Liu, P., Wakamiya, M., Shea, M. J., Albrecht, U., Behringer, R. R., & Bradley, A. (1999). Requirement for Wnt3 in vertebrate axis formation. *Nature Genetics, 22*, 361–365.

Lutz, H. (1949). Sur la production expérimentale de la polyembryonie et de la monstruosité double chez les oiseaux. *Archives d'Anatomie Microscopique et de Morphologie Expérimentale, 39*, 79–144.

Massague, J., & Chen, Y. G. (2000). Controlling TGF-beta signaling. *Genes & Development, 14*, 627–644.

Merrill, B. J., Pasolli, H. A., Polak, L., Rendl, M., Garcia-Garcia, M. J., Anderson, K. V., et al. (2004). Tcf3: A transcriptional regulator of axis induction in the early embryo. *Development, 131*, 263–274. https://doi.org/10.1242/dev.00935.

Mitrani, E., & Shimoni, Y. (1990). Induction by soluble factors of organized axial structures in chick epiblasts. *Science, 247*, 1092–1094.

Mitrani, E., Shimoni, Y., & Eyal-Giladi, H. (1983). Nature of the hypoblastic influence on the chick embryo epiblast. *Journal of Embryology and Experimental Morphology, 75*, 21–30.

Montague, T. G., Gagnon, J. A., & Schier, A. F. (2018). Conserved regulation of Nodal-mediated left-right patterning in zebrafish and mouse. *Development, 145*. https://doi.org/10.1242/dev.171090.

Montague, T. G., & Schier, A. F. (2017). Vg1-Nodal heterodimers are the endogenous inducers of mesendoderm. *eLife, 6*. https://doi.org/10.7554/eLife.28183.

Nishita, M., Hashimoto, M. K., Ogata, S., Laurent, M. N., Ueno, N., Shibuya, H., et al. (2000). Interaction between Wnt and TGF-b signalling pathways during formation of Spemann's Organizer. *Nature, 403*, 781–785.

Pasteels, J. (1940). Un aperçu comparitif de la gastrulation chez les chordés. *Biological Reviews of the Cambridge Philosophical Society, 15*, 59–106.

Pelliccia, J. L., Jindal, G. A., & Burdine, R. D. (2017). Gdf3 is required for robust Nodal signaling during germ layer formation and left-right patterning. *eLife, 6*. https://doi.org/10.7554/eLife.28635.

Perea-Gómez, A., Vella, F. D., Shawlot, W., Oulad-Abdelghani, M., Chazaud, C., Meno, C., et al. (2002). Nodal antagonists in the anterior visceral endoderm prevent the formation of multiple primitive streaks. *Developmental Cell, 3*, 745–756.

Peter, K. (1938). Untersuchungen über die Entwicklung des Dotterentoderms. 1. Die Entwicklung des Entoderms beim Hühnchen. *Zeitschrift für Mikroskopisch-Anatomische Forschung, 43*, 362–415.

Read, E. M., Rodaway, A. R., Neave, B., Brandon, N., Holder, N., Patient, R. K., et al. (1998). Evidence for non-axial A/P patterning in the nonneural ectoderm of Xenopus and zebrafish pregastrula embryos. *The International Journal of Developmental Biology, 42*, 763–774.

Rhinn, M., Dierich, A., Shawlot, W., Behringer, R. R., Le Meur, M., & Ang, S. L. (1998). Sequential roles for Otx2 in visceral endoderm and neuroectoderm for forebrain and midbrain induction and specification. *Development, 125*, 845–856.

Rivera-Perez, J. A., & Magnuson, T. (2005). Primitive streak formation in mice is preceded by localized activation of Brachyury and Wnt3. *Developmental Biology, 288*, 363–371. https://doi.org/10.1016/j.ydbio.2005.09.012.

Rozbicki, E., Chuai, M., Karjalainen, A. I., Song, F., Sang, H. M., Martin, R., et al. (2015). Myosin-II-mediated cell shape changes and cell intercalation contribute to primitive streak formation. *Nature Cell Biology, 17*, 397–408. https://doi.org/10.1038/ncb3138.

Schmidt, J. E., von Dassow, G., & Kimelman, D. (1996). Regulation of dorsal-ventral patterning: The ventralizing effects of the novel Xenopus homeobox gene Vox. *Development, 122*, 1711–1721.

Schulte-Merker, S., Lee, K. J., McMahon, A. P., & Hammerschmidt, M. (1997). The zebrafish organizer requires chordino. *Nature, 387*, 862–863.

Seleiro, E. A., Connolly, D. J., & Cooke, J. (1996). Early developmental expression and experimental axis determination by the chicken Vg1 gene. *Current Biology, 6*, 1476–1486.

Shah, S. B., Skromne, I., Hume, C. R., Kessler, D. S., Lee, K. J., Stern, C. D., et al. (1997). Misexpression of chick Vg1 in the marginal zone induces primitive streak formation. *Development, 124*, 5127–5138.

Sheng, G., & Stern, C. D. (1999). Gata2 and Gata3: Novel markers for early embryonic polarity and for non-neural ectoderm in the chick embryo. *Mechanisms of Development, 87*, 213–216.

Skromne, I., & Stern, C. D. (2001). Interactions between Wnt and Vg1 signalling pathways initiate primitive streak formation in the chick embryo. *Development, 128*, 2915–2927.

Skromne, I., & Stern, C. D. (2002). A hierarchy of gene expression accompanying induction of the primitive streak by Vg1 in the chick embryo. *Mechanisms of Development, 114*, 115–118.

Sokol, S. Y., & Melton, D. A. (1992). Interaction of Wnt and activin in dorsal mesoderm induction in Xenopus. *Developmental Biology, 154*, 348–355.

Spemann, H. (1919). Experimentelle Forschungen zum Determinations- und Individualitätsproblem. *Naturwissenschaften, 7*, 581–591.

Spratt, N. T. (1942). Location of organ-specific regions and their relationship to the development of the premitive streak in the early chick blastoderm. *The Journal of Experimental Zoology*, *89*, 69–101.

Spratt, N. T. (1946). Formation of the primitive streak in the explanted chick blastoderm marked with carbon particles. *The Journal of Experimental Zoology*, *103*, 259–304.

Spratt, N. T. (1947). Regression and shortening of the primitve streak in the explanted chick blastoderm. *The Journal of Experimental Zoology*, *104*, 69–100.

Spratt, N. T., & Haas, H. (1960a). Integrative mechanisms in development of the early chick blastoderm. I. Regulative potentiality of separated parts. *The Journal of Experimental Zoology*, *145*, 97–137.

Spratt, N. T., & Haas, H. (1960b). Morphogenetic movements in the lower surface of the unincubated and early chick blastoderm. *The Journal of Experimental Zoology*, *144*, 139–157.

Srivastava, R. A., & Srivastava, N. (2000). High density lipoprotein, apolipoprotein A-I, and coronary artery disease. *Molecular and Cellular Biochemistry*, *209*, 131–144.

Stern, C. D. (1990). The marginal zone and its contribution to the hypoblast and primitive streak of the chick embryo. *Development*, *109*, 667–682.

Stern, C. D., & Canning, D. R. (1990). Origin of cells giving rise to mesoderm and endoderm in chick embryo. *Nature*, *343*, 273–275.

Stern, C. D., & Downs, K. M. (2012). The hypoblast (visceral endoderm): An evo-devo perspective. *Development*, *139*, 1059–1069. https://doi.org/10.1242/dev.070730.

Stern, C. D., & Ireland, G. W. (1981). An integrated experimental study of endoderm formation in avian embryos. *Anatomy and Embryology*, *163*, 245–263.

Stern, C. D., Yu, R. T., Kakizuka, A., Kintner, C. R., Mathews, L. S., Vale, W. W., et al. (1995). Activin and its receptors during gastrulation and the later phases of mesoderm development in the chick embryo. *Developmental Biology*, *172*, 192–205.

Streit, A., Berliner, A. J., Papanayotou, C., Sirulnik, A., & Stern, C. D. (2000). Initiation of neural induction by FGF signalling before gastrulation. *Nature*, *406*, 74–78.

Streit, A., Lee, K. J., Woo, I., Roberts, C., Jessell, T. M., & Stern, C. D. (1998). Chordin regulates primitive streak development and the stability of induced neural cells, but is not sufficient for neural induction in the chick embryo. *Development*, *125*, 507–519.

Streit, A., & Stern, C. D. (1999). Mesoderm patterning and somite formation during node regression: Differential effects of chordin and noggin. *Mechanisms of Development*, *85*, 85–96.

Sykes, T. G., Rodaway, A. R., Walmsley, M. E., & Patient, R. K. (1998). Suppression of GATA factor activity causes axis duplication in Xenopus. *Development*, *125*, 4595–4605.

Takaoka, K., Nishimura, H., & Hamada, H. (2017). Both Nodal signalling and stochasticity select for prospective distal visceral endoderm in mouse embryos. *Nature Communications*, *8*, 1492. https://doi.org/10.1038/s41467-017-01625-x.

Thomas, P., & Beddington, R. (1996). Anterior primitive endoderm may be responsible for patterning the anterior neural plate in the mouse embryo. *Current Biology*, *6*, 1487–1496.

Thomsen, G. H., & Melton, D. A. (1993). Processed Vg1 protein is an axial mesoderm inducer in Xenopus. *Cell*, *74*, 433–441.

Torlopp, A., Khan, M. A., Oliveira, N. M., Lekk, I., Soto-Jimenez, L. M., Sosinsky, A., et al. (2014). The transcription factor Pitx2 positions the embryonic axis and regulates twinning. *eLife*, *3*. https://doi.org/10.7554/eLife.03743.

Ulshafer, R. J., & Clavert, A. (1979). The use of avian double monsters in studies on induction of the nervous system. *Journal of Embryology and Experimental Morphology*, *53*, 237–243.

Vakaet, L. (1967). Contribution a l'étude de la pregastrulation et de la gastrulation de l'embryon de poulet en culture 'in vitro'. *Mémoires de l'Académie Royale de Médecine de Belgique*, *5*, 235–257, 2e ser.

Vakaet, L. (1970). Cinephotomicrographic investigations of gastrulation in the chick blastoderm. *Archives de Biologie*, *81*, 387–426.

Varlet, I., Collignon, J., & Robertson, E. J. (1997). Nodal expression in the primitive endoderm is required for specification of the anterior axis during mouse gastrulation. *Development*, *124*, 1033–1044.

Voiculescu, O., Bertocchini, F., Wolpert, L., Keller, R. E., & Stern, C. D. (2007). The amniote primitive streak is defined by epithelial cell intercalation before gastrulation. *Nature*, *449*, 1049–1052.

Voiculescu, O., Bodenstein, L., Lau, I. J., & Stern, C. D. (2014). Local cell interactions and self-amplifying individual cell ingression drive amniote gastrulation. *eLife*, *3*, e01817. https://doi.org/10.7554/eLife.01817.

Waddington, C. H. (1932). Experiments on the development of chick and duck embryos cultivated in vitro. *Philosophical Transactions of the Royal Society of London. Series B, Biological Sciences*, *221*, 179–230.

Waddington, C. H. (1933). Induction by the endoderm in birds. *Wilhelm Roux' Archiv für Entwicklungsmechanik der Organismen*, *128*, 502–521.

Walmsley, M. E., Guille, M. J., Bertwistle, D., Smith, J. C., Pizzey, J. A., & Patient, R. K. (1994). Negative control of Xenopus GATA-2 by activin and noggin with eventual expression in precursors of the ventral blood islands. *Development*, *120*, 2519–2529.

Wei, Y., & Mikawa, T. (2000). Formation of the avian primitive streak from spatially restricted blastoderm: Evidence for polarized cell division in the elongating streak. *Development*, *127*, 87–96.

Willnow, T. E., Hammes, A., & Eaton, S. (2007). Lipoproteins and their receptors in embryonic development: More than cholesterol clearance. *Development*, *134*, 3239–3249.

Yamamoto, M., Saijoh, Y., Perea-Gomez, A., Shawlot, W., Behringer, R. R., Ang, S. L., et al. (2004). Nodal antagonists regulate formation of the anteroposterior axis of the mouse embryo. *Nature*, *428*, 387–392.

CHAPTER FOUR

Comparative analysis of human and mouse development: From zygote to pre-gastrulation

Matteo A. Molè†, Antonia Weberling†, Magdalena Zernicka-Goetz*
University of Cambridge, Cambridge, United Kingdom
*Corresponding author: e-mail address: mz205@cam.ac.uk

Contents

Abstract

Development of the mammalian embryo begins with formation of the totipotent zygote during fertilization. This initial cell is able to give rise to every embryonic tissue of the developing organism as well as all extra-embryonic lineages, such as the placenta and the yolk sac, which are essential for the initial patterning and support growth of the fetus until birth. As the embryo transits from pre- to post-implantation, major structural and transcriptional changes occur within the embryonic lineage to set up the basis for the subsequent phase of gastrulation. Fine-tuned coordination of cell division, morphogenesis and differentiation is essential to ultimately promote assembly of the future fetus. Here, we review the current knowledge of mammalian development of both mouse and human focusing on morphogenetic processes leading to the onset of gastrulation, when the embryonic anterior-posterior axis becomes established and the three germ layers start to be specified.

† Both authors are contributed equally:

Current Topics in Developmental Biology, Volume 136
ISSN 0070-2153
https://doi.org/10.1016/bs.ctdb.2019.10.002

1. Introduction

Mammalian embryogenesis initiates with the establishment of the zygote upon fertilization, during which the maternal (oocyte) and paternal (sperm) gametes fuse to establish the initial embryonic cell, the diploid zygote. Its ability to produce every cell of the future organism, alongside all extra-embryonic structures, defines the initial unrestricted totipotent character of this founder cell. Pre-implantation development, encompassing the period from the zygote to blastocyst formation, is characterized by striking conservation across species, enabling the study of these initial stages in model organisms such as the mouse. At the end of the pre-implantation period, the embryo begins to establish interactions with the maternal uterine endometrium while undergoing major structural changes that will set up the base for the following phase of development, gastrulation. However, as mouse and human embryogenesis diverge significantly during this successive stage, it becomes necessary to investigate post-implantation development on the human embryo directly.

Recent developments of *ex vivo* culture systems for the mouse (Bedzhov & Zernicka-Goetz, 2014; Morris et al., 2012) and human (Deglincerti et al., 2016; Shahbazi et al., 2016) have enabled culture of mouse and human embryos beyond implantation, opening the possibility to directly characterize peri-implantation and early phases of post-implantation development up to initiation of gastrulation. Analysis of these critical stages will be important not only to increase our understanding of embryo development but also to elucidate the underlying molecular mechanisms to eventually uncover the causes behind low efficiency of human reproduction as failure during pre- to post-implantation development represents one of the major causes of pre-clinical miscarriages, accounting for up to 30% of conception losses prior to live birth (Macklon, Geraedts, & Fauser, 2002; Wilcox, Baird, & Weinberg, 1999; Wilcox et al., 1988).

2. Pre-implantation development: From zygote to blastocyst formation

The onset of pre-implantation development shows a high degree of conservation between mouse and human embryogenesis (Fig. 1). Following fertilization, the zygote undergoes a series of cleavage divisions leading to an increase in cell number, while an overall constant volume

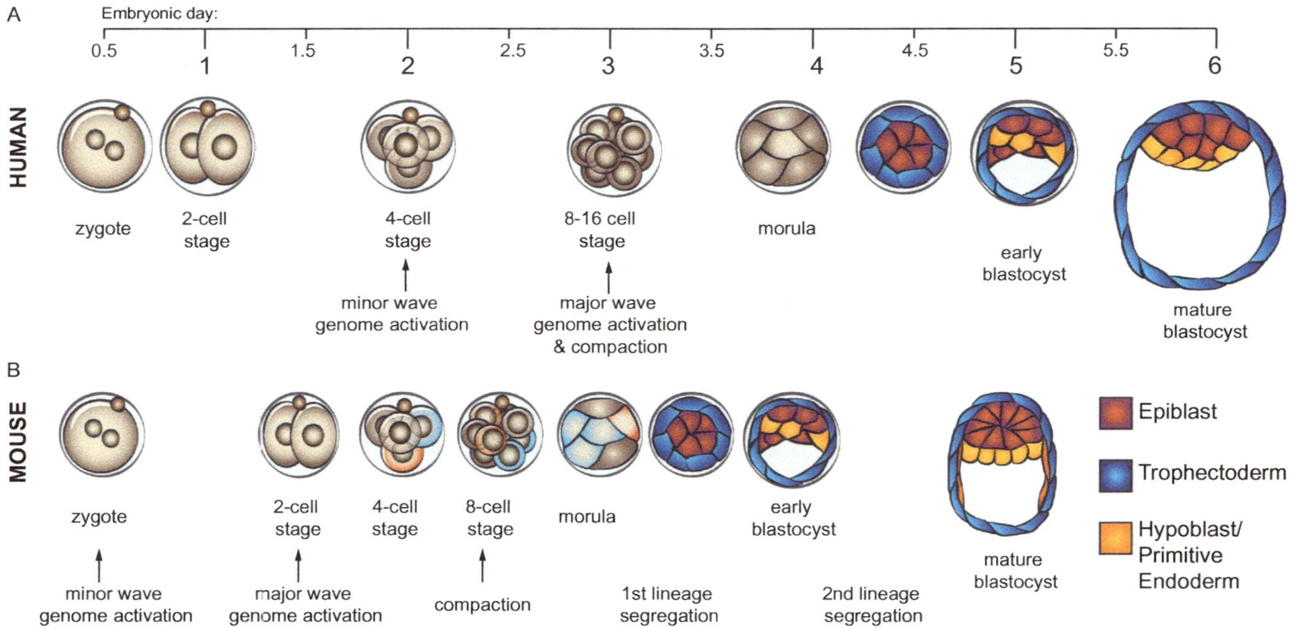

Fig. 1 Pre-implantation development: from zygote to blastocyst. (A) Human embryogenesis starts with formation of the zygote at fertilization, followed by a series of cleavage divisions. The major wave of embryonic genome activation occurs at eight-cell stage (embryonic day 3) alongside initiation of compaction and polarization. At embryonic day 4.5, the outer cells, precursors of the future trophectoderm (blue) segregate from inner cells, precursors of the epiblast and hypoblast lineages (red). At embryonic day 5, the human embryo forms a blastocyst consisting of the inner cell mass (red and yellow, salt and pepper), the spherical trophectoderm (blue) and the blastocoelic cavity. Before hatching, the size of the human blastocyst is estimated to be of ~200 μm in diameter (Richter, Harris, Daneshmand, & Shapiro, 2001) while following hatching from zona pellucida human blastocyst expands to reach an estimated diameter ~600–800 μm (Deglincerti et al., 2016; Niakan & Eggan, 2013). (B) Mouse embryogenesis starts with formation of the zygote at fertilization, followed by cleavage divisions. The major wave of embryonic (zygotic) genome activation occurs at two-cell stage (embryonic day 1.5). Compaction and polarization take place at eight-cell stage (embryonic day 2.5). The first lineage segregation between outer cells (future trophectoderm, blue) and inner cells (epiblast and primitive endoderm precursors, red) occurs at embryonic day 3.5. At embryonic day 3.75–4.0, the mouse embryo has matured into a blastocyst consisting of the inner cell mass (red and yellow, salt and pepper), the spherical trophectoderm (blue) and the blastocoelic cavity. Upon hatching from zona pellucida and further expansion, the size of the mouse blastocyst is estimated to be of ~110–120 μm in diameter. The second lineage segregation between epiblast (red) and primitive endoderm (yellow) is fully completed before initiation of implantation while epiblast acquires epithelial morphology and adopts a 3D rosette configuration (embryonic day E4.75).

is retained. During the initial developmental steps of this process, the embryo remains transcriptionally silent and relies on inherited maternal mRNAs and proteins until initiation of the maternal to zygotic transition during which the embryonic genome is activated replacing the maternal RNA. Activation of the zygotic genome encompasses two waves: a minor and major wave. In the mouse, the minor wave takes place at late zygote stage, whereas the major one occurs at two-cell stage (a major wave; Fig. 1B) (Schulz & Harrison, 2019). In humans, the minor wave of zygotic genome activation takes place at four-cell stage while the major occurs at eight-cell stage (Fig. 1A) (Braude, Bolton, & Moore, 1988). Detailed reviews of the sequence of events of zygotic genome activation can be found elsewhere (Jukam, Shariati, & Skotheim, 2017; Lee, Bonneau, & Giraldez, 2014; Schulz & Harrison, 2019; Tadros & Lipshitz, 2009).

Despite the fact that individual blastomeres appear morphologically identical at early pre-implantation stages, heterogeneities in cell fate and potential have been discovered as early as at the two-cell stage (Casser et al., 2017; Gardner, 2001; Piotrowska, Wianny, Pedersen, & Zernicka-Goetz, 2001; Piotrowska & Zernicka-Goetz, 2001) and four-cell stage (Piotrowska-Nitsche, 2005; Piotrowska-Nitsche & Zernicka-Goetz, 2005). This early heterogeneity has been shown to relate to changes at the epigenetic level (Torres-Padilla, Parfitt, Kouzarides, & Zernicka-Goetz, 2007), which leads to differential division- and gene expression patterns that can bias cell fate specification (Goolam et al., 2016; Hupalowska et al., 2018; Wang et al., 2018; White, Angiolini, et al., 2016; White, Bissiere, et al., 2016). Despite these differences, early blastomeres as well as the inner cell mass (ICM) of the early blastocyst still have capacity to form both embryonic and extra-embryonic lineages (Kelly, Mulnard, & Graham, 1978; Posfai et al., 2017; Rossant & Lis, 1979; Suwińska, Czołowska, Ożdżeński, & Tarkowski, 2008). Whether early blastomeres' heterogenity is exhibited by the human embryo (Piras, Tomita, & Selvarajoo, 2015) remains a fascinating question currently unaddressed.

Initially, the embryo consists of loosely adhering blastomeres, but during the process known as compaction, apposition between individual blastomeres increases with concurrent decrease of intercellular space (Ducibella, 1977; Zenker et al., 2018). The process of compaction in mouse occurs in a synchronous manner at the eight-cell stage (Fig. 1B). In contrast to this, compaction in human embryos takes place asynchronously at the transition between the 8- and 16-cell stage (Fig. 1A) (Nikas, Ao, Winston, & Handyside, 1996). Simultaneously, the blastomeres acquire a defined outside-in polarity through accumulation of apical polarity complexes on the outside surface of the cells, under the influence of the contractile actomyosin cytoskeleton (Reeve, 1981; Zhu, Leung, Shahbazi, & Zernicka-Goetz, 2017).

In the course of the following cleavage division, the first lineage segregation takes place, discriminating the inner cells (ICM) from the outer blastomeres with inner cells maintaining pluripotency and outer cells initiating differentiation into trophectoderm, which will mediate apposition of the embryo to the maternal endometrium and later will generate the placenta. In the mouse embryo, this process depends on the differential activation of the Hippo/YAP signaling pathway between inner and outer blastomeres, leading to the expression of trophectoderm-specific genes in the outer cells and maintenance of the core pluripotency markers *Nanog*, *Oct4*, and *Sox2* in the ICM (Chazaud & Yamanaka, 2016; White, Zenker, Bissiere, & Plachta, 2018). Specifically, YAP localizes in the nucleus of the outside cells, converging with Notch signaling (Watanabe et al., 2017) to induce the expression of the trophectoderm marker *Cdx2*. Conversely, YAP in the inner cells is sequestered in the cytoplasm leading to maintenance of expression of the key pluripotency genes. Whether a similar molecular mechanism underlies the first lineage segregation during human development remains unknown. However, recently it has been shown that OCT4 is critical not only for epiblast specification but also for the expression of the trophectoderm marker CDX2 (Fogarty et al., 2017), which is in contrast to mouse embryogenesis. Furthermore, specification of the human trophectoderm initiates independently of CDX2 expression, which becomes upregulated only after formation of the blastocyst (Niakan & Eggan, 2013), suggesting that a differential mechanism from the one in the mouse may govern the first lineage segregation in the human embryo. *In vitro* differentiation of human ESCs suggests instead GATA3 as potential central regulator of trophectoderm specification during this first lineage segregation step (Krendl et al., 2017).

As development progresses, the trophectoderm undergoes epithelialization, which leads to the maturation of tight junctions at the cell-cell interfaces. Concurrently, active transport of Na^+ ions at the basolateral side of the trophectoderm generates an osmotic flow of water toward the inside of the embryo. Multiple intercellular lumens first form and then fuse into a single fluid filled cavity known as blastocoel (Dumortier et al., 2019) (Fig. 1). This blastocyst structure is conserved between human and mouse embryos and marks the stage of development when the free-floating embryo reaches the uterus to implant. *In vitro* structures resembling this initial stages of development have been recently generated by the combination of mouse ESC and trophoblast stem cells (TSC) in microwell arrays (Kime et al., 2019; Rivron et al., 2018). However, these blastocyst-like structures are currently unable to progress through implantation and post-implantation development *in vitro* and *in vivo*.

Despite close similarities between mouse and human blastocysts at this early stage, a major difference exists in respect to the timing of the second lineage segregation. In the mouse embryo, the ICM shows heterogeneity between the individual cells (Fig. 1B) (Chazaud, Yamanaka, Pawson, & Rossant, 2006) that becomes established through the sequential waves of cell internalization (Morris et al., 2010): the cells internalized early express higher level of FGF4 ligands whereas the cells internalized at later stages exhibit higher levels of FGF receptors. Secretion of FGF4 ligands activates FGF/ERK signaling in cells expressing FGFR1 and FGFR2, thus specifying progenitors of the primitive endoderm, leading to the expression of primitive endoderm markers such as *Gata6* (Morris et al., 2010; Yamanaka, Lanner, & Rossant, 2010). This causes separation between the group of *Gata6* expressing cells and *Nanog* expressing cells, with the latter becoming precursors of the epiblast instead (Bassalert, Valverde-Estrella, & Chazaud, 2018; Kang, Garg, & Hadjantonakis, 2017; Molotkov, Mazot, Brewer, Cinalli, & Soriano, 2017). While initially dispersed within the ICM, these two groups of cells resolve by active cell movement and apoptosis of incorrectly positioned cells to form a sheet of primitive endoderm cells toward the blastocoel cavity (Meilhac et al., 2009; Plusa, Piliszek, Frankenberg, Artus, & Hadjantonakis, 2008; Schrode, Saiz, Di Talia, & Hadjantonakis, 2014).

In the human embryo, distinct molecular profiles between epiblast, trophectoderm and hypoblast, equivalent of the mouse primitive endoderm, lineages appear to arise only after blastocyst formation (Petropoulos et al., 2016; Stirparo et al., 2018). While clear separation between the epiblast and hypoblast lineage is achieved, a mature hypoblast layer becomes evident only at the late blastocyst upon initiation of implantation, by downregulation of OCT4 and upregulation of the hypoblast marker SOX17 (Fig. 1A) (Deglincerti et al., 2016, Niakan & Eggan, 2013, Shahbazi et al., 2016). Even though, FGF/ERK signaling has been shown to be essential for mouse primitive endoderm specification, a different signaling pathway may underlie this second lineage segregation in the human embryo. In fact, despite the conservation of the key lineage specific markers, inhibition of FGF/ERK signaling in humans does not appear to prevent hypoblast specification (Kuijk et al., 2012; Roode et al., 2012). Therefore, the molecular mechanism driving this second step of lineage segregation in human embryogenesis remains still unclear (Boroviak et al., 2018).

3. Implantation and the role of trophectoderm

Upon implantation, the blastocyst establishes primary contacts with the maternal tissue to become eventually attached to the endometrium. Attachment and adhesion to the maternal endometrium is mediated by the trophectoderm. At this stage of development, two major subpopulations of trophectoderm can be distinguished: the region in direct contact with the ICM located at the embryonic pole of the embryo, the polar trophectoderm, and the region surrounding the blastocoel cavity localized at the abembryonic pole, the mural trophectoderm. Despite having equal potency at the beginning, they eventually undertake distinct behavior due to the differential exposure to inductive signaling emanating from the ICM (Copp, 1978, 1979; Gardner, Papaioannou, & Barton, 1973; Papaioannou, 1982).

In the mouse, the process of adhesion is mediated by the mural trophectoderm and takes place at embryonic day 4.5–4.75 (Fig. 2D). Prior to implantation, the trophectoderm marker CDX2 becomes downregulated in the mural in contrast to the polar trophectoderm. Loss of *CDX2* expression results in differential cell fate acquisition and the establishment of a tissue boundary physically separating the polar from the mural compartment (Christodoulou et al., 2019). As a result of multipotency loss, cells of the mural trophectoderm undergo an epithelial to mesenchymal-like transition (Mobley et al., 2017) and acquire an invasive character, differentiating into trophoblast giant cells. Initial contacts to the uterine extracellular matrix are established via integrin αVβ3 receptors (Schultz, Mayernik, Rout, & Armant, 1997). This results in activation of intracellular Ca^{2+} signaling, which recruits further integrin receptors to the site of the initial contact reinforcing adhesion for successful implantation of the embryo. Implantation in the mouse therefore represents a two-step process, where initial weak adhesions are followed by a second phase to strengthen the primary contacts (Wang, Mayernik, & Armant, 2007).

In humans, the process of adhesion and implantation is driven by the polar compartment of the trophectoderm instead and occurs between embryonic day 7 and 8 (Fig. 2A) (Hertig & Rock, 1944; Lindenberg, Hyttel, Sjøgren, & Greve, 1989). When the human embryo comes in

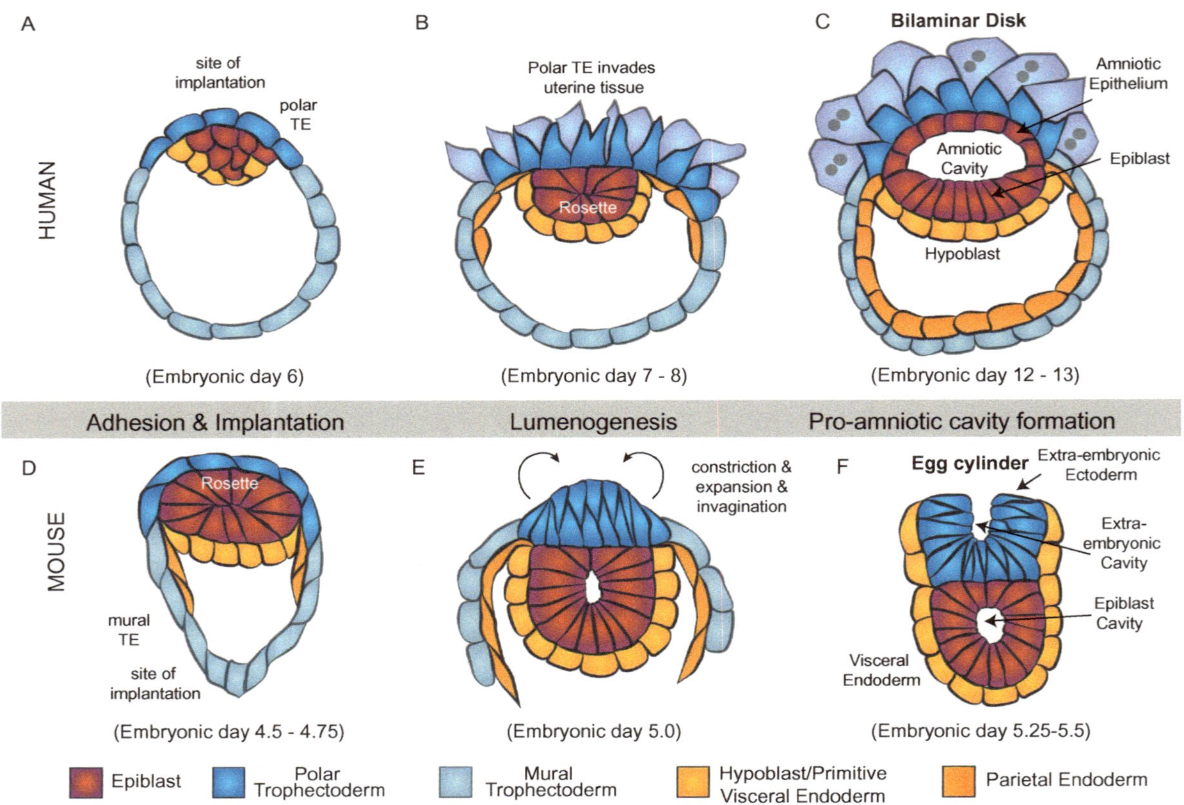

Fig. 2 See legend on opposite page.

contact with the maternal endometrium, the polar trophectoderm proliferates and differentiates into cytotrophoblast and multinucleated syncytiotrophoblast cells, to eventually mediate penetration of the blastocyst inside the decidual tissue (Fig. 2B) (Aplin & Ruane, 2017; Weimar, Post Uiterweer, Teklenburg, Heijnen, & Macklon, 2013). Recent transcriptome dynamics identified TBX3 as a novel key regulator of trophoblast differentiation from cytotrophoblast to syncytiotrophoblast (Lv et al., 2019).

In contrast to the human, the polar trophectoderm in the mouse does not participate in the process of implantation *per se* and maintains its multipotent character due to activation of FGF signaling secreted by the epiblast (Christodoulou et al., 2019; Gardner et al., 1973; Goldin & Papaioannou, 2003). Its high mitotic index (Copp, 1978) together with the contractility imposed by the tissue boundary (Christodoulou et al., 2019) lead the polar trophectoderm epithelium to acquire a highly columnar shape (Fig. 2E). Actomyosin contractility and invagination of this tissue eventually give rise to the extra-embryonic ectoderm (ExE) with acquisition of the so-called egg cylinder structure in the post-implantation mouse embryo (Fig. 2F).

4. Epiblast epithelization and pro-amniotic cavity formation

As embryo implants, major transcriptional and morphogenetic changes take place within the embryonic compartment. The epiblast transitions from a previously unrestricted state of naïve pluripotency toward a

Fig. 2 Implantation and tissue morphogenesis. (A–C) Human implantation and morphogenesis. (A and B) The human embryo mediates attachment and invasion of the maternal endometrial tissue at embryonic day 7 via the polar trophectoderm. Physical segregation between epiblast and hypoblast lineages occurs in the period between late blastocyst and initiation of implantation (embryonic day 6–7). The epiblast establishes apico-basal polarity and adopt a rosette-like configuration surrounding a central opening. (C) Expansion of the latter results into formation of the amniotic cavity surrounded by the squamous amniotic epithelium and the columnar-shaped epiblast. The embryo acquires a bilaminar disk morphology with epiblast on top and hypoblast on the bottom of the discoidal structure. (D–F) Mouse implantation and morphogenesis. (D) In the mouse embryo, implantation is initiated at embryonic day 4.75 via the mural trophectoderm located at the abembryonic pole of the blastocyst. Lineage segregation between epiblast and primitive endoderm occurs before implantation. The epiblast initiates epithelization forming a rosette with a central opening. (E) The epiblast cavity expands. The polar trophectoderm acquire a columnar shape due to contractility of the tissue boundary leading to invagination and formation of the extra-embryonic trophectoderm. (F) Formation of the egg cylinder configuration.

state of primed pluripotency (Nichols & Smith, 2009). This is coincident with the maturation of the epiblast into a fully structured and polarized epithelium surrounding a cavity at its center (Bedzhov & Zernicka-Goetz, 2014; Wallingford, Angelo, & Mager, 2013). Examination of early human embryos from the Carnegie collection provided the first description of this striking tissue remodeling event (Hertig, Rock, & Adams, 1956). At embryonic day 7.5, the implanted epiblast was shown to consist of a group of radially oriented cells surrounding a small primordial cavity at its center (Luckett, 1975). This represents the earliest signature of epithelialization in the epiblast, which transforms the initially nonpolar group of cells into an epithelium with defined polarity (Fig. 2B). Expansion of this primordial opening at the center of the developing epiblast into the future amniotic cavity marks the initiation of amniogenesis (Fig. 2C).

Recent development of *ex vivo* culture conditions has allowed further investigation of the molecular events underlying these early stages of epiblast morphogenesis by mimicking implantation and post-implantation growth of the human embryo in a dish (Deglincerti et al., 2016; Shahbazi et al., 2016). Accumulation of major polarity complexes such as aPKC and actomyosin at the apical site confirmed transition of the epiblast into a mature epithelium, during which epiblast cells arrange around a pivotal point of apical constriction giving rise to a fully structured 3D rosette (Fig. 2B) (Shahbazi et al., 2016; Wallingford et al., 2013). Signals originating from the basement membrane drive polarity establishment and subsequent lumenogenesis in the mouse epiblast (Bedzhov & Zernicka-Goetz, 2014). It was shown that the loss of integrin β1 receptor prevents the remodeling of the mouse ESCs into polarized structures (Bedzhov & Zernicka-Goetz, 2014). This result could provide an explanation for the peri-implantation lethality of the mouse embryo when integrin signaling is defective (Fassler & Meyer, 1995; Li et al., 2005; Sakai et al., 2003; Stephens et al., 1995), illustrating the importance of polarization and lumen formation for early embryonic development. Following the establishment of the future apical site, a central cavity forms via Rab11-dependent secretion of glycosylated sialomucin proteins such as Podocalyxin (Shahbazi et al., 2017). The high negative charge of these proteins results in charge repulsion of the apical cell membranes and it is thought to facilitates opening of the lumen at the center of the epiblast. These major morphogenetic changes could be further recapitulated *in vitro* by culturing individual mouse ESCs (Bedzhov & Zernicka-Goetz, 2014) and human ESCs (Shahbazi et al., 2017; Taniguchi et al., 2015) in 3D matrices. Under the influence of signals from the surrounding extracellular matrix, the ESCs initiate polarization, acquire a multicellular rosette configuration and ultimately undertake lumenogenesis closely mimicking morphogenesis of the mouse and human epiblast *in vitro*.

The morphogenetic transition from a disorganized group of cells to a structured epithelium is also accompanied by major pluripotency changes within the epiblast during which cells lose their state of naïve pluripotency and commit to a more differentiated primed state. This pluripotency state transition is a necessary requirement for induction of epithelization and lumen formation (Shahbazi et al., 2017), highlighting the close interplay between morphogenesis and pluripotency transition for development of the epiblast epithelium in both mouse and human embryos.

In humans, the central lumen in the center of the rosette-like configuration represents the primordium of the future amniotic cavity, which expands further to separate the thick pseudostratified epiblast epithelium from the overlaying squamous amnion layer (Fig. 2C) (Luckett, 1975). This additional lineage segregation is instrumental to discriminate the pre-gastrulation epiblast disk from the thin amnion layer, which will cover the entire conceptus with the amniotic sac, preventing dehydration and protecting the fetus from physical stresses until completion of gestation. However, no clear difference between the pseudostratified columnar epiblast and the squamous amnion epithelium could be observed in *ex vivo* culture of human embryos (Deglincerti et al., 2016; Shahbazi et al., 2016). *In vitro* instead, human ESCs either plated in 3D substrata or via controllable microfluidic devices can recapitulate major aspects of human amniogenesis by giving rise to asymmetric cysts with a disk-like epiblast structure on one pole and a squamous epithelium on the opposite regions, induced via BMP-SMAD signaling (Shao, Taniguchi, Gurdziel, et al., 2017; Shao, Taniguchi, Townshend, et al., 2017; Zheng et al., 2019). This putative amnion epithelium appears to differ from the epiblast disk by downregulation of the core pluripotency markers SOX2, NANOG and OCT4 while upregulating of the putative amnion marker TFAP2A. Studies on monkey embryos (*Macaca fascicularis*) have also revealed robust activation of BMP and WNT signaling within the amnion, as shown by the expression of *BMP4* and *WNT3* ligands together with key immediate BMP (*ID2 and MSX2*) and WNT signaling targets (*AXIN2*) within the amnion epithelium (Sasaki et al., 2016). Further improvement of *ex vivo* culture conditions would be instrumental to elucidate the central signaling pathways underlying morphogenesis of the human amnion.

Formation of the pro-amniotic cavity in the mouse differs significantly from the human embryo and entails a two-step process instead. The first step is conserved between the mouse and the human and involves formation of the rosette configuration within the epiblast (Fig. 2B, human and 2D, mouse) (Bedzhov & Zernicka-Goetz, 2014; Wallingford et al., 2013). The second step occurs only in the mouse and involves the formation of a second cavity through apical constriction driven invagination of the extra-embryonic ectoderm at the proximal site of the embryo (Fig. 2F). These two cavities, one in the epiblast and one in the extra-embryonic ectoderm, expand further through integrin-mediated podocalyxin secretion toward their apical domains to become eventually united into a single pro-amniotic cavity by the resolution of hybrid rosettes at the boundary between the two tissues (Christodoulou et al., 2018). The synergistic interactions of embryonic and extra-embryonic tissues during early post-implantation development could be replicated by the assembly of mouse ESC and TSC (trophoblast-stem cells) in a 3D matrix leading to formation of embryo-like structures (Harrison, Sozen, Christodoulou, Kyprianou, & Zernicka-Goetz, 2017). These structures not only resemble the post-implantation mouse egg cylinder but also recapitulate formation of the pro-amniotic cavity by fusion of the epiblast and extra-embryonic ectoderm cavities.

These different patterns lead to striking divergence in the overall configuration between the post-implantation mouse and human embryo (Fig. 2). The human epiblast adopts a flat disk configuration with a single cavity covered by a thin layer of amniotic epithelium (Fig. 2C). Below the epiblast disk, the hypoblast epithelium is believed to induce key signaling for the epiblast patterning while giving rise to the future yolk sac cavity. In contrast to the human, the mouse embryo gives rise to a concave configuration of an egg cylinder (Fig. 2F). We hypothesize that this is due to the physical pressure imposed on the epiblast by the invagination of the polar trophectoderm to form the ExE. This process does not take place in humans where the polar trophectoderm mediates implantation instead, in contrast to formation of the ExE in the mouse. Thus, as the epiblast is pushed inside the blastocoel cavity, the mouse embryo adopts a cup-shaped configuration resembling a cylindrical structure rather than a flat disk as in humans.

5. Pre-gastrulation patterning and establishment of the anterior-posterior axis

The post-implantation mouse embryo is composed of the epiblast and ExE, which surround the pro-amniotic cavity (following the fusion between the epiblast and ExE cavities), which in turn are enveloped by the visceral endoderm (Fig. 3A). Before gastrulation commences, the epiblast undergoes a process of regional differentiation to lay the foundation of the future anterior-posterior axis. This process is essential to determine the posterior side of the epiblast where the primitive streak will form and the process of gastrulation initiates.

Major insights into the molecular and cellular dynamics underlying the process of epiblast patterning were obtained through studies of the mouse embryo between embryonic day 5.5 and 6.5, just before the start of gastrulation (reviewed in Rivera-Pérez & Hadjantonakis, 2015). During this stage, the extra-embryonic lineages (ExE and visceral endoderm) provide essential paracrine signaling to specify the anterior-posterior axis. At E5.0, the ExE secretes BMP4 ligands that instruct the epiblast to express *Wnt3* and *Nodal* within the cup-shaped epithelium (Fig. 3A) (Arnold & Robertson, 2009; Robertson, 2014). Expression of *Nodal* again is maintained through signaling by the visceral endoderm (Kumar et al., 2015). In the following 12h, an important signaling center arises at the distal part of the visceral endoderm, the distal visceral endoderm (DVE) (Thomas & Beddington, 1996). This structure might be the first to break the initially radial symmetry by demarcating the future anterior region of the epiblast via unilateral migration of the DVE cells toward the anterior of the embryo until reaching the border between epiblast and ExE, forming the anterior-visceral endoderm (AVE) (Antonica, Orietti, Mort, & Zernicka-Goetz, 2019; Bloomekatz, Grego-Bessa, Migeotte, & Anderson, 2012; Migeotte, Omelchenko, Hall, & Anderson, 2010; Omelchenko et al., 2014; Srinivas, 2004; Stower & Srinivas, 2018). Early expression of anterior markers in primitive endoderm cells prior to implantation have suggested the possibility that the DVE may originate in the primitive endoderm of the late blastocyst already (Morris et al., 2012; Takaoka, Yamamoto, & Hamada, 2011; Takaoka et al., 2006; Torres-Padilla et al., 2007). It has been suggested that not only the DVE could be specified at these stages, but that also the future anterior could already be specified through a characteristic tilt of the epiblast along the proximo-distal axis of the embryo (Takaoka et al., 2011).

The AVE acts as a signaling center producing key antagonists such as LEFTY1, CER1 (Belo et al., 2000; Belo et al., 1997) and DKK1 (Kemp, Willems, Abdo, Lambiv, & Leyns, 2005), which suppress activation of NODAL (Perea-Gomez et al., 2002; Yamamoto et al., 2004), BMP and WNT pathways, respectively, in close proximity to the AVE, specifying this epiblast region as the future anterior side (Rivera-Pérez & Hadjantonakis, 2015). On the opposite side, distant from the AVE and thereby exposed

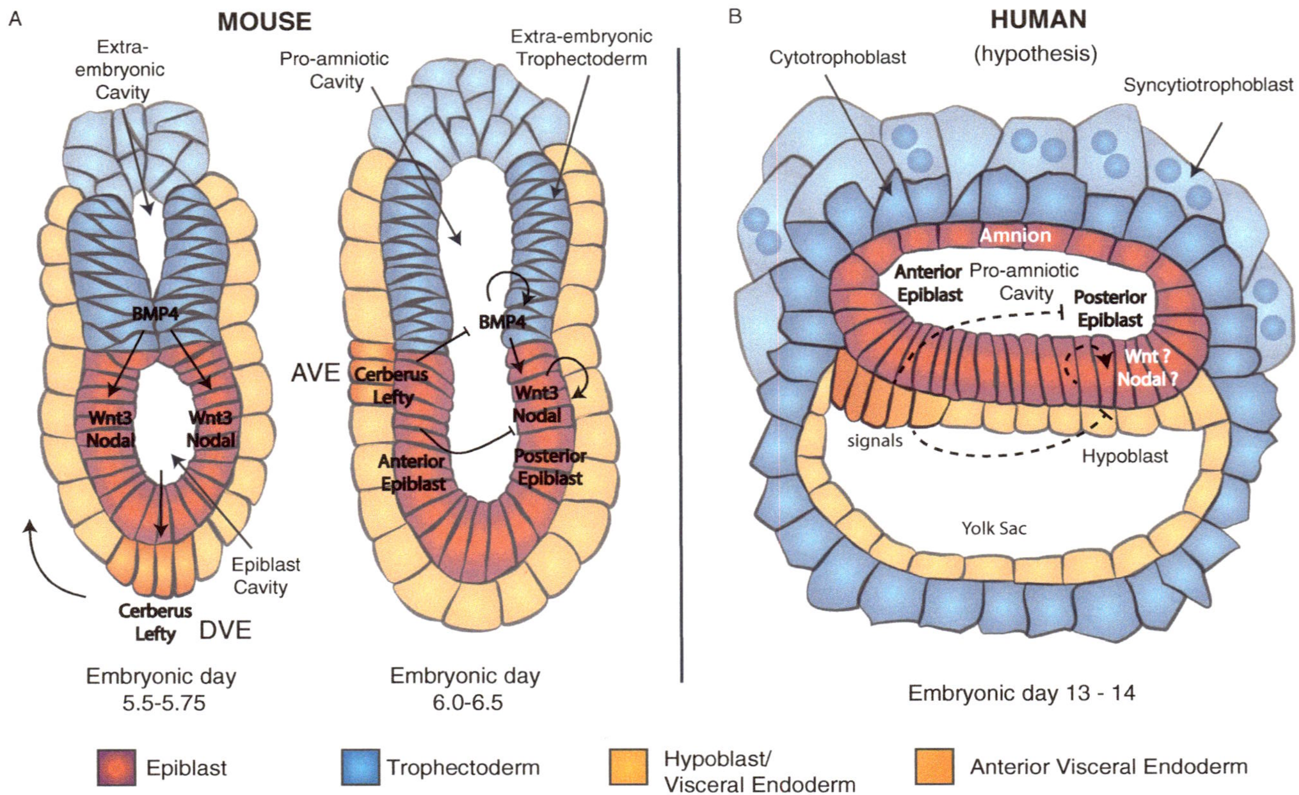

Fig. 3 See legend on opposite page.

to low concentration of the above inhibitors, WNT, BMP and NODAL remain active. NODAL (Brennan et al., 2001) and WNT3 (Liu et al., 1999) have been shown to regulate the induction of gastrulation. This is achieved through inducing expression of the master regulatory gene of gastrulation Brachyury (*Tbxt).* Activation of this gene expression in the posterior epiblast defines the site of primitive streak formation where the gastrulation will occur (Rivera-Pérez & Magnuson, 2005). Recent analysis of the monkey embryo confirmed expression of similarly conserved AVE antagonists in the primate VE, with *CER1* widely expressed along two-third of the VE length and *DKK1* confined to a small region (Sasaki et al., 2016). Whether a similar signaling center as the mouse AVE exists in human embryos is currently unknown. Molecular and morphological characterization of a potential human AVE signaling center would be essential to elucidate mechanisms behind symmetry breaking and anterior-posterior axis specification in the human embryo preceding initiation of gastrulation (Fig. 3B).

Another crucial event at this developmental stage is the specification and segregation of primordial germ cells (PGCs), the precursors of the future gametes. The origin of PGCs in the human embryo remains currently unknown but the observations in mouse, primate, and porcine embryos led to two opposing hypotheses. The first hypothesis supports specification of PGCs from the posterior region of the epiblast, based on mouse (Ohinata et al., 2009) and porcine studies (Kobayashi et al., 2017). The second model suggests the amnion as a potential source of primate PGCs, based on

Fig. 3 Pre-gastrulation patterning. (A) The extra-embryonic and embryonic cavity fuse together into the pro-amniotic cavity. BMP4, secreted by the extra-embryonic ectoderm, induces activation of *Nodal* and *Wnt3* uniformly in the epiblast. The visceral endoderm cells localized at the distal tip acquire columnar morphology, start to express Nodal and Wnt inhibitors (*Cer1* and *Lefty1*) forming the distal visceral endoderm (DVE). The DVE migrates unilaterally toward the proximal site of the epiblast until reaching the epiblast-extra-embryonic ectoderm boundary. The resulting AVE inhibits Nodal and Wnt signaling in the epiblast region in close proximity, defining the future anterior side of the epiblast. The epiblast distal from the AVE keeps activating Nodal and Wnt signaling, resulting in the specification of the posterior side of the epiblast where gastrulation initiates. (B) Hypothetical model of anterior-posterior patterning in human embryo. BMP ligands may be secreted by the amniotic epithelium leading to expression of Nodal and Wnt signaling in the epiblast disk uniformly. Specification of an AVE like structure in the hypoblast epithelium underlying the epiblast could cause secretion Nodal and Wnt inhibitors to demarcate the anterior epiblast from the posterior epiblast where the activity of Nodal and Wnt is preserved promoting initiation of gastrulation.

cynomolgus monkey embryos (*M. fascicularis*) (Sasaki et al., 2016). Despite differences in their tissue of origin, PGCs appear to maintain a conserved regulatory gene network among different model organisms based on the co-expression of the master PGCs transcription factors BLIMP1, also known as PRDM1, and TFAP2c in addition to markers of the pluripotent epiblast lineage NANOG and OCT4. However, expression of SOX17 (Irie et al., 2015) and the naïve pluripotency factor KLF4, together with repression of SOX2 have been shown to be essential markers unique to human PGC specification (reviewed in Kobayashi & Surani, 2018). The difference in the origin between primate and murine PGCs may also reflect species specific differences in the morphogenesis of the epiblast lineage at post-implantation. The epiblast of cynomolgus monkey embryos develops into a discoidal structure similar to the post-implantation human embryo with a pseudostratified epiblast and a squamous amnion epithelium surrounding a central amniotic cavity. In contrast to that, the amnion epithelium in the mouse is formed toward the end of gastrulation, due to tissue folding and reconnection of the proximal parts of the cup-shaped epiblast, which were previously separated (Pereira et al., 2011). In conclusion, it will be important to trace the tissue of origin of human PGCs and determine the gene regulatory network underlying their specification in the human embryo (Kobayashi & Surani, 2018; Sasaki et al., 2015), before their migration toward the yolk sac tissue where they finally reach the gonads (McLaren, 2003).

Upon establishment of the anterior-posterior axis through signaling originating from the extra-embryonic tissues, the process of gastrulation commences at the posterior end of the epiblast epithelium with the formation of the primitive streak, demarcated anteriorly by the node. Gastrulation leads to the formation of the three germ layers ectoderm, mesoderm and endoderm, which represent the basic building blocks for the assembly of every tissue of the developing fetus. Primitive streak initiates *in situ* by progressive epithelial-to-mesenchymal transition (EMT) in the posterior epiblast (Williams, Burdsal, Periasamy, Lewandoski, & Sutherland, 2012), followed by ingression through the primitive streak of epiblast cells (Ichikawa et al., 2013) that acquire full mesenchymal migratory behavior (Lamouille, Xu, & Derynck, 2014; Nieto, Huang, Jackson, & Thiery, 2016; Przybyla, Muncie, & Weaver, 2016). These cells eventually form the endoderm, precursor of all internal tissues as well as organs of the digestive and respiratory systems, and the middle mesoderm layer, precursor of the muscular, skeletal, and connective tissues. The remaining epiblast cells, which fail to ingress through the primitive streak, constitute the ectoderm epithelium from which the central nervous system and the epidermis layer originate. The process of EMT together with formation of mesoderm and definitive endoderm layers of gastrulation has been recently mimicked *in vitro* by generation of complete embryo-like structures through promoting self-assembly of stem cells for all three tissues: ESC, TSC and extra-embryonic endoderm—XEN—cells (Sozen et al., 2018).

Gastrulation in the human embryo appears to initiate at embryonic day 14 based on histological studies of the Carnegie Institution, where formation of the primitive streak and the emergence of embryonic mesoderm was observed to start at this stage (Hertig et al., 1956a; Heuser, ROCK, & Hertig, 1945; Luckett, 1978). The widely accepted 14-day rule (Hurlbut et al., 2017; Pera, 2017), which has developed into legislation in a number of jurisdictions such as in the UK, prohibits the culture of human embryos beyond day 14 when the primitive streak starts to appear. Further investigation is required to understand the basis of human embryo patterning up to day 14. But to elucidate the mechanism governing the molecular and cellular events of human gastrulation we rely on alternative *in vitro* stem cell-based models. Differentiation of the three lineages could be obtained taking advantage of engineered micropatterns for both mouse (Morgani, Metzger, Nichols, Siggia, & Hadjantonakis, 2018) and human ESCs (Simunovic et al., 2018). A 3D model of the human epiblast by using human ESCs revealed the intrinsic capability of human ESCs to break symmetry when exposed to homogenous concentrations of BMP4 (Simunovic et al., 2019). Despite the absence of signaling from extra-embryonic tissues, cells break symmetry by establishing a mutually exclusive gradient between SOX2, defining the anterior side, and BRA (T), defining the posterior region where epithelial-to-mesenchymal transition is initiated.

6. Conclusions

Mouse and human morphogenesis shows remarkable similarities before implantation despite differences in the timing of lineage segregation. Upon implantation, their developmental trajectories diverge substantially with the mouse giving rise to a cup-shaped embryo while the human embryo forms a flat bilaminar disk structure. Therefore, the principles governing human morphogenesis at this stage cannot be extrapolated by studies in mouse solely but necessitate direct investigation of the embryonic development. In addition to our fundamental knowledge originated from initial histological studies of the human specimens at the Carnegie Institution, recent advances in embryo culture systems have enabled the analysis of implantation and early post-implantation stages of the previously inaccessible human embryo. However, the processes of epiblast patterning and germ layer formation during gastrulation still presents a challenge, as the 14-day rule prohibits the culture of human embryos beyond this point. Implementation of stem cell-based models alongside direct investigation of the human embryo will be therefore instrumental to provide further insights into the morphogenetic, molecular and signaling mechanisms underlying human embryogenesis in order to build a new organism.

References

Antonica, F., Orietti, L. C., Mort, R. L., & Zernicka-Goetz, M. (2019). Concerted cell divisions in embryonic visceral endoderm guide anterior visceral endoderm migration. *Developmental Biology*, *450*(2), 132–140. https://doi.org/10.1016/j.ydbio.2019.03.016.

Aplin, J. D., & Ruane, P. T. (2017). Embryo–epithelium interactions during implantation at a glance. *Journal of Cell Science*, *130*(1), 15–22. https://doi.org/10.1242/jcs.175943.

Arnold, S. J., & Robertson, E. J. (2009). Making a commitment: Cell lineage allocation and axis patterning in the early mouse embryo. *Nature Reviews. Molecular Cell Biology*, *10*(2), 91–103. https://doi.org/10.1038/nrm2618.

Bassalert, C., Valverde-Estrella, L., & Chazaud, C. (2018). Primitive endoderm differentiation: From specification to epithelialization. *Current Topics in Developmental Biology*, *128*, 81–104. https://doi.org/10.1016/bs.ctdb.2017.12.001.

Bedzhov, I., & Zernicka-Goetz, M. (2014). Self-organizing properties of mouse pluripotent cells initiate morphogenesis upon implantation. *Cell*, *156*(5), 1032–1044. https://doi.org/10.1016/j.cell.2014.01.023.

Belo, J. A., Bachiller, D., Agius, E., Kemp, C., Borges, A. C., Marques, S., et al. (2000). Cerberus-like is a secreted BMP and nodal antagonist not essential for mouse development. *Genesis (New York, N.Y.: 2000)*, *26*(4), 265–270.

Belo, J. A., Bouwmeester, T., Leyns, L., Kertesz, N., Gallo, M., Follettie, M., et al. (1997). Cerberus-like is a secreted factor with neuralizing activity expressed in the anterior primitive endoderm of the mouse gastrula. *Mechanisms of Development*, *68*(1–2), 45–57. https://doi.org/10.1016/S0925-4773(97)00125-1.

Bloomekatz, J., Grego-Bessa, J., Migeotte, I., & Anderson, K. V. (2012). Pten regulates collective cell migration during specification of the anterior–posterior axis of the mouse embryo. *Developmental Biology*, *364*(2), 192–201. https://doi.org/10.1016/j.ydbio.2012.02.005.

Boroviak, T., Stirparo, G. G., Dietmann, S., Hernando-Herraez, I., Mohammed, H., Reik, W., et al. (2018). Single cell transcriptome analysis of human, marmoset and mouse embryos reveals common and divergent features of preimplantation development. *Development*, *145*(21), dev167833. https://doi.org/10.1242/dev.167833.

Braude, P., Bolton, V., & Moore, S. (1988). Human gene expression first occurs between the four- and eight-cell stages of preimplantation development. *Nature*, *332*(6163), 459–461. https://doi.org/10.1038/332459a0.

Brennan, J., Lu, C. C., Norris, D. P., Rodriguez, T. A., Beddington, R. S. P., & Robertson, E. J. (2001). Nodal signaling in the epiblast patterns the early mouse embryo. *Nature*, *411*, 965–969. https://doi.org/10.1038/35082103.

Casser, E., Israel, S., Witten, A., Schulte, K., Schlatt, S., Nordhoff, V., et al. (2017). Totipotency segregates between the sister blastomeres of two-cell stage mouse embryos. *Scientific Reports*, 7(1), 8299. https://doi.org/10.1038/s41598-017-08266-6.

Chazaud, C., & Yamanaka, Y. (2016). Lineage specification in the mouse preimplantation embryo. *Development*, *143*(7), 1063–1074. https://doi.org/10.1242/dev.128314.

Chazaud, C., Yamanaka, Y., Pawson, T., & Rossant, J. (2006). Early lineage segregation between epiblast and primitive endoderm in mouse blastocysts through the Grb2-MAPK pathway. *Developmental Cell*, *10*(5), 615–624. https://doi.org/10.1016/j.devcel.2006.02.020.

Christodoulou, N., Kyprianou, C., Weberling, A., Wang, R., Cui, G., Peng, G., et al. (2018). Sequential formation and resolution of multiple rosettes drive embryo remodelling after implantation. *Nature Cell Biology*, *20*(11), 1278–1289. https://doi.org/10.1038/s41556-018-0211-3.

Christodoulou, N., Weberling, A., Strathdee, D., Anderson, K. I., Timpson, P., & Zernicka-Goetz, M. (2019). Morphogenesis of extra-embryonic tissues directs the remodelling of the mouse embryo at implantation. *Nature Communications*, *10*(1), 3557. https://doi.org/10.1038/s41467-019-11482-5.

Copp, A. J. (1978). Interaction between inner cell mass and trophectoderm of the mouse blastocyst. I. A study of cellular proliferation. *Journal of Embryology and Experimental Morphology*, *48*, 109–125. Retrieved from http://www.ncbi.nlm.nih.gov/pubmed/744943.

Copp, A. J. (1979). Interaction between inner cell mass and trophectoderm of the mouse blastocyst. II. The fate of the polar trophectoderm. *Journal of Embryology and Experimental Morphology*, *51*, 109–120. Retrieved from http://www.ncbi.nlm.nih.gov/pubmed/479740.

Deglincerti, A., Croft, G. F., Pietila, L. N., Zernicka-Goetz, M., Siggia, E. D., & Brivanlou, A. H. (2016). Self-organization of the in vitro attached human embryo. *Nature*, *533*, 251. Retrieved from https://doi.org/10.1038/nature17948.

Ducibella, T. (1977). Changes in cell surface and cortical cytoplasmic organization during early embryogenesis in the preimplantation mouse embryo. *The Journal of Cell Biology*, *74*(1), 153–167. https://doi.org/10.1083/jcb.74.1.153.

Dumortier, J. G., Le Verge-Serandour, M., Tortorelli, A. F., Mielke, A., de Plater, L., Turlier, H., et al. (2019). Hydraulic fracturing and active coarsening position the lumen of the mouse blastocyst. *Science*, *365*(6452), 465–468. https://doi.org/10.1126/science.aaw7709.

Fassler, R., & Meyer, M. (1995). Consequences of lack of beta 1 integrin gene expression in mice. *Genes & Development*, *9*(15), 1896–1908.

Fogarty, N. M. E., McCarthy, A., Snijders, K. E., Powell, B. E., Kubikova, N., Blakeley, P., et al. (2017). Genome editing reveals a role for OCT4 in human embryogenesis. *Nature*, *550*(7674), 67–73. https://doi.org/10.1038/nature24033.

Gardner, R. L. (2001). Specification of embryonic axes begins before cleavage in normal mouse development. *Development*, *128*, 839–847, PMID:11222139.

Gardner, R. L., Papaioannou, V. E., & Barton, S. C. (1973). Origin of the ectoplacental cone and secondary giant cells in mouse blastocysts reconstituted from isolated trophoblast and inner cell mass. *Journal of Embryology and Experimental Morphology*, *30*(3), 561–572. https://doi.org/4772385.

Goldin, S. N., & Papaioannou, V. E. (2003). Paracrine action of FGF4 during periimplantation development maintains trophectoderm and primitive endoderm. *Genesis*, *36*(1), 40–47. https://doi.org/10.1002/gene.10192.

Goolam, M., Scialdone, A., Graham, S. J. L., Macaulay, I. C., Jedrusik, A., Hupalowska, A., et al. (2016). Heterogeneity in Oct4 and Sox2 targets biases cell fate in 4-cell mouse embryos. *Cell*, *165*(1), 61–74. https://doi.org/10.1016/j.cell.2016.01.047.

Harrison, S. E., Sozen, B., Christodoulou, N., Kyprianou, C., & Zernicka-Goetz, M. (2017). Assembly of embryonic and extraembryonic stem cells to mimic embryogenesis in vitro. *Science*, *356*(6334), eaal1810. https://doi.org/10.1126/science.aal1810.

Hertig, A. T., & Rock, J. (1944). On the development of the early human ovum, with special reference to the trophoblast of the previllous stage: A description of a normal and 5 pathologic human ova. *American Journal of Obstetrics and Gynecologyss*, *47*, 149–184.

Hertig, A. T., Rock, J., & Adams, E. C. (1956). A description of 34 human ova within the first 17 days of development. *The American Journal of Anatomy*, *98*(3), 435–493. Retrieved from http://www.ncbi.nlm.nih.gov/pubmed/13362122.

Heuser, C., Rock, J., & Hertig, A. T. (1945). Two human embryos showing early stages of the definitive yolk sac. *Contributions to embryology*, *557*(31), 85–99. Carnegie Institute of Washington Publisher.

Hupalowska, A., Jedrusik, A., Zhu, M., Bedford, M. T., Glover, D. M., & Zernicka-Goetz, M. (2018). CARM1 and paraspeckles regulate pre-implantation mouse embryo development. *Cell*, *175*(7), 1902–1916.e13. https://doi.org/10.1016/j.cell.2018.11.027.

Hurlbut, J. B., Hyun, I., Levine, A. D., Lovell-Badge, R., Lunshof, J. E., Matthews, K. R. W., et al. (2017). Revisiting the Warnock rule. *Nature Biotechnology*, *35*(11), 1029–1042. https://doi.org/10.1038/nbt.4015.

Ichikawa, T., Nakazato, K., Keller, P. J., Kajiura-Kobayashi, H., Stelzer, E. H. K., Mochizuki, A., et al. (2013). Live imaging of whole mouse embryos during gastrulation: Migration analyses of epiblast and mesodermal cells. *PLoS One*, *8*(7), e64506. https://doi.org/10.1371/journal.pone.0064506.

Irie, N., Weinberger, L., Tang, W. W. C., Kobayashi, T., Viukov, S., Manor, Y. S., et al. (2015). SOX17 is a critical specifier of human primordial germ cell fate. *Cell*, *160*(1–2), 253–268. https://doi.org/10.1016/j.cell.2014.12.013.

Jukam, D., Shariati, S. A. M., & Skotheim, J. M. (2017). Zygotic genome activation in vertebrates. *Developmental Cell*, *42*(4), 316–332. https://doi.org/10.1016/j.devcel.2017.07.026.

Kang, M., Garg, V., & Hadjantonakis, A.-K. (2017). Lineage establishment and progression within the inner cell mass of the mouse blastocyst requires FGFR1 and FGFR2. *Developmental Cell*, *41*(5), 496–510.e5. https://doi.org/10.1016/j.devcel.2017.05.003.

Kelly, S. J., Mulnard, J. G., & Graham, C. F. (1978). Cell division and cell allocation in early mouse development. *Journal of Embryology and Experimental Morphology*, *48*, 37–51.

Kemp, C., Willems, E., Abdo, S., Lambiv, L., & Leyns, L. (2005). Expression of all Wnt genes and their secreted antagonists during mouse blastocyst and postimplantation development. *Developmental Dynamics*, *233*(3), 1064–1075. https://doi.org/10.1002/dvdy.20408.

Kime, C., et al. (2019). Induced 2C Expression and implantation-competent blastocyst-like cysts from primed pluripotent stem cells. *Stem Cell Reports*, *13*, 485–498. https://doi.org/10.1016/j.stemcr.2019.07.011.

Kobayashi, T., & Surani, M. A. (2018). On the origin of the human germline. *Development*, *145*(16), dev150433. https://doi.org/10.1242/dev.150433.

Kobayashi, T., Zhang, H., Tang, W. W. C., Irie, N., Withey, S., Klisch, D., et al. (2017). Principles of early human development and germ cell program from conserved model systems. *Nature*, *546*, 416. Retrieved from https://doi.org/10.1038/nature22812.

Krendl, C., Shaposhnikov, D., Rishko, V., Ori, C., Ziegenhain, C., Sass, S., et al. (2017). GATA2/3-TFAP2A/C transcription factor network couples human pluripotent stem cell differentiation to trophectoderm with repression of pluripotency. *Proceedings of the National Academy of Sciences of the United Sates of America*, *114*(45), E9579–E9588. https://doi.org/10.1073/pnas.1708341114.

Kuijk, E. W., van Tol, L. T., Van de Velde, H., Wubbolts, R., Welling, M., Geijsen, N., et al. (2012). The roles of FGF and MAP kinase signaling in the segregation of the epiblast and hypoblast cell lineages in bovine and human embryos. *Development*, *139*(5), 871–882. https://doi.org/10.1242/dev.071688.

Kumar, A., Lualdi, M., Lyozin, G. T., Sharma, P., Loncarek, J., Fu, X.-Y., et al. (2015). Nodal signaling from the visceral endoderm is required to maintain Nodal gene expression in the epiblast and drive DVE/AVE migration. *Developmental Biology*, *400*(1), 1–9. https://doi.org/10.1016/j.ydbio.2014.12.016.

Lamouille, S., Xu, J., & Derynck, R. (2014). Molecular mechanisms of epithelial–mesenchymal transition. *Nature Reviews Molecular Cell Biology*, *15*(3), 178–196. https://doi.org/10.1038/nrm3758.

Lee, M. T., Bonneau, A. R., & Giraldez, A. J. (2014). Zygotic genome activation during the maternal-to-zygotic transition. *Annual Review of Cell and Developmental Biology*, *30*, 581–613. https://doi.org/10.1146/annurev-cellbio-100913-013027.

Li, S., Bordoy, R., Stanchi, F., Moser, M., Braun, A., Kudlacek, O., et al. (2005). PINCH1 regulates cell-matrix and cell-cell adhesions, cell polarity and cell survival during the peri-implantation stage. *Journal of Cell Science*, *118*(Pt. 13), 2913–2921. https://doi.org/10.1242/jcs.02422.

Lindenberg, S., Hyttel, P., Sjøgren, A., & Greve, T. (1989). A comparative study of attachment of human, bovine and mouse blastocysts to uterine epithelial monolayer. *Human Reproduction*, *4*(4), 446–456. https://doi.org/10.1093/oxfordjournals.humrep.a136925.

Liu, P., Wakamiya, M., Shea, M. J., Albrecht, U., Behringer, R. R., & Bradley, A. (1999). Requirement for Wnt3 in vertebrate axis formation. *Nature Genetics*, *22*(4), 361–365. https://doi.org/10.1038/11932.

Luckett, W. P. (1975). The development of primordial and definitive amniotic cavities in early Rhesus monkey and human embryos. *The American Journal of Anatomy*, *144*(2), 149–167. https://doi.org/10.1002/aja.1001440204.

Luckett, W. P. (1978). Origin and differentiation of the yolk sac and extraembryonic mesoderm in presomite human and rhesus monkey embryos. *The American Journal of Anatomy*, *152*(1), 59–97. https://doi.org/10.1002/aja.1001520106.

Lv, B., An, Q., Zeng, Q., Lu, P., Zhu, X., Ji, Y., et al. (2019). Single-cell RNA sequencing reveals regulatory mechanism for trophoblast cell-fate divergence in human Peri-implantation embryo. *bioRxiv*, 567362. https://doi.org/10.1101/567362.

Macklon, N. S., Geraedts, J. P. M., & Fauser, B. C. J. M. (2002). Conception to ongoing pregnancy: The "black box" of early pregnancy loss. *Human Reproduction Update*, *8*(4), 333–343. https://doi.org/10.1093/humupd/8.4.333.

McLaren, A. (2003). Primordial germ cells in the mouse. *Developmental Biology*, *262*(1), 1–15. https://doi.org/10.1016/S0012-1606(03)00214-8.

Meilhac, S. M., Adams, R. J., Morris, S. A., Danckaert, A., Le Garrec, J.-F., & Zernicka-Goetz, M. (2009). Active cell movements coupled to positional induction are involved in lineage segregation in the mouse blastocyst. *Developmental Biology*, *331*(2), 210–221. https://doi.org/10.1016/j.ydbio.2009.04.036.

Migeotte, I., Omelchenko, T., Hall, A., & Anderson, K. V. (2010). Rac1-dependent collective cell migration is required for specification of the anterior-posterior body axis of the mouse. *PLoS Biology*, *8*(8), e1000442. https://doi.org/10.1371/journal.pbio.1000442.

Mobley, R. J., Raghu, D., Duke, L. D., Abell-Hart, K., Zawistowski, J. S., Lutz, K., et al. (2017). MAP3K4 controls the chromatin modifier HDAC6 during trophoblast stem cell epithelial-to-mesenchymal transition. *Cell Reports*, *18*(10), 2387–2400. https://doi.org/10.1016/j.celrep.2017.02.030.

Molotkov, A., Mazot, P., Brewer, J. R., Cinalli, R. M., & Soriano, P. (2017). Distinct requirements for FGFR1 and FGFR2 in primitive endoderm development and exit from pluripotency. *Developmental Cell*, *41*(5), 511–526.e4. https://doi.org/10.1016/j.devcel.2017.05.004.
Morgani, S. M., Metzger, J. J., Nichols, J., Siggia, E. D., & Hadjantonakis, A.-K. (2018). Micropattern differentiation of mouse pluripotent stem cells recapitulates embryo regionalized cell fate patterning. *eLife*, 7, e32839. https://doi.org/10.7554/eLife.32839.
Morris, S. A., et al. (2010). Origin and formation of the first two distinct cell types of the inner cell mass in the mouse embryo. *Proceedings of the National Academy of Sciences*, *107*, 6364–6369. https://doi.org/10.1073/pnas.0915063107.
Morris, S. A., Grewal, S., Barrios, F., Patankar, S. N., Strauss, B., Buttery, L., et al. (2012). Dynamics of anterior-posterior axis formation in the developing mouse embryo. *Nature Communications*, *3*, 673. https://doi.org/10.1038/ncomms1671.
Niakan, K. K., & Eggan, K. (2013). Analysis of human embryos from zygote to blastocyst reveals distinct gene expression patterns relative to the mouse. *Developmental Biology*, *375*(1), 54–64. https://doi.org/10.1016/j.ydbio.2012.12.008.
Nichols, J., & Smith, A. (2009). Naive and primed pluripotent states. *Cell Stem Cell*, *4*(6), 487–492. https://doi.org/10.1016/j.stem.2009.05.015.
Nieto, M. A., Huang, R. Y.-J., Jackson, R. A., & Thiery, J. P. (2016). EMT: 2016. *Cell*, *166*(1), 21–45. https://doi.org/10.1016/j.cell.2016.06.028.
Nikas, G., Ao, A., Winston, R. M., & Handyside, A. H. (1996). Compaction and surface polarity in the human embryo in vitro. *Biology of Reproduction*, *55*(1), 32–37. Retrieved from https://www.ncbi.nlm.nih.gov/pubmed/8793055.
Ohinata, Y., Ohta, H., Shigeta, M., Yamanaka, K., Wakayama, T., & Saitou, M. (2009). A signaling principle for the specification of the germ cell lineage in mice. *Cell*, *137*(3), 571–584. https://doi.org/10.1016/j.cell.2009.03.014.
Omelchenko, T., Rabadan, M. A., Hernández-Martínez, R., Grego-Bessa, J., Anderson, K. V., & Hall, A. (2014). β-Pix directs collective migration of anterior visceral endoderm cells in the early mouse embryo. *Genes & Development*, *28*(24), 2764–2777. https://doi.org/10.1101/gad.251371.114.
Papaioannou, V. E. (1982). Lineage analysis of inner cell mass and trophectoderm using microsurgically reconstituted mouse blastocysts. *Journal of Embryology and Experimental Morphology*, *68*, 199–209. Retrieved from http://www.ncbi.nlm.nih.gov/pubmed/7108422.
Pera, M. F. (2017). Human embryo research and the 14-day rule. *Development*, *144*(11), 1923–1925. https://doi.org/10.1242/dev.151191.
Perea-Gomez, A., Vella, F. D. J., Shawlot, W., Oulad-Abdelghani, M., Chazaud, C., Meno, C., et al. (2002). Nodal antagonists in the anterior visceral endoderm prevent the formation of multiple primitive streaks. *Developmental Cell*, *3*(5), 745–756.
Pereira, P. N., Dobreva, M. P., Graham, L., Huylebroeck, D., Lawson, K. A., & Zwijsen, A. (2011). Amnion formation in the mouse embryo: The single amniochorionic fold model. *BMC Developmental Biology*, *11*(1), 48. https://doi.org/10.1186/1471-213X-11-48.
Petropoulos, S., Edsgärd, D., Reinius, B., Deng, Q., Panula, S. P., Codeluppi, S., et al. (2016). Single-cell RNA-Seq reveals lineage and X chromosome dynamics in human preimplantation embryos. *Cell*, *165*(4), 1012–1026. https://doi.org/10.1016/j.cell.2016.03.023.
Piotrowska, K., Wianny, F., Pedersen, R. A., & Zernicka-Goetz, M. (2001). Blastomeres arising from the first cleavage division have distinguishable fates in normal mouse development. *Development (Cambridge, England)*, *128*(19), 3739–3748. Retrieved from http://www.ncbi.nlm.nih.gov/pubmed/11585800.

Piotrowska, K., & Zernicka-Goetz, M. (2001). Role for sperm in spatial patterning of the early mouse embryo. *Nature, 409*(6819), 517–521. https://doi.org/10.1038/35054069.

Piotrowska-Nitsche, K. (2005). Four-cell stage mouse blastomeres have different developmental properties. *Development, 132*(3), 479–490. https://doi.org/10.1242/dev.01602.

Piotrowska-Nitsche, K., & Zernicka-Goetz, M. (2005). Spatial arrangement of individual 4-cell stage blastomeres and the order in which they are generated correlate with blastocyst pattern in the mouse embryo. *Mechanisms of Development, 122*(4), 487–500. https://doi.org/10.1016/j.mod.2004.11.014.

Piras, V., Tomita, M., & Selvarajoo, K. (2015). Transcriptome-wide variability in single embryonic development cells. *Scientific Reports, 4*(1), 7137. https://doi.org/10.1038/srep07137.

Plusa, B., Piliszek, A., Frankenberg, S., Artus, J., & Hadjantonakis, A.-K. (2008). Distinct sequential cell behaviours direct primitive endoderm formation in the mouse blastocyst. *Development, 135*(18), 3081–3091. https://doi.org/10.1242/dev.021519.

Posfai, E., Petropoulos, S., de Barros, F. R. O., Schell, J. P., Jurisica, I., Sandberg, R., et al. (2017). Position- and Hippo signaling-dependent plasticity during lineage segregation in the early mouse embryo. *eLife, 6*, e22906. https://doi.org/10.7554/eLife.22906.

Przybyla, L., Muncie, J. M., & Weaver, V. M. (2016). Mechanical control of epithelial-to-mesenchymal transitions in development and cancer. *Annual Review of Cell and Developmental Biology, 32*(1), 527–554. https://doi.org/10.1146/annurev-cellbio-111315-125150.

Reeve, W. J. (1981). Cytoplasmic polarity develops at compaction in rat and mouse embryos. *Journal of Embryology and Experimental Morphology, 62*, 351–367. Retrieved from http://www.ncbi.nlm.nih.gov/pubmed/7276818.

Richter, K. S., Harris, D. C., Daneshmand, S. T., & Shapiro, B. S. (2001). Quantitative grading of a human blastocyst: Optimal inner cell mass size and shape. *Fertility and Sterility, 76*(6), 1157–1167. https://doi.org/10.1016/S0015-0282(01)02870-9.

Rivera-Pérez, J. A., & Hadjantonakis, A.-K. (2015). The dynamics of morphogenesis in the early mouse embryo. *Cold Spring Harbor Perspectives in Biology*, 7(11), a015867. https://doi.org/10.1101/cshperspect.a015867.

Rivera-Pérez, J. A., & Magnuson, T. (2005). Primitive streak formation in mice is preceded by localized activation of Brachyury and Wnt3. *Developmental Biology, 288*(2), 363–371. https://doi.org/10.1016/j.ydbio.2005.09.012.

Rivron, N. C., Frias-Aldeguer, J., Vrij, E. J., Boisset, J.-C., Korving, J., Vivié, J., et al. (2018). Blastocyst-like structures generated solely from stem cells. *Nature, 557*(7703), 106–111. https://doi.org/10.1038/s41586-018-0051-0.

Robertson, E. J. (2014). Dose-dependent Nodal/Smad signals pattern the early mouse embryo. *Seminars in Cell & Developmental Biology, 32*, 73–79. https://doi.org/10.1016/j.semcdb.2014.03.028.

Roode, M., Blair, K., Snell, P., Elder, K., Marchant, S., Smith, A., et al. (2012). Human hypoblast formation is not dependent on FGF signalling. *Developmental Biology, 361*(2), 358–363. https://doi.org/10.1016/j.ydbio.2011.10.030.

Rossant, J., & Lis, W. T. (1979). Potential of isolated mouse inner cell masses to form trophectoderm derivatives in vivo. *Developmental Biology, 70*(1), 255–261. https://doi.org/10.1016/0012-1606(79)90022-8.

Sakai, T., Li, S., Docheva, D., Grashoff, C., Sakai, K., Kostka, G., et al. (2003). Integrin-linked kinase (ILK) is required for polarizing the epiblast, cell adhesion, and controlling actin accumulation. *Genes & Development, 17*(7), 926–940. https://doi.org/10.1101/gad.255603.

Sasaki, K., Nakamura, T., Okamoto, I., Yabuta, Y., Iwatani, C., Tsuchiya, H., et al. (2016). The germ cell fate of Cynomolgus monkeys is specified in the nascent amnion. *Developmental Cell, 39*(2), 169–185. https://doi.org/10.1016/j.devcel.2016.09.007.

Sasaki, K., Yokobayashi, S., Nakamura, T., Okamoto, I., Yabuta, Y., Kurimoto, K., et al. (2015). Robust in vitro induction of human germ cell fate from pluripotent stem cells. *Cell Stem Cell, 17*(2), 178–194. https://doi.org/10.1016/j.stem.2015.06.014.

Schrode, N., Saiz, N., Di Talia, S., & Hadjantonakis, A. K. (2014). GATA6 levels modulate primitive endoderm cell fate choice and timing in the mouse blastocyst. *Developmental Cell, 29*, 454–467. https://doi.org/10.1016/j.devcel.2014.04.011.

Schultz, J. F., Mayernik, L., Rout, U. K., & Armant, D. R. (1997). Integrin trafficking regulates adhesion to fibronectin during differentiation of mouse peri-implantation blastocysts. *Developmental Genetics, 21*(1), 31–43. https://doi.org/10.1002/(SICI)1520-6408(1997)21:1<31::AID-DVG4>3.0.CO;2-9.

Schulz, K. N., & Harrison, M. M. (2019). Mechanisms regulating zygotic genome activation. *Nature Reviews. Genetics, 20*(4), 221–234. https://doi.org/10.1038/s41576-018-0087-x.

Shahbazi, M. N., Jedrusik, A., Vuoristo, S., Recher, G., Hupalowska, A., Bolton, V., et al. (2016). Self-organization of the human embryo in the absence of maternal tissues. *Nature Cell Biology, 18*(6), 700–708. https://doi.org/10.1038/ncb3347.

Shahbazi, M. N., Scialdone, A., Skorupska, N., Weberling, A., Recher, G., Zhu, M., et al. (2017). Pluripotent state transitions coordinate morphogenesis in mouse and human embryos. *Nature, 552*(7684), 239–243. https://doi.org/10.1038/nature24675.

Shao, Y., Taniguchi, K., Gurdziel, K., Townshend, R. F., Xue, X., Yong, K. M. A., et al. (2017). Self-organized amniogenesis by human pluripotent stem cells in a biomimetic implantation-like niche. *Nature Materials, 16*(4), 419–425. https://doi.org/10.1038/nmat4829.

Shao, Y., Taniguchi, K., Townshend, R. F., Miki, T., Gumucio, D. L., & Fu, J. (2017). A pluripotent stem cell-based model for post-implantation human amniotic sac development. *Nature Communications, 8*(1), 208. https://doi.org/10.1038/s41467-017-00236-w.

Simunovic, M., Metzger, J. J., Etoc, F., Yoney, A., Ruzo, A., Martyn, I., et al. (2018). Molecular mechanism of symmetry breaking in a 3D model of a human epiblast. *bioRxiv*, 330704. https://doi.org/10.1101/330704.

Simunovic, M., Metzger, J. J., Etoc, F., Yoney, A., Ruzo, A., Martyn, I., et al. (2019). A 3D model of a human epiblast reveals BMP4-driven symmetry breaking. *Nature Cell Biology, 21*(7), 900–910. https://doi.org/10.1038/s41556-019-0349-7.

Sozen, B., Amadei, G., Cox, A., Wang, R., Na, E., Czukiewska, S., et al. (2018). Self-assembly of embryonic and two extra-embryonic stem cell types into gastrulating embryo-like structures. *Nature Cell Biology, 20*(8), 979–989. https://doi.org/10.1038/s41556-018-0147-7.

Srinivas, S. (2004). Active cell migration drives the unilateral movements of the anterior visceral endoderm. *Development, 131*(5), 1157–1164. https://doi.org/10.1242/dev.01005.

Stephens, L. E., Sutherland, A. E., Klimanskaya, I. V., Andrieux, A., Meneses, J., Pedersen, R. A., et al. (1995). Deletion of beta 1 integrins in mice results in inner cell mass failure and peri-implantation lethality. *Genes & Development, 9*(15), 1883–1895. Retrieved from http://www.ncbi.nlm.nih.gov/pubmed/7544312.

Stirparo, G. G., Boroviak, T., Guo, G., Nichols, J., Smith, A., & Bertone, P. (2018). Integrated analysis of single-cell embryo data yields a unified transcriptome signature for the human pre-implantation epiblast. *Development, 145*(3), dev158501. https://doi.org/10.1242/dev.158501.

Stower, M. J., & Srinivas, S. (2018). The head's tale: Anterior-posterior axis formation in the mouse embryo. *Current Topics in Developmental Biology, 128*, 365–390. https://doi.org/10.1016/bs.ctdb.2017.11.003.

Suwińska, A., Czołowska, R., Ożdżeński, W., & Tarkowski, A. K. (2008). Blastomeres of the mouse embryo lose totipotency after the fifth cleavage division: Expression of Cdx2 and Oct4 and developmental potential of inner and outer blastomeres of 16- and 32-cell embryos. *Developmental Biology*, *322*(1), 133–144. https://doi.org/10.1016/j.ydbio.2008.07.019.

Tadros, W., & Lipshitz, H. D. (2009). The maternal-to-zygotic transition: A play in two acts. *Development*, *136*(18), 3033–3042. https://doi.org/10.1242/dev.033183.

Takaoka, K., Yamamoto, M., & Hamada, H. (2011). Origin and role of distal visceral endoderm, a group of cells that determines anterior–posterior polarity of the mouse embryo. *Nature Cell Biology*, *13*(7), 743–752. https://doi.org/10.1038/ncb2251.

Takaoka, K., Yamamoto, M., Shiratori, H., Meno, C., Rossant, J., Saijoh, Y., et al. (2006). The mouse embryo autonomously acquires anterior-posterior polarity at implantation. *Developmental Cell*, *10*(4), 451–459. https://doi.org/10.1016/j.devcel.2006.02.017.

Taniguchi, K., Shao, Y., Townshend, R. F., Tsai, Y.-H., DeLong, C. J., Lopez, S. A., et al. (2015). Lumen formation is an intrinsic property of isolated human pluripotent stem cells. *Stem Cell Reports*, *5*(6), 954–962. https://doi.org/10.1016/j.stemcr.2015.10.015.

Thomas, P., & Beddington, R. (1996). Anterior primitive endoderm may be responsible for patterning the anterior neural plate in the mouse embryo. *Current Biology*, *6*(11), 1487–1496. https://doi.org/10.1016/S0960-9822(96)00753-1.

Torres-Padilla, M.-E., Parfitt, D.-E., Kouzarides, T., & Zernicka-Goetz, M. (2007). Histone arginine methylation regulates pluripotency in the early mouse embryo. *Nature*, *445*(7124), 214–218. https://doi.org/10.1038/nature05458.

Wallingford, M. C., Angelo, J. R., & Mager, J. (2013). Morphogenetic analysis of peri-implantation development. *Developmental Dynamics*, *242*(9), 1110–1120. https://doi.org/10.1002/dvdy.23991.

Wang, J., Mayernik, L., & Armant, D. R. (2007). Trophoblast adhesion of the peri-implantation mouse blastocyst is regulated by integrin signaling that targets phospholipase C. *Developmental Biology*, *302*(1), 143–153. https://doi.org/10.1016/j.ydbio.2006.09.015.

Wang, J., Wang, L., Feng, G., Wang, Y., Li, Y., Li, X., et al. (2018). Asymmetric expression of LincGET biases cell fate in two-cell mouse embryos. *Cell*, *175*(7), 1887–1901.e18. https://doi.org/10.1016/j.cell.2018.11.039.

Watanabe, Y., Miyasaka, K. Y., Kubo, A., Kida, Y. S., Nakagawa, O., Hirate, Y., et al. (2017). Notch and Hippo signaling converge on strawberry notch 1 (Sbno1) to synergistically activate Cdx2 during specification of the trophectoderm. *Scientific Reports*, 7(1), 46135. https://doi.org/10.1038/srep46135.

Weimar, C. H. E., Post Uiterweer, E. D., Teklenburg, G., Heijnen, C. J., & Macklon, N. S. (2013). In-vitro model systems for the study of human embryo–endometrium interactions. *Reproductive Biomedicine Online*, *27*(5), 461–476. https://doi.org/10.1016/j.rbmo.2013.08.002.

White, M. D., Angiolini, J. F., Alvarez, Y. D., Kaur, G., Zhao, Z. W., Mocskos, E., et al. (2016). Long-lived binding of Sox2 to DNA predicts cell fate in the four-cell mouse embryo. *Cell*, *165*(1), 75–87. https://doi.org/10.1016/j.cell.2016.02.032.

White, M. D., Bissiere, S., Alvarez, Y. D., & Plachta, N. (2016). Mouse embryo compaction. *Current Topics in Developmental Biology*, *120*, 235–258. https://doi.org/10.1016/bs.ctdb.2016.04.005.

White, M. D., Zenker, J., Bissiere, S., & Plachta, N. (2018). Instructions for assembling the early mammalian embryo. *Developmental Cell*, *45*(6), 667–679. https://doi.org/10.1016/j.devcel.2018.05.013.

Wilcox, A. J., Baird, D. D., & Weinberg, C. R. (1999). Time of implantation of the conceptus and loss of pregnancy. *New England Journal of Medicine*, *340*(23), 1796–1799. https://doi.org/10.1056/NEJM199906103402304.
Wilcox, A. J., Weinberg, C. R., O'Connor, J. F., Baird, D. D., Schlatterer, J. P., Canfield, R. E., et al. (1988). Incidence of early loss of pregnancy. *New England Journal of Medicine*, *319*(4), 189–194. https://doi.org/10.1056/NEJM198807283190401.
Williams, M., Burdsal, C., Periasamy, A., Lewandoski, M., & Sutherland, A. (2012). Mouse primitive streak forms in situ by initiation of epithelial to mesenchymal transition without migration of a cell population. *Developmental Dynamics*, *241*(2), 270–283. https://doi.org/10.1002/dvdy.23711.
Yamamoto, M., Saijoh, Y., Perea-Gomez, A., Shawlot, W., Behringer, R. R., Ang, S.-L., et al. (2004). Nodal antagonists regulate formation of the anteroposterior axis of the mouse embryo. *Nature*, *428*(6981), 387–392. https://doi.org/10.1038/nature02418.
Yamanaka, Y., Lanner, F., & Rossant, J. (2010). FGF signal-dependent segregation of primitive endoderm and epiblast in the mouse blastocyst. *Development*, *137*(5), 715–724. https://doi.org/10.1242/dev.043471.
Zenker, J., White, M. D., Gasnier, M., Alvarez, Y. D., Lim, H. Y. G., Bissiere, S., et al. (2018). Expanding actin rings zipper the mouse embryo for blastocyst formation. *Cell*, *173*(3), 776–791.e17. https://doi.org/10.1016/j.cell.2018.02.035.
Zheng, Y., Xue, X., Shao, Y., Wang, S., Esfahani, S. N., Li, Z., et al. (2019). Controlled modelling of human epiblast and amnion development using stem cells. *Nature*, *573*, 421–425. https://doi.org/10.1038/s41586-019-1535-2.
Zhu, M., Leung, C. Y., Shahbazi, M. N., & Zernicka-Goetz, M. (2017). Actomyosin polarisation through PLC-PKC triggers symmetry breaking of the mouse embryo. *Nature Communications*, *8*(1), 921. https://doi.org/10.1038/s41467-017-00977-8.

Further reading

Kang, M., Piliszek, A., Artus, J., & Hadjantonakis, A.-K. (2013). FGF4 is required for lineage restriction and salt-and-pepper distribution of primitive endoderm factors but not their initial expression in the mouse. *Development*, *140*(2), 267–279. https://doi.org/10.1242/dev.084996.

PART II

Many ways to gastrulate

CHAPTER FIVE

The cellular and molecular mechanisms that establish the mechanics of *Drosophila* gastrulation

Clint S. Ko, Adam C. Martin*

Department of Biology, Massachusetts Institute of Technology, Cambridge, MA, United States
*Corresponding author: e-mail address: acmartin@mit.edu

Contents

Abstract

In this review, we cover advances in the field that have contributed to our mechanistic understanding of how tissues internalize during *Drosophila melanogaster* gastrulation. The changes in tissue shape and architecture that are associated with mesoderm and endoderm invagination in the early *Drosophila* embryo are accompanied by cell shape changes which are driven by actomyosin contractility. The activation of signal transduction pathways is patterned by embryonic transcription factors, which define distinct geometries of gene expression and the tissue contractile domains. At the subcellular level, outputs from signaling pathways that activate actomyosin contractility are highly polarized and their behavior is fine-tuned by a balance of both positive and negative regulation. Cells are mechanically linked through adherens junctions, allowing forces that are generated by cells to be integrated across the tissue, ensuring coordinated cell behavior during tissue invagination. The transmission of force between cells also enables mechanical feedback whereby force generation influences both cell and cytoskeletal behavior. Finally, after tissue invagination, mesoderm cells undergo an epithelial-to-mesenchymal transition and cell spreading. We highlight studies that have utilized this model system to uncover fundamental principles at molecular-, cell-, and tissue-levels, which have contributed to our understanding of similar tissue morphogenetic processes across different organisms.

Current Topics in Developmental Biology, Volume 136
ISSN 0070-2153
https://doi.org/10.1016/bs.ctdb.2019.08.003

1. Overview

Gastrulation involves dramatic rearrangements of tissue architecture, leading to the internalization of the mesoderm and endoderm into the embryo to generate multiple germ layers. During early embryonic development in *Drosophila melanogaster*, the blastula is comprised of a single epithelial sheet that surrounds a central yolk. At the onset of gastrulation, presumptive mesoderm cells that comprise a region of about 70 cells long (along the anterior-posterior axis) and 18 cells wide (along the dorsal-ventral axis) internalize through a process of fast cell shape changes (i.e., apical constriction) that fold the tissue and form a furrow along the ventral midline, called the ventral furrow (Leptin & Grunewald, 1990; Sweeton, Parks, Costa, & Wieschaus, 1991) (Fig. 1A). This process is followed by a cup-like tissue invagination on the posterior end during posterior midgut (PMG) invagination, which internalizes endoderm cells (Leptin & Grunewald, 1990; Sweeton et al., 1991) (Fig. 1A). Mesoderm and endoderm fates are specified by patterns of differential gene expression that become active during oocyte formation, which lead to the establishment of anterior-posterior and dorsal-ventral polarity in the embryo (Belvin & Anderson, 1996; Govind & Steward, 1991; Nüsslein-Volhard, Frohnhöfer, & Lehmann, 1987). Gene expression then activates signal transduction pathways, which regulate cytoskeletal activities in precise geometries across the embryo. The cytoskeleton, most prominently the actomyosin cytoskeleton, generates forces that drive apical constriction and tissue internalization. Cellular forces are subsequently transmitted and integrated across a tissue, introducing tissue-level stresses that influence tissue shape. These processes are spatiotemporally regulated and coordinated through an interplay of gene expression, cell signaling, and mechanical feedback. This review will begin with the molecular mechanisms of cell shape change and will build up to larger scales, explaining how cells interact and how tissue invagination occurs.

2. Apical constriction: Force generation at the molecular and cellular level

The blastoderm of the *Drosophila* embryo is an epithelium with defined apical (outer) and basal (inner) polarity and with apically localized adherens junctions (Fig. 1A and C). In many developmental contexts, cell shape changes drive epithelial morphogenesis (Heisenberg & Bellaïche, 2013;

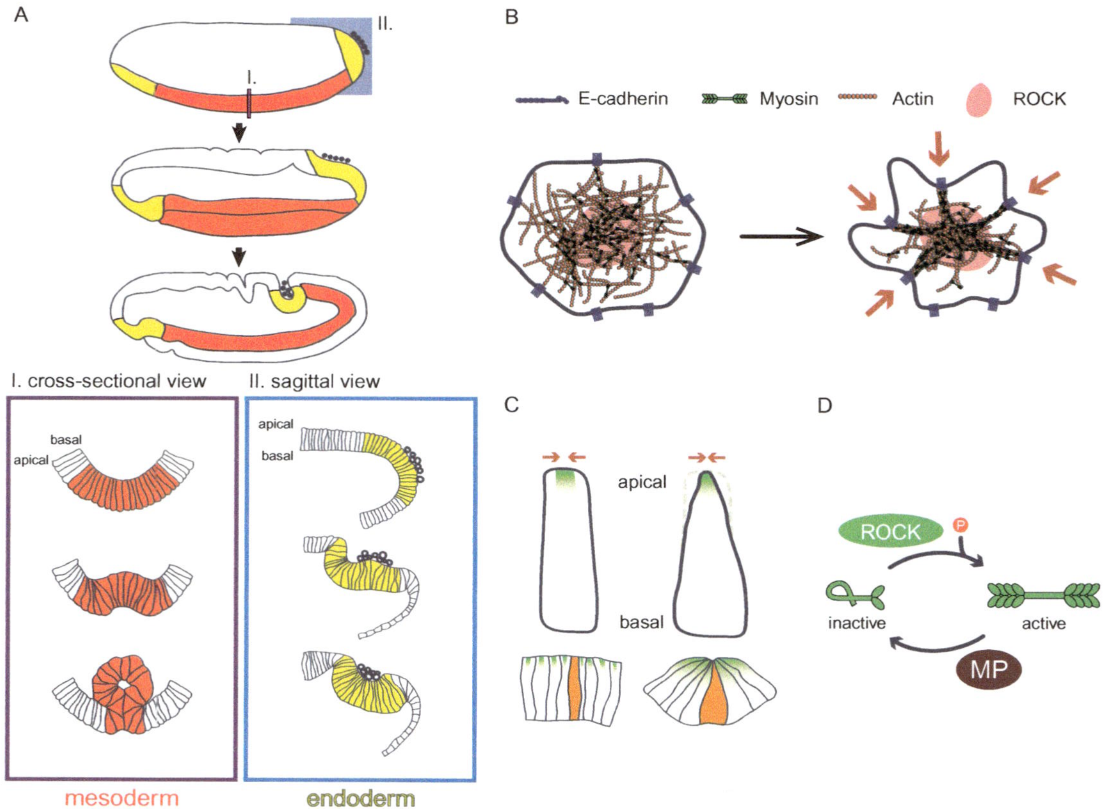

Fig. 1 See figure legend on next page.

Lecuit, Lenne, & Munro, 2011). During gastrulation, invagination of the mesoderm and posterior midgut primordium is associated with apical constriction, a cell shape change whereby the apex of the cell constricts (Costa, Wilson, & Wieschaus, 1994; Leptin & Grunewald, 1990; Sweeton et al., 1991) (Fig. 1B and C). Apical constriction shrinks the outer surface of the epithelium relative to the basal surface. Because the two surfaces are connected by the lateral domain, these cell shape changes lead to inward tissue curvature (Fig. 1A and C).

Apical constriction is driven by apical actomyosin contractility (Dawes-Hoang et al., 2005; Franke, Montague, & Kiehart, 2005; Levayer & Lecuit, 2012; Martin, Kaschube, & Wieschaus, 2009; Young, Pesacreta, & Kiehart, 1991). In *Drosophila*, nonmuscle myosin II (henceforth referred to as myosin) is activated in a similar manner to vertebrate myosin (Jordan & Karess, 1997; Vasquez, Heissler, Billington, & Martin, 2016). Phosphorylation of the regulatory light chain of myosin at highly conserved residues (T18 and S19) by Rho-associated coiled-coil kinase (ROCK), for instance, activates the motor and promotes the formation of myosin assemblies termed bipolar minifilaments (Amano et al., 1996; Jordan & Karess, 1997; Matsumura, 2005; Mizuno, Amano, Kaibuchi, & Nishida, 1999; Sellers, 1991; Vasquez et al., 2016; Winter et al., 2001) (Fig. 1D). This assembly that contains motor domains at both ends can pull on antiparallel arrays of filamentous actin (F-actin) with plus ends facing outward to generate contractile force. During apical constriction in *Drosophila* gastrulation and other tissue morphogenetic contexts, myosin activity is enriched in the middle

Fig. 1 Apical constriction drives internalization of the mesoderm and endoderm during *Drosophila* gastrulation. (A) Diagram of embryos at various stages of gastrulation. The mesoderm (red) and endoderm (yellow) internalize into the embryo through coordinated cell shape changes. Bottom insets show cross-section views of mesoderm invagination (left) and a sagittal view of endoderm invagination (right). (B) A polarized actomyosin cortex promotes apical constriction. Medioapical ROCK activity activates myosin, pulling on a polarized F-actin meshwork that is attached to adherens junctions. (C) Apical constriction changes tissue architecture. Diagram demonstrates the constriction of the cell apex at the apical side (arrows), which is associated with medioapically polarized myosin (green). Top shows the cell highlighted in orange in the bottom cartoon of a contractile epithelium. (D) Myosin undergoes cycles of activation and inactivation. ROCK phosphorylates the regulatory light chain of myosin, leading to its activation and promoting the formation of bipolar minifilaments. Myosin phosphatase (MP) removes the phosphates and inactivates myosin. *Panel (A): Modified from Leptin, M., (2005). Gastrulation movements: The logic and the nuts and bolts.* Developmental Cell, 8, *305–320.*

of the apical domain (medioapical) (Booth, Blanchard, Adams, & Röper, 2014; Chanet et al., 2017; Chung, Kim, & Andrew, 2017; Coravos & Martin, 2016; Mason, Tworoger, & Martin, 2013). This medioapical localization is thought to increase the efficiency of cell apex constriction as active medial myosin pulls centripetally on a polarized F-actin network, where F-actin plus ends are often enriched at junctions (Coravos & Martin, 2016; Kovacs, Goodwin, Ali, Paterson, & Yap, 2002; Tang & Brieher, 2012) (Fig. 1B). The microtubule cytoskeleton also exhibits a striking polarity within the apical cortex. Apically constricting mesoderm and endoderm cells exhibit a prominent noncentrosomal microtubule-organizing center near the center of the cell apex (Ko, Tserunyan, & Martin, 2019). Thus, the apical cortex of apically constricting cells exhibits striking organization and polarity.

Myosin activity levels are dynamic, exhibiting a pulsatile pattern of myosin assembly and disassembly (Martin et al., 2009). These contractile pulses are regulated by cycles of myosin phosphorylation and dephosphorylation by ROCK and myosin phosphatase, respectively (Munjal, Philippe, Munro, & Lecuit, 2015; Vasquez, Tworoger, & Martin, 2014) (Fig. 1D). During early stages of invagination, cells relax their apical area after constriction, undergoing a so-called unratcheted pulse. Over time, cells switch to stabilizing their apical area after constriction, a process referred to as a ratcheted pulse (Martin et al., 2009; Xie & Martin, 2015). This behavior has been attributed to a persistence of myosin structures at the medioapical cortex during pulse disassembly, allowing cell shape to be stabilized (Coravos & Martin, 2016; Xie & Martin, 2015).

Much as how the spatiotemporal regulation of myosin phosphorylation is integral for cell shape change, the regulation of actin polymer dynamics also plays a key role in modulating apical constriction (Fox & Peifer, 2007; Homem & Peifer, 2008). Actin filaments (F-actin) are polymerized from monomers by different classes of nucleators, such as formins and actin-related protein 2/3 (Arp2/3), which polymerize linear- and branched-actin networks, respectively (Campellone & Welch, 2010; Goode & Eck, 2007; Grikscheit & Grosse, 2016). In vitro studies have demonstrated that the organization and connectivity of F-actin in a network determines the response to myosin-induced contraction (Ennomani et al., 2016). In turn, the architecture and dynamics of F-actin networks at the cell cortex can influence cell surface mechanics, which has important implications for how cells deform and change shape (Chugh et al., 2017; Clark, Wartlick, Salbreux, & Paluch, 2014; Clément, Dehapiot, Collinet,

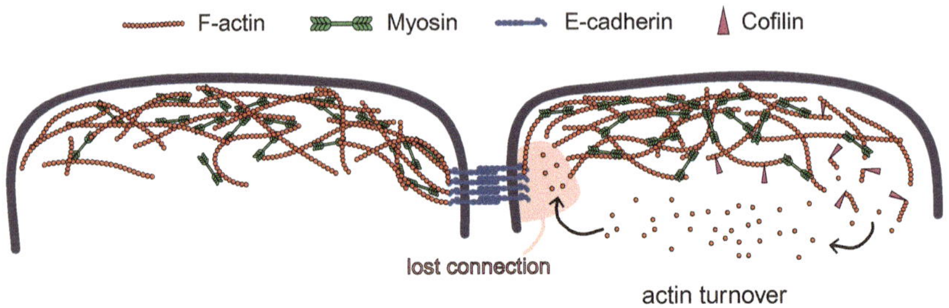

Fig. 2 Contractile force is integrated across the tissue by actomyosin attachment to adherens junctions. Actomyosin networks in neighboring cells can form supracellular connections by attaching to E-cadherin-based adherens junctions. During apical constriction, lost attachments are repaired by actin turnover. Depolymerization of F-actin, promoted by cofilin, can remodel contracted networks, providing monomers (G-actin) that can serve as building blocks for new F-actin growth around the AJ to promote reattachment after a lost connection (pink region).

Lecuit, & Lenne, 2017). During apical constriction of the mesoderm, the Abelson nonreceptor tyrosine kinase (Abl), which regulates the actin modulator Enabled (Ena) (Gertler et al., 1995), is important for the proper organization and localization of apical actin (Fox & Peifer, 2007). In addition, the sole Diaphanous-related formin in *Drosophila*, Diaphanous (Dia) plays an important role during apical constriction in stabilizing adherens junctions and promoting the attachment between medioapical contractile networks and junctions (Homem & Peifer, 2008; Mason et al., 2013). Tight regulation of F-actin polymerization and depolymerization has been shown to be critical for other contractile processes, such as cytokinesis (Engl, Arasi, Yap, Thiery, & Viasnoff, 2014; Vavylonis, Wu, Hao, O'Shaughnessy, & Pollard, 2008). Recently, actin turnover at the apical cortex of constricting mesoderm cells was shown to be required to disperse contracted F-actin networks to maintain a coherent apical F-actin meshwork that spans the apical domain, which is necessary for proper tissue invagination (Jodoin et al., 2015) (Fig. 2).

3. Tissue invagination: Integrating cells across the tissue

Actomyosin contractility at the apical cortex generates forces that lead to apical constriction, but these cellular forces are also integrated across the tissue during tissue invagination (Harris & Tepass, 2010; Heisenberg & Bellaïche, 2013; Lecuit et al., 2011). Force transmission requires actomyosin to be mechanically linked to E-cadherin-containing adherens junctions (AJs) (Hoffman & Yap, 2015). In epithelia, AJs mediate cell adhesion, which is necessary to maintain the integrity of the tissue (Baum & Georgiou, 2011; Leckband & de Rooij, 2014). Cell adhesion is mediated by homotypic interactions of the extracellular cadherin domains (Leckband & Prakasam, 2006; Leckband & Sivasankar, 2012). The cytoplasmic tail of cadherins bind to α-, β-, and p120 catenins and the cadherin-catenin complexes mediate force-sensitive binding to the cytoskeleton with other accessory proteins (Harris & Tepass, 2010; Hoffman & Yap, 2015; Leckband & de Rooij, 2014; Lecuit & Yap, 2015). Furthermore, regulation of actomyosin at AJs plays an important role in maintaining strong cell adhesion and mechanosensitivity during cell rearrangements and tissue morphogenesis (Engl et al., 2014; Kovacs et al., 2002; Leerberg et al., 2014; Priya et al., 2015; Ratheesh & Yap, 2012).

At the onset of *Drosophila* mesoderm invagination, AJs shift from an initially sub-apical position to the apical surface in the mesoderm (Dawes-Hoang et al., 2005; Kölsch, Seher, Fernandez-Ballester, Serrano, & Leptin, 2007). The apical shift of E-cadherin clusters requires Traf4, a member of the tumor necrosis factor receptor-associated factor (TRAF) family (Mathew, Rembold, & Leptin, 2011), and also depends on apical actomyosin accumulation (Weng & Wieschaus, 2016). During mesoderm invagination, apical actomyosin networks connect with neighboring cells by attaching to apical E-cadherin clusters, forming a supracellular actomyosin network that can transmit forces across the tissue (Martin, Gelbart, Fernandez-Gonzalez, Kaschube, & Wieschaus, 2010; Sánchez-Corrales & Röper, 2018; Yevick, Miller, Dunkel, & Martin, 2019) (Fig. 2). Actomyosin at the apical cortex can form connections with AJs through direct physical interactions with α-catenin and vinculin (Buckley et al., 2014; Desai et al., 2013; Homem & Peifer, 2008; le Duc et al., 2010; Sako, Nagafuchi, Tsukita, Takeichi, & Kusumi, 1998; Yamada, Pokutta, Drees, Weis, & Nelson, 2005; Yonemura, Wada, Watanabe, Nagafuchi, & Shibata, 2010). In addition, the Afadin homologue Canoe (Cno) has been shown to localize to AJs and to mediate linkage to the actin cytoskeleton, a function that is required for apical constriction and convergent extension movements during *Drosophila* gastrulation (Sawyer et al., 2011; Sawyer, Harris, Slep, Gaul, & Peifer, 2009).

The importance of physically linking actomyosin to AJs is exemplified in embryos that are depleted for β-catenin (Armadillo, Arm) where myosin that is no longer tethered to intact AJs contracts at the apical cortex with no associated reduction in apical cell area (Dawes-Hoang et al., 2005; Martin et al., 2010). More recently, it was demonstrated that the connection between medioapical actin and the AJ is dynamic and that actin turnover repairs fractures between actomyosin and AJs as tension builds, which stabilizes actomyosin attachments to AJs over long time scales (Jodoin et al., 2015) (Fig. 2). Interestingly, without an organized microtubule cytoskeleton, separations of the apical F-actin meshwork from AJs are not repaired efficiently and lead to large apical actomyosin network fractures, suggesting that microtubules promote actomyosin reattachment to AJs (Ko et al., 2019).

The integration of contractile forces at the tissue surface is concomitant with the formation of a supracellular actomyosin network that leads to a buildup of epithelial tension. In the *Drosophila* mesoderm, tension is highest along the anterior-posterior axis, an anisotropy that is reflected in how ventral cells constrict (Chanet et al., 2017; Martin et al., 2010). Mesoderm cells constrict primarily along the dorsal-ventral axis and remain elongated along the anterior-posterior axis. Thus, tissue mechanics can bias the directionality of cell shape changes and the resulting deformations and curvatures that are associated with these cell shape changes (Fig. 3). In addition to active forces at the apical surface, passive mechanical forces also contribute to mesoderm invagination and internalization. Critically, during mesoderm invagination, ventral cells conserve their volume as they apically constrict, which is

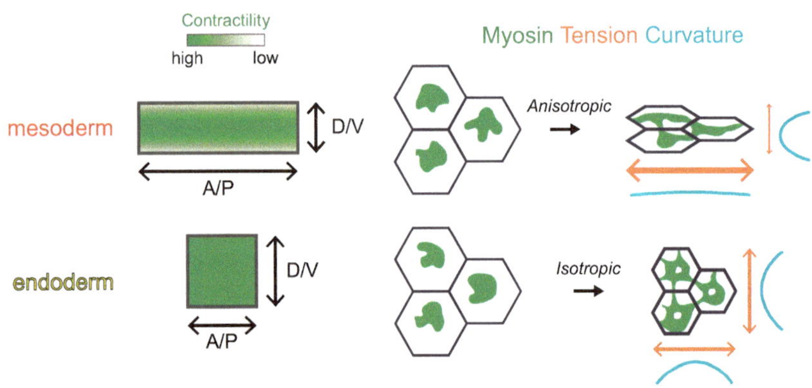

Fig. 3 Tissue geometry influences tissue mechanics, cell shape change, and cytoskeletal organization. The contractile domain of the mesoderm is elongated along the anterior-posterior axis (A/P) and graded along the dorsal-ventral axis (D/V). As a result, cells constrict more along the dorsal-ventral axis (D/V), leading to anisotropic apical constriction. The contractile domain of the endoderm is more isotropic, leading to isotropic apical constriction and myosin ring structures. The effect of tissue geometry on the magnitude of tension and degree of curvature are shown for both the A/P and D/V axis.

thought as being sufficient to generate invagination (Gelbart et al., 2012; Izquierdo, Quinkler, & De Renzis, 2018; Polyakov et al., 2014). Volume conservation results in a lengthening of lateral surfaces along the apical-basal axis (Costa et al., 1994; Gelbart et al., 2012). It has been suggested that apical-basal shortening and basal expansion after cell lengthening at later stages of ventral furrow formation comprises the final force that drives full mesoderm internalization (Conte et al., 2012; Sweeton et al., 1991). The mesoderm-specific disassembly of basal myosin coincides with basal expansion and invagination and is necessary for invagination (Krueger, Tardivo, Nguyen, & De Renzis, 2018; Polyakov et al., 2014). Finally, either pushing forces or mechanical compliance of the lateral and dorsal ectodermal tissues may play an important role during mesoderm internalization as increasing stiffness in these tissues prevents tissue invagination (Conte, Muñoz, Baum, & Miodownik, 2009; Perez-Mockus et al., 2017; Rauzi et al., 2015).

Cellular actomyosin contractility and force transduction through connections to AJs create a complex interplay of active and passive forces within the tissue that promotes tissue internalization. An additional level of regulation is conferred through mechanical feedback that fine tunes cellular response to force, such as by modulation of junctional strength. (Gomez, McLachlan, & Yap, 2011; Lecuit & Yap, 2015; Leerberg et al., 2014; Ratheesh & Yap, 2012). During mesoderm invagination, actomyosin contractility is responsible for protecting AJs against Snail-mediated disassembly, suggesting that contractility reinforces the coupling between cells (Weng & Wieschaus, 2016). Tissue mechanics can also influence cytoskeletal organization with actomyosin aligning with the axis that most greatly resists contraction (Chanet et al., 2017; Yevick et al., 2019) (Fig. 3). Finally, mechanical deformation of the embryo by a micromanipulation needle has been shown to be sufficient to induce apical myosin accumulation and to promote mesoderm invagination, suggesting the possibility that mechanical tissue deformations that occur as cells initiate constriction can feed back and further activate myosin, leading to more collective cell constrictions (Pouille, Ahmadi, Brunet, & Farge, 2009). Overall, studies of *Drosophila* gastrulation have illustrated how cell force generation can be modulated by mechanical and geometric properties of the tissue.

4. RhoA signaling: Activating contractility and coordinating cell behavior

The activation and regulation of actomyosin contractility plays an important role in many cell and tissue behaviors (Agarwal & Zaidel-Bar, 2018).

Across many systems, including *Drosophila* gastrulation, precise and dynamic control of the small GTPase RhoA plays an integral part in modulating actomyosin contractility (Jaffe & Hall, 2005) (Fig. 4A). RhoA acts as an upstream activator of contractility by activating ROCK and Dia (Jaffe & Hall, 2005). Thus, the spatiotemporal regulation of RhoA dictates the organization and behavior of actomyosin-based cytoskeletal structures in the cell (Bement et al., 2015; Bement, Miller, & von Dassow, 2006; Etienne-Manneville & Hall, 2002; Georgiou & Baum, 2010; Izquierdo et al., 2018; Maddox & Burridge, 2003; Matthews et al., 2012; Su, Takaki, &

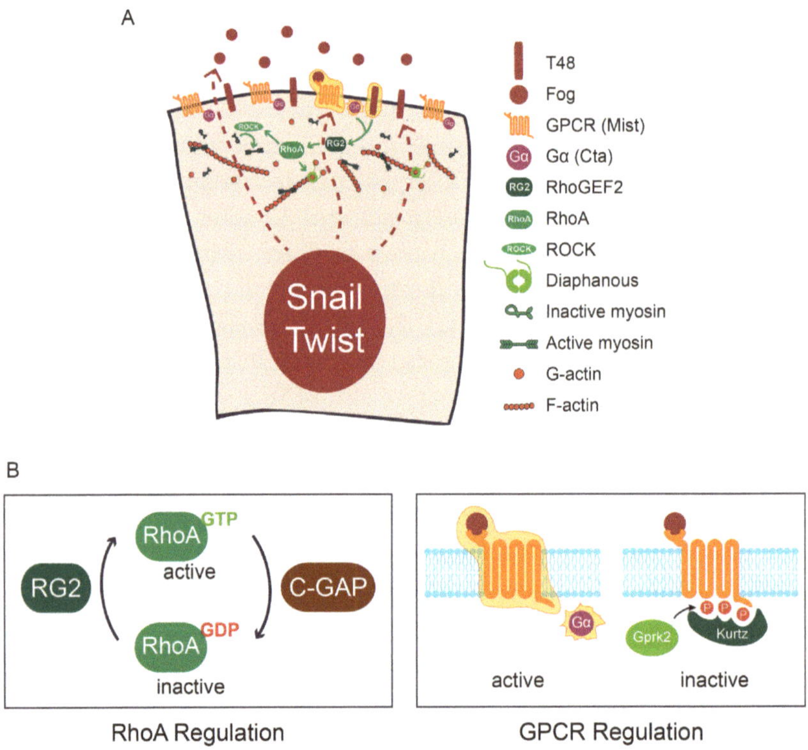

Fig. 4 RhoA activation at the apical surface is downstream of embryonic transcription factors, Snail and Twist. (A) Transcriptional targets of Snail and Twist include Fog, an extracellular ligand for a GPCR, and T48, a transmembrane protein, that act together to enrich the cell apex with active RhoGEF2. RhoA activation occurs apically, leading to apical actomyosin contractility. (B) Different modes of dynamic signal regulation. RhoA activity is regulated by GEFs and GAPs. RhoA is activated by RhoGEF2 (RG2) and inactivated by Cumberland-GAP (C-GAP). The GPCR (Mist and Smog) is activated by binding to Fog, transducing the signal through the $G_{\alpha12/13}$ protein Cta. Signaling can be inhibited through phosphorylation by Gprk2 and the β-arrestin Kurtz.

Petronczki, 2011; Wagner & Glotzer, 2016). Here, we focus on work done in the early *Drosophila* embryo during mesoderm invagination that has delineated the importance of tight spatiotemporal regulation of RhoA activity in promoting cell shape change and tissue invagination.

The activity of RhoA is regulated by guanine nucleotide exchange factors (GEFs) which switch RhoA to an active state and GTPase activating proteins (GAPs) that switch RhoA to an inactive state (Hodge & Ridley, 2016) (Fig. 4B). During gastrulation, cell shape change and subsequent mesoderm and endoderm invagination require the activity of a RhoA GEF called RhoGEF2 in order to form active actomyosin networks at the apical surface (Barrett, Leptin, & Settleman, 1997; Dawes-Hoang et al., 2005; Fox & Peifer, 2007; Häcker & Perrimon, 1998; Izquierdo et al., 2018; Kölsch et al., 2007). RhoGEF2 is maternally supplied, but in constricting cells, RhoGEF2 is enriched at the apical side of the cells, in a mechanism that could rely, in part, on the association of EB1 in a polarized microtubule network (Kölsch et al., 2007; Rogers, Wiedemann, Häcker, Turck, & Vale, 2004). However, RhoGEF2 does not remain localized across the apical surface, but becomes enriched at the medial region of the apical cortex, as well as junctions (Mason et al., 2013; Mason, Xie, Vasquez, Tworoger, & Martin, 2016), which reflects the localization of RhoA itself (Coravos & Martin, 2016; Mason et al., 2013). In the case of *Drosophila* mesoderm invagination, negative regulation by a RhoA GAP, Cumberland-GAP (C-GAP or RhoGAP71E), is critical for polarizing RhoA activity and also coordinating cell contractility at the tissue level (Mason et al., 2016). In *C-GAP* mutants, the shape of the ventral furrow as it invaginates is abnormal, suggesting that negative RhoA regulation has a tissue-scale impact by influencing the curvature of the internalizing tissue (Mason et al., 2016). Indeed, the importance of cycling through active and inactive states of RhoA in orchestrating and organizing contractile arrays of actomyosin as well as maintaining an excitable cell cortex has been demonstrated in a wide variety of systems (Bement et al., 2015; Maddox & Burridge, 2003; Mason et al., 2016; Michaux, Robin, McFadden, & Munro, 2018; Miller & Bement, 2008; Munjal et al., 2015).

Finely-tuned regulation of RhoA, particularly through cycles of activation and inhibition, might also play an important role in defining the type of contractile cell behavior. Ratcheted pulses, which require myosin persistence at the medioapical cortex, have been shown to depend on the embryonic transcription factor Twist (Martin et al., 2009; Mason et al., 2013; Xie & Martin, 2015). More recently, it has been demonstrated that the level of myosin persistence depends on the relative ratio of RhoA GEF and GAP levels in the cell or the net RhoA activity (Izquierdo et al., 2018; Mason et al., 2016). Thus, RhoA regulation is critical for coordinating apical constriction across the tissue during mesoderm invagination. In the next section, we review the components that are upstream of RhoA signaling.

5. Gene regulation: Cell signaling centers and tissue geometry

In order to coordinate cell behaviors for effective tissue internalization, an interplay of cell fate and cell signaling regulates the spatiotemporal activation of RhoA, leading to controlled apical constriction and proper tissue invagination. The mesoderm and endoderm both express cell fate-determining transcription factors. During mesoderm invagination, the contractile domain in the ventral region of the embryo is set and regulated by transcriptional targets of Twist and Snail (Seher, Narasimha, Vogelsang, & Leptin, 2007). Twist and Snail regulate genes that modulate cell signaling to control cell shape change (Fig. 4A). In mesoderm and endoderm invagination, cell shape changes are associated with the expression of *folded gastrulation* (*fog*), a ligand for G-protein coupled receptors (GPCR) Mist and Smog, which lie upstream of RhoGEF2 activation (Costa et al., 1994; Dawes-Hoang et al., 2005; Fuse, Yu, & Hirose, 2013; Jha, van Zanten, Philippe, Mayor, & Lecuit, 2018; Kerridge et al., 2016; Manning, Peters, Peifer, & Rogers, 2013; Sweeton et al., 1991; Urbansky, González Avalos, Wosch, & Lemke, 2016; Xie & Martin, 2015; Zusman & Wieschaus, 1985). In addition, Twist expression in the mesoderm is associated with the expression of T48, a transmembrane protein that acts in parallel with Fog signaling to activate apical actomyosin contractility (Kölsch et al., 2007; Urbansky et al., 2016) (Fig. 4A).

Downstream of GPCR activation is a heterotrimeric G-protein, which transduces the signal. A $G_{\alpha 12/13}$ protein named Concertina (Cta) functions during mesoderm invagination to promote contractility. In *cta* mutants, apical constriction across the mesoderm is delayed and uncoordinated because apical cell size heterogeneity is increased with disrupted apical cortex organization and contractile pulse initiation (Costa et al., 1994; Parks & Wieschaus, 1991; Xie, Mason, & Martin, 2016).

Apical constriction is associated with a signaling center in the center of the apical domain. In both the *Drosophila* mesoderm and endoderm cells, RhoA activation and downstream signaling such as ROCK recruitment are polarized within the apex (Chanet et al., 2017; Mason et al., 2013). As a result, signaling processes are radially polarized in apically constricting mesoderm and endoderm cells. This leads to myosin activation being predominantly localized to the medioapical surface. Signaling processes upstream of RhoA also exhibit spatial organization, where mesoderm cells have large apical plasma membrane invaginations that partition the active GPCR, Smog (Jha et al., 2018). The size of these plasma membrane invaginations and the level of GPCR signaling are attenuated by GPCR endocytosis mediated by a GPCR kinase Gprk2 (Fuse et al., 2013) and the β-arrestin Kurtz (Jha et al., 2018; Roman, He, & Davis, 2000; Tipping et al., 2010) (Fig. 4B).

In addition to spatial organization of signaling at the subcellular level, multicellular patterns of gene expression and contractility influence cell and tissue shape changes. In the mesoderm, for example, the pattern of Snail/Twist expression establishes a domain of contractile cells that is longer along the anterior-posterior axis than the dorsal-ventral axis. However, the asymmetry is even more extreme because there is graded Twist expression and contractility along the dorsal-ventral axis (Heer et al., 2017; Lim, Levine, & Yamazaki, 2017) (Fig. 3), meaning that cells will experience significantly more contractile force along the anterior-posterior axis than the dorsal-ventral axis. This geometry can be manipulated by affecting upstream Toll signaling which establishes dorsal-ventral polarity, such as through the depletion of the *Drosophila* serine protease inhibitor, Serpin-27A (Spn27A) (Ligoxygakis, Roth, & Reichhart, 2003). When Spn27A is depleted, the Snail domain is expanded along the dorsal-ventral axis and the Twist gradient is flattened, resulting in a contractile domain that is more square-like and less graded (Chanet et al., 2017; Heer et al., 2017; Ligoxygakis et al., 2003). Normally mesoderm cells constrict anisotropically along the dorsal-ventral axis, such that they are longer in the anterior-posterior direction (Fig. 3). Expanding the width of the contractile domain along the dorsal-ventral axis results in isotropic cell apex constrictions and a loss of anisotropic tension (Chanet et al., 2017). The organization of apical actomyosin is also affected as actomyosin rings form at the cortex, suggesting that tension feeds back on actomyosin to align it with the long axis of the embryo, making the tissue straight along the axis of the furrow. During invagination of the endoderm, the tissue geometry of the contractile domain is more symmetrical and cells constrict isotropically, leading to isotropic epithelial tension (Chanet et al., 2017) (Fig. 3). This is consistent with the posterior midgut domain being more symmetrical in shape than the primordial mesoderm domain. Further support for the conclusion that tissue shape influences cell shape comes from the optogenetic activation of contractility in different patterns (Guglielmi, Barry, Huber, & De Renzis, 2015).

Proper invagination of the mesoderm also requires tight coordination between apical constriction and other cell processes, such as cell division. Regions of the early embryo have a stereotyped pattern of mitotic entry, termed mitotic domains, which is regulated by the expression of the cyclin-dependent kinase-activating phosphatase Cdc25 (String, Stg) (Edgar & O'Farrell, 1989; Foe, 1989). Typically, mitotic entry tightly correlates with *stg* expression; however, in the mesoderm, despite being the first cells to express *stg*, mitosis is delayed until after the tissue has internalized into the embryo as mitotic domain 10. This mitotic delay is primarily regulated by the zygotically expressed Tribbles, a putative serine/threonine kinase that is active specifically in the mesoderm to promote the degradation of Stg (Grosshans & Wieschaus, 2000; Mata, Curado, Ephrussi, & Rørth, 2000; Seher & Leptin, 2000) and Frühstart, a cytoplasmic protein that binds to the hydrophobic patch of cyclins to block interactions with Stg (Gawliński et al., 2007; Grosshans, Müller, & Wieschaus, 2003; Grosshans & Wieschaus, 2000). Inhibition of Stg activity has also been shown to be regulated post-transcriptionally by an RNA-binding repressor, Held Out Wing (HOW) (Nabel-Rosen, Toledano-Katchalski, Volohonsky, & Volk, 2005). When Stg activity is not properly repressed during mesoderm invagination, the formation of the ventral furrow is inhibited and the tissue is not internalized normally. However, it has been reported that mesoderm cells that remain at the surface are eventually internalized in a slower manner after divisions have completed (Seher & Leptin, 2000). Nevertheless, it is clear that fast, coordinated internalization of the mesoderm requires coordination with the cell cycle by the expression of zygotic repressors of mitotic entry.

6. Late gastrulation: Subsequent mesoderm EMT and spreading

The mesoderm is internalized as a coherent epithelial sheet of cells that maintain intercellular E-cadherin adhesions and apical-basal polarity (Clark, Muha, Klingseisen, Leptin, & Müller, 2011; McMahon, Supatto, Fraser, & Stathopoulos, 2008). Once internalized, the cells undergo cell division and after ~50–60 min eventually undergo an epithelial-to-mesenchymal transition (EMT) and spreading (Clark et al., 2011; McMahon et al., 2008)

(Fig. 5A). The EMT of the *Drosophila* mesoderm is characterized by a decrease in adherens junction number and loss of apical E-cadherin polarity as mesoderm cells spread onto the basal surface of the ectoderm (Clark et al., 2011; Sun & Stathopoulos, 2018). As they spread, mesoderm cells exhibit plasma membrane protrusions (Clark et al., 2011; Murray & Saint, 2007; Smallhorn, Murray, & Saint, 2004), however, it is not clear whether these membrane protrusions have molecules that are characteristic of lamellipodia or filopodia, such as Arp2/3 complex/PIP_3 or Ena/Dia, respectively (Ridley, 2011; Small, Stradal, Vignal, & Rottner, 2002).

The mechanism of mesoderm cell migration is still unclear. However, spreading, in part, depends on fibroblast growth factor (FGF) signaling (Michaelson, Gisselbrecht, Zhou, Baek, & Buff, 1998; Vincent, Wilson,

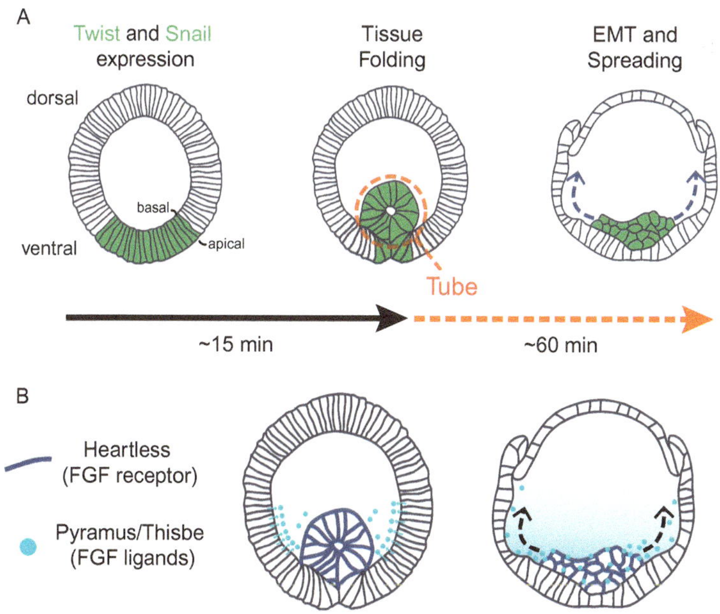

Fig. 5 After tissue folding, the mesoderm undergoes EMT and cell spreading. (A) Diagram showing cross-section views of embryos from early to late stages of gastrulation. At the conclusion of ventral furrow formation (~15 min after the onset of apical constriction), the mesoderm is maintained as a coherent epithelial sheet, forming a tube-like structure. After EMT, the tube collapses and mesoderm cells spread dorsally along the ectoderm. (B) FGF signaling promotes proper mesoderm cell migration. The ectoderm secretes the FGF ligands Pyramus/Thisbe while the mesoderm cells express the FGF receptor *heartless*. *Panel (A): Modified from Leptin, M., (2005). Gastrulation movements: The logic and the nuts and bolts.* Developmental Cell, *8, 305–320.*

Coelho, Affolter, & Leptin, 1998; Wilson, Vogelsang, & Leptin, 2005). Mesoderm cells express the FGF receptor *heartless*, which responds to FGF ligands secreted by the ectoderm (Gryzik & Müller, 2004; Stathopoulos, Tam, Ronshaugen, Frasch, & Levine, 2004) (Fig. 5B). FGF signaling and mitotic entry appear to function in parallel to accelerate EMT and spreading (Clark et al., 2011; Schaeffer, Somarelli, Hanna, Palmer, & Garcia-Blanco, 2014). However, while inhibiting FGF signaling and mitotic entry delays EMT and spreading, spreading still occurs (Clark et al., 2011; Schaeffer et al., 2014). These results suggest that FGF signaling is necessary to ensure that mesoderm cells migrate to their proper locations, but that FGF signaling is not necessary for EMT. Another factor involved in mesoderm spreading is the RhoA GEF, Pebble (Ect2 in mammals). Pebble/Ect2 is most well-known for its role in promoting cytokinetic contractile ring assembly (Prokopenko et al., 1999; Su et al., 2011; Yüce, Piekny, & Glotzer, 2005), however, *pebble* mutants exhibit defects in mesoderm spreading and lack the protrusions normally displayed by migrating mesoderm cells (Smallhorn et al., 2004).

EMT is often associated with the loss of intercellular adhesion mediated by transcription factors of the Twist and Snail family (Nieto, 2011). This change can involve a switch in cadherin type expression, with E-cadherin being repressed and N-cadherin being induced (Hatta & Takeichi, 1986; Nandadasa, Tao, Menon, Heasman, & Wylie, 2009; Oda, Tsukita, & Takeichi, 1998). Snail transcriptionally represses *E*-cadherin gene expression in many contexts (Cano et al., 2000; Oda et al., 1998; Wu & McClay, 2007). Despite the fact that a switch from E- to N-cadherin expression is seen in *Drosophila* mesoderm cells (Oda et al., 1998), it now seems clear that changes in cadherin transcription have no functional consequence on *Drosophila* mesoderm spreading. E-cadherin expression with a ubiquitous promoter that cannot be repressed by Snail rescues the function of E-cadherin null mutants, suggesting that mesoderm formation is not perturbed (Oda & Tsukita, 2001). More recently, research has shown that differential expression of E- and N-cadherin in ectoderm and mesoderm, respectively, is not important for germ layer formation (Schäfer, Narasimha, Vogelsang, & Leptin, 2014). Consistent with the cadherin switch not being important for EMT and migration, E-cadherin is not completely lost from the plasma membrane of spreading mesoderm cells. In fact, E-cadherin is enriched at sites of mesoderm-ectoderm contact, suggesting that E-cadherin may provide traction for migrating mesoderm cells (McMahon et al., 2008).

Given that Twist and Snail are expressed more than 1 hour before EMT occurs, an interesting question is what dictates the timing and duration of EMT. In addition, apical constriction is often associated with epithelial cell delamination/extrusion (Simões, Oh, Wang, Fernandez-Gonzalez, & Tepass, 2017). Thus, how mesoderm cells remain a coherent epithelial tissue during the invagination process and yet undergo EMT ~50 min later is unclear. One insight regarding the timing of EMT is the finding that the Abelson kinase (Abl) suppresses premature cell delamination. In fact, loss of Abl results in reduced apical-basal polarity and the delamination of individual cells during the inward folding of the epithelium, 20 min prior to when this normally happens (Jodoin & Martin, 2016). These premature EMT-like behaviors depend on the activity of Ena, which suggests that precise regulation of the actin cytoskeleton is key to properly time EMT (Jodoin & Martin, 2016). The mesenchymal state of the mesoderm is transient; the mesoderm undergoes a mesenchymal-to-epithelial transition (MET) after spreading, with the apical domain of mesoderm cells facing the yolk and the basal domain contacting the ectoderm (Sun & Stathopoulos, 2018). FGF appears to regulate the establishment of epithelial polarity during the MET (Sun & Stathopoulos, 2018). How FGF promotes EMT at one time and MET at a later time is an interesting and unanswered question in the field.

7. Concluding remarks

Our review highlights the power of *Drosophila melanogaster* gastrulation as a model system to uncover the molecular mechanisms that change cell shape and coordinate cell behavior across a precisely patterned tissue. Genetic and molecular approaches have demonstrated how processes that establish the polarity of the body axes in the early embryo promote cell-type specific expression of transcription factors that activate signal transduction, which produces coordinated force generation. This integration transduces dorsal-ventral and anterior-posterior patterning processes into contractile domains with specific geometries, which sets the stage for mesoderm and endoderm invagination. Importantly, activation of actomyosin contractility is polarized within a cell to promote efficient constriction and also remains dynamic to maintain a coherent apical actomyosin network during tissue invagination. Indeed, both positive and negative regulation is important for the cytoskeleton and components of the upstream signal transduction pathway, demonstrating the importance of dynamics/turnover. Active forces generated by actomyosin contraction are integrated across the tissue through adherens junctions, which in conjunction with passive forces in the tissue, promote proper tissue shape. Finally, in the mesoderm, cell and tissue invagination is tightly coordinated with other cell behaviors to ensure that cell divisions and EMT do not occur until after the mesoderm has internalized. Much of the current progress in our understanding of the mechanics of tissue internalization during *Drosophila* gastrulation can be attributed to advances in microscopy and the development of molecular and biophysical tools, such as optogenetics, which has allowed researchers to more specifically probe gene function and cell/tissue mechanics. Thus, as more powerful tools are developed, we look forward to seeing further progress in the years to come.

References

Agarwal, P., & Zaidel-Bar, R. (2018). Principles of actomyosin regulation in vivo. *Trends in Cell Biology*, *29*, 1–14.

Amano, M., Ito, M., Kimura, K., Fukata, Y., Chihara, K., Nakano, T., et al. (1996). Phosphorylation and activation of myosin by Rho-associated Kinase (Rho-kinase). *The Journal of Biological Chemistry*, *271*, 20246–20249.

Barrett, K., Leptin, M., & Settleman, J. (1997). The Rho GTPase and a putative RhoGEF mediate a signaling pathway for the cell shape changes in Drosophila gastrulation. *Cell*, *91*, 905–915.

Baum, B., & Georgiou, M. (2011). Dynamics of adherens junctions in epithelial establishment, maintenance, and remodeling. *The Journal of Cell Biology*, *192*, 907–917.

Belvin, M. P., & Anderson, K. V. (1996). A conserved signaling pathway: The Drosophila Toll-dorsal pathway. *Annual Review of Cell and Developmental Biology*, *12*, 393–416.

Bement, W. M., Leda, M., Moe, A. M., Kita, A. M., Larson, M. E., Golding, A. E., et al. (2015). Activator–inhibitor coupling between Rho signalling and actin assembly makes the cell cortex an excitable medium. *Nature Cell Biology*, *17*, 1471–1483.

Bement, W. M., Miller, A. L., & von Dassow, G. (2006). Rho GTPase activity zones and transient contractile arrays. *BioEssays*, *28*, 983–993.

Booth, A. J. R., Blanchard, G. B., Adams, R. J., & Röper, K. (2014). A dynamic microtubule cytoskeleton directs medial actomyosin function during tube formation. *Developmental Cell*, *29*, 562–576.

Buckley, C. D., Tan, J., Anderson, K. L., Hanein, D., Volkmann, N., Weis, W. I., et al. (2014). The minimal cadherin-catenin complex binds to actin filaments under force. *Science*, *346*, 1254211.

Campellone, K. G., & Welch, M. D. (2010). A nucleator arms race: Cellular control of actin assembly. *Nature Reviews Molecular Cell Biology*, *11*, 237–251.

Cano, A., Pérez-Moreno, M. A., Rodrigo, I., Locascio, A., Blanco, M. J., del Barrio, M. G., et al. (2000). The transcription factor Snail controls epithelial–mesenchymal transitions by repressing E-cadherin expression. *Nature Cell Biology*, *2*, 76–83.

Chanet, S., Miller, C. J., Vaishnav, E. D., Ermentrout, B., Davidson, L. A., & Martin, A. C. (2017). Actomyosin meshwork mechanosensing enables tissue shape to orient cell force. *Nature Communications*, *8*, 1–13.

Chugh, P., Clark, A. G., Smith, M. B., Cassani, D. A. D., Dierkes, K., Ragab, A., et al. (2017). Actin cortex architecture regulates cell surface tension. *Nature Cell Biology*, *19*, 689–697.

Chung, S., Kim, S., & Andrew, D. J. (2017). Uncoupling apical constriction from tissue invagination. *eLife*, *6*, e22235. https://doi.org/10.7554/eLife.22235.
Clark, I. B. N., Muha, V., Klingseisen, A., Leptin, M., & Müller, H.-A. J. (2011). Fibroblast growth factor signalling controls successive cell behaviours during mesoderm layer formation in Drosophila. *Development*, *138*, 2705–2715.
Clark, A. G., Wartlick, O., Salbreux, G., & Paluch, E. K. (2014). Stresses at the cell surface during animal cell review morphogenesis. *Current Biology*, *24*, R484–R494.
Clément, R., Dehapiot, B., Collinet, C., Lecuit, T., & Lenne, P.-F. (2017). Viscoelastic dissipation stabilizes cell shape changes during tissue morphogenesis. *Current Biology*, *27*, 1–16.
Conte, V., Muñoz, J. J., Baum, B., & Miodownik, M. (2009). Robust mechanisms of ventral furrow invagination require the combination of cellular shape changes. *Physical Biology*, *6*, 016010–016013.
Conte, V., Ulrich, F., Baum, B., Muñoz, J., Veldhuis, J., Brodland, W. G., et al. (2012). A biomechanical analysis of ventral furrow formation in the Drosophila melanogaster embryo. *PLoS One*, *7*, e34473.
Coravos, J. S., & Martin, A. C. (2016). Apical sarcomere-like actomyosin contracts nonmuscle Drosophila epithelial cells. *Developmental Cell*, *39*, 1–14.
Costa, M., Wilson, E. T., & Wieschaus, E. F. (1994). A putative cell signal encoded by the folded gastrulation gene coordinates cell shape changes during Drosophila gastrulation. *Cell*, *76*, 1075–1089.
Dawes-Hoang, R. E., Parmar, K. M., Christiansen, A. E., Phelps, C. B., Brand, A. H., & Wieschaus, E. F. (2005). Folded gastrulation, cell shape change and the control of myosin localization. *Development*, *132*, 4165–4178.
Desai, R., Sarpal, R., Ishiyama, N., Pellikka, M., Ikura, M., & Tepass, U. (2013). Monomeric alpha-catenin links cadherin to the actin cytoskeleton. *Nature Cell Biology*, *15*, 261–273.
Edgar, B. A., & O'Farrell, P. H. (1989). Genetic control of cell division patterns in the Drosophila embryo. *Cell*, *57*, 177–187.
Engl, W., Arasi, B., Yap, L. L., Thiery, J. P., & Viasnoff, V. (2014). Actin dynamics modulate mechanosensitive immobilization of E-cadherin at adherens junctions. *Nature Cell Biology*, *16*, 587–594.
Ennomani, H., Letort, G., Guérin, C., Martiel, J.-L., Cao, W., Nédélec, F., et al. (2016). Architecture and connectivity govern actin network contractility. *Current Biology*, *26*, 616–626.
Etienne-Manneville, S., & Hall, A. (2002). Rho GTPases in cell biology. *Nature*, *420*, 629–635.
Foe, V. E. (1989). Mitotic domains reveal early commitment of cells in Drosophila embryos. *Development*, *107*, 1–22.
Fox, D. T., & Peifer, M. (2007). Abelson kinase (Abl) and RhoGEF2 regulate actin organization during cell constriction in Drosophila. *Development*, *134*, 567–578.
Franke, J. D., Montague, R. A., & Kiehart, D. P. (2005). Nonmuscle myosin II generates forces that transmit tension and drive contraction in multiple tissues during dorsal closure. *Current Biology*, *15*, 2208–2221.
Fuse, N., Yu, F., & Hirose, S. (2013). Gprk2 adjusts Fog signaling to organize cell movements in Drosophila gastrulation. *Development*, *140*, 4246–4255.
Gawliński, P., Nikolay, R., Goursot, C., Lawo, S., Chaurasia, B., Herz, H.-M., et al. (2007). The Drosophila mitotic inhibitor Frühstart specifically binds to the hydrophobic patch of cyclins. *EMBO Reports*, *8*, 490–496.
Gelbart, M. A., He, B., Martin, A. C., Thiberge, S. Y., Wieschaus, E. F., & Kaschube, M. (2012). Volume conservation principle involved in cell lengthening and nucleus movement during tissue morphogenesis. *Proceedings of the National Academy of Sciences of the United States of America*, *109*, 19298–19303.

Georgiou, M., & Baum, B. (2010). Polarity proteins and Rho GTPases cooperate to spatially organise epithelialactin-based protrusions. *Journal of Cell Science, 123*, 1383.

Gertler, F. B., Comer, A. R., Juang, J.-L., Ahern, S. M., Clark, M. J., Liebl, E. C., et al. (1995). Enabled, a dosage-sensitive suppressor of mutations in the Drosophila Abl tyrosine kinase, encodes an Abl substrate with SH3 domain-binding properties. *Genes & Development, 9*, 521–533.

Gomez, G. A., McLachlan, R. W., & Yap, A. S. (2011). Productive tension: Force-sensing and homeostasis of cell-cell junctions. *Trends in Cell Biology, 21*, 499–505.

Goode, B. L., & Eck, M. J. (2007). Mechanism and function of formins in the control of actin assembly. *Annual Review of Biochemistry, 76*, 593–627.

Govind, S., & Steward, R. (1991). Dorsoventral pattern formation in Drosophila: Signal transduction and nuclear targeting. *Trends in Genetics, 7*, 119–125.

Grikscheit, K., & Grosse, R. (2016). Formins at the Junction. *Trends in Biochemical Sciences, 41*, 148–159.

Grosshans, J., Müller, H. A. J., & Wieschaus, E. F. (2003). Control of cleavage cycles in Drosophila embryos by Frühstart. *Developmental Cell, 5*, 285–294.

Grosshans, J., & Wieschaus, E. F. (2000). A genetic link between morphogenesis and cell division during formation of the ventral furrow in Drosophila. *Cell, 101*, 523–531.

Gryzik, T., & Müller, H. A. J. (2004). FGF8-like1 and FGF8-like2 encode putative ligands of the FGF receptor Htl and are required for mesoderm migration in the Drosophila gastrula. *Current Biology, 14*, 659–667.

Guglielmi, G., Barry, J. D., Huber, W., & De Renzis, S. (2015). An optogenetic method to modulate cell contractility during tissue morphogenesis. *Developmental Cell, 35*, 646–660.

Häcker, U., & Perrimon, N. (1998). DRhoGEF2 encodes a member of the Dbl family of oncogenes and controls cell shape changes during gastrulation in Drosophila. *Genes & Development, 12*, 274–284.

Harris, T. J. C., & Tepass, U. (2010). Adherens junctions: From molecules to morphogenesis. *Nature Reviews Molecular Cell Biology, 11*, 1–13.

Hatta, K., & Takeichi, M. (1986). Expression of N-cadherin adhesion molecules associated with early morphogenetic events in chick development. *Nature, 320*, 447–449.

Heer, N. C., Miller, P. W., Chanet, S., Stoop, N., Dunkel, J., & Martin, A. C. (2017). Actomyosin-based tissue folding requires a multicellular myosin gradient. *Development, 144*, 1876–1886.

Heisenberg, C.-P., & Bellaïche, Y. (2013). Forces in tissue morphogenesis and patterning. *Cell, 153*, 948–962.

Hodge, R. G., & Ridley, A. J. (2016). Regulating Rho GTPases and their regulators. *Nature Publishing Group, 17*, 496–510.

Hoffman, B. D., & Yap, A. S. (2015). Towards a dynamic understanding of cadherin-based mechanobiology. *Trends in Cell Biology, 25*, 803–814.

Homem, C. C. F., & Peifer, M. (2008). Diaphanous regulates myosin and adherens junctions to control cell contractility and protrusive behavior during morphogenesis. *Development, 135*, 1005–1018.

Izquierdo, E., Quinkler, T., & De Renzis, S. (2018). Guided morphogenesis through optogenetic activation of Rho signalling during early Drosophila embryogenesis. *Nature Communications, 9*, 1–13.

Jaffe, A. B., & Hall, A. (2005). Rho GTPases: Biochemistry and biology. *Annual Review of Cell and Developmental Biology, 21*, 247–269.

Jha, A., van Zanten, T. S., Philippe, J.-M., Mayor, S., & Lecuit, T. (2018). Quantitative control of GPCR organization and signaling by endocytosis in epithelial morphogenesis. *Current Biology, 28*, 1–22.

Jodoin, J. N., Coravos, J. S., Chanet, S., Vasquez, C. G., Tworoger, M., Kingston, E. R., et al. (2015). Stable force balance between epithelial cells arises from F-actin turnover. *Developmental Cell*, *35*, 685–697.

Jodoin, J. N., & Martin, A. C. (2016). Abl suppresses cell extrusion and intercalation during epithelium folding. *Molecular Biology of the Cell*, *27*, 2822–2832.

Jordan, P., & Karess, R. (1997). Myosin light chain-activating phosphorylation sites are required for oogenesis in Drosophila. *The Journal of Cell Biology*, *139*, 1805–1819.

Kerridge, S., Munjal, A., Philippe, J.-M., Jha, A., de las Bayonas, A. G., Saurin, A. J., et al. (2016). Modular activation of Rho1 by GPCR signalling imparts polarized myosin II activation during morphogenesis. *Nature Cell Biology*, *18*, 1–28.

Ko, C. S., Tserunyan, V., & Martin, A. C. (2019). Microtubules promote intercellular contractile force transmission during tissue folding. *The Journal of Cell Biology*, *218*(8), 2726–2742. jcb.201902011-201902017.

Kölsch, V., Seher, T. C., Fernandez-Ballester, G. J., Serrano, L., & Leptin, M. (2007). Control of Drosophila gastrulation by apical localization of adherens junctions and RhoGEF2. *Science*, *315*, 384–386.

Kovacs, E. M., Goodwin, M., Ali, R. G., Paterson, A. D., & Yap, A. S. (2002). Cadherin-directed actin assembly: E-cadherin physically associates with the Arp2/3 complex to direct actin assembly in nascent adhesive contacts. *Current Biology*, *12*, 379–382.

Krueger, D., Tardivo, P., Nguyen, C., & De Renzis, S. (2018). Downregulation of basal myosin-II is required for cell shape changes and tissue invagination. *The EMBO Journal*, *37*. e100170-100116.

le Duc, Q., Shi, Q., Blonk, I., Sonnenberg, A., Wang, N., Leckband, D., et al. (2010). Vinculin potentiates E-cadherin mechanosensing and is recruited to actin-anchored sites within adherens junctions in a myosin II-dependent manner. *The Journal of Cell Biology*, *189*, 1107–1115.

Leckband, D. E., & de Rooij, J. (2014). Cadherin adhesion and mechanotransduction. *Annual Review of Cell and Developmental Biology*, *30*, 291–315.

Leckband, D., & Prakasam, A. (2006). Mechanism and dynamics of cadherin adhesion. *Annual Review of Biomedical Engineering*, *8*, 259–287.

Leckband, D., & Sivasankar, S. (2012). Cadherin recognition and adhesion. *Current Opinion in Cell Biology*, *24*, 620–627.

Lecuit, T., Lenne, P.-F., & Munro, E. M. (2011). Force generation, transmission, and integration during cell and tissue morphogenesis. *Annual Review of Cell and Developmental Biology*, *27*, 157–184.

Lecuit, T., & Yap, A. S. (2015). E-cadherin junctions as active mechanical integrators in tissue dynamics. *Nature Publishing Group*, *17*, 533–539.

Leerberg, J. M., Gomez, G. A., Verma, S., Moussa, E. J., Wu, S. K., Priya, R., et al. (2014). Tension-sensitive actin assembly supports contractility at the epithelial zonula adherens. *Current Biology*, *24*, 1689–1699.

Leptin, M., & Grunewald, B. (1990). Cell shape changes during gastrulation in Drosophila. *Development*, *110*, 73–84.

Levayer, R., & Lecuit, T. (2012). Biomechanical regulation of contractility: Spatial control and dynamics. *Trends in Cell Biology*, *22*, 61–81.

Ligoxygakis, P., Roth, S., & Reichhart, J.-M. (2003). A serpin regulates dorsal-ventral axis formation in the Drosophila embryo. *Current Biology*, *13*, 2097–2102.

Lim, B., Levine, M., & Yamazaki, Y. (2017). Transcriptional pre-patterning of Drosophila gastrulation. *Current Biology*, *27*, 286–290.

Maddox, A. S., & Burridge, K. (2003). RhoA is required for cortical retraction and rigidity during mitotic cell rounding. *The Journal of Cell Biology*, *160*, 255–265.

Manning, A. J., Peters, K. A., Peifer, M., & Rogers, S. L. (2013). Regulation of epithelial morphogenesis by the G protein-coupled receptor mist and Its ligand fog. *Science Signaling*, *6*, ra98.

Martin, A. C., Gelbart, M., Fernandez-Gonzalez, R., Kaschube, M., & Wieschaus, E. F. (2010). Integration of contractile forces during tissue invagination. *The Journal of Cell Biology*, *188*, 735–749.

Martin, A. C., Kaschube, M., & Wieschaus, E. F. (2009). Pulsed contractions of an actin-myosin network drive apical constriction. *Nature*, *457*, 495–501.

Mason, F. M., Tworoger, M., & Martin, A. C. (2013). Apical domain polarization localizes actin–myosin activity to drive ratchet-like apical constriction. *Nature Publishing Group*, *15*, 926–936.

Mason, F. M., Xie, S., Vasquez, C. G., Tworoger, M., & Martin, A. C. (2016). RhoA GTPase inhibition organizes contraction during epithelial morphogenesis. *The Journal of Cell Biology*, *214*, 603–617.

Mata, J., Curado, S., Ephrussi, A., & Rørth, P. (2000). Tribbles coordinates mitosis and morphogenesis in Drosophila by regulating string/CDC25 proteolysis. *Cell*, *101*, 511–522.

Mathew, S. J., Rembold, M., & Leptin, M. (2011). Role for Traf4 in polarizing adherens junctions as a prerequisite for efficient cell shape changes. *Molecular and Cellular Biology*, *31*, 4978–4993.

Matsumura, F. (2005). Regulation of myosin II during cytokinesis in higher eukaryotes. *Trends in Cell Biology*, *15*, 371–377.

Matthews, H. K., Delabre, U., Rohn, J. L., Guck, J., Kunda, P., & Baum, B. (2012). Changes in Ect2 localization couple actomyosin-dependent cell shape changes to mitotic progression. *Developmental Cell*, *23*, 371–383.

McMahon, A., Supatto, W., Fraser, S. E., & Stathopoulos, A. (2008). Dynamic analyses of Drosophila gastrulation provide insights into collective cell migration. *Science*, *322*, 1546–1550.

Michaelson, A. M., Gisselbrecht, S., Zhou, Y., Baek, K. H., & Buff, E. M. (1998). Dual functions of the heartless fibroblast growth factor receptor in development of the Drosophila embryonic mesoderm. *Developmental Genetics*, *22*, 212–229.

Michaux, J. B., Robin, F. B., McFadden, W. M., & Munro, E. M. (2018). Excitable RhoA dynamics drive pulsed contractions in the early C. elegans embryo. *The Journal of Cell Biology*, *115*, 4230–4252. https://doi.org/10.1083/jcb.201806161.

Miller, A. L., & Bement, W. M. (2008). Regulation of cytokinesis by Rho GTPase flux. *Nature Cell Biology*, *11*, 71–77.

Mizuno, T., Amano, M., Kaibuchi, K., & Nishida, Y. (1999). Identification and characterization of Drosophila homolog of Rho-kinase. *Gene*, *238*, 437–444.

Munjal, A., Philippe, J.-M., Munro, E. M., & Lecuit, T. (2015). A self-organized biomechanical network drives shape changes during tissue morphogenesis. *Nature*, *524*, 351–355.

Murray, M. J., & Saint, R. (2007). Photoactivatable GFP resolves Drosophila mesoderm migration behaviour. *Development*, *134*, 3975–3983.

Nabel-Rosen, H., Toledano-Katchalski, H., Volohonsky, G., & Volk, T. (2005). Cell divisions in the Drosophila embryonic mesoderm are repressed via posttranscriptional regulation of string/cdc25 by HOW. *Current Biology*, *15*, 295–302.

Nandadasa, S., Tao, Q., Menon, N. R., Heasman, J., & Wylie, C. (2009). N- and E-cadherins in Xenopus are specifically required in the neural and non-neural ectoderm, respectively, for F-actin assembly and morphogenetic movements. *Development*, *136*, 1327–1338.

Nieto, M. A. (2011). The ins and outs of the epithelial to mesenchymal transition in health and disease. *Annual Review of Cell and Developmental Biology*, *27*, 347–376.

Nüsslein-Volhard, C., Frohnhöfer, H. G., & Lehmann, R. (1987). Determination of anteroposterior polarity in Drosophila. *Science*, *238*, 1675–1681.

Oda, H., & Tsukita, S. (2001). Real-time imaging of cell-cell adherens junctions reveals that Drosophila mesoderm invagination begins with two phases of apical constriction of cells. *Journal of Cell Science*, *114*, 493–501.

Oda, H., Tsukita, S., & Takeichi, M. (1998). Dynamic behavior of the cadherin-based cell–cell adhesion system during Drosophila gastrulation. *Developmental Biology, 203*, 435–450.

Parks, S., & Wieschaus, E. F. (1991). The Drosophila gastrulation gene concertina encodes a Ga-like protein. *Cell, 64*, 447–458.

Perez-Mockus, G., Mazouni, K., Roca, V., Corradi, G., Conte, V., & Schweisguth, F. (2017). Spatial regulation of contractility by neuralized and bearded during furrow invagination in Drosophila. *Nature Communications, 8*, 1–15.

Polyakov, O., He, B., Swan, M., Shaevitz, J. W., Kaschube, M., & Wieschaus, E. F. (2014). Passive mechanical forces control cell-shape change during Drosophila ventral furrow formation. *Biophysical Journal, 107*, 998–1010.

Pouille, P.-A., Ahmadi, P., Brunet, A.-C., & Farge, E. (2009). Mechanical signals trigger myosin II redistribution and mesoderm invagination in Drosophila embryos. *Science Signaling, 2*, 1–9.

Priya, R., Gomez, G. A., Budnar, S., Verma, S., Cox, H. L., Hamilton, N. A., et al. (2015). Feedback regulation through myosin II confers robustness on RhoA signalling at E-cadherin junctions. *Nature Cell Biology, 17*, 1282–1293.

Prokopenko, S. N., Brumby, A., OKeefe, L., Prior, L., He, Y., Saint, R., et al. (1999). A putative exchange factor for Rho1 GTPase is required for initiation of cytokinesis in Drosophila. *Genes & Development, 13*, 2301–2314.

Ratheesh, A., & Yap, A. S. (2012). A bigger picture: Classical cadherins and the dynamic actin cytoskeleton. *Nature Publishing Group, 13*, 673–679.

Rauzi, M., Krzic, U., Saunders, T. E., Krajnc, M., Ziherl, P., Hufnagel, L., et al. (2015). Embryo-scale tissue mechanics during Drosophila gastrulation movements. *Nature Communications, 6*, 1–12.

Ridley, A. J. (2011). Life at the leading edge. *Cell, 145*, 1012–1022.

Rogers, S. L., Wiedemann, U., Häcker, U., Turck, C., & Vale, R. D. (2004). Drosophila RhoGEF2 associates with microtubule plus ends in an EB1-dependent manner. *Current Biology, 14*, 1827–1833.

Roman, G., He, J., & Davis, R. L. (2000). Kurtz, a novel nonvisual arrestin, is an essential neural gene in Drosophila. *Genetics, 155*, 1281–1295.

Sako, Y., Nagafuchi, A., Tsukita, S., Takeichi, M., & Kusumi, A. (1998). Cytoplasmic regulation of the movement of E-cadherin on the free cell surface as studied by optical tweezers and single particle tracking: Corralling and tethering by the membrane skeleton. *The Journal of Cell Biology, 140*, 1227–1240.

Sánchez-Corrales, Y. E., & Röper, K. (2018). Alignment of cytoskeletal structures across cell boundaries generates tissue cohesion during organ formation. *Current Opinion in Cell Biology, 55*, 104–110.

Sawyer, J. K., Choi, W., Jung, K.-C., He, L., Harris, N. J., & Peifer, M. (2011). A contractile actomyosin network linked to adherens junctions by Canoe/afadin helps drive convergent extension. *Molecular Biology of the Cell, 22*, 2423–2659.

Sawyer, J. K., Harris, N. J., Slep, K. C., Gaul, U., & Peifer, M. (2009). The Drosophila afadin homologue Canoe regulates linkage of the actin cytoskeleton to adherens junctions during apical constriction. *The Journal of Cell Biology, 186*, 57–73.

Schaeffer, D., Somarelli, J. A., Hanna, G., Palmer, G. M., & Garcia-Blanco, M. A. (2014). Cellular migration and invasion uncoupled: Increased migration is not an inexorable consequence of epithelial-to-mesenchymal transition. *Molecular and Cellular Biology, 34*, 3486–3499.

Schäfer, G., Narasimha, M., Vogelsang, E., & Leptin, M. (2014). Cadherin switching during the formation and differentiation of the Drosophila mesoderm—Implications for epithelial-to-mesenchymal transitions. *Journal of Cell Science, 127*, 1511–1522.

Seher, T. C., & Leptin, M. (2000). Tribbles, a cell-cycle brake that coordinates proliferation and morphogenesis during Drosophila gastrulation. *Current Biology, 10*, 623–629.

Seher, T. C., Narasimha, M., Vogelsang, E., & Leptin, M. (2007). Analysis and reconstitution of the genetic cascade controlling early mesoderm morphogenesis in the Drosophila embryo. *Mechanisms of Development, 124*, 167–179.
Sellers, J. R. (1991). Regulation of cytoplasmic and smooth muscle myosin. *Current Opinion in Cell Biology, 3*, 98–104.
Simões, S., Oh, Y., Wang, M. F. Z., Fernandez-Gonzalez, R., & Tepass, U. (2017). Myosin II promotes the anisotropic loss of the apical domain during Drosophila neuroblast ingression. *The Journal of Cell Biology, 216*, 1387–1404.
Small, J. V., Stradal, T., Vignal, E., & Rottner, K. (2002). The lamellipodium: Where motility begins. *Trends in Cell Biology, 12*, 112–120.
Smallhorn, M., Murray, M. J., & Saint, R. (2004). The epithelial-mesenchymal transition of the Drosophila mesoderm requires the Rho GTP exchange factor Pebble. *Development, 131*, 2641–2651.
Stathopoulos, A., Tam, B., Ronshaugen, M., Frasch, M., & Levine, M. (2004). Pyramus and Thisbe: FGF genes that pattern the mesoderm of Drosophila embryos. *Genes & Development, 18*, 687–699.
Su, K.-C., Takaki, T., & Petronczki, M. (2011). Targeting of the RhoGEF Ect2 to the equatorial membrane controls cleavage furrow formation during cytokinesis. *Developmental Cell, 21*, 1104–1115.
Sun, J., & Stathopoulos, A. (2018). FGF controls epithelial-mesenchymal transitions during gastrulation by regulating cell division and apicobasal polarity. *Development, 145*, 1–15.
Sweeton, D., Parks, S., Costa, M., & Wieschaus, E. F. (1991). Gastrulation in Drosophila: The formation of the ventral furrow and posterior midgut invaginations. *Development, 112*, 775–789.
Tang, V. W., & Brieher, W. M. (2012). α-Actinin-4/FSGS1 is required for Arp2/3-dependent actin assembly at the adherens junction. *The Journal of Cell Biology, 196*, 115–130.
Tipping, M., Kim, Y., Kyriakakis, P., Tong, M., Shvartsman, S. Y., & Veraksa, A. (2010). Beta-arrestin Kurtz inhibits MAPK and Toll signalling in Drosophila development. *The EMBO Journal, 29*, 3222–3235.
Urbansky, S., González Avalos, P., Wosch, M., & Lemke, S. (2016). Folded gastrulation and T48 drive the evolution of coordinated mesoderm internalization in flies. *eLife, 5*, 1–20.
Vasquez, C. G., Heissler, S. M., Billington, N., & Martin, A. C. (2016). Drosophila non-muscle myosin II motor activity determines the rate of tissue folding. *eLife, 5*, e20828.
Vasquez, C. G., Tworoger, M., & Martin, A. C. (2014). Dynamic myosin phosphorylation regulates contractile pulses and tissue integrity during epithelial morphogenesis. *The Journal of Cell Biology, 206*, 435–450.
Vavylonis, D., Wu, J.-Q., Hao, S., O'Shaughnessy, B., & Pollard, T. D. (2008). Assembly mechanism of the contractile ring for cytokinesis by fission yeast. *Science, 319*, 97–100.
Vincent, S., Wilson, R., Coelho, C., Affolter, M., & Leptin, M. (1998). The Drosophila protein Dof is specifically required for FGF signaling. *Molecular Cell, 2*, 515–525.
Wagner, E., & Glotzer, M. (2016). Local RhoA activation induces cytokinetic furrows independent of spindle position and cell cycle stage. *The Journal of Cell Biology, 213*, 641–649.
Weng, M., & Wieschaus, E. F. (2016). Myosin-dependent remodeling of adherens junctions protects junctions from Snail-dependent disassembly. *The Journal of Cell Biology, 212*, 219–229.
Wilson, R., Vogelsang, E., & Leptin, M. (2005). FGF signalling and the mechanism of mesoderm spreading in Drosophila embryos. *Development, 132*, 491–501.
Winter, C. G., Wang, B., Ballew, A., Royou, A., Karess, R., Axelrod, J. D., et al. (2001). Drosophila Rho-associated kinase (Drok) links frizzled-mediated planar cell polarity signaling to the actin cytoskeleton. *Cell, 105*, 81–91.

Wu, S.-Y., & McClay, D. R. (2007). The Snail repressor is required for PMC ingression in the sea urchin embryo. *Development, 134*, 1061–1070.
Xie, S., & Martin, A. C. (2015). Intracellular signalling and intercellular coupling coordinate heterogeneous contractile events to facilitate tissue folding. *Nature Communications, 6*, 1–13.
Xie, S., Mason, F. M., & Martin, A. C. (2016). Loss of Ga12/13 exacerbates apical area dependence of actomyosin contractility. *Molecular Biology of the Cell, 27*, 3526–3536.
Yamada, S., Pokutta, S., Drees, F., Weis, W. I., & Nelson, W. J. (2005). Deconstructing the cadherin-catenin-actin complex. *Cell, 123*, 889–901.
Yevick, H. G., Miller, P. W., Dunkel, J., & Martin, A. C. (2019). Structural redundancy in supracellular actomyosin networks enables robust tissue folding. *Developmental Cell* https://www.ncbi.nlm.nih.gov/pubmed/31353314. (Epub ahead of print).
Yonemura, S., Wada, Y., Watanabe, T., Nagafuchi, A., & Shibata, M. (2010). α-Catenin as a tension transducer that induces adherens junction development. *Nature Cell Biology, 12*, 533–542.
Young, P. E., Pesacreta, T. C., & Kiehart, D. P. (1991). Dynamic changes in the distribution of cytoplasmic myosin during Drosophila embryogenesis. *Development, 111*, 1–14.
Yüce, Ö., Piekny, A., & Glotzer, M. (2005). An ECT2-centralspindlin complex regulates the localization and function of RhoA. *The Journal of Cell Biology, 170*, 571–582.
Zusman, S. B., & Wieschaus, E. F. (1985). Requirements for zygotic gene activity during gastrulation in Drosophila melanogaster. *Developmental Biology, 111*, 359–371.

CHAPTER SIX

Cellular, molecular, and biophysical control of epithelial cell intercalation

Adam C. Paré[a,*], Jennifer A. Zallen[b,*]
[a]Department of Biological Sciences, University of Arkansas, Fayetteville, AR, United States
[b]Howard Hughes Medical Institute and Developmental Biology Program, Sloan Kettering Institute, New York, NY, United States
*Corresponding authors: e-mail address: pare@uark.edu; zallenj@mskcc.org

Contents

Abstract

Convergent extension is a conserved mechanism for elongating tissues. In the *Drosophila* embryo, convergent extension is driven by planar polarized cell intercalation and is a paradigm for understanding the cellular, molecular, and biophysical mechanisms that establish tissue structure. Studies of convergent extension in *Drosophila* have provided key insights into the force-generating molecules that promote convergent extension in epithelial tissues, as well as the global systems of spatial information that systematically organize these cell behaviors. A general framework has emerged in which asymmetrically localized proteins involved in cytoskeletal tension and cell adhesion direct oriented cell movements, and spatial signals provided by the Toll, Tartan, and Teneurin receptor families break planar symmetry to establish and coordinate planar cell polarity throughout the tissue. In this chapter, we describe the cellular, molecular, and biophysical mechanisms that regulate cell intercalation in the *Drosophila* embryo, and discuss how research in this system has revealed conserved biological principles that control the organization of multicellular tissues and animal body plans.

Current Topics in Developmental Biology, Volume 136
ISSN 0070-2153
https://doi.org/10.1016/bs.ctdb.2019.11.014

1. Convergent extension: A conserved mechanism for shaping epithelia

A conserved feature of animal development is the formation of a body axis that is elongated from head to tail. This characteristic structure of the body plan is generated by dynamic cell movements that cause the embryo to narrow along one axis and elongate along a perpendicular axis, a process known as convergent extension. Convergent extension can occur through a wide range of cell behaviors, including polarized cell divisions, stereotyped changes in cell shape, and oriented cell movements. In particular, cell intercalation is an essential mechanism for convergent extension during body axis elongation in frogs, fish, flies, chicks, worms, and mice (Keller et al., 2000; Kong, Wolf, & Grosshans, 2017; Takeichi, 2014; Walck-Shannon & Hardin, 2014; Wallingford, Fraser, & Harland, 2002). First described in migratory mesenchymal cells in the *Xenopus* embryo (Keller et al., 2000; Wallingford et al., 2002), it is now recognized that even highly adherent epithelial cells can dynamically remodel to drive tissue elongation. Studies of convergent extension in *Drosophila* have uncovered the force-generating machinery and tissue-wide spatial information systems that link axial patterning to polarized cell behavior. In this chapter, we describe the cellular, molecular, and biophysical mechanisms that regulate epithelial cell intercalation in the *Drosophila* germband, and we discuss outstanding questions in the field regarding the molecular basis of planar symmetry breaking and how biochemical and biophysical processes converge to control epithelial organization.

2. Cell rearrangements during convergent extension in the *Drosophila* embryo

A premier model for studying convergent extension is elongation of the *Drosophila* germband epithelium (Fig. 1A). Pioneering mutagenesis screens carried out by Eric Wieschaus and Christiane Nüsslein-Volhard shed light on the upstream molecular cues that pattern the germband and surrounding tissues (Nüsslein-Volhard & Wieschaus, 1980; Wieschaus & Nüsslein-Volhard, 2016). Since then, elegant studies by many labs have revealed how networks of transcriptional enhancers and repressors generate complex patterns of gene expression from relatively simple starting inputs. However, a major challenge in the field has been to understand how these gene expression patterns induce changes in cell shape and behavior to yield proper embryo morphology.

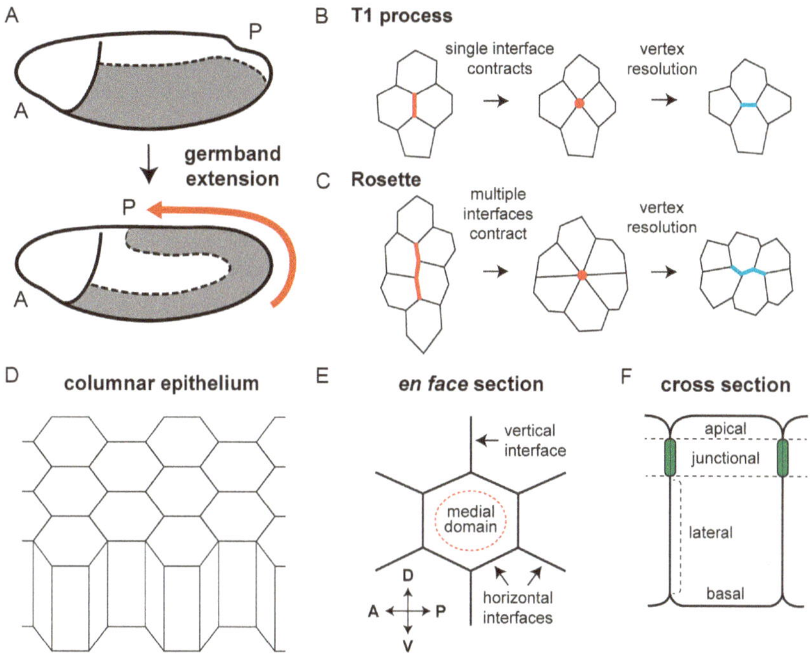

Fig. 1 Cell rearrangements during *Drosophila* germband extension. (A) Convergent extension in the germband ectoderm (gray) elongates the anterior–posterior (AP) body axis. (B) T1 processes occur through the contraction of a single vertical cell interface. (C) Multicellular rosettes form through the contraction of 2 interfaces (for a 5-cell rosette), 3 interfaces (for a 6-cell rosette), or 4 or more interfaces (for rosettes containing 7 or more cells). (D) The germband ectoderm is a simple columnar epithelium. (E) An *en face* view showing one hexagonal cell and its neighbors. (F) Cross-sectional view showing different cellular domains along the apical–basal axis.

The germband ectoderm is a single-layered columnar epithelium on the ventrolateral surface of the developing embryo (Fig. 1A and D–F) that differentiates into a wide range of tissues, including the nervous system, epidermis, airways, and imaginal discs. Several characteristics make the germband ectoderm an attractive system for studying the cellular basis of convergent extension, including (1) the relatively simple columnar organization of the epithelium, (2) the absence of cell division during most of this process, and (3) the ease with which cell behaviors can be visualized by microscopy. The *Drosophila* germband more than doubles in length along the AP axis over the course of approximately 2 h (Fig. 1A), with most of this elongation achieved during an initial, 30-min fast phase. Time-lapse epi-illumination movies of ectodermal cells demonstrated that the fast phase of elongation occurs primarily through cell intercalation, in which cell movements oriented along the dorsal-ventral (DV) axis drive tissue elongation along the anterior-posterior (AP) axis (Irvine & Wieschaus, 1994). Importantly, mutations that eliminate mesodermal cell fates do not alter the rate or extent of elongation, indicating that axis elongation is driven solely by cell behaviors in the ectoderm (Irvine & Wieschaus, 1994).

Direct visualization of cell shapes during germband extension using fluorescent markers revealed several unexpected cell behaviors. Unlike mesenchymal cells, which intercalate largely by crawling on neighboring cells using spatially regulated protrusions (see chapters "Convergent extension in the amphibian, *Xenopus laevis*" by Keller and Sutherland; "Cellular and molecular mechanisms of convergence and extension in zebrafish" by Williams and Solnica-Krezel), intercalating epithelial cells in the germband rearrange by remodeling cell–cell junctions in a spatially regulated manner (Fig. 1B and C). Cell–cell interfaces oriented perpendicular to the direction of tissue elongation (vertical interfaces) contract to form four-cell vertices, which resolve to produce a new cell–cell interface oriented parallel to the AP axis (horizontal interfaces) (Bertet, Sulak, & Lecuit, 2004) (Fig. 1B). During each rearrangement, a pair of adjacent cells along the AP axis becomes separated, and two cells that were previously separated come into contact. These local cell rearrangements are referred to as neighbor exchange events or T1 processes, following the convention first introduced for two-dimensional foams (Weaire & Rivier, 1984).

Early models proposed that simple T1 processes within an ordered, hexagonal cell array are sufficient to account for tissue elongation (Bertet et al., 2004). However, it is now known that germband cells are not strictly hexagonal and display more complex behaviors (Zallen & Zallen, 2004). In particular, germband cells assemble into multicellular rosette structures that form and resolve directionally and are necessary for tissue elongation (Blankenship, Backovic, Sanny, Weitz, & Zallen, 2006). During rosette formation, connected vertical interfaces contract to bring five or more cells into contact at a single point or vertex. This vertex then resolves in the perpendicular direction, transforming two columns of cells into two rows of cells (Fig. 1C) (Blankenship et al., 2006). Mutants that assemble fewer rosettes, with no change in the number of T1 processes, display significantly reduced elongation, demonstrating that rosette behaviors are required for convergent extension (Tamada, Farrell, & Zallen, 2012). Since the discovery of the rosette mechanism in *Drosophila*, rosette-based intercalation has now been shown to be a general mechanism for convergent extension in flies, worms, chicks, frogs, and mice (Chacon-Heszele, Ren, Reynolds, Chi, & Chen, 2012; Lau et al., 2015; Lienkamp et al., 2012; McGreevy,

Vijayraghavan, Davidson, & Hildebrand, 2015; Nishimura & Takeichi, 2008; Rozbicki et al., 2015; Shah et al., 2017; Sun et al., 2017; Williams, Yen, Lu, & Sutherland, 2014).

3. The molecular basis of epithelial cell intercalation

Intercalary behaviors must ultimately be propelled by proteins that are asymmetrically localized within the plane of the tissue, a property known as planar cell polarity, but the molecular nature of these asymmetries has long been mysterious. Key findings in the *Drosophila* germband demonstrate that cell intercalation is driven by a conserved set of planar polarized cytoskeletal and junctional proteins that modulate cortical tension and cell adhesion in distinct cellular domains (Fig. 2), directly linking cellular force-generating machineries to polarized cell behavior.

The contractile domain of intercalating cells is established by the localized recruitment of nonmuscle myosin II (referred to here as myosin), a motor protein that binds to, crosslinks, and translocates along actin filaments (Bertet et al., 2004; Zallen & Wieschaus, 2004). In addition, a complementary adhesive domain is defined by Par-3, a PDZ protein that associates with and stabilizes adherens junctions (Simões et al., 2010; Zallen & Wieschaus, 2004). The loss of either protein leads to severe defects in axis elongation, demonstrating that these molecules are essential for polarized cell behavior (Bertet et al., 2004; Simões et al., 2010; Zallen & Wieschaus, 2004). Subsequent studies revealed that these patterns reflect a general segregation

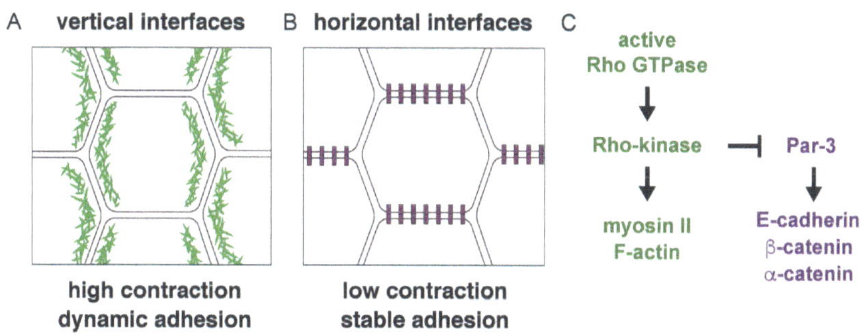

Fig. 2 Planar polarity in the *Drosophila* germband. (A) Vertical cell interfaces in the germband display increased localization or activity of proteins involved in promoting cortical tension, adherens junction turnover, and endocytosis. (B) Horizontal interfaces are enriched for proteins involved in cell adhesion. (C) Rho-kinase activity at vertical interfaces directs multiple aspects of planar polarity.

of the contractile and adhesive machinery within cells (Fig. 2), as actin filaments colocalize with myosin at vertical interfaces (Blankenship et al., 2006), and the adherens junction proteins E-cadherin, α-catenin, and β-catenin are enriched at horizontal interfaces (Blankenship et al., 2006; Levayer & Lecuit, 2013; Levayer, Pelissier-Monier, & Lecuit, 2011; Simões et al., 2010; Tamada et al., 2012; Warrington, Strutt, & Strutt, 2013). These results suggest a model in which cell intercalation is driven by complementary cellular domains specialized for contraction or adhesion, a molecular framework that has now been shown to be generalizable to many epithelial contexts.

The proof that planar polarized actomyosin contractility is necessary for cell intercalation came from the use of laser ablation methods to analyze cortical tension (Farhadifar, Röper, Aigouy, Eaton, & Jülicher, 2007; Hutson et al., 2003). Biophysical studies using this approach revealed that vertical interfaces are under higher tension than horizontal interfaces, proving that planar polarized myosin localization translates into planar polarized contractile forces (Fernandez-Gonzalez, Simões, Röper, Eaton, & Zallen, 2009; Rauzi, Verant, Lecuit, & Lenne, 2008). Since its discovery in *Drosophila*, planar polarized actomyosin contractility is now recognized to drive convergent extension in many tissues, including the chick and mouse neural plate (Nishimura, Honda, & Takeichi, 2012; Nishimura & Takeichi, 2008; Williams et al., 2014), the mouse cochlea and limb bud (Chacon-Heszele et al., 2012; Lau et al., 2015), the mouse and frog kidney (Lienkamp et al., 2012), the *Xenopus* and zebrafish notochord (Shindo & Wallingford, 2014; Williams et al., 2018), the chick primitive streak (Rozbicki et al., 2015), and the *C. elegans* nervous system (Shah et al., 2017).

The critical role of actomyosin contractility in driving cell intercalation raises the question of how planar polarized patterns of myosin localization and activity are established in the tissue. The complementary distributions of contractile and adhesive proteins in the germband are generated by a single serine-threonine protein kinase, Rho-kinase, which is asymmetrically localized to vertical interfaces and is essential for intercalary behavior (Simões et al., 2010). Rho-kinase regulates planar polarity in two ways. First, Rho-kinase phosphorylates the myosin regulatory light chain, enhancing myosin localization and activity at vertical edges (Kasza, Farrell, & Zallen, 2014; Simões et al., 2010). In addition, Rho-kinase phosphorylates Par-3 and inhibits Par-3 association with the cell cortex at vertical edges (Simões et al., 2010), likely by preventing its interaction with phosphoinositide membrane lipids (Krahn, Klopfenstein, Fischer, & Wodarz, 2010). Par-3 in turn enhances cell-cell adhesion at horizontal interfaces (Simões et al., 2010). Thus, Rho-kinase is a critical regulator of planar polarity that has direct, opposing effects on contraction and adhesion at vertical interfaces.

The upstream signals that regulate Rho-kinase localization during cell intercalation are less well understood, but evidence points to an essential role for the small GTPase Rho, a conserved activator of Rho-kinase activity and actomyosin contraction in many contexts (Jaffe & Hall, 2005). The effects of Rho are difficult to test genetically, due to its pleiotropic requirements in many aspects of embryogenesis, but dominant-negative Rho abolishes planar polarity in the germband (Simões et al., 2014) and activity probes designed to detect active Rho are enriched at contracting cell interfaces during germband elongation (Garcia De Las Bayonas, Philippe, Lellouch, & Lecuit, 2019; Munjal, Philippe, Munro, & Lecuit, 2015; Simões et al., 2014). Elucidating the upstream mechanisms that direct localized Rho-kinase activity to establish planar cell polarity in the germband remains an important question in the field.

Several other mechanisms contribute to the planar polarized distribution of adhesion complexes during cell intercalation. The Abl tyrosine kinase localizes to vertical interfaces and enhances junctional turnover by phosphorylating β-catenin (Tamada et al., 2012). Proper localization of E-cadherin requires the glucosyltransferase Xiantuan (Zhang et al., 2014). Finally, components of the membrane trafficking machinery, including clathrin and dynamin (Levayer et al., 2011) and the Rab GTPase Rab35 (Jewett et al., 2017), promote the localized endocytosis of membrane and adherens junction complexes at vertical edges, resulting in edge shortening and cell separation.

Once planar polarized contraction and adhesion are established, positive and negative interactions dynamically reinforce planar polarity. For example, the F-actin-binding protein Shroom is required to maintain Rho-kinase planar polarity at later stages of intercalation (Simões et al., 2014), and Par-3 is required for the planar polarized localization of myosin (Simões et al., 2010). In addition, dynamic myosin flows along the medial (apical) cell cortex are oriented by fluctuations in E-cadherin levels (Levayer & Lecuit, 2013) and are associated with changes in apical cell area (Fernandez-Gonzalez & Zallen, 2011; Sawyer et al., 2011), edge contraction (Levayer & Lecuit, 2013; Munjal et al., 2015; Rauzi, Lenne, & Lecuit, 2010), and new edge formation (Collinet, Rauzi, Lenne, & Lecuit, 2015; Yu & Fernandez-Gonzalez, 2016). Finally, cells that undergo changes in angular orientation as they rearrange can reorganize Par-3 and myosin polarity to match their new orientation, revealing that planar polarity mechanisms display an unexpected plasticity (Farrell, Weitz, Magnasco, & Zallen, 2017). In addition, vertex sliding (Vanderleest et al., 2018) and basolateral cellular protrusions (Sun et al., 2017) also contribute to intercalation and may be controlled by distinct molecular mechanisms. How these diverse cell biological processes are dynamically and coordinately controlled to establish and maintain planar polarized behaviors in the apical and basal domains of the cell is an important area for future study.

4. Biophysical control of epithelial cell intercalation

Epithelial cells are connected by adherens junctions that facilitate the transmission of mechanical forces from cell to cell. Therefore, a unique feature of intercalation in epithelia is that actomyosin forces generated in one cell can propagate to neighboring cells and influence their behavior. Indeed, laser ablation studies demonstrate that germband cells can respond to physical changes that occur several cell diameters away (Fernandez-Gonzalez et al., 2009; Wang et al., 2017), and genetic and biophysical experiments show that cells respond to mechanical forces generated in neighboring tissues (Collinet et al., 2015; Lye et al., 2015). An emerging picture is that physical forces provide an additional tier of mechanical regulation that influences cell behavior during intercalation.

Myosin dynamics are sensitive to mechanical forces in single cells (Effler et al., 2006) and multicellular epithelia (Fernandez-Gonzalez et al., 2009; Pouille, Ahmadi, Brunet, & Farge, 2009). During cell intercalation, cortical myosin turnover is selectively decreased in regions of high tension, resulting in the localized stabilization and accumulation of myosin (Fernandez-Gonzalez et al., 2009). Moreover, an applied force can recruit myosin to the cortex within tens of seconds (Fernandez-Gonzalez et al., 2009). These results demonstrate a positive feedback loop in which myosin generates cortical tension, which in turn recruits and stabilizes more myosin at the cortex. This mechanical feedback loop provides a mechanism by which tension from neighboring cells influences myosin contractility in a non-cell-autonomous manner, contributing to the formation of supracellular myosin cables that contract to form rosettes (Blankenship et al., 2006; Fernandez-Gonzalez et al., 2009).

In contrast to the planar polarized contractile structures that mediate cell interface contraction during intercalation, the mechanisms that control the direction of vertex resolution to promote productive cell rearrangement are less well understood. Pulsatile myosin flows along the medial cell surface are often observed in epithelial cells (Martin, Kaschube, & Wieschaus, 2009), and during cell intercalation, these flows are biased along the AP axis (Levayer & Lecuit, 2013; Rauzi et al., 2010). It has been proposed that medial myosin flows create local, nonautonomous AP-oriented pulling forces that influence the rate and direction of new interface formation (Collinet et al., 2015; Lan, Wang, Fernandez-Gonzalez, & Feng, 2015; Yu & Fernandez-Gonzalez, 2016). Consistent with this possibility, regulated actomyosin activity is necessary for directional vertex resolution (Kasza et al., 2014; Yu & Fernandez-Gonzalez, 2016), and laser ablation experiments demonstrate that nonautonomous forces influence the rate and orientation of new interface formation (Collinet et al., 2015; Yu & Fernandez-Gonzalez, 2016). These results raise the intriguing possibility that myosin-dependent mechanical forces directly regulate multiple steps of cell intercalation.

In addition to the transmission of mechanical forces between cells, forces that are transmitted between tissues can also influence cell intercalation. Notably, invagination of the posterior midgut induces a strong pulling force in the direction of tissue extension that contributes to axis elongation (Collinet et al., 2015; Lye et al., 2015), and mesoderm invagination imparts an orthogonal pulling force that causes transient cell-shape deformations (Butler et al., 2009; Farrell et al., 2017). Conversely, forces generated in the germband can affect cell behaviors in surrounding tissues, such as the anisotropic apical constriction of cells in the mesoderm (Martin, Gelbart, Fernandez-Gonzalez, Kaschube, & Wieschaus, 2010) and the orientation of cell divisions in the mesectoderm (Wang et al., 2017). The ability to independently modulate forces in distinct tissues using genetic patterning mutations and laser ablation makes the *Drosophila* embryo a particularly attractive model for addressing the biomechanical relationships between cells and tissues in vivo.

5. Breaking planar symmetry

Despite significant strides in understanding the cellular, molecular, and biophysical processes that drive cell intercalation, the spatial cues that orient and coordinate planar polarity across the hundreds of cells of the germband had long been elusive. An important clue came from the discovery that germband extension requires inputs from the AP-patterning system (Irvine & Wieschaus, 1994). In particular, mutants lacking transcription factors encoded by gap or pair-rule genes exhibit a strong reduction in cell intercalation in discrete regions where these factors are normally present (Irvine & Wieschaus, 1994). By contrast, embryos with mutations that disrupt DV patterning, segment polarity genes (e.g., Wingless), or classical planar cell polarity signaling (through Frizzled and Dishevelled) undergo robust cell intercalation and convergent extension (Irvine & Wieschaus, 1994; Zallen & Wieschaus, 2004). In addition, local differences in the expression of the pair-rule genes *eve* and *runt* are necessary and sufficient for planar cell polarity (Zallen & Wieschaus, 2004). These results indicate that striped patterns of gene expression perpendicular to the AP axis orient cell intercalation (Wieschaus, Sweeton, & Costa, 1991). However, it was two decades before the targets of the AP patterning system that break planar symmetry and align cell polarity with the body axes were identified.

Studies of striped proteins in the germband, combined with genome engineering approaches to knock out multiple genes simultaneously, revealed that two independent cell-surface receptor systems act downstream of the pair-rule genes to direct planar polarity (Fig. 3) (Paré et al., 2014, 2019). The first system consists of three members of the Toll receptor family, Toll-2, Toll-6, and Toll-8, that direct actomyosin and adherens junction polarity and cell intercalation throughout the germband (Paré et al., 2014). The second system,

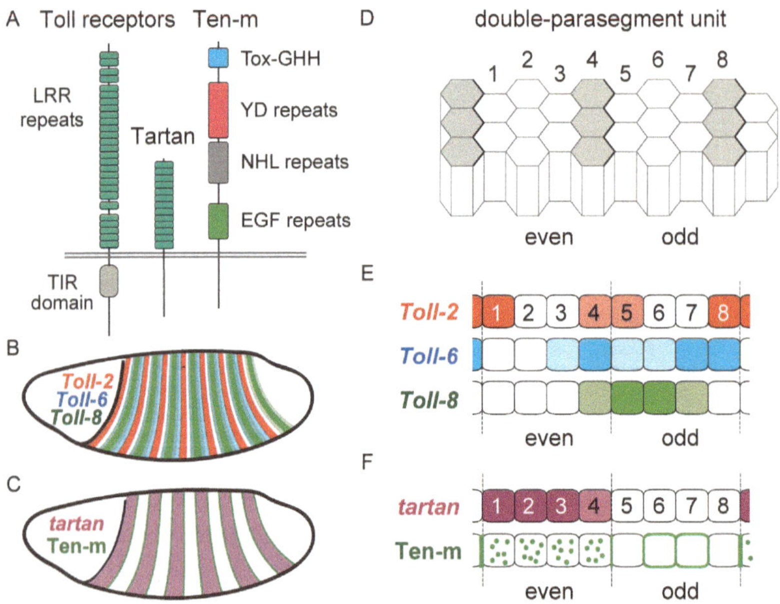

Fig. 3 Receptor systems mediating planar polarity in the germband. (A) Schematics of Toll receptors, Tartan, and Ten-m (not to scale). (B) *Toll-2*, *Toll-6*, and *Toll-8* are expressed in staggered striped patterns in the germband. (C) *tartan* (purple) is expressed in stripes and Ten-m protein (green) is enriched at the stripe borders. (D) The germband is organized into double-parasegment units, each made up of seven or eight columns of cells. Even parasegments (cells 1-4), odd parasegments (cells 5-8). Wingless-expressing cells are shown in gray. (E) Toll receptor expression within the double-parasegment unit. (F) *tartan* expression and Ten-m localization within the double-parasegment unit. Ten-m is enriched at compartment boundaries and is absent from the membrane in Tartan-positive cells.

involving the leucine-rich repeat (LRR) receptor Tartan and the teneurin Ten-m, controls planar polarity specifically at compartment (parasegment) boundaries (Paré et al., 2019). The discovery of these two systems has yielded a surprisingly complex answer to how patterned gene expression organizes cell behavior during convergent extension in the *Drosophila* germband.

6. Toll receptors direct planar polarity and cell intercalation

The discovery that Toll-2, Toll-6, and Toll-8 are required for planar polarity and intercalary behavior during germband extension revealed the long-sought link between the transcriptional events that pattern the AP axis of the embryo and the morphogenetic mechanisms that convert this pattern into changes in tissue structure. The absence of all three Toll-related receptors causes strong defects in planar polarity, cell intercalation, and axis elongation (Paré et al., 2014). This new morphogenetic role for Toll receptors in convergent extension was subsequently shown to be conserved in other arthropods (Benton et al., 2016). Toll family receptors are transmembrane proteins that contain extracellular LRRs and an intracellular TIR (Toll/interleukin-1 receptor) domain that signals to downstream effectors (Fig. 3A). The founding member of this family, Toll, was originally discovered as a critical component of the DV patterning system in *Drosophila* (Anderson, Bokla, & Nüsslein-Volhard, 1985; Anderson, Jürgens, & Nüsslein-Volhard, 1985; Hashimoto, Hudson, & Anderson, 1988). This receptor family was subsequently shown to be essential for innate immunity in *Drosophila* and vertebrates (Anderson, 2000). The roles of *Drosophila* Toll and vertebrate Toll-like receptors in innate immunity have been the subject of intense study (Kawasaki & Kawai, 2014; Leulier & Lemaitre, 2008), but the developmental roles of Toll family receptors, beyond DV patterning, have received comparably less attention.

Toll-2, Toll-6, and Toll-8 are expressed in non-uniform patterns in the germband, placing them in a position to act as the symmetry-breaking molecules that organize planar cell polarity. The *Drosophila* germband is transcriptionally subdivided into a series of repeating, double-parasegment units along the AP axis that each span approximately eight columns of cells (Fig. 3D) (Akam, 1987; Clark, 2017). *Toll-2*, *Toll-6*, and *Toll-8* are expressed in distinct patterns within this double-parasegment unit (Fig. 3B and E) under the control of pair-rule genes (Chiang & Beachy, 1994; Eldon et al., 1994; Graham, Anderson, Brandt, Xiang, & Pick, 2019; Paré et al., 2014). *Toll-2* is expressed in alternating major and minor stripes, whereas

Toll-8 is expressed in fewer, broader stripes. *Toll-6* has the most complex pattern, largely overlapping with Toll-8 but with differing regions of low and high expression. Thus, every column of cells within each unit has a distinct complement of Toll receptor types and expression levels (Paré et al., 2014, 2019). Notably, the loss of one or two Toll receptors disrupts planar polarity in different regions of the double-parasegment unit, and ectopic stripes of Toll receptor expression are sufficient to induce myosin planar polarity locally (Paré et al., 2014, 2019). These observations suggest that the juxtaposition of cells that express different levels or combinations of Toll receptors is a critical symmetry-breaking event in the establishment of planar polarity.

The expression of Toll receptors in striped patterns acts as a bridge between axial patterning information and polarized cell behavior. However, the mechanisms that convert local differences in receptor expression into planar polarity are currently unknown. Here we consider two classes of models that could explain how striped Toll receptor expression leads to planar polarity, either through interactions between different types of Toll receptors (heterotypic models) or independent Toll receptor functions (single-component models). In a *heterotypic activation* model, *trans* interactions between different Toll receptors expressed in neighboring columns of cells could activate Toll receptors and myosin contractility at vertical cell interfaces (Fig. 4A). In this model, Toll receptors serve as both ligand and receptor, eliminating the requirement for other patterned activators. Consistent with a heterotypic activation model, cells expressing Toll-2 interact heterotypically with cultured *Drosophila* cells expressing Toll-6 or Toll-8 (Paré et al., 2014). However, the heterotypic model does not explain why receptor activation does not occur at horizontal interfaces, as most cells in the germband express multiple receptor types and the subcellular localization of Toll receptors has not been determined. In addition, there is a strong genetic argument against the heterotypic model. Namely, if heterotypic interactions are required for receptor activity, then a double mutant lacking two Toll receptors should be just as defective as a triple-mutant embryo lacking all three. However, triple mutants are generally more defective than double-mutant combinations (Paré et al., 2014). Thus, although heterotypic interactions may play some role in Toll receptor signaling during convergent extension, they cannot be strictly required for receptor function.

By contrast, *single-component* models propose that each Toll receptor type functions independently of the others. Single-component models could work in a variety of ways. In a *homotypic inactivation* model, inhibitory *trans* interactions between Toll receptors expressed in the same stripe could

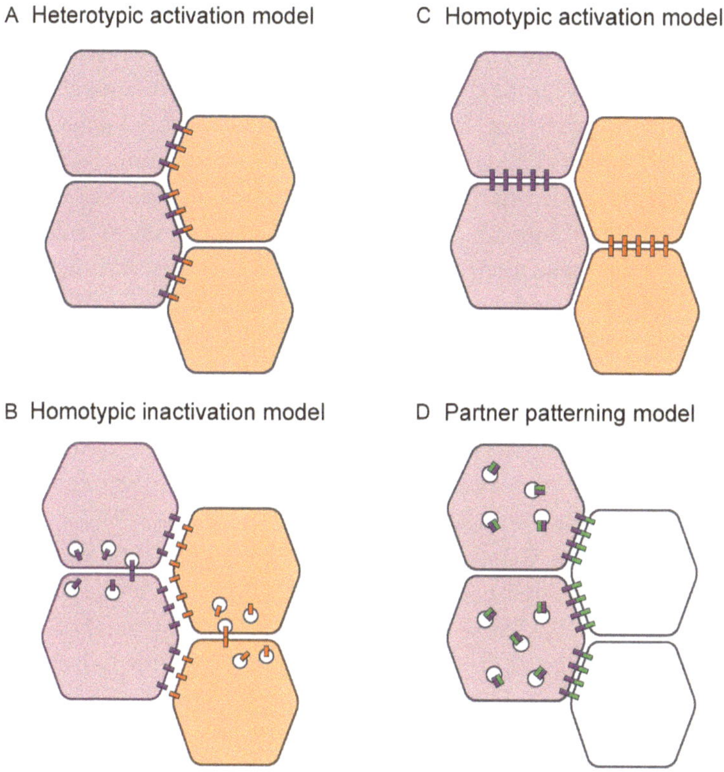

Fig. 4 Models for how striped Toll receptors generate planar polarity. (A) In a heterotypic activation model, *trans* interactions between cells expressing different Toll receptors (purple and orange) in adjacent stripes could stabilize Toll receptors at vertical interfaces. (B) In a homotypic inactivation model, *trans* interactions between cells expressing the same Toll receptor type could destabilize Toll receptors at horizontal interfaces. (C) In a homotypic activation model, *trans* interactions between cells expressing the same Toll receptor type could stabilize receptors at horizontal interfaces. (D) In a partner patterning model, a patterned Toll receptor (purple) and an unpatterned interaction partner (green) could undergo inhibitory *cis* interactions and stabilizing *trans* interactions, promoting the enrichment of the partner at vertical interfaces along the Toll stripe border.

downregulate Toll receptors at horizontal cell interfaces (Fig. 4B). Alternatively, in a *homotypic activation* model, activating *trans* interactions between Toll receptors in the same stripe could stabilize Toll receptors at horizontal cell interfaces (Fig. 4C). In both models, Toll receptors would become planar polarized through homotypic interactions, and would go on to promote planar polarity by influencing either actomyosin contractility (Fig. 4B) or cell adhesion (Fig. 4C). Finally, in a *partner-patterning* model, Toll receptors could pattern the localization or activity of a ubiquitous interaction partner, causing it to become differentially localized in Toll-expressing and Toll-nonexpressing cells (Fig. 4D). Such a mechanism was recently demonstrated for Tartan and Ten-m in the *Drosophila* germband (see below) (Paré et al., 2019). A significant advantage of single-component models is that they do not require pre-patterned interaction partners to trigger planar polarity. Instead, the capacity of Toll receptors to induce planar polarity is inherent in their striped expression patterns. Although there is currently no direct evidence for single-component models of Toll receptor activity, genetic evidence supports the idea of independent receptor functions (Paré et al., 2014). Insight into where Toll receptors are localized and active within cells will help to distinguish between these models.

Toll receptors display complex expression patterns in a variety of tissues throughout *Drosophila* development, including the embryonic head and germband (Chiang & Beachy, 1994; Eldon et al., 1994; Kambris, Hoffmann, Imler, & Capovilla, 2002; Paré et al., 2014) and the wing and leg imaginal discs (Kim, Chung, Yoon, Choi, & Yim, 2006; Yagi, Nishida, & Ip, 2010). These results raise the possibility that Toll receptors could influence cell polarity and behavior in other tissues. Consistent with their widespread expression, Toll receptor mutants have defects in the structure and organization of *Drosophila* appendages (Eldon et al., 1994; Yagi et al., 2010), follicle cells (Kleve, Siler, Syed, & Eldon, 2006), salivary glands (Kolesnikov & Beckendorf, 2007), and nervous system (Ballard, Miller, & Ganetzky, 2014; Foldi et al., 2017; McIlroy et al., 2013; Ward, Hong, Favaloro, & Luo, 2015). In addition to their roles in development, Toll is required for wound repair in the late embryonic epidermis (Carvalho, Jacinto, & Matova, 2014), and Toll-2, Toll-3, Toll-8, and Toll-9 regulate cell competition in wing imaginal discs (Meyer et al., 2014). A better understanding of how Toll receptors mediate cell polarity and behavior in the germband will provide insight into their roles in diverse biological processes involving cell and tissue remodeling.

7. Regulation of planar polarity at compartment boundaries

Compartment boundaries are physical structures that delineate distinct domains within epithelial sheets. Originally discovered in insects (Garcia-Bellido, Ripoll, & Morata, 1973; Lawrence, 1973), compartment boundaries are now recognized to be important for tissue organization in both vertebrates and invertebrates (Batlle & Wilkinson, 2012; Dahmann & Basler, 1999; Dahmann, Oates, & Brand, 2011; Fagotto, 2014). These boundaries not only prevent mixing between different cell lineages, they also provide natural, defined locations that position the source of diffusible morphogens (Lawrence & Struhl, 1996). As a consequence, compartment boundaries have profound effects on cell identity and tissue architecture. Compartment boundaries between parasegments in the germband are established early in embryogenesis (Vincent & O'Farrell, 1992), and they act as barriers to cell crossing throughout embryonic and larval development, as cells undergo dramatic displacements resulting from epithelial remodeling and tissue growth (Landsberg et al., 2009; Monier, Pélissier-Monier, Brand, & Sanson, 2010; Scarpa, Finet, Blanchard, & Sanson, 2018). While the transcriptional inputs that establish compartment boundaries have been well-characterized, the cellular mechanisms that create physical barriers between cell populations are less well understood.

Compartment boundaries act as barriers, at least in part, by locally increasing actomyosin contractility to create a contractile cable that resists cell crossing (Aliee et al., 2012; Landsberg et al., 2009; Major & Irvine, 2005, 2006; Monier et al., 2010; Tetley, Blanchard, Fletcher, Adams, & Sanson, 2016; Umetsu, Dunst, & Dahmann, 2014). Myosin is still enriched at compartment boundaries in Toll-deficient embryos, indicating that other signals regulate boundary formation in the germband (Paré et al., 2019). A key insight into the mechanisms that establish compartment boundaries was the discovery that the segmentally expressed LRR protein Tartan and its binding partner Ten-m direct planar polarity specifically at compartment boundaries in the germband (Fig. 3A, C and F) (Paré et al., 2019). Tartan influences cell shape in *Drosophila* epithelial tissues (Krause, Wolf, Hemphälä, Samakovlis, & Schuh, 2006; Mao, Kerr, & Freeman, 2008; Milán, Pérez, & Cohen, 2005; Milán, Weihe, Pérez, & Cohen, 2001; Sakurai, Kojima, Aigaki, & Hayashi, 2007) and neurons (Kurusu et al., 2008), and teneurins regulate neural development in *Drosophila* (Baumgartner & Wides, 2019; Hong et al., 2009; Hong & Luo, 2014; Mosca, 2015; Mosca, Hong, Dani, Favaloro, & Luo, 2012) and vertebrates (Antinucci, Nikolaou, Meyer, & Hindges, 2013; Berns, DeNardo, Pederick, & Luo, 2018; Dharmaratne et al., 2012; Leamey et al., 2008). *Drosophila* embryos defective for either Tartan or Ten-m display decreased myosin and increased Par-3 levels specifically at compartment boundaries, whereas non-boundary regions are not affected (Paré et al., 2019). These functions appear to involve direct receptor–receptor interactions, as Tartan and Ten-m can interact *in trans* when expressed in cultured cells *in vitro*, and Tartan recruits Ten-m protein to compartment boundaries, corresponding to the borders of the Tartan stripes, *in vivo* (Paré et al., 2019). Notably, either reducing Tartan or Ten-m expression or overexpressing either protein ubiquitously disrupts cell alignment at boundaries, indicating that local differences in Tartan and Ten-m are important for planar polarity and boundary structure (Paré et al., 2019).

These results identify a novel receptor interaction that regulates planar polarity at compartment boundaries during germband extension. Tartan is expressed in even parasegments, providing a clear source of differences in Tartan activity at compartment boundaries (Fig. 3C and F) (Chang et al., 1993; Paré et al., 2019). By contrast, the Ten-m protein appears to be present in all cells, raising the question of how Ten-m is specifically enriched at compartment boundaries (Paré et al., 2019). Loss- and gain-of-function experiments show that Tartan inhibits Ten-m membrane localization when these proteins are present in the same cell (*in cis*), but stabilizes Ten-m at the membrane when these proteins are present in neighboring cells (*in trans*). These dual interactions involving *cis*-inhibition and *trans*-activation result in the enrichment of Ten-m at the borders of Tartan stripes (Fig. 3C and F) (Paré et al., 2019), and may represent a general mechanism for patterning the activity of uniformly expressed receptors (Fig. 4D). Consistent with this possibility, there is evidence for similar mechanisms regulating Notch signaling (del Álamo, Rouault, & Schweisguth, 2011) and the localized activities of proteins in the Frizzled/Disheveled planar cell polarity (PCP) system (Peng & Axelrod, 2012; Yang & Mlodzik, 2015).

Together, the Tartan/Ten-m and Toll receptor systems represent spatially overlapping yet independent mechanisms for establishing planar polarity in distinct regions of the germband. However, how these signals influence actomyosin organization and cell polarity is not known. An intriguing idea is that the downstream effectors of Toll receptors may induce dynamic contractile structures that promote cell intercalation, whereas the effectors of Tartan and Ten-m may confer distinct contractile and adhesive properties at compartment boundaries that allow them to serve as stable barriers to cell crossing.

8. Control of junctional and medial myosin by G protein-coupled receptors

Apical constriction in the mesoderm, like cell intercalation in the ectoderm, relies on actomyosin contractility to produce spatially regulated

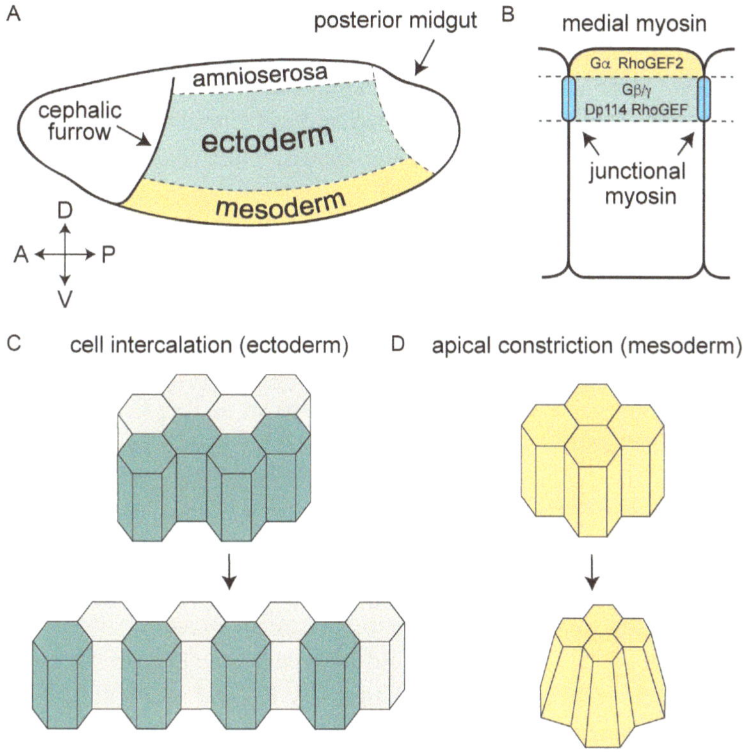

Fig. 5 Region-specific cell behaviors in the *Drosophila* embryo. (A) Embryo schematic showing the germband ectoderm (blue) and mesoderm (yellow). (B) GPCR signaling through the Gα alpha subunit and RhoGEF2 activates myosin in the medial (apical) cellular domain. GPCR signaling through the Gβ/Gγ subunits and the RhoGEF Dp114/Cysts recruits myosin to adherens junctions in the ectoderm to promote cell intercalation. (C) Schematic of intercalating cells in the ectoderm. (D) Schematic of apically constricting cells in the mesoderm.

and coordinated changes in cell shape (Fig. 5A, C and D) (Dawes-Hoang et al., 2005; Gorfinkiel & Blanchard, 2011; Kasza & Zallen, 2011; Martin & Goldstein, 2014; Munjal & Lecuit, 2014; Sweeton, Parks, Costa, & Wieschaus, 1991). Recent studies demonstrate that these two cell behaviors also share common molecular regulators. In particular, signaling through distinct sets of upstream regulators in the G protein-coupled receptor (GPCR) and RhoGEF protein families plays a critical role in activating apical constriction in mesodermal cells and cell intercalation in the germband. These studies provide insight into how myosin activity is differentially activated in the ectoderm and mesoderm to mediate distinct cell behaviors in these neighboring tissues.

In the mesoderm, apical constriction is induced by the ventral expression of specific factors under the control of the DV-patterning system, including the secreted ligand Fog (Costa, Wilson, & Wieschaus, 1994), the transmembrane protein T48 (Kölsch, Seher, Fernandez-Ballester, Serrano, & Leptin, 2007), and the GPCR Mist (Manning, Peters, Peifer, & Rogers, 2013). Fog binds to Mist (Manning et al., 2013), and perhaps to the ubiquitously expressed GPCR Smog (Kerridge et al., 2016), leading to activation of the Gα protein Concertina (Parks & Wieschaus, 1991). T48 and Concertina localize to and activate RhoGEF2 at the apical cell cortex of mesodermal cells (Kölsch et al., 2007; Rogers, Wiedemann, Häcker, Turck, & Vale, 2004), and RhoGEF2 in turn activates Rho GTPase by converting the inactive GDP-bound form to an active, GTP-bound form (Jaffe & Hall, 2005). Rho then activates Rho-kinase to trigger myosin contractility in the medial cell cortex, a process referred to as radial cell polarity (Martin & Goldstein, 2014; Mason, Tworoger, & Martin, 2013). Myosin contractility decreases the apical surface area of mesodermal cells in a pulsatile fashion through physical linkages to the actin cytoskeleton and the cellular junctions, driving mesoderm invagination (Martin et al., 2009) (see chapter "The cellular and molecular mechanisms that establish the mechanics of *Drosophila* gastrulation" by Ko and Martin).

In the germband, actomyosin networks are organized into medial and junctional myosin populations (Levayer & Lecuit, 2013; Rauzi et al., 2010). Both populations of myosin are responsive to GPCR signaling, but they appear to be separately controlled by Gα and RhoGEF2 in the medial domain and by Gβ13F/Gγ1 and the RhoGEF Dp114/Cysts at junctions (Fig. 5B) (Garcia De Las Bayonas et al., 2019; Kerridge et al., 2016). Disruption of RhoGEF2 or the Gα protein Concertina selectively affects medial myosin, without altering junctional myosin (Garcia De Las Bayonas et al., 2019; Kerridge et al., 2016). By contrast, mutations affecting Smog and Dp114/Cysts specifically reduce junctional myosin (Garcia De Las Bayonas et al., 2019; Kerridge et al., 2016; Silver et al., 2019). Dp114/Cysts specifically localizes to and enhances Rho GTPase activity at adherens junctions in response to the apical polarity protein Crumbs (Silver et al., 2019) and Gβ13F/Gγ1 activity (Garcia De Las Bayonas et al., 2019). Importantly, junctional Dp114/Cysts localization is observed in the ectoderm, but not in the mesoderm, suggesting a mechanism for restricting junctional myosin localization to the ectoderm (Garcia De Las Bayonas et al., 2019). The identification of an ectoderm-specific RhoGEF that promotes junctional myosin is a significant step forward in understanding how distinct cell behaviors are differentially activated in the ectoderm and mesoderm. It is interesting to speculate that cell intercalation may only occur in regions where Dp114/Cysts-mediated junctional myosin recruitment and Toll receptor-mediated planar polarity overlap. In this model, the ectoderm-specific junctional localization of Dp114/Cysts would render cells competent to support junctional myosin contractility, whereas patterned Toll receptors would be the triggering event that activates planar polarized myosin contractility to induce cell intercalation. Whether Toll receptors and GPCR signaling pathways function in parallel to mediate different aspects of cell polarity, or if these pathways converge at a molecular level to regulate Rho GTPase signaling, remains to be determined.

9. Current questions and future challenges

Significant advances have been made in understanding the cell behaviors, mechanical forces, and spatial cues that control convergent extension in epithelial tissues. Four general principles of epithelial remodeling have emerged. First, the molecular basis of planar polarity during cell intercalation ultimately involves the spatially regulated localization and activities of contractile and adhesive proteins. Second, mechanical forces transmitted between cells and tissues provide important spatial inputs that modulate protein dynamics and cell behavior. Third, biochemical and mechanical signals are actively integrated to promote collective cell behaviors such as rosette assembly and boundary formation that profoundly influence tissue structure. Finally, cell polarity is oriented to the body axes by a high-resolution network of locally acting spatial cues that establish planar polarity, position compartment boundaries, and restrict intercalary behavior to the ectoderm. Together, these spatial cues provide the structural foundation for multicellular organization in the *Drosophila* embryo, linking intrinsic cell polarities to tissue-scale organization.

Despite significant progress, much remains to be discovered. First, the discovery that stripes of Toll receptor and Tartan expression are necessary for planar polarity suggests that neighboring epithelial cells compute axial patterning information by interpreting local differences in receptor levels. While heterotypic interactions between Tartan and Ten-m provide a plausible mechanism for demarcating compartment boundaries, the molecular logic of the Toll receptor code is less clear. Insight into how cells detect qualitative and quantitative receptor differences in neighboring cells will provide fundamental insights into how patterned gene expression is converted into tissue structure.

Second, although several cell-surface receptors that regulate planar polarity have now been identified, the molecular connections between these upstream inputs and the downstream effectors of polarity are not understood. Many receptors, in addition to the Toll receptor and Tartan/Ten-m systems, have been shown to induce myosin accumulation at interfaces between cells that express different levels or types of receptors (Bielmeier, 2016). The signaling pathways that communicate to the actomyosin cytoskeleton downstream of these diverse receptor families are not known. How molecularly distinct classes of receptors in the germband all converge on actomyosin contractility, yet elicit different cell behaviors in intercalating and boundary domains, will be a fascinating avenue of further investigation.

Finally, these studies of dynamically reorganizing tissues reveal that mechanical forces play important roles in controlling contractile dynamics and cell behavior. However, the mechanosensors that detect these forces, as well as the downstream mechanotransduction pathways that convert mechanical forces into biochemical changes in protein localization and activity, largely remain to be discovered. A better understanding of how mechanical forces and biochemical signals are integrated during convergent extension will provide insight into how cell polarity and behavior are dynamic coordinated across multicellular tissues to produce robust and stereotyped tissue structures.

Acknowledgments

The authors thank all members of the Zallen and Paré labs, past and present, for helpful discussions, and Eric Brooks, Rodrigo Fernandez-Gonzalez, Karen Kasza, Sergio Simões, Masako Tamada, Athea Vichas, and Richard Zallen for comments on the manuscript. Research in ACP's lab is funded by the Vice Provost for Research and Economic Development and the Department of Biological Sciences at University of Arkansas. Research in JAZ's lab is funded by NIH/NIGMS grant GM079340. JAZ is an investigator of the Howard Hughes Medical Institute.

References

Akam, M. (1987). The molecular basis for metameric pattern in the *Drosophila* embryo. *Development*, *101*, 1–22.

Aliee, M., Röper, J.-C., Landsberg, K. P., Pentzold, C., Widmann, T. J., Jülicher, F., et al. (2012). Physical mechanisms shaping the *Drosophila* dorsoventral compartment boundary. *Current Biology*, *22*, 967–976.

Anderson, K. V. (2000). Toll signaling pathways in the innate immune response. *Current Opinion in Immunology, 12*, 13–19.

Anderson, K. V., Bokla, L., & Nüsslein-Volhard, C. (1985). Establishment of dorsal-ventral polarity in the *Drosophila* embryo: The induction of polarity by the Toll gene product. *Cell, 42*, 791–798.

Anderson, K. V., Jürgens, G., & Nüsslein-Volhard, C. (1985). Establishment of dorsal-ventral polarity in the *Drosophila* embryo: Genetic studies on the role of the Toll gene product. *Cell, 42*, 779–789.

Antinucci, P., Nikolaou, N., Meyer, M. P., & Hindges, R. (2013). Teneurin-3 specifies morphological and functional connectivity of retinal ganglion cells in the vertebrate visual system. *Cell Reports, 5*, 582–592.

Ballard, S. L., Miller, D. L., & Ganetzky, B. (2014). Retrograde neurotrophin signaling through Tollo regulates synaptic growth in *Drosophila*. *The Journal of Cell Biology, 204*, 1157–1172.

Batlle, E., & Wilkinson, D. G. (2012). Molecular mechanisms of cell segregation and boundary formation in development and tumorigenesis. *Cold Spring Harbor Perspectives in Biology, 4*, a008227.

Baumgartner, S., & Wides, R. (2019). Discovery of teneurins. *Frontiers in Neuroscience, 13*, 230.

Benton, M. A., Pechmann, M., Frey, N., Stappert, D., Conrads, K. H., Chen, Y. T., et al. (2016). Toll genes have an ancestral role in axis elongation. *Current Biology, 26*, 1609–1615.

Berns, D. S., DeNardo, L. A., Pederick, D. T., & Luo, L. (2018). Teneurin-3 controls topographic circuit assembly in the hippocampus. *Nature, 554*, 328–333.

Bertet, C., Sulak, L., & Lecuit, T. (2004). Myosin-dependent junction remodelling controls planar cell intercalation and axis elongation. *Nature, 429*, 667–671.

Bielmeier, C. (2016). Interface contractility between differently fated cells drives cell elimination and cyst formation. *Current Biology, 26*, 563–574.

Blankenship, J. T., Backovic, S. T., Sanny, J. S. S. P., Weitz, O., & Zallen, J. A. (2006). Multicellular rosette formation links planar cell polarity to tissue morphogenesis. *Developmental Cell, 11*, 459–470.

Butler, L. C., Blanchard, G. B., Kabla, A. J., Lawrence, N. J., Welchman, D. P., Mahadevan, L., et al. (2009). Cell shape changes indicate a role for extrinsic tensile forces in *Drosophila* germ-band extension. *Nature Cell Biology, 11*, 859–864.

Carvalho, L., Jacinto, A., & Matova, N. (2014). The Toll/NF-kappaB signaling pathway is required for epidermal wound repair in *Drosophila*. *Proceedings of the National Academy of Sciences of the United States of America, 111*, E5373–E5382.

Chacon-Heszele, M. F., Ren, D., Reynolds, A. B., Chi, F., & Chen, P. (2012). Regulation of cochlear convergent extension by the vertebrate planar cell polarity pathway is dependent on p120-catenin. *Development, 139*, 968–978.

Chang, Z., Price, B. D., Bockheim, S., Boedigheimer, M. J., Smith, R., & Laughon, A. (1993). Molecular and genetic characterization of the *Drosophila tartan* gene. *Developmental Biology, 160*, 315–332.

Chiang, C., & Beachy, P. A. (1994). Expression of a novel Toll-like gene spans the parasegment boundary and contributes to hedgehog function in the adult eye of *Drosophila*. *Mechanisms of Development, 47*, 225–239.

Clark, E. (2017). Dynamic patterning by the *Drosophila* pair-rule network reconciles long-germ and short-germ segmentation. *PLoS Biology, 15*, e2002439.

Collinet, C., Rauzi, M., Lenne, P.-F., & Lecuit, T. (2015). Local and tissue-scale forces drive oriented junction growth during tissue extension. *Nature Cell Biology, 17*, 1247–1258.

Costa, M., Wilson, E. T., & Wieschaus, E. (1994). A putative cell signal encoded by the folded gastrulation gene coordinates cell shape changes during *Drosophila* gastrulation. *Cell, 76*, 1075–1089.

Dahmann, C., & Basler, K. (1999). Compartment boundaries at the edge of development. *Trends in Genetics*, *15*, 320–326.
Dahmann, C., Oates, A. C., & Brand, M. (2011). Boundary formation and maintenance in tissue development. *Nature Reviews. Genetics*, *12*, 43–55.
Dawes-Hoang, R. E., Parmar, K. M., Chistiansen, A. E., Phelps, C. B., Brand, A. H., & Wieschaus, E. F. (2005). Folded gastrulation, cell shape change and the control of myosin localization. *Development*, *132*, 4165–4178.
del Álamo, D., Rouault, H., & Schweisguth, F. (2011). Mechanism and significance of cis-inhibition in notch signalling. *Current Biology*, *21*, R40–R47.
Dharmaratne, N., Glendining, K. A., Young, T. R., Tran, H., Sawatari, A., & Leamey, C. A. (2012). Ten-m3 is required for the development of topography in the ipsilateral retinocollicular pathway. *Plos One*, *7*, e43083.
Effler, J. C., Kee, Y.-S., Berk, J. M., Tran, M. N., Iglesias, P. A., & Robinson, D. N. (2006). Mitosis-specific mechanosensing and contractile-protein redistribution control cell shape. *Current Biology*, *16*, 1962–1967.
Eldon, E., Kooyer, S., D'Evelyn, D., Duman, M., Lawinger, P., Botas, J., et al. (1994). The *Drosophila* 18 wheeler is required for morphogenesis and has striking similarities to Toll. *Development*, *120*, 885–899.
Fagotto, F. (2014). The cellular basis of tissue separation. *Development*, *141*, 3303–3318.
Farhadifar, R., Röper, J.-C., Aigouy, B., Eaton, S., & Jülicher, F. (2007). The influence of cell mechanics, cell-cell interactions, and proliferation on epithelial packing. *Current Biology*, *17*, 2095–2104.
Farrell, D. L., Weitz, O., Magnasco, M. O., & Zallen, J. A. (2017). SEGGA: a toolset for rapid automated analysis of epithelial cell polarity and dynamics. *Development*, *144*, 1725–1734.
Fernandez-Gonzalez, R., Simões, S. d. M., Röper, J.-C., Eaton, S., & Zallen, J. A. (2009). Myosin II dynamics are regulated by tension in intercalating cells. *Developmental Cell*, *17*, 736–743.
Fernandez-Gonzalez, R., & Zallen, J. A. (2011). Oscillatory behaviors and hierarchical assembly of contractile structures in intercalating cells. *Physical Biology*, *8*, 045005.
Foldi, I., Anthoney, N., Harrison, N., Gangloff, M., Verstak, B., Nallasivan, M. P., et al. (2017). Three-tier regulation of cell number plasticity by neurotrophins and Tolls in *Drosophila*. *The Journal of Cell Biology*, *216*, 1421–1438.
Garcia De Las Bayonas, A., Philippe, J.-M., Lellouch, A. C., & Lecuit, T. (2019). Distinct RhoGEFs activate apical and junctional contractility under control of G proteins during epithelial morphogenesis. *Current Biology*, *29*, 3370–3385.e7.
Garcia-Bellido, A., Ripoll, P., & Morata, G. (1973). Developmental compartmentalisation of the wing disk of *Drosophila*. *Nature: New Biology*, *245*, 251–253.
Gorfinkiel, N., & Blanchard, G. B. (2011). Dynamics of actomyosin contractile activity during epithelial morphogenesis. *Current Opinion in Cell Biology*, *23*, 531–539.
Graham, P. L., Anderson, W. R., Brandt, E. A., Xiang, J., & Pick, L. (2019). Dynamic expression of *Drosophila* segmental cell surface-encoding genes and their pair-rule regulators. *Developmental Biology*, *447*, 147–156.
Hashimoto, C., Hudson, K. L., & Anderson, K. V. (1988). The Toll gene of *Drosophila*, required for dorsal-ventral embryonic polarity, appears to encode a transmembrane protein. *Cell*, *52*, 269–279.
Hong, W., & Luo, L. (2014). Genetic control of wiring specificity in the fly olfactory system. *Genetics*, *196*, 17–29.
Hong, W., Zhu, H., Potter, C. J., Barsh, G., Kurusu, M., Zinn, K., et al. (2009). Leucine-rich repeat transmembrane proteins instruct discrete dendrite targeting in an olfactory map. *Nature Neuroscience*, *12*, 1542–1550.

Hutson, M. S., Tokutake, Y., Chang, M. S., Bloor, J. W., Venakides, S., Kiehart, D. P., et al. (2003). Forces for morphogenesis investigated with laser microsurgery and quantitative modeling. *Science*, *300*, 145–149.

Irvine, K. D., & Wieschaus, E. (1994). Cell intercalation during *Drosophila* germband extension and its regulaton by pair- rule segmentation genes. *Development*, *120*, 827–841.

Jaffe, A. B., & Hall, A. (2005). Rho GTPases: biochemistry and biology. *Annual Review of Cell and Developmental Biology*, *21*, 247–269.

Jewett, C. E., Vanderleest, T. E., Miao, H., Xie, Y., Madhu, R., Loerke, D., et al. (2017). Planar polarized Rab35 functions as an oscillatory ratchet during cell intercalation in the *Drosophila* epithelium. *Nature Communications*, *8*, 476.

Kambris, Z., Hoffmann, J. A., Imler, J. L., & Capovilla, M. (2002). Tissue and stage-specific expression of the Tolls in *Drosophila* embryos. *Gene Expression Patterns*, *2*, 311–317.

Kasza, K. E., Farrell, D. L., & Zallen, J. A. (2014). Spatiotemporal control of epithelial remodeling by regulated myosin phosphorylation. *Proceedings of the National Academy of Sciences of the United States of America*, *111*, 11732–11737.

Kasza, K. E., & Zallen, J. A. (2011). Dynamics and regulation of contractile actin-myosin networks in morphogenesis. *Current Opinion in Cell Biology*, *23*, 30–38.

Kawasaki, T., & Kawai, T. (2014). Toll-like receptor signaling pathways. *Frontiers in Immunology*, *5*, 461.

Keller, R., Davidson, L., Edlund, A., Elul, T., Ezin, M., Shook, D., et al. (2000). Mechanisms of convergence and extension by cell intercalation. *Philosophical Transactions of the Royal Society of London. Series B, Biological Sciences*, *355*, 897–922.

Kerridge, S., Munjal, A., Philippe, J.-M., Jha, A., de las Bayonas, A. G., Saurin, A. J., et al. (2016). Modular activation of Rho1 by GPCR signalling imparts polarized myosin II activation during morphogenesis. *Nature Cell Biology*, *18*, 261–270.

Kim, S., Chung, S., Yoon, J., Choi, K.-W., & Yim, J. (2006). Ectopic expression of Tollo/Toll-8 antagonizes Dpp signaling and induces cell sorting in the *Drosophila* wing. *Genesis*, *44*, 541–549.

Kleve, C. D., Siler, D. A., Syed, S. K., & Eldon, E. D. (2006). Expression of *18-wheeler* in the follicle cell epithelium affects cell migration and egg morphology in *Drosophila*. *Developmental Dynamics*, *235*, 1953–1961.

Kolesnikov, T., & Beckendorf, S. K. (2007). 18 wheeler regulates apical constriction of salivary gland cells via the Rho-GTPase-signaling pathway. *Developmental Biology*, *307*, 53–61.

Kölsch, V., Seher, T., Fernandez-Ballester, G. J., Serrano, L., & Leptin, M. (2007). Control of *Drosophila* gastrulation by apical localization of adherens junctions and RhoGEF2. *Science*, *315*, 384–386.

Kong, D., Wolf, F., & Grosshans, J. (2017). Forces directing germ-band extension in *Drosophila* embryos. *Mechanisms of Development*, *144*, 11–22.

Krahn, M. P., Klopfenstein, D. R., Fischer, N., & Wodarz, A. (2010). Membrane targeting of Bazooka/PAR-3 is mediated by direct binding to phosphoinositide lipids. *Current Biology*, *20*, 636–642.

Krause, C., Wolf, C., Hemphälä, J., Samakovlis, C., & Schuh, R. (2006). Distinct functions of the leucine-rich repeat transmembrane proteins capricious and tartan in the *Drosophila* tracheal morphogenesis. *Developmental Biology*, *296*, 253–264.

Kurusu, M., Cording, A., Taniguchi, M., Menon, K., Suzuki, E., & Zinn, K. (2008). A screen of cell-surface molecules identifies leucine-rich repeat proteins as key mediators of synaptic target selection. *Neuron*, *59*, 972–985.

Lan, H., Wang, Q., Fernandez-Gonzalez, R., & Feng, J. J. (2015). A biomechanical model for cell polarization and intercalation during *Drosophila* germband extension. *Physical Biology*, *12*, 056011.

Landsberg, K. P., Farhadifar, R., Ranft, J., Umetsu, D., Widmann, T. J., Bittig, T., et al. (2009). Increased cell bond tension governs cell sorting at the *Drosophila* anteroposterior compartment boundary. *Current Biology, 19*, 1950–1955.

Lau, K., Tao, H., Liu, H., Wen, J., Sturgeon, K., Sorfazlian, N., et al. (2015). Anisotropic stress orients remodelling of mammalian limb bud ectoderm. *Nature Cell Biology, 17*, 569–579.

Lawrence, P. A. (1973). A clonal analysis of segment development in *Oncopeltus* (Hemiptera). *Journal of Embryology and Experimental Morphology, 30*, 681–699.

Lawrence, P. A., & Struhl, G. (1996). Morphogens, compartments, and pattern: Lessons from *Drosophila? Cell, 85*, 951–961.

Leamey, C. A., Glendining, K. A., Kreiman, G., Kang, N.-D., Wang, K. H., Fassler, R., et al. (2008). Differential gene expression between sensory neocortical areas: Potential roles for Ten_m3 and Bcl6 in patterning visual and somatosensory pathways. *Cerebral Cortex, 18*, 53–66.

Leulier, F., & Lemaitre, B. (2008). Toll-like receptors—Taking an evolutionary approach. *Nature Reviews. Genetics, 9*, 165.

Levayer, R., & Lecuit, T. (2013). Oscillation and polarity of E-cadherin asymmetries control actomyosin flow patterns during morphogenesis. *Developmental Cell, 26*, 162–175.

Levayer, R., Pelissier-Monier, A., & Lecuit, T. (2011). Spatial regulation of Dia and Myosin-II by RhoGEF2 controls initiation of E-cadherin endocytosis during epithelial morphogenesis. *Nature Cell Biology, 13*, 529–542.

Lienkamp, S. S., Liu, K., Karner, C. M., Carroll, T. J., Ronneberger, O., Wallingford, J. B., et al. (2012). Vertebrate kidney tubules elongate using a planar cell polarity-dependent, rosette-based mechanism of convergent extension. *Nature Genetics, 44*, 1382–1387.

Lye, C. M., Blanchard, G. B., Naylor, H. W., Muresan, L., Huisken, J., Adams, R. J., et al. (2015). Mechanical coupling between endoderm invagination and axis extension in *Drosophila. PLoS Biology, 13*, e1002292.

Major, R. J., & Irvine, K. D. (2005). Influence of Notch on dorsoventral compartmentalization and actin organization in the *Drosophila* wing. *Development, 132*, 3823–3833.

Major, R. J., & Irvine, K. D. (2006). Location and requirement for Myosin II at the dorsal-ventral compartment boundary of the *Drosophila* wing. *Developmental Dynamics, 235*, 3051–3058.

Manning, A. J., Peters, K. A., Peifer, M., & Rogers, S. L. (2013). Regulation of epithelial morphogenesis by the G protein-coupled receptor mist and its ligand fog. *Science Signaling, 6*, ra98.

Mao, Y., Kerr, M., & Freeman, M. (2008). Modulation of *Drosophila* retinal epithelial integrity by the adhesion proteins capricious and tartan. *Plos One, 3*, e1827.

Martin, A. C., Gelbart, M., Fernandez-Gonzalez, R., Kaschube, M., & Wieschaus, E. F. (2010). Integration of contractile forces during tissue invagination. *The Journal of Cell Biology, 188*, 735–749.

Martin, A. C., & Goldstein, B. (2014). Apical constriction: Themes and variations on a cellular mechanism driving morphogenesis. *Development, 141*, 1987–1998.

Martin, A. C., Kaschube, M., & Wieschaus, E. F. (2009). Pulsed contractions of an actin-myosin network drive apical constriction. *Nature, 457*, 495–499.

Mason, F. M., Tworoger, M., & Martin, A. C. (2013). Apical domain polarization localizes actin-myosin activity to drive ratchet-like apical constriction. *Nature Cell Biology, 15*, 926–936.

McGreevy, E. M., Vijayraghavan, D., Davidson, L. A., & Hildebrand, J. D. (2015). Shroom3 functions downstream of planar cell polarity to regulate myosin II distribution and cellular organization during neural tube closure. *Biology Open, 4*, 186–196.

McIlroy, G., Foldi, I., Aurikko, J., Wentzell, J. S., Lim, M. A., Fenton, J. C., et al. (2013). Toll-6 and Toll-7 function as neurotrophin receptors in the *Drosophila melanogaster* CNS. *Nature Neuroscience*, *16*, 1248–1256.

Meyer, S. N., Amoyel, M., Bergantinos, C., de la Cova, C., Schertel, C., Basler, K., et al. (2014). An ancient defense system eliminates unfit cells from developing tissues during cell competition. *Science*, *346*, 1258236.

Milán, M., Pérez, L., & Cohen, S. M. (2005). Boundary formation in the *Drosophila* wing: Functional dissection of capricious and tartan. *Developmental Dynamics*, *233*, 804–810.

Milán, M., Weihe, U., Pérez, L., & Cohen, S. M. (2001). The LRR proteins capricious and Tartan mediate cell interactions during DV boundary formation in the *Drosophila* wing. *Cell*, *106*, 785–794.

Monier, B., Pélissier-Monier, A., Brand, A. H., & Sanson, B. (2010). An actomyosin-based barrier inhibits cell mixing at compartmental boundaries in *Drosophila* embryos. *Nature Cell Biology*, *12*, 60–69.

Mosca, T. J. (2015). On the Teneurin track: A new synaptic organization molecule emerges. *Frontiers in Cellular Neuroscience*, *9*, 204.

Mosca, T. J., Hong, W., Dani, V. S., Favaloro, V., & Luo, L. (2012). Trans-synaptic Teneurin signalling in neuromuscular synapse organization and target choice. *Nature*, *484*, 237–241.

Munjal, A., & Lecuit, T. (2014). Actomyosin networks and tissue morphogenesis. *Development*, *141*, 1789–1793.

Munjal, A., Philippe, J.-M., Munro, E., & Lecuit, T. (2015). A self-organized biomechanical network drives shape changes during tissue morphogenesis. *Nature*, *524*, 351–355.

Nishimura, T., Honda, H., & Takeichi, M. (2012). Planar cell polarity links axes of spatial dynamics in neural-tube closure. *Cell*, *149*, 1084–1097.

Nishimura, T., & Takeichi, M. (2008). Shroom3-mediated recruitment of Rho kinases to the apical cell junctions regulates epithelial and neuroepithelial planar remodeling. *Development*, *135*, 1493–1502.

Nüsslein-Volhard, C., & Wieschaus, E. (1980). Mutations affecting segment number and polarity in *Drosophila*. *Nature*, *287*, 795–801.

Paré, A. C., Naik, P., Shi, J., Mirman, Z., Palmquist, K. H., & Zallen, J. A. (2019). An LRR receptor-Teneurin system directs planar polarity at compartment boundaries. *Developmental Cell*, *51*, 208–221.

Paré, A. C., Vichas, A., Fincher, C. T., Mirman, Z., Farrell, D. L., Mainieri, A., et al. (2014). A positional Toll receptor code directs convergent extension in *Drosophila*. *Nature*, *515*, 523–527.

Parks, S., & Wieschaus, E. (1991). The *Drosophila* gastrulation gene *concertina* encodes a G alpha-like protein. *Cell*, *64*, 447–458.

Peng, Y., & Axelrod, J. D. (2012). Asymmetric protein localization in planar cell polarity: Mechanisms, puzzles, and challenges. *Current Topics in Developmental Biology*, *101*, 33–53.

Pouille, P.-A., Ahmadi, P., Brunet, A.-C., & Farge, E. (2009). Mechanical signals trigger Myosin II redistribution and mesoderm invagination in *Drosophila* embryos. *Science Signaling*, *2*, ra16.

Rauzi, M., Lenne, P.-F., & Lecuit, T. (2010). Planar polarized actomyosin contractile flows control epithelial junction remodelling. *Nature*, *468*, 1110–1114.

Rauzi, M., Verant, P., Lecuit, T., & Lenne, P.-F. (2008). Nature and anisotropy of cortical forces orienting *Drosophila* tissue morphogenesis. *Nature Cell Biology*, *10*, 1401–1410.

Rogers, S. L., Wiedemann, U., Häcker, U., Turck, C., & Vale, R. D. (2004). *Drosophila* RhoGEF2 associates with microtubule plus ends in an EB1-dependent manner. *Current Biology*, *14*, 1827–1833.

Rozbicki, E., Chuai, M., Karjalainen, A. I., Song, F., Sang, H. M., Martin, R., et al. (2015). Myosin-II-mediated cell shape changes and cell intercalation contribute to primitive streak formation. *Nature Cell Biology*, *17*, 397–408.

Sakurai, K. T., Kojima, T., Aigaki, T., & Hayashi, S. (2007). Differential control of cell affinity required for progression and refinement of cell boundary during *Drosophila* leg segmentation. *Developmental Biology*, *309*, 126–136.

Sawyer, J. K., Choi, W., Jung, K. C., He, L., Harris, N. J., & Peifer, M. (2011). A contractile actomyosin network linked to adherens junctions by Canoe/afadin helps drive convergent extension. *Molecular Biology of the Cell*, *22*, 2491–2508.

Scarpa, E., Finet, C., Blanchard, G. B., & Sanson, B. (2018). Actomyosin-driven tension at compartmental boundaries orients cell division independently of cell geometry in vivo. *Developmental Cell*, *47*, 727–740.e6.

Shah, P. K., Tanner, M. R., Kovacevic, I., Rankin, A., Marshall, T. E., Noblett, N., et al. (2017). PCP and SAX-3/Robo pathways cooperate to regulate convergent extension-based nerve cord assembly in *C. elegans*. *Developmental Cell*, *41*, 195–203.e3.

Shindo, A., & Wallingford, J. B. (2014). PCP and septins compartmentalize cortical actomyosin to direct collective cell movement. *Science*, *343*, 649–652.

Silver, J. T., Wirtz-Peitz, F., Simões, S., Pellikka, M., Yan, D., Binari, R., et al. (2019). Apical polarity proteins recruit the RhoGEF Cysts to promote junctional myosin assembly. *The Journal of Cell Biology*, *218*, 3397–3414.

Simões, S. M., Blankenship, J. T., Weitz, O., Farrell, D. L., Tamada, M., Fernandez-Gonzalez, R., et al. (2010). Rho-kinase directs Bazooka/Par-3 planar polarity during *Drosophila* axis elongation. *Developmental Cell*, *19*, 377–388.

Simões, S. M., Mainieri, A., & Zallen, J. A. (2014). Rho GTPase and Shroom direct planar polarized actomyosin contractility during convergent extension. *The Journal of Cell Biology*, *204*, 575–589.

Sun, Z., Amourda, C., Shagirov, M., Hara, Y., Saunders, T. E., & Toyama, Y. (2017). Basolateral protrusion and apical contraction cooperatively drive *Drosophila* germ-band extension. *Nature Cell Biology*, *19*, 375–383.

Sweeton, D., Parks, S., Costa, M., & Wieschaus, E. (1991). Gastrulation in *Drosophila:* The formation of the ventral furrow and posterior midgut invaginations. *Development*, *112*, 775–789.

Takeichi, M. (2014). Dynamic contacts: Rearranging adherens junctions to drive epithelial remodeling. *Nature Reviews. Molecular Cell Biology*, *15*, 397–410.

Tamada, M., Farrell, D. L., & Zallen, J. A. (2012). Abl regulates planar polarized junctional dynamics through β-catenin tyrosine phosphorylation. *Developmental Cell*, *22*, 309–319.

Tetley, R. J., Blanchard, G. B., Fletcher, A. G., Adams, R. J., & Sanson, B. (2016). Unipolar distributions of junctional Myosin II identify cell stripe boundaries that drive cell intercalation throughout *Drosophila* axis extension. *eLife*, *5*, e12094.

Umetsu, D., Dunst, S., & Dahmann, C. (2014). An RNA interference screen for genes required to shape the anteroposterior compartment boundary in *Drosophila* identifies the Eph receptor. *Plos One*, *9*, e114340.

Vanderleest, T. E., Smits, C. M., Xie, Y., Jewett, C. E., Blankenship, J. T., & Loerke, D. (2018). Vertex sliding drives intercalation by radial coupling of adhesion and actomyosin networks during *Drosophila* germband extension. *eLife*, 7, e34586.

Vincent, J. P., & O'Farrell, P. H. (1992). The state of *engrailed* expression is not clonally transmitted during early *Drosophila* development. *Cell*, *68*, 923–931.

Walck-Shannon, E., & Hardin, J. (2014). Cell intercalation from top to bottom. *Nature Reviews. Molecular Cell Biology*, *15*, 34–48.

Wallingford, J. B., Fraser, S. E., & Harland, R. M. (2002). Convergent extension: the molecular control of polarized cell movement during embryonic development. *Developmental Cell*, *2*, 695–706.

Wang, M. F. Z., Hunter, M. V., Wang, G., McFaul, C., Yip, C. M., & Fernandez-Gonzalez, R. (2017). Automated cell tracking identifies mechanically oriented cell divisions during *Drosophila* axis elongation. *Development, 144*, 1350–1361.

Ward, A., Hong, W., Favaloro, V., & Luo, L. (2015). Toll receptors instruct axon and dendrite targeting and participate in synaptic partner matching in a *drosophila* olfactory circuit. *Neuron, 85*, 1013–1028.

Warrington, S. J., Strutt, H., & Strutt, D. (2013). The Frizzled-dependent planar polarity pathway locally promotes E-cadherin turnover via recruitment of RhoGEF2. *Journal of Cell Science, 140*, 1045–1054.

Weaire, D., & Rivier, N. (1984). Soap, cells and statistics—Random patterns in two dimensions. *Contemporary Physics, 25*, 59–99.

Wieschaus, E., & Nüsslein-Volhard, C. (2016). The Heidelberg screen for pattern mutants of *Drosophila*: A personal account. *Annual Review of Cell and Developmental Biology, 32*, 1–46.

Wieschaus, E., Sweeton, D., & Costa, M. (1991). Convergence and extension during germband elongation in *Drosophila* embryos. In R. Keller (Ed.), *Gastrulation* (pp. 213–223). New York: Plenum Press.

Williams, M. L. K., Sawada, A., Budine, T., Yin, C., Gontarz, P., & Solnica-Krezel, L. (2018). Gon4l regulates notochord boundary formation and cell polarity underlying axis extension by repressing adhesion genes. *Nature Communications, 9* 1319.

Williams, M., Yen, W., Lu, X., & Sutherland, A. (2014). Distinct apical and basolateral mechanisms drive planar cell polarity-dependent convergent extension of the mouse neural plate. *Developmental Cell, 29*, 34–46.

Yagi, Y., Nishida, Y., & Ip, Y. T. (2010). Functional analysis of Toll-related genes in *Drosophila. Development, Growth & Differentiation, 52*, 771–783.

Yang, Y., & Mlodzik, M. (2015). Wnt-Frizzled/planar cell polaritiy signaling: Cellular orientation by facing the Wing (Wnt). *Annual Review of Cell and Developmental Biology, 31*, 623–646.

Yu, J. C., & Fernandez-Gonzalez, R. (2016). Local mechanical forces promote polarized junctional assembly and axis elongation in *Drosophila. eLife, 5*, e10757.

Zallen, J. A., & Wieschaus, E. (2004). Patterned gene expression directs bipolar planar polarity in *Drosophila. Developmental Cell, 6*, 343–355.

Zallen, J. A., & Zallen, R. (2004). Cell-pattern disordering during convergent extension in *Drosophila. Journal of Physics. Condensed Matter, 16*, S5073.

Zhang, Y., Kong, D., Reichl, L., Vogt, N., Wolf, F., & Grosshans, J. (2014). The glucosyltransferase Xiantuan of the endoplasmic reticulum specifically affects E-Cadherin expression and is required for gastrulation movements in *Drosophila. Developmental Biology, 390*, 208–220.

Further reading

Finegan, T. M., Hervieux, N., Nestor-Bergmann, A., Fletcher, A. G., Blanchard, G. B., & Sanson, B. (2019). The tricellular vertex-specific adhesion molecule Sidekick facilitates polarised cell intercalation during *Drosophila* axis extension. *PLoS Biology, 17*, e3000522.

Razzell, W., Bustillo, M. E., & Zallen, J. A. (2018). The force-sensitive protein Ajuba regulates cell adhesion during epithelial morphogenesis. *The Journal of Cell Biology, 217*, 3715–3730.

CHAPTER SEVEN

Gastrulation in the sea urchin

David R. McClay[a,*], Jacob Warner[b], Megan Martik[c], Esther Miranda[a], Leslie Slota[a]

[a]Department of Biology, Duke University, Durham, NC, United States
[b]Department of Biology, University of North Carolina, Wilmington, NC, United States
[c]Division of Biology and Biological Engineering, California institute of Technology, Pasadena, CA, United States
*Corresponding author: e-mail address: dmcclay@duke.edu

Contents

Abstract

Gastrulation is arguably the most important evolutionary innovation in the animal kingdom. This process provides the basic embryonic architecture, an inner layer separated from an outer layer, from which all animal forms arise. An extraordinarily simple and elegant process of gastrulation is observed in the sea urchin embryo. The cells participating in sea urchin gastrulation are specified early during cleavage. One outcome of that specification is the expression of transcription factors that control each of the many subsequent morphogenetic changes. The first of these movements is an epithelial-mesenchymal transition (EMT) of skeletogenic mesenchyme cells, then EMT of pigment cell progenitors. Shortly thereafter, invagination of the archenteron occurs. At the end of archenteron extension, a second wave of EMT occurs to release immune cells into the blastocoel and primordial germ cells that will home to the coelomic pouches. The archenteron then remodels to establish the three parts of the gut, and at the anterior end, the gut fuses with the stomodaeum to form the through-gut. As part of the anterior remodeling, mesodermal coelomic pouches bud off the lateral sides of the archenteron tip. Multiple cell biological processes conduct each of these movements and in some

Current Topics in Developmental Biology, Volume 136
ISSN 0070-2153
https://doi.org/10.1016/bs.ctdb.2019.08.004

cases the upstream transcription factors controlling this process have been identified. Remarkably, each event seamlessly occurs at the right time to orchestrate formation of the primitive body plan. This review covers progress toward understanding many of the molecular mechanisms underlying this sequence of morphogenetic events.

1. A sequential overview of sea urchin gastrulation

As a model system for studying gastrulation the sea urchin embryo is valuable for several properties. Those properties include the ability to perform live whole-embryo imaging with many fluorescent probes to reveal molecular dynamics of the process. The system is tractable for experiments that enable assembly of gene regulatory networks, and these provide insight into the control of morphogenetic events. Classically, the system is known for its availability for cell transplantation experiments which open avenues for testing hypotheses about how each of the many events work. These approaches have led to a growing depth of understanding of the steps that lead up to and through gastrulation. As such, sea urchin gastrulation is an excellent model for the study of morphogenesis in the animal kingdom.

To understand sea urchin gastrulation, several species have been utilized including *Strongylocentrotus purpuratus* and *Lytechinus variegatus* in the United States, *Paracentrotus lividus* in Europe, and *Hemicentrotus pulcherrimus* in Japan. Each of these species has its own advantages for studying morphogenetic events of gastrulation, and for the most part the sequence of events is highly conserved. In the description that follows, the timing refers to hours post fertilization (hpf) in *L. variegatus* embryos grown at 23 °C.

In preparation for gastrulation, a number of important steps occur beginning with oogenesis. Egg-localized maternal factors establish the anterior-posterior axis. After fertilization the dorsal-ventral (aboral-oral) axis is established by Nodal signaling early in cleavage (Duboc, Rottinger, Besnardeau, & Lepage, 2004) and the left-right axis is established later during gastrulation, again by Nodal signaling (Duboc, Rottinger, Lapraz, Besnardeau, & Lepage, 2005). During cleavage and throughout the blastula stage, molecular specification establishes each of the three germ layers by about hatched blastula stage (7.5 hpf) and specification continues after that time to program several subtypes of cells within those germ layers. By the mesenchyme blastula stage (9 hpf) specification has diversified cells into about 15 different fates, and each lineage has been well documented (Cameron, Hough-Evans, Britten, & Davidson, 1987; Logan & McClay, 1997). Prior to initiation of gastrulation, the embryo is a simple spherical monolayer. A large number of experiments over the last two decades have established models of gene regulatory networks that approximate the molecular sequence in specification of each of the cell types in the embryo as it approaches gastrulation (Davidson & Erwin, 2006; Davidson et al., 2002a; Li, Cui, Peter, & Davidson, 2014; Peter & Davidson, 2017; Su et al., 2009).

At the hatched blastula stage (7.5 hpf), the specified cell types present in the embryo can be broadly classified into anterior (animal) cells that will give rise to the ectoderm and neurons and the posterior (vegetal) cells which will give rise to the endoderm and mesoderm (Fig. 1). As the first sign of impending gastrulation the posterior cells change shape and become more columnar in an apical-basal orientation. At about 9 hpf the mesenchyme blastula stage is initiated with an epithelial-mesenchyme transition (EMT) and ingression of the mesodermal skeletogenic cells into the blastocoel (Fig. 1A). The 32 ingressed skeletogenic cells become motile, change shape with the cell body moving basally and through the basal lamina. The cells de-adhere from the adherens junction (Saunders & McClay, 2014), and immediately endocytose cadherin from the membrane and coincidentally exocytose proteins needed for mesenchymal migration (Fig. 2). After entering the blastocoel, the skeletogenic cells divide once mitotically, then migrate on the matrix substrate to encircle the base of the invaginating archenteron. The skeletogenic cells then fuse to form a syncytium and begin synthesizing the calcium carbonate skeleton at about 12 hpf beginning at two lateral sites in the embryo (Fig. 1B) (McIntyre, Lyons, Martik, & McClay, 2014). The biomineralization sequence is initiated in response to VEGF and FGF signaling from two sites in the ectoderm (Duloquin, Lhomond, & Gache, 2007; Rottinger et al., 2008).

Pigment cells, the second mesodermal cell type to undergo an EMT, enter the blastocoel between 12 and 13 hpf. Upon arrival in the blastocoel the pigment cells migrate to the dorsal side where they re-enter the ectoderm via a mesoderm to epithelial transition. Shortly after the pigment cells initiate their EMT, invagination of the archenteron begins. Non-skeletal mesoderm cells (NSM) at the center of the vegetal plate change shape as the first evidence of that invagination. Some of these NSM cells extend their cell bodies inward in a maneuver that caused these cells to be called "bottle cells" based on their shape (Kimberly & Hardin, 1998). The bottle cells initiate an inward bending of the archenteron and the surrounding NSM cells adopt a keystone shape which establishes the inbending dome. Mechanical properties involving the extracellular matrix are thought to participate in this

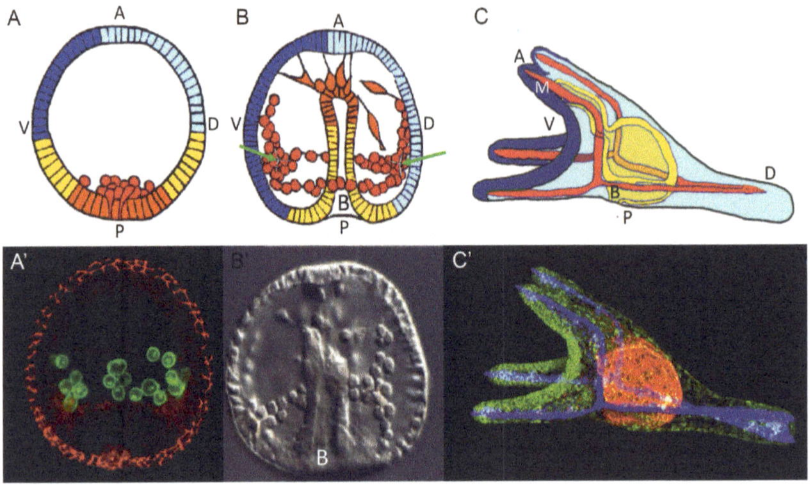

Fig. 1 Major steps in gastrulation. At the mesenchyme blastula stage (A and A′), the skeletogenic cells (red in (A), green in (A′)) undergo an epithelial-mesenchymal transition (EMT) to begin the gastrulation movements. Several hours later as shown in (B and B′), invagination of the archenteron occurs at the posterior end of the embryo (P), beginning with the inbending of the non-skeletal mesoderm (orange) to form the blastopore (B) which remains at the posterior end, and followed by invagination of the endoderm (yellow). As invagination occurs the skeletogenic cells move to surround the base of the archenteron, fuse to form a syncytium, and begin synthesizing the $CaCO_3$ skeleton (green arrows in (B)). Following completion of invagination (C and C′) the anterior end of the archenteron bends toward the ventral side (V) to contact and then fuse with the stomodaeum. That anterior opening becomes the mouth (M), the archenteron differentiates into a three-part gut, and the shape of the embryo changes to conform to the shape of the growing skeletal rods (red in (C), blue in (C′)). This stage is known as the pluteus larva. In the diagrams blue = ectoderm, yellow = endoderm, red and orange = mesoderm, (A) anterior (approximately the animal pole), (P) posterior (approximately the vegetal pole), (V) ventral (also known as oral), (D) dorsal (also known as aboral), (B) blastopore (which later becomes the site of the anus), (M) mouth (derived from the invagination of the stomodaeum). In (A′) the ectoderm is stained with an antibody to β-catenin (red) and an antibody that recognizes skeletogenic cells (green). The pluteus larva in (C′) is stained with a membrane marker (green), a marker specific for the gut (red), and a marker specific for skeletogenic cells (blue).

bending process (Davidson, Koehl, Keller, & Oster, 1995; Davidson, Oster, Keller, & Koehl, 1999). The archenteron then advances, first with the invagination of additional NSM cells (cells that will later form the coelomic pouches, two immune cell types, primordial germ cells and muscle cell progenitors). Endoderm invagination follows and begins at about 13 hpf. These cells push the mesoderm cells upward through the blastocoel (Fig. 1B and B′).

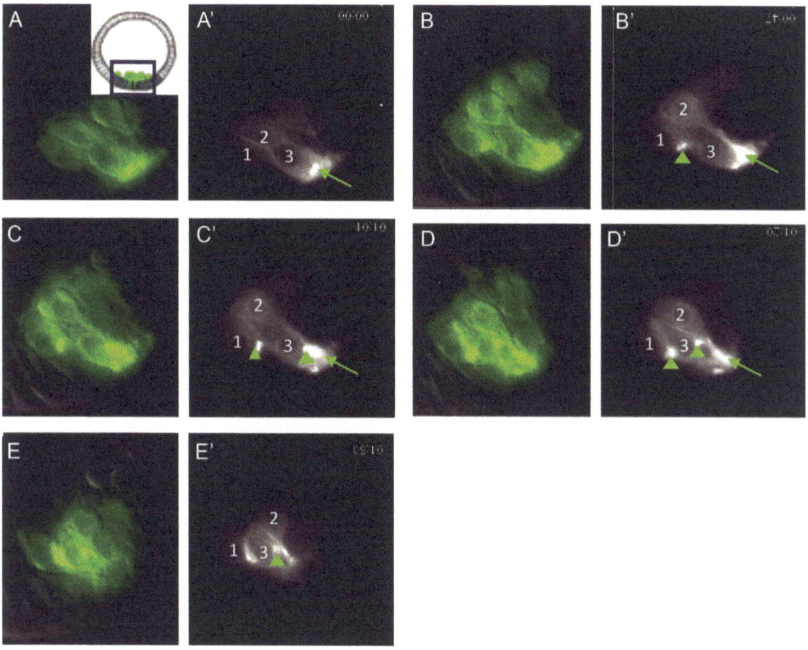

Fig. 2 Sequence of skeletogenic EMT. Shown from top to bottom are single images from a time lapse movie of an EMT (images of the region shown in the diagrammed insert in (A)). A single micromere at the 16-cell stage, previously injected with cadherin-GFP RNA, expresses cadherin GFP in adherens junctions (green arrow). From top to bottom cells 1, 2, and 3 were tracked. Cell 1 is the first to pull its cytoplasmic tail away from the adherens junctions and when that happens an intracellular vesicle of cadherin GFP appears (arrowhead in frame (B′)). Shortly thereafter an intracellular vesicle appears in cell 2 (right arrowhead in frame C′). Cell 3 is last to leave the adherens junction and internalize the fluorescent cadherin (arrowhead in frame (D′)). Finally, the adherens junction is no longer present when the skeletogenic cells from the transplant complete the EMT (frame (E, E′)).

Several types of cell movements are necessary to conduct the invagination. The arrival of endoderm cells from lateral positions on the vegetal plate results, at least in part, from the force derived from the spreading cell sheet of the vegetal plate. This spreading force is not globally provided as demonstrated by Ettensohn (1985b), who observed that isolated vegetal half embryos continued to invaginate, demonstrating that the mechanical force had to be local. Presumably, that force is generated in part by the resolution of a previously thickened vegetal plate which undergoes cell shape changes, and these cause a localized spreading, creating the force allowing cells to move through the blastopore and into the lengthening archenteron.

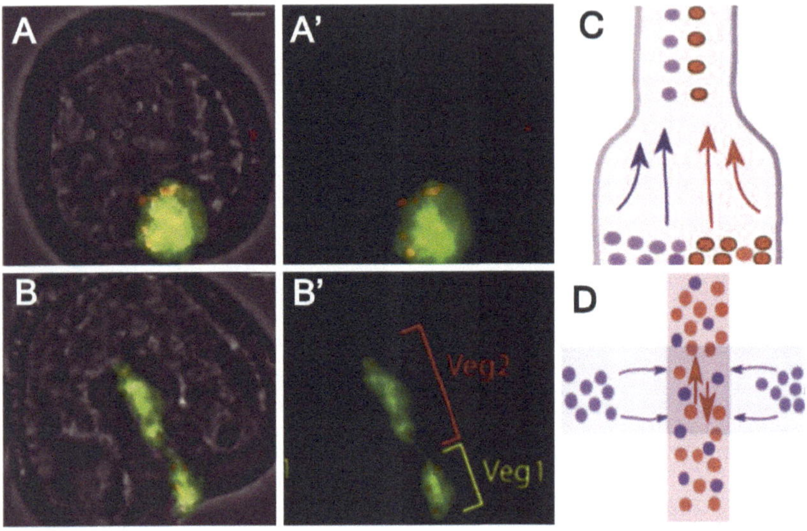

Fig. 3 Invagination of the archenteron. Shown are an early frame from a time lapse movie (A, A′) and a later frame of the same movie (B, B′). In the experiment, a single cell at the 32-cell stage was replaced by a cell expressing Histone RFP and membrane GFP. Progeny of that cell are seen to be entering the archenteron at the beginning of gastrulation. Later in gastrulation convergent-extension aligns the cells of that cluster into a single file. The diagram in (C) and (D) illustrate the two extreme forms of convergent extension. In (C), modeled after sea urchins, cells converge from a patch of cells as those cells move in the direction of extension and establish single files. In (D), modeled after zebrafish and frog, cells approach from 90 degrees and randomly merge to extend the tissue. *From Martik, M.L., & McClay, D.R. (2017). New insights from a high-resolution look at gastrulation in the sea urchin, Lytechinus variegatus.* Mechanisms of Development 148, *3–10.*

Additional forces for lengthening the archenteron are generated by repeated cycles of filopodial extension and contraction by NSM cells, and by convergent-extension movements of the endoderm (Fig. 3). Filopodial extension occurs cyclically; NSM cell filopodia make contact with the basal lamina on the sides of the blastocoel, briefly exert a pulling force on the archenteron, then the filopodia are released and retracted. This extension, pull, and retraction cycle continues throughout archenteron extension. As the endoderm cells invaginate, the diameter of the gut narrows from about 32 cells surrounding the blastopore to about eight cells per diameter along the archenteron (Ettensohn, 1985a). That narrowing is the result of convergent-extension (Martik & McClay, 2017). From time lapse movies, the convergent-extension operates as if each cluster of cells around the base of the blastopore converges about twice (Fig. 3A, A′, B, B′) (Martik & McClay, 2017). Though convergent-extension accounts for much of the lengthening of the archenteron, that morphogenetic movement occurs somewhat differently than the process seen in frogs and fish (contrast Fig. 3C with D). Further aiding archenteron elongation, as invagination proceeds, the endoderm cells divide on average one time each and the orientation of the divisions are biased along the long axis of the archenteron. The divisions therefore add to the lengthening of the archenteron though quantitative analyzes established that the oriented divisions contribute only about 10% of the total length of the archenteron (Martik & McClay, 2017).

The final 1/3 of archenteron extension relies heavily on filopodial contraction. In a series of experiments filopodia were severed with a laser each time they were extended. As a consequence, the archenteron did not proceed to more than 2/3 its final length (Hardin & McClay, 1990). Normally, the filopodia cause the archenteron to stretch, so the cells along the gut become anisotropic to account for much of the final third of the archenteron extension (Fig. 4D, inset). The filopodia continue to pull the gut toward the anterior end of the embryo until a "target" is reached at the anterior end. Experiments showed that if the target site is moved out of reach by distorting the ectoderm into a sausage shape, the filopodia continue to pull the archenteron for hours longer than the usual endpoint (Hardin & McClay, 1990). If the sausage shape of the embryo is relaxed at any time to allow the tip of the archenteron to contact the target, the filopodia attach and terminate their cycle of extension, attachment, contraction, and release (Hardin & McClay, 1990). Following the termination of archenteron extension, the anisotropic shape of the cells along the archenteron is resolved to be more isotropic due at least in part to the late addition of Veg1 hindgut cells at the posterior end (Fig. 4A–F and F inset) (Logan & McClay, 1997; Miller & McClay, 1997). Archenteron extension lasts for a total of about 5 h from 12 to 17 hpf. The target is reached at about 15 hpf and during the next 2 h the hindgut cells move into the archenteron.

After the archenteron reaches its final length the anterior end is the first to be remodeled. Several types of cells leave the archenteron tip via EMT to invade the blastocoel. Among these, eight primordial germ cells go through a typical EMT, migrate a short distance, and enter the coelomic pouches (Campanale et al., 2014; Martik & McClay, 2015). About 16 presumptive immune cells also go through an EMT at the tip of the archenteron and migrate through the blastocoel until they attach themselves to the extracellular matrix surrounding the growing skeleton and to the outside of the gut tube. These cells will then engage in immunosurveillance during the larval stage

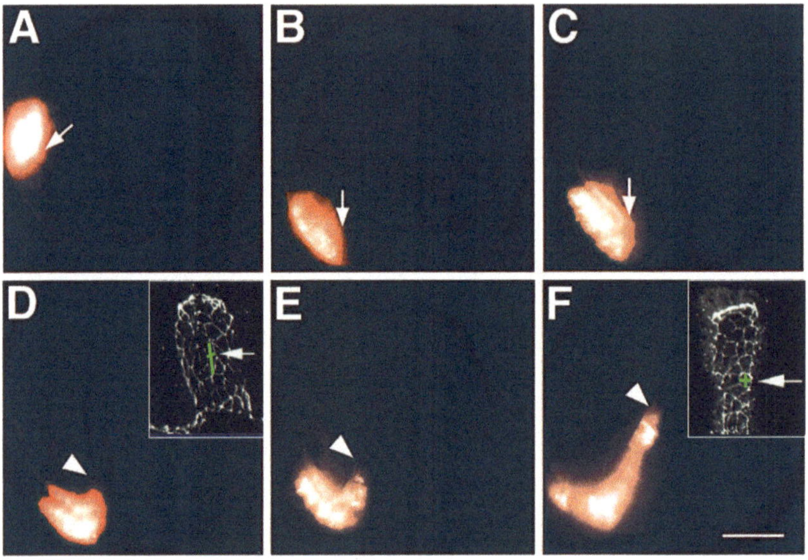

Fig. 4 Veg1 cells enter the archenteron late in gastrulation. A single Veg1 blastomere was labeled with diI at the 60-cell stage. In the sequence shown, the progeny of that blastomere eventually enter the archenteron late in gastrulation, after the Veg2 endoderm and mesoderm produce much of the length of the archenteron. Upon entering the archenteron, the Veg1 progeny align in a file (as in Fig. 3B), and their entry allows the stretched anisotropic Veg2 cells to return to isotropy. The inserts in (D) and (F) show embryos stained with β-catenin at the same stages shown in (D) and (F), respectively. In the (D) inset many of the cells are stretched and anisotropic (arrow) while later, as Veg1 cells enter the archenteron, most cells return to isotropy (arrow). The green lines in the insets mark length vs width of single cells. *From Logan, C. Y., & McClay, D. R. (1997). The allocation of early blastomeres to the ectoderm and endoderm is variable in the sea urchin embryo.* Development, 124, *2213–2223; Miller, J. R., & McClay, D. R. (1997). Characterization of the role of cadherin in regulating cell adhesion during sea urchin development.* Developmental Biology, 192, *323–339.*

(Ch Ho et al., 2016). Later, muscle cells delaminate and spread to wrap around the foregut and are patterned into circumesophageal rings by Hedgehog signaling (Walton, Warner, Hertzler, & McClay, 2009).

Meanwhile, coelomic pouches form as outpockets from the mesoderm at the tip of the archenteron and later "pinch off" to form lumens separated from the gut. These structures eventually house the primordial germ cells, the left of which becomes incorporated into the rudiment that will form the adult following metamorphosis. Finally, at about 23 hpf the anterior end of the endodermal gut contacts and fuses with the invaginating stomodaeum to establish the through-gut (Gustafson & Wolpert, 1963). Little is currently understood about the mechanisms of this deuterostome fusion event, though it is known that the stomodaeum is specified by oral ectoderm autonomously long before the stomodaeum contacts the foregut (Gross & McClay, 2001; Oliveri, Walton, Davidson, & McClay, 2006). Once the anterior end of the gut is remodeled, the stomodaeum-anterior gut fusion involves loss of the basement membranes separating the two invaginations, and then fusion of the two epithelia. Further morphogenetic events establish sphincters between the foregut and midgut, between the midgut and hindgut, and at the anal opening. Fig. 1C and C′ illustrate a 24 hpf pluteus larva showing how the skeleton growth distorts the A-P and D-V axis, and also the appearance of the differentiated gut within the blastocoel.

Much of the above was established by observation, experiments using classic and modern imaging techniques, or classical embryological experiments. Over the past two decades a strong effort by many in the sea urchin community have added insights into some of the molecular mechanisms necessary for the gastrulation process. Below, we review advances toward understanding those molecular mechanisms.

2. Setting the stage for gastrulation: Specification of the vegetal plate

Most of the vegetal half of the embryo (the progeny of the micromeres and the macromeres of the 16-cell stage embryo) will later be internalized during gastrulation movements. Between the 16-cell stage and mesenchyme blastula stage (2.5–9 hpf) the vegetal half of the embryo is specified to produce cells that will later differentiate into a number of cell types prior to gastrulation: primordial germ cells (8 cells at the beginning of gastrulation), skeletogenic cells (32 cells), non-skeletal mesoderm cells (~64 cells), and endoderm cell precursors (~160 cells). The non-skeletal mesoderm cells further separate into pigment cell precursors (~20 cells), immune cell precursors (~20 cells), coelomic pouch cell precursors (~18 cells), and pharyngeal muscle precursors (~6 cells). Each of these cell types divides once more during or just after their gastrulation movements. Specification of these several cell types has been examined in some detail. Gene regulatory networks operating during cleavage have been extensively characterized over the past 20 years (Croce & McClay, 2010; Davidson et al., 2002a, 2002b; Duboc et al., 2010; Li et al., 2014; Li, Materna, & Davidson, 2012; Oliveri, Tu, & Davidson, 2008; Peter & Davidson, 2010; Saudemont et al., 2010; Sethi, Angerer, & Angerer, 2009; Sethi, Wikramanayake, Angerer, Range, & Angerer, 2012). Those studies and others established

GRN models of the sequence of specification and signaling that programs each of the cell types listed above for their future differentiation, and for their specific role in gastrulation movements (Figs. 5 and 6). For some of the cell types the specification is known in enough detail to know how particular gastrulation movements are programmed. Two examples are illustrated. Without going into detail on all the steps of specification there are several important steps that establish the several cell types in the vegetal plate that will later engage in the gastrulation movements. Fig. 5 illustrates the specification sequence of the skeletogenic cells. The Wnt pathway initiates this lineage by the asymmetrical fourth cleavage, so the skeletogenic cell progenitors are initially programmed toward their eventual fate by activation of a repressor (Pmar1) that, in turn, represses a repressor (HesC) (Logan, Miller, Ferkowicz, & McClay, 1999; Oliveri, Davidson, & McClay, 2003; Revilla-i-Domingo et al., 2007). Repression of HesC expression results in activation of skeletogenic cell specification. The GRN model in Fig. 5 shows the progression of that network activation with time reflected (HPF) from top to bottom at the right. At about 9 hpf, two events happen. A number of differentiation genes begin to be activated by network transcription factors, and subsets of those factors direct the epithelial-mesenchymal transition which consists of (1) acquisition of motility, (2) cell shape change, (3) cell polarity change, (4) invasion of the cell body through the basement membrane, and (5) de-adhesion from the adherens junction. Those movements, also illustrated above in Fig. 2, are the morphogenetic output driven by the transcription factors of the specified skeletogenic cells.

The epithelial-mesenchymal transition of the skeletogenic cells begins 3 h before all other gastrulation movements allowing analysis of that event to be studied independent of other gastrulation movements. As revealed in the GRN analysis, in the 2 h prior to skeletogenic cell EMT (7–9 hpf) 10 transcription factors are uniquely activated in the skeletogenic cell GRN, with three others upstream of those (Fig. 5) (Oliveri et al., 2008; Wu & McClay, 2007; Wu, Yang, & McClay, 2008; and unpublished comparisons of the *S. purpuratus GRN* connections in *L. variagatus*). Those transcription factors were systematically knocked down and assays were developed to ask if absence of that transcription had an impact on the EMT, and if so, how (Saunders & McClay, 2014). Using this approach, several distinct transcriptional subcircuits illustrated in Fig. 5 were found to separately control (1) onset of motility, (2) cell shape change, (3) polarity change, (4) invasion through the basement membrane, and (5) de-adhesion of the skeletogenic cells from the adherens junction. Together these several cell biological functions

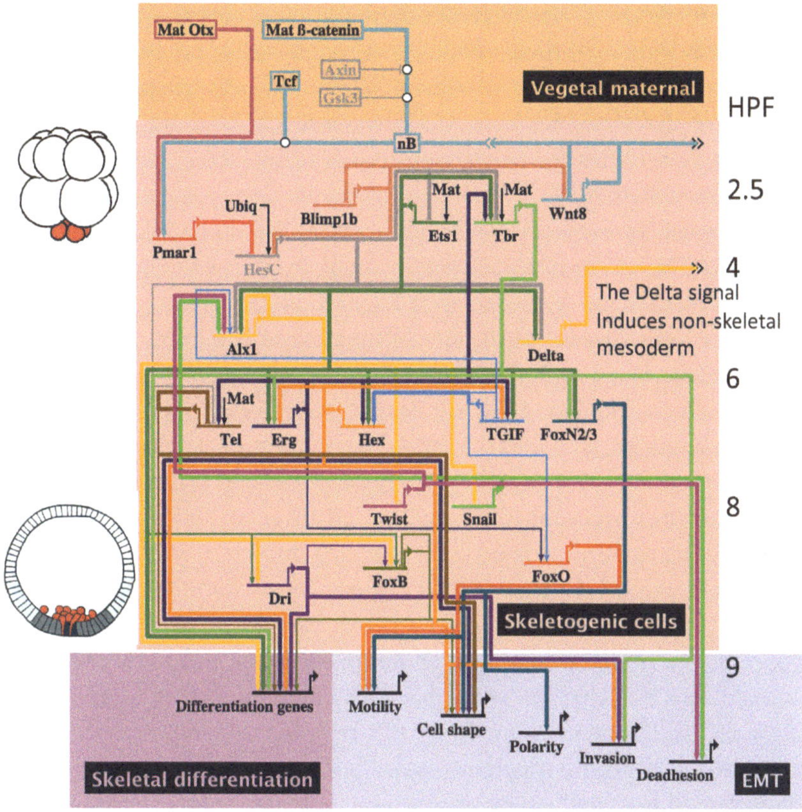

Fig. 5 Skeletogenic cell gene regulatory network (GRN). Shown is a model of the GRN experimentally established in *Lytechinus variegatus*. It is very similar to the skeletogenic cell GRN established for *Strongylocentrotus purpuratus* (Oliveri et al., 2008). In the Biotapestry diagram shown, specification begins in the micromeres at fourth cleavage (embryo to upper left). Genes are depicted as horizontal lines and their activities are depicted as lines with arrows (activators) or blocks (repressors) on the other genes in the network. The genes are drawn from top to bottom based on time of first expression as seen by in situ hybridization and/or QPCR, with approximate times in hours. Post fertilization times provided on the right side of the diagram. Activities directed downward advance the GRN while activities directed upward provide feedback in the system. At about 9 hpf EMT begins and at the same time the skeletogenic cells begin to differentiate. The model shows the collective group of transcription factors that activate expression of a number of known differentiation genes, and separately, those transcription factors that control each of the EMT component activities. Thick lines represent activities that have been confirmed or established experimentally in *L. variegatus*. Thin lines are drawn where genes are present in *L. variegatus*, but the connections have been established only for *S. purpuratus*. Several genes are missing from the *Lytechinus* model that are in the *S. purpuratus* GRN model. Those genes are present but not expressed in the *L. variegatus* skeletogenic cells. Two genes are in *L. variegatus* model

(Continued)

constitute the EMT. Thus, of the 13 transcription factors that are upstream of the EMT, at least 10 of them control component parts of the EMT morphogenetic event (Saunders & McClay, 2014). The transcription factors are upstream of the actual cell biological processes that conduct the EMT, but they provide a means for identifying those cell biological components that are driven by the GRN.

At the cell biological level, a number of EMT properties have been identified. The emerging skeletogenic cells invade through a remodeled basement membrane. If stained with an antibody to laminin, the basement membrane loses laminin at the site of invasion at the time the cells move through. That loss is cell autonomous in that if micromeres are moved to ectopic positions the laminin is eliminated from that ectopic site at the time of EMT (Saunders & McClay, 2014). A number of experiments utilized time-lapse movies to assess the progress of the EMT using transplanted micromeres that were fluorescently labeled in unlabeled host embryos so when the micromere progeny (the skeletogenic cells) went through the EMT it was easy to visualize detailed aspects of that morphogenetic movement. The cells first become motile while still part of the blastula monolayer. At about the same time the laminin hole appears in the basement membrane. It should be noted that the entire basement membrane is not eliminated as the "hole" still contains some basement membrane components. The cell then changes shape and much of the cell body breaches the hole and moves to the other side of the basement membrane. At this point the tail of the cell is still attached to the adherens junction distorting the cell shape. Finally, the tail of the cell releases from the adherens junction and the tail withdraws into the now mesenchymal skeletogenic cell. In cells expressing fluorescent Cadherin, strong staining is present in the adherens junction until that release (Miller & McClay, 1997; Saunders & McClay, 2014). As soon as the cell is released the Cadherin is endocytosed and the fluorescence quickly lost, presumably via lysosomal hydrolysis (Fig. 2). When Cadherin-GFP-expressing skeletogenic cells release their cytoplasmic tail from the adherens junction,

Fig. 5—Cont'd are not present in, or yet found, in *S. purpuratus*. Ubiq are assumed ubiquitously expressed drivers of genes where no other activator has been discovered. Mat are maternal inputs of early expressed genes. Circles represent formation of complexes or activities along a signal transduction pathway. Slashes represent intercellular signal transmission. Delta and Wnt8 are produced by the early skeletogenic cells. These signals act on the macromeres and their progeny as modeled in Fig. 6. *From Saunders, L.R., & McClay, D.R. (2014). Sub-circuits of a gene regulatory network control a developmental epithelial-mesenchymal transition.* Development, 141, *1503–1513.*

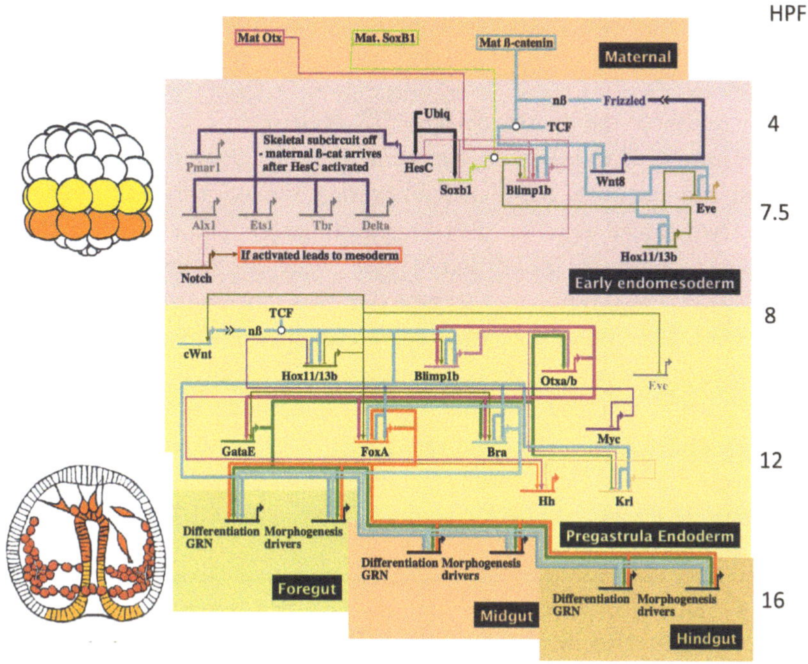

Fig. 6 Endoderm specification leading to gastrulation. Shown is a simplified model of the GRN that was identified through a number of perturbation experiments. It reflects experiments conducted in *L. variegatus* and is similar to the GRN identified in *S. purpuratus* (Peter & Davidson, 2011). Known maternal inputs are shown at the top. Endomesoderm specification occurs between fourth and eighth cleavages (the embryo diagrammed at the upper left is at sixth cleavage. The yellow cells are the Veg1 cells and the orange cells are Veg2 cells). A late arrival of β-catenin, after the HesC repressor is activated in macromeres, excludes macromeres from a skeletogenic fate (Revilla-i-Domingo, Oliveri, & Davidson, 2007). Later, macromere progeny remaining in direct contact with the Delta-signaling micromeres for more than 3 h (Croce & McClay, 2010), become the non-skeletogenic mesoderm. Macromere progeny that escape the Delta signal (first Veg1 at sixth cleavage, and later upper Veg2 at eighth cleavage) become the endoderm. The lower Veg2 progeny resulting from eighth cleavage become the non-skeletal mesoderm (specification not shown). Specification of endoderm cells leads to activation of a number of differentiation genes, and to proteins necessary for morphogenesis of the three parts of the gut. Genes are represented as horizontal bars with output arrows. Gene activities are depicted as line with arrows representing activating activities, and bars representing repression activity. Thick lines were established experimentally in *L. variegatus* and largely agree with *S. purpuratus* GRN models. Thin lines were established experimentally in *S. purpuratus* but both genes involved are expressed in endoderm in *L. variegatus*. In the pregastrula endoderm a several Wnts are present so cWnt is used to represent the activity of one or more of these.

all the fluorescence is taken up by the ingressed skeletal cell indicating that no Cadherin and no pinched off part of a cell is left behind (Saunders & McClay, 2014).

Once inside the blastocoel the 32 skeletogenic cells divide once, then rearrange themselves to surround the invaginating archenteron at the posterior end of the blastocoel. They then fuse to form into a syncytium, and at two lateral sites of the syncytium nuclei gather into a cluster, prompted by release of VEGF and FGF by the ectoderm just above the clusters. Skeletogenesis is initiated at these two sites (arrows in Fig. 1B). Several reviews and accounts of skeletogenesis document this process in some detail (Ettensohn & Dey, 2017; Guss & Ettensohn, 1997; Hodor & Ettensohn, 1998; Lyons, Martik, Saunders, & McClay, 2014; Malinda & Ettensohn, 1994; McIntyre et al., 2014; Piacentino et al., 2016; Piacentino, Ramachandran, & Bradham, 2015; Rafiq, Cheers, & Ettensohn, 2012; Rafiq, Shashikant, McManus, & Ettensohn, 2014; Sharma & Ettensohn, 2010).

3. The pigment cells invade the blastocoel shortly after the skeletogenic cells

The Delta-Notch pathway with signaling from the micromeres induces the non-skeletogenic mesoderm starting at sixth cleavage a process that will later result in mesoderm being separated from endomesoderm. Endoderm specification continues in endomesoderm cells that lose contact with the Delta-expressing skeletogenic cells between the sixth and eighth cleavage, thereby subdividing the mesoderm from specification of the endoderm. Delta from micromeres (Fig. 5) activates Notch in cells that will become the endomesoderm, and later some of those cells will be further specified as mesoderm (Fig. 6) (Sherwood & McClay, 1997, 1999; Sweet, Gehring, & Ettensohn, 2002). Endomesoderm specification also requires Wnt signaling (Fig. 6).

Three hours after the EMT of skeletogenic cells is initiated the pigment cell progenitors undergo an EMT and enter the blastocoel. The timing of this event is coincident with the earliest visual evidence of invagination of the archenteron (primary invagination), and by the time the primary invagination has established a dome at the vegetal plate, the pigment cells have completed their EMT. Once inside the blastocoel most pigment cells immediately migrate toward the posterior end of the embryo where they undergo a mesenchymal-to-epithelial transition and re-invade the ectoderm. They selectively invade the dorsal ectoderm somehow using

Eph and ephrin proteins as a mechanism to provide that selectivity (Krupke, Zysk, Mellott, & Burke, 2016). The EMT of the pigment cells looks similar to the EMT of skeletogenic cells in time-lapse movies; however, most of the transcription factors found to participate in controlling skeletogenic cell EMT are not expressed by pigment cells (D. McClay Unpublished Data). Thus, this EMT must occur with a largely different set of controlling transcription factors.

4. Mechanistic studies of primary invagination

At 12 hpf, at the same time the pigment cells undergo their EMT, the remaining non-skeletogenic cells undergo cell shape changes to produce the initial inbending of the archenteron. The site of the inbending varies with species, being at the middle of the vegetal plate in *L. variegatus*, and offset toward the oral side in *S. purpuratus*. Veg2 cells that receive the Notch signal continuously for about 3 h become the non-skeletogenic mesoderm cells, and of all macromere progeny only those from the lower tier of Veg2 after eighth cleavage satisfy this requirement (Croce & McClay, 2010; Materna, Ransick, Li, & Davidson, 2013; Ransick & Davidson, 2006, 2012). A separate signaling input, this time from Nodal, separates the ventral (oral) from dorsal (aboral) non-skeletogenic mesoderm (Duboc et al., 2010). Much of the dorsal NSM gives rise to the pigment cells. Pigment cells are not major participants in the primary inbending of the archenteron because by that time the pigment cells have largely removed themselves from the vegetal plate by EMT. Several mechanisms have been proposed to cause the initial inbending, one involving bottle cells (Hardin & Keller, 1988), and others involving extraembryonic matrix (Lane, Koehl, Wilt, & Keller, 1993), RhoA (Beane, Gross, & McClay, 2006) or by several other proposed mechanisms involving mechanical properties (Davidson et al., 1995). Details of the initial inbending sequence likely involve processes known in other systems to control cell shape change but there is much to learn about the process that initiates this conserved deuterostome event.

5. Specification of endoderm

Following the initial inbending by the NSM, the archenteron lengthens by addition of endoderm cells from the vegetal plate. Those endoderm cells move toward the blastopore, pass through it, and contribute to the lengthening archenteron. At a molecular level, much has been learned about how the endoderm cells are specified prior to and during invagination. Fig. 6 is a GRN model of endoderm specification. Initially, macromeres and their progeny (yellow and orange cells in the diagram at top left of Fig. 6) begin specification as endomesoderm through a combination of Wnt and Delta-Notch signaling. At sixth cleavage the Veg1 cells (the yellow cells in Fig. 6, and the upper progeny of division of macromeres at sixth cleavage), are removed from the source of Delta signaling and proceed to become endoderm (though some Veg1 progeny also become ectoderm). The lower macromere progeny, the Veg2 cells (orange cells in Fig. 6), continue to receive Delta for two more cleavages and at eighth cleavage the upper Veg2 cells are removed from Delta as a consequence of a meridional cleavage, and because these cells receive fewer than 3 h of continuous Delta signaling, these cells become endoderm, while the lower tier of Veg2 cells continue to receive Delta for 3 h and as a consequence become the non-skeletogenic mesoderm (Croce & McClay, 2010). Fig. 6 is a general model for events leading to endoderm specification and morphogenesis. For simplification, parts of specification leading to mesoderm specification are excluded.

Endoderm-activating genes of the endomesoderm (the genes illustrated in the early endomesoderm portion of the model), direct the upper Veg2 cells toward an endoderm fate once those cells are removed from the Delta signal. Among the first genes activated in the definitive endoderm is *brachyury (bra)* which, in invertebrates, is expressed in the endoderm (Fig. 6). In the sea urchin *bra* is expressed 4 h prior to the onset of invagination in a ring that will later become the blastopore (Fig. 7) (Croce, Lhomond, & Gache, 2001; Gross & McClay, 2001). Shortly thereafter, *foxA* is expressed in the same region (Oliveri et al., 2006). These two transcription factors eventually are expressed in all endoderm cells with *foxA* expression remaining on, once activated, while *bra* expression is restricted to the ring of cells surrounding the blastopore throughout gastrulation (Fig. 7). Since all endoderm cells pass through the blastopore this means that presumptive endoderm cells activate *bra* expression as they approach the blastopore and extinguish *bra* expression as they enter the archenteron (Gross & McClay, 2001). Recently, it was learned that Wnts also display dynamic expression changes throughout gastrulation (Fig. 7) (also see Robert, Lhomond, Schubert, & Croce, 2014). As with the skeletogenic GRN, endoderm specification leads to two outcomes: activation of genes involved in differentiation and activation of genes expressing proteins involved in morphogenesis. Specification events even later in gastrulation include several

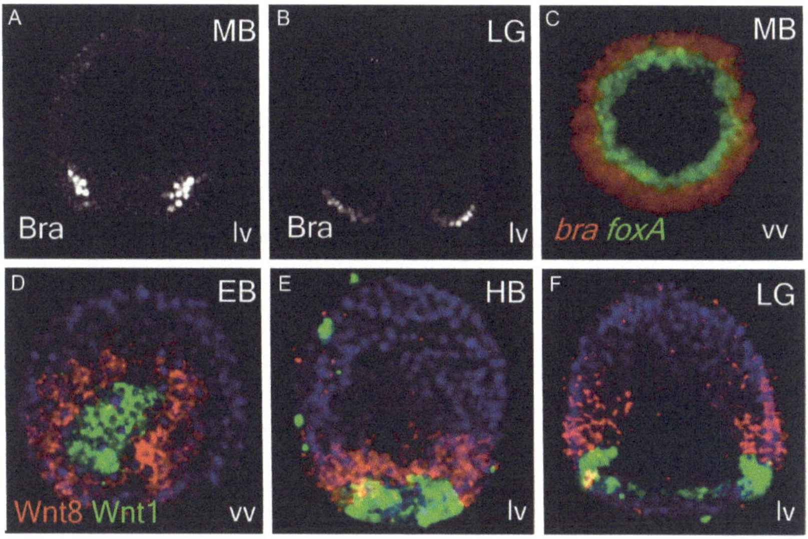

Fig. 7 The blastopore is a dynamic structure both for specification and morphogenesis. (A) Brachyury is present in cells just outside the blastopore from mesenchyme blastula, through gastrulation (B), as seen with an antibody to Brachyury. (C) In a double in situ of an embryo at late mesenchyme blastula, cells just outside the blastopore express *brachyury* (red), a few cells express *bra* and *foxA* (yellow), and still other cells express only *foxA* (green). The *foxA*-expressing cells previously had expressed *bra,* then both *bra* and *foxA,* and then *bra* expression was extinguished. This pattern is followed by all endoderm cells sequentially as they approach and move through the blastopore. (D–F) Expression of Wnts is dynamic at the blastopore. Shown here are *wnt1* (green) and *wnt8* (red) expression at three stages, before, at the beginning and at the end of gastrulation. Here, as the cells move into the blastopore and archenteron, *wnt1* and *wnt8* expression moves from cell to cell, counter to the direction to cell movement toward the blastopore. lv, lateral view; vv, vegetal view; MB, mesenchyme blastula; EB, early blastula; HB, hatched blastula; LG, late gastrula. *Panels (A) and (B): From Gross, J. M., & McClay, D. R. (2001). The role of Brachyury (T) during gastrulation movements in the sea urchin Lytechinus variegatus.* Developmental Biology, 239, *132–147.)*

evolutionarily conserved pathways with expression of transcription factors that are also expressed in vertebrate guts and pancreas (Cole, Rizzo, Martinez, Fernandez-Serra, & Arnone, 2009; Perillo, Wang, Leach, & Arnone, 2016).

6. Cells at the tip of the advancing archenteron are necessary for establishing right-left asymmetry

Nodal is responsible for establishing the dorsal-ventral axis in all three germ layers in the embryo (Duboc et al., 2010, 2004). Nodal signaling is also responsible for establishing right-left asymmetry (Duboc et al., 2005). Nodal signaling from the tip of the archenteron, beginning late in gastrulation, activates Nodal signaling on the right side to establish an asymmetry in specification of the left and right coelomic pouches (Fig. 8) (Bessodes et al., 2012; Warner et al., 2016). Misexpression of Hedgehog compromises that left-right asymmetry (Walton et al., 2009; Warner, Lyons, & McClay, 2012; Warner et al., 2016). This raised the question as to which signal was upstream, Nodal or Hedgehog. A careful study showed that Nodal can initiate right-left in the absence of Hedgehog but the asymmetry cannot be maintained without Hedgehog (Warner et al., 2016). Thus, it appears that Nodal is the symmetry-breaking signal, but, perdurance of Nodal for a period of time is required to stabilize the asymmetry and Hedgehog is necessary for maintenance of that signal. In other deuterostomes, Nodal signaling establishes the left side, while in the sea urchin, Nodal operates on the right side to prevent the right coelomic pouch from becoming part of the rudiment (Duboc et al., 2005).

7. Homing of the primordial germ cells to the coelomic pouches

At fifth cleavage, the cells at the vegetal pole of the embryo divide asymmetrically to produce four large and four small micromeres. The small micromeres divide once more before gastrulation and are fated to become the primordial germ cells (PGCs) (Yajima & Wessel, 2011). These cells remain attached to the archenteron during invagination. Then, when the archenteron reaches its full length, the PGCs undergo an EMT, migrate in the blastocoel briefly, then enter the coelomic pouches via a MET (Campanale et al., 2014; Martik & McClay, 2015). Experiments with PGCs revealed several properties of these cells. First, the PGCs are actively retained in the archenteron cell sheet until gastrulation is completed. If PGCs are transplanted to their normal position in a second host embryo this remains true. If, however, the PGCs are transplanted to an ectopic position in the host embryo they undergo an EMT precociously suggesting that in their normal position they are actively restrained from undergoing an EMT until gastrulation is completed (Martik & McClay, 2015). Additionally, if the PGCs are transplanted into an ectopic position, they go through the EMT, then demonstrate a remarkably robust ability to find their way to the coelomic pouches indicating a highly efficient homing mechanism (Martik & McClay, 2015) (Fig. 8). To home correctly to the coelomic

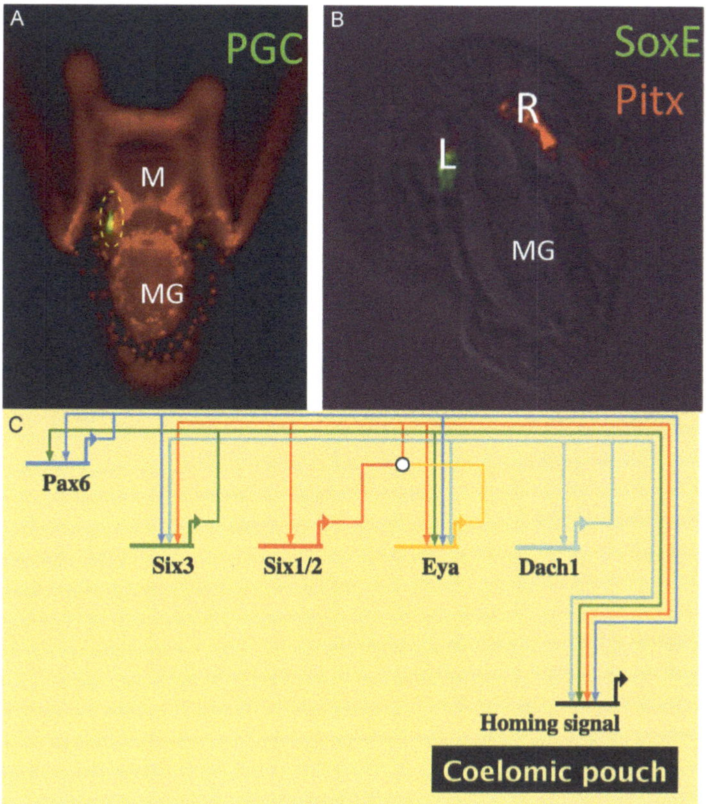

Fig. 8 The coelomic pouches are specified asymmetrically and are the targets for primordial germ cell homing. (A) A single primordial germ cell (PGC), labeled green with FITC, was placed in an embryo stained with Rhodamine-dextran at the 60-cell stage. During gastrulation the PGC successfully homed to the left coelomic pouch where its two progeny are seen at the pluteus larval stage. (B) During gastrulation left-right asymmetry is established as evidenced by a double in situ of *soxE* (green) and *pitx* (red) showing that the left and right coelomic pouches express different molecular markers. The left coelomic pouch will contribute to the rudiment of the adult. (C) Experiments used the homing of PGCs as an assay. Perturbation of transcription factors expressed by the coelomic pouches asked which factors were necessary for homing. Then, experiments asked how necessary factors interacted with each other. Results identified a GRN circuit necessary for production of the homing signal. Shown is that circuit which is both a feed forward and a feedback circuit. *Panel (A): From Martik, M. L., & McClay, D. R. (2015). Deployment of a retinal determination gene network drives directed cell migration in the sea urchin embryo.* Elife, 4, *e08827; Panel (B): From Warner, J. F., Miranda, E. L., & McClay, D. R. (2016). Contribution of hedgehog signaling to the establishment of left-right asymmetry in the sea urchin.* Developmental Biology, 411, *314–324; Panel (C): From Martik, M. L., & McClay, D. R. (2015). Deployment of a retinal determination gene network drives directed cell migration in the sea urchin embryo.* Elife 4, *e08827.*

pouch, a group of five transcription factors expressed by the coelomic pouch cells were found to establish an upstream control circuit that is necessary for the coelomic pouch cells to release the presumed homing signal. Perturbation of those five transcription factors established that the control circuit is both a classic feed forward and a positive feedback device (Martik & McClay, 2015). In *Lytechinus,* the PGCs preferentially home to the left coelomic pouch, and that pouch later becomes an important component of the embryonic rudiment which becomes the juvenile at metamorphosis. Thus, like PGCs of many species in the animal kingdom, sea urchin PGCs actively home to a specific target. Presumably, the left coelomic pouch, containing the PGCs, later contributes to the development of the adult gonad after metamorphosis.

8. Summary

Gastrulation in the sea urchin is a beautifully orchestrated sequence of morphogenetic movements. After over a century of study the molecular underpinnings of these processes of specification, morphogenesis, and patterning are beginning to be understood at a molecular level. The lessons we have learned so far have applications and implications across many phyla and biological processes from cancer metastasis to cell specification. Recent advances in high throughput omics and advances in quantitative microscopy will further provide the foundation for future mechanistic studies.

Acknowledgments

The authors thank Esther Miranda for her technical support in all aspects of the research. We also thank the members of the McClay lab for their advice and input. Support for this project was provided by: NIH RO1 HD14483 (to D.R.M.), PO1 HD37105 (to D.R.M.), and NSF fellowship support (to L.S.).

References

Beane, W. S., Gross, J. M., & McClay, D. R. (2006). RhoA regulates initiation of invagination, but not convergent extension, during sea urchin gastrulation. *Developmental Biology*, *292*, 213–225.

Bessodes, N., Haillot, E., Duboc, V., Rottinger, E., Lahaye, F., & Lepage, T. (2012). Reciprocal signaling between the ectoderm and a mesendodermal left-right organizer directs left-right determination in the sea urchin embryo. *PLoS Genetics*, *8*, e1003121.

Cameron, R. A., Hough-Evans, B. R., Britten, R. J., & Davidson, E. H. (1987). Lineage and fate of each blastomere of the eight-cell sea urchin embryo. *Genes & Development*, *1*, 75–85.

Campanale, J. P., Gokirmak, T., Espinoza, J. A., Oulhen, N., Wessel, G. M., & Hamdoun, A. (2014). Migration of sea urchin primordial germ cells. *Developmental Dynamics, 243*, 917–927.

Ch Ho, E., Buckley, K. M., Schrankel, C. S., Schuh, N. W., Hibino, T., Solek, C. M., et al. (2016). Perturbation of gut bacteria induces a coordinated cellular immune response in the purple sea urchin larva. *Immunology and Cell Biology, 94*, 861–874.

Cole, A. G., Rizzo, F., Martinez, P., Fernandez-Serra, M., & Arnone, M. I. (2009). Two ParaHox genes, SpLox and SpCdx, interact to partition the posterior endoderm in the formation of a functional gut. *Development, 136*, 541–549.

Croce, J., Lhomond, G., & Gache, C. (2001). Expression pattern of Brachyury in the embryo of the sea urchin *Paracentrotus lividus*. *Development Genes and Evolution, 211*, 617–619.

Croce, J. C., & McClay, D. R. (2010). Dynamics of Delta/Notch signaling on endomesoderm segregation in the sea urchin embryo. *Development, 137*, 83–91.

Davidson, E. H., & Erwin, D. H. (2006). Gene regulatory networks and the evolution of animal body plans. *Science, 311*, 796–800.

Davidson, L. A., Koehl, M. A., Keller, R., & Oster, G. F. (1995). How do sea urchins invaginate? Using biomechanics to distinguish between mechanisms of primary invagination. *Development, 121*, 2005–2018.

Davidson, L. A., Oster, G. F., Keller, R. E., & Koehl, M. A. (1999). Measurements of mechanical properties of the blastula wall reveal which hypothesized mechanisms of primary invagination are physically plausible in the sea urchin *Strongylocentrotus purpuratus*. *Developmental Biology, 209*, 221–238.

Davidson, E. H., Rast, J. P., Oliveri, P., Ransick, A., Calestani, C., Yuh, C. H., et al. (2002a). A genomic regulatory network for development. *Science, 295*, 1669–1678.

Davidson, E. H., Rast, J. P., Oliveri, P., Ransick, A., Calestani, C., Yuh, C. H., et al. (2002b). A provisional regulatory gene network for specification of endomesoderm in the sea urchin embryo. *Developmental Biology, 246*, 162–190.

Duboc, V., Lapraz, F., Saudemont, A., Bessodes, N., Mekpoh, F., Haillot, E., et al. (2010). Nodal and BMP2/4 pattern the mesoderm and endoderm during development of the sea urchin embryo. *Development, 137*, 223–235.

Duboc, V., Rottinger, E., Besnardeau, L., & Lepage, T. (2004). Nodal and BMP2/4 signaling organizes the oral-aboral axis of the sea urchin embryo. *Developmental Cell, 6*, 397–410.

Duboc, V., Rottinger, E., Lapraz, F., Besnardeau, L., & Lepage, T. (2005). Left-right asymmetry in the sea urchin embryo is regulated by nodal signaling on the right side. *Developmental Cell, 9*, 147–158.

Duloquin, L., Lhomond, G., & Gache, C. (2007). Localized VEGF signaling from ectoderm to mesenchyme cells controls morphogenesis of the sea urchin embryo skeleton. *Development, 134*, 2293–2302.

Ettensohn, C. A. (1985a). Gastrulation in the sea urchin embryo is accompanied by the rearrangement of invaginating epithelial cells. *Developmental Biology, 112*, 383–390.

Ettensohn, C. A. (1985b). Mechanisms of epithelial invagination. *The Quarterly Review of Biology, 60*, 289–307.

Ettensohn, C. A., & Dey, D. (2017). Kirrel L, a member of the Ig-domain superfamily of adhesion proteins, is essential for fusion of primary mesenchyme cells in the sea urchin embryo. *Developmental Biology, 421*, 258–270.

Gross, J. M., & McClay, D. R. (2001). The role of Brachyury (T) during gastrulation movements in the sea urchin *Lytechinus variegatus*. *Developmental Biology, 239*, 132–147.

Guss, K. A., & Ettensohn, C. A. (1997). Skeletal morphogenesis in the sea urchin embryo: Regulation of primary mesenchyme gene expression and skeletal rod growth by ectoderm-derived cues. *Development, 124*, 1899–1908.

Gustafson, T., & Wolpert, L. (1963). Studies on the cellular basis of morphogenesis in the sea urchin embryo. Formation of the coelom, the mouth, and the primary pore canal. *Experimental Cell Research, 29*, 561–582.
Hardin, J., & Keller, R. (1988). The behaviour and function of bottle of cells in gastrulation of *Xenopus laevis. Development, 103*, 211–230.
Hardin, J., & McClay, D. R. (1990). Target recognition by the archenteron during sea urchin gastrulation. *Developmental Biology, 142*, 86–102.
Hodor, P. G., & Ettensohn, C. A. (1998). The dynamics and regulation of mesenchymal cell fusion in the sea urchin embryo. *Developmental Biology, 199*, 111–124.
Kimberly, E. L., & Hardin, J. (1998). Bottle cells are required for the initiation of primary invagination in the sea urchin embryo. *Developmental Biology, 204*, 235–250.
Krupke, O. A., Zysk, I., Mellott, D. O., & Burke, R. D. (2016). Eph and Ephrin function in dispersal and epithelial insertion of pigmented immunocytes in sea urchin embryos. *Elife, 5,* e16000.
Lane, M. C., Koehl, M. A., Wilt, F., & Keller, R. (1993). A role for regulated secretion of apical extracellular matrix during epithelial invagination in the sea urchin. *Development, 117*, 1049–1060.
Li, E., Cui, M., Peter, I. S., & Davidson, E. H. (2014). Encoding regulatory state boundaries in the pregastrular oral ectoderm of the sea urchin embryo. *Proceedings of the National Academy of Sciences of the United States of America, 111*, E906–E913.
Li, E., Materna, S. C., & Davidson, E. H. (2012). Direct and indirect control of oral ectoderm regulatory gene expression by Nodal signaling in the sea urchin embryo. *Developmental Biology, 369*, 377–385.
Logan, C. Y., & McClay, D. R. (1997). The allocation of early blastomeres to the ectoderm and endoderm is variable in the sea urchin embryo. *Development, 124*, 2213–2223.
Logan, C. Y., Miller, J. R., Ferkowicz, M. J., & McClay, D. R. (1999). Nuclear beta-catenin is required to specify vegetal cell fates in the sea urchin embryo. *Development, 126*, 345–357.
Lyons, D. C., Martik, M. L., Saunders, L. R., & McClay, D. R. (2014). Specification to biomineralization: Following a single cell type as it constructs a skeleton. *Integrative and Comparative Biology, 54*, 723–733.
Malinda, K. M., & Ettensohn, C. A. (1994). Primary mesenchyme cell migration in the sea urchin embryo: Distribution of directional cues. *Developmental Biology, 164*, 562–578.
Martik, M. L., & McClay, D. R. (2015). Deployment of a retinal determination gene network drives directed cell migration in the sea urchin embryo. *Elife, 4*, e08827.
Martik, M. L., & McClay, D. R. (2017). New insights from a high-resolution look at gastrulation in the sea urchin, Lytechinus variegatus. *Mechanisms of Development, 148*, 3–10.
Materna, S. C., Ransick, A., Li, E., & Davidson, E. H. (2013). Diversification of oral and aboral mesodermal regulatory states in pregastrular sea urchin embryos. *Developmental Biology, 375*, 92–104.
McIntyre, D. C., Lyons, D. C., Martik, M., & McClay, D. R. (2014). Branching out: Origins of the sea urchin larval skeleton in development and evolution. *Genesis, 52*, 173–185.
Miller, J. R., & McClay, D. R. (1997). Characterization of the role of cadherin in regulating cell adhesion during sea urchin development. *Developmental Biology, 192*, 323–339.
Oliveri, P., Davidson, E. H., & McClay, D. R. (2003). Activation of pmar1 controls specification of micromeres in the sea urchin embryo. *Developmental Biology, 258*, 32–43.
Oliveri, P., Tu, Q., & Davidson, E. H. (2008). Global regulatory logic for specification of an embryonic cell lineage. *Proceedings of the National Academy of Sciences of the United States of America, 105*, 5955–5962.
Oliveri, P., Walton, K. D., Davidson, E. H., & McClay, D. R. (2006). Repression of mesodermal fate by foxa, a key endoderm regulator of the sea urchin embryo. *Development, 133*, 4173–4181.

Perillo, M., Wang, Y. J., Leach, S. D., & Arnone, M. I. (2016). A pancreatic exocrine-like cell regulatory circuit operating in the upper stomach of the sea urchin Strongylocentrotus purpuratus larva. *BMC Evolutionary Biology*, *16*, 117.

Peter, I. S., & Davidson, E. H. (2010). The endoderm gene regulatory network in sea urchin embryos up to mid-blastula stage. *Developmental Biology*, *340*, 188–199.

Peter, I. S., & Davidson, E. H. (2011). A gene regulatory network controlling the embryonic specification of endoderm. *Nature*, *474*, 635–639.

Peter, I. S., & Davidson, E. H. (2017). Assessing regulatory information in developmental gene regulatory networks. *Proceedings of the National Academy of Sciences of the United States of America*, *114*, 5862–5869.

Piacentino, M. L., Chung, O., Ramachandran, J., Zuch, D. T., Yu, J., Conaway, E. A., et al. (2016). Zygotic LvBMP5-8 is required for skeletal patterning and for left-right but not dorsal-ventral specification in the sea urchin embryo. *Developmental Biology*, *412*, 44–56.

Piacentino, M. L., Ramachandran, J., & Bradham, C. A. (2015). Late Alk4/5/7 signaling is required for anterior skeletal patterning in sea urchin embryos. *Development*, *142*, 943–952.

Rafiq, K., Cheers, M. S., & Ettensohn, C. A. (2012). The genomic regulatory control of skeletal morphogenesis in the sea urchin. *Development*, *139*, 579–590.

Rafiq, K., Shashikant, T., McManus, C. J., & Ettensohn, C. A. (2014). Genome-wide analysis of the skeletogenic gene regulatory network of sea urchins. *Development*, *141*, 950–961.

Ransick, A., & Davidson, E. H. (2006). Cis-regulatory processing of Notch signaling input to the sea urchin glial cells missing gene during mesoderm specification. *Developmental Biology*, *297*, 587–602.

Ransick, A., & Davidson, E. H. (2012). Cis-regulatory logic driving glial cells missing: Self-sustaining circuitry in later embryogenesis. *Developmental Biology*, *364*, 259–267.

Revilla-i-Domingo, R., Oliveri, P., & Davidson, E. H. (2007). A missing link in the sea urchin embryo gene regulatory network: hesC and the double-negative specification of micromeres. *Proceedings of the National Academy of Sciences of the United States of America*, *104*, 12383–12388.

Robert, N., Lhomond, G., Schubert, M., & Croce, J. C. (2014). A comprehensive survey of wnt and frizzled expression in the sea urchin *Paracentrotus lividus*. *Genesis*, *52*, 235–250.

Rottinger, E., Saudemont, A., Duboc, V., Besnardeau, L., McClay, D., & Lepage, T. (2008). FGF signals guide migration of mesenchymal cells, control skeletal morphogenesis [corrected] and regulate gastrulation during sea urchin development. *Development*, *135*, 353–365.

Saudemont, A., Haillot, E., Mekpoh, F., Bessodes, N., Quirin, M., Lapraz, F., et al. (2010). Ancestral regulatory circuits governing ectoderm patterning downstream of Nodal and BMP2/4 revealed by gene regulatory network analysis in an echinoderm. *PLoS Genetics*, *6*, e1001259.

Saunders, L. R., & McClay, D. R. (2014). Sub-circuits of a gene regulatory network control a developmental epithelial-mesenchymal transition. *Development*, *141*, 1503–1513.

Sethi, A. J., Angerer, R. C., & Angerer, L. M. (2009). Gene regulatory network interactions in sea urchin endomesoderm induction. *PLoS Biology*, *7*, e1000029.

Sethi, A. J., Wikramanayake, R. M., Angerer, R. C., Range, R. C., & Angerer, L. M. (2012). Sequential signaling crosstalk regulates endomesoderm segregation in sea urchin embryos. *Science*, *335*, 590–593.

Sharma, T., & Ettensohn, C. A. (2010). Activation of the skeletogenic gene regulatory network in the early sea urchin embryo. *Development*, *137*, 1149–1157.

Sherwood, D. R., & McClay, D. R. (1997). Identification and localization of a sea urchin Notch homologue: Insights into vegetal plate regionalization and Notch receptor regulation. *Development*, *124*, 3363–3374.

Sherwood, D. R., & McClay, D. R. (1999). LvNotch signaling mediates secondary mesenchyme specification in the sea urchin embryo. *Development*, *126*, 1703–1713.

Su, Y. H., Li, E., Geiss, G. K., Longabaugh, W. J., Kramer, A., & Davidson, E. H. (2009). A perturbation model of the gene regulatory network for oral and aboral ectoderm specification in the sea urchin embryo. *Developmental Biology*, *329*, 410–421.

Sweet, H. C., Gehring, M., & Ettensohn, C. A. (2002). LvDelta is a mesoderm-inducing signal in the sea urchin embryo and can endow blastomeres with organizer-like properties. *Development*, *129*, 1945–1955.

Walton, K. D., Warner, J., Hertzler, P. H., & McClay, D. R. (2009). Hedgehog signaling patterns mesoderm in the sea urchin. *Developmental Biology*, *331*, 26–37.

Warner, J. F., Lyons, D. C., & McClay, D. R. (2012). Left-right asymmetry in the sea urchin embryo: BMP and the asymmetrical origins of the adult. *PLoS Biology*, *10,* e1001404.

Warner, J. F., Miranda, E. L., & McClay, D. R. (2016). Contribution of hedgehog signaling to the establishment of left-right asymmetry in the sea urchin. *Developmental Biology*, *411*, 314–324.

Wu, S. Y., & McClay, D. R. (2007). The snail repressor is required for PMC ingression in the sea urchin embryo. *Development*, *134*, 1061–1070.

Wu, S. Y., Yang, Y. P., & McClay, D. R. (2008). Twist is an essential regulator of the skeletogenic gene regulatory network in the sea urchin embryo. *Developmental Biology*, *319*, 406–415.

Yajima, M., & Wessel, G. M. (2011). Small micromeres contribute to the germline in the sea urchin. *Development*, *138*, 237–243.

CHAPTER EIGHT

Tunicate gastrulation

Konner M. Winkley[a], Matthew J. Kourakis[b], Anthony W. DeTomaso[b], Michael T. Veeman[a], William C. Smith[b,*]
[a]Division of Biology, Kansas State University, Manhattan, KS, United States
[b]Department of Molecular, Cellular and Developmental Biology, University of California, Santa Barbara, CA, United States
*Corresponding author: e-mail address: w_smith@lifesci.ucsb.edu

Contents

Abstract

Tunicates are a diverse group of invertebrate marine chordates that includes the larvaceans, thaliaceans, and ascidians. Because of their unique evolutionary position as the sister group of the vertebrates, tunicates are invaluable as a comparative model and hold the promise of revealing both conserved and derived features of chordate gastrulation. Descriptive studies in a broad range of tunicates have revealed several important unifying traits that make them unique among the chordates, including invariant cell lineages through gastrula stages and an overall morphological simplicity. Gastrulation has only been studied in detail in ascidians such as *Ciona* and *Phallusia*, where it involves a simple cup-shaped gastrula driven primarily by endoderm invagination. This appears to differ significantly from vertebrate models, such as *Xenopus*, in which mesoderm convergent extension and epidermal epiboly are major contributors to involution. These differences may reflect the cellular simplicity of the ascidian embryo.

Current Topics in Developmental Biology, Volume 136
ISSN 0070-2153
https://doi.org/10.1016/bs.ctdb.2019.09.001

1. Introduction: Tunicates—their place on the evolutionary tree and their contribution to our understanding of embryology

Tunicates were among the earliest experimental models for embryology. Embryologists were attracted to the ascidian embryo, with its regular cleavage program and small, simple embryonic body plan. Chabry performed blastomere separation experiments in early embryos of the ascidian *Ascidiella aspersa* that are regarded as foundational to the discipline of experimental embryology (Chabry, 1887; Fischer, 1992). He found that isolated blastomeres showed predetermined fates, dividing as if they were still in the intact embryo. Although not obvious in their diverse adult forms, tunicates embryos are unmistakably chordate with a notochord and dorsal hollow nerve cord. The close evolutionary relationship of ascidians to vertebrates was well appreciated by this time (Darwin, 1871; Haeckel, 1874; Kowalevsky, 1866). Conklin (1905a) built on Chabry's work and mapped the complete lineage of cells through and beyond gastrulation, with illustrations by embryonic stage and a nomenclature still in use. Conklin's work included descriptions of cleavage planes, cell-cell contacts, nuclear positions, distribution of cytoplasmic determinants, cell fates, polar body location, and spindle dynamics, as well as comparisons of gastrulation and other aspects of embryogenesis between ascidians and other animals. Satoh's SEM studies of *Halocynthia roretzi* confirmed and expanded on these early descriptions of ascidian development, bolstering inferences concerning the coordination of cell movements during gastrulation and neurulation and revealing additional surface features, such as the presence of filopodia from interior membranes over the blastocoel (Satoh, 1978).

Since that time, molecular approaches in *Halocynthia*, *Ciona* and *Phallusia* have revealed much about the mechanisms driving ascidian patterning and morphogenesis, but little is known about tunicate gastrulation and other developmental processes outside these ascidian models. The tunicates are thought to have diverged from their last common ancestor with the vertebrates more than 500 million years ago, and are a large and diverse group. They are usually regarded as a chordate sub-phylum, but some authors have argued that they should be elevated to phylum status (Satoh, Rokhsar, & Nishikawa, 2014). The ascidians (class *Ascidiacea*) are sessile as adults and are the most speciose group of tunicates, represented by nearly 3000 separate types (WoRMS, 2019) across three orders, Phlebobranchia, Stolidobranchia and Aplousobranchia, each characterized by differences in gill-slit morphology. Many of these are solitary, but colonialism has evolved independently in multiple ascidian lineages (Brown & Swalla, 2012), with varying degrees of physiological integration. The thaliaceans (class *Thaliacea*), with about 80 known species, are pelagic as adults and move through the water column by pulsating muscular contractions, using incurrent and excurrent siphons at opposite ends of the body. Thaliaceans, including pyrosomes, salps and doliolids, can be seen singly, or as chains of asexually budding animals. They can show alternation of sexual and asexual generations, and can reproduce asexually at an explosive rate during plankton blooms (Holland, 2016). The larvaceans (class *Appendicularia*) are also pelagic and represented by about 70 species (WoRMS, 2019). Larvaceans are so-called because they never lose their tails and resemble larvae into adulthood. They live inside buoyant glycoprotein 'houses' that aid in filter feeding and are secreted by specialized epidermal organs.

Molecular phylogenies using concatenated sequence datasets reveal that tunicates, and not cephalochordates, are the sister taxon to the vertebrates (Fig. 1) (Bourlat, Rota-Stabelli, Lanfear, & Telford, 2009; Delsuc, Brinkmann, Chourrout, & Philippe, 2006). Two recent phylogenies, largely congruent, go further to resolve relations among the tunicates (Delsuc et al., 2018;

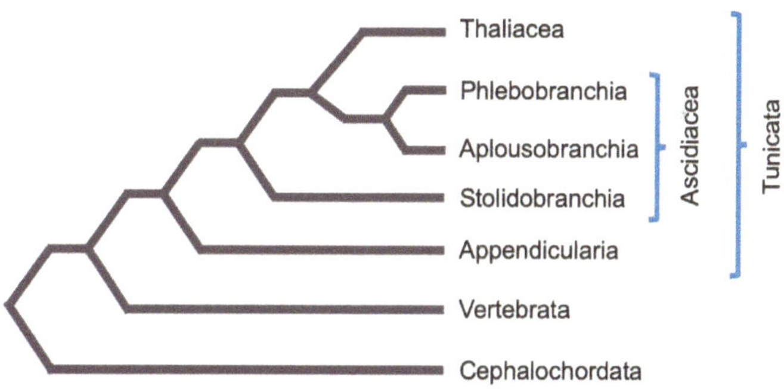

Fig. 1 Phylogenetic relationships between the chordates, with emphasis on the tunicate clade. Tunicates and vertebrates share a more recent common ancestor than either do with cephalochordates. Because larvaceans (Appendicularia) are basal in the Tunicata, the ancestral tunicate may have been pelagic. The Ascidiacea are revealed to be paraphyletic due to the placement of the Thaliacea as sister to Phlebobranchia/Aplousiobranchia. Branch lengths are not to scale. *Based on Kocot, K. M., Tassia, M. G., Halanych, K. M., & Swalla, B. J. (2018). Phylogenomics offers resolution of major tunicate relationships.* Molecular Phylogenetics and Evolution, *121, 166–173. https://doi.org/10.1016/j.ympev.2018.01.005; Delsuc, F., Philippe, H., Tsagkogeorga, G., Simion, P., Tilak, M.K., Turon, X., et al. A phylogenomic framework and timescale for comparative studies of tunicates.* BMC Biology. *https://doi.org/10.1186/s12915-018-0499-2.*

Kocot, Tassia, Halanych, & Swalla, 2018) as summarized in Fig. 1. Larvaceans are the basal lineage, pointing to a possible pelagic ancestor for the tunicates as a whole. The ascidian order Stolidobranchia is sister to a group that includes the Thaliacea plus an Aplousiobranchia/Phlebobranchia pairing (the phlebobranch ascidians may themselves be paraphyletic, however). The placement of thaliaceans *within* the Ascidiacea is contrary to most older classification schemes of the last century (see, e.g., Brusca & Brusca, 1990) but is well-supported and suggests this free-swimming group evolved from a benthic ascidian ancestor, underscoring the potential for extensive morphological innovation in these lesser-studied tunicates (Kocot et al., 2018).

2. Events leading up to gastrulation

2.1 Cleavage patterns

Only minor differences in embryonic cleavage patterns have been described between different ascidian species. *Ciona* cell lineages were first described in detail by Conklin (1905a), who also took advantage of the highly pigmented *Styela* egg to identify a cytoplasmic component he called the 'myoplasm' that segregates with muscle fate (Conklin, 1905b). This was arguably the first cytoplasmic determinant of cell fate to be identified, and provided the foundation for the mosaic theory of development (Conklin, 1905c). Conklin's cell lineages were refined and extended by Nishida who performed blastomere labeling experiments in *Halocynthia*, which has a larger embryo better suited to single-cell microinjection than *Ciona* (Nishida, 1987; Nishida & Satoh, 1983, 1985). Most cells are restricted to a single tissue fate by the start of gastrulation at the 112-cell stage (Nishida, 1987), though there is extensive subsequent refinement of cell identity in the neural lineages in particular (Hudson, Lotito, & Yasuo, 2007; Racioppi et al., 2014; Stolfi, Wagner, Taliaferro, Chou, & Levine, 2011). The fate map has not been deciphered with single cell resolution all the way to the hatched larval stage, but several specific lineages have been traced well past gastrulation (Carlson, Reeves, & Veeman, 2015; Nicol & Meinertzhagen, 1988a; Stolfi & Levine, 2011; Stolfi, Ryan, Meinertzhagen, & Christiaen, 2015; Tokuoka, Imai, Satou, & Satoh, 2004; Wang et al., 2019). Many markers of tissue differentiation are expressed normally even in embryos that have been cleavage arrested prior to gastrulation (Nishikata, Mita-Miyazawa, Deno, & Satoh, 1987; Okado & Takahashi, 1988; Whittaker, 1973).

The first two rounds of division are at right angles and give rise to four similarly sized daughter cells. The first division splits the left and right halves of the embryo, and the second division nominally separates anterior from posterior (Conklin, 1905a). The third round of division is again orthogonal to the prior two and separates the animal and vegetal hemispheres of the embryo. These cleavages are slightly asymmetric in volume, with the animal daughters all somewhat smaller than their vegetal siblings (Conklin, 1905a). Cell divisions remain synchronous through the fourth and fifth cell cycles, but many of the vegetal cell divisions are highly asymmetric with major differences in size and fate between anterior and posterior daughters (Conklin, 1905a; Dumollard et al., 2017; Hibino, Nishikata, & Nishida, 1998; Iseto & Nishida, 1999; Nishikata, Hibino, & Nishida, 1999; Patalano et al., 2006; Tassy, Daian, Hudson, Bertrand, & Lemaire, 2006). After the 32-cell stage, cell cycle times slow in the vegetal hemisphere to give distinct 44 and 76-cell stages (Dumollard, Hebras, Besnardeau, & McDougall, 2013). Animal cells continue to divide relatively symmetrically to give rise to most of the ectoderm, whereas many of the vegetal divisions are asymmetric in fate and/or size to give rise to a variety of largely mesodermal and endodermal cell types (Conklin, 1905a; Nishida, 1987).

2.2 Ooplasmic segregation/PEM

Early embryonic patterning depends on two successive cytoplasmic reorganizations in the first cell cycle known as the first and second ooplasmic segregations. Ascidian oocytes are polarized along the animal-vegetal axis prior to fertilization (Sardet et al., 1992), and the first ooplasmic segregation helps to reinforce this polarity. The second ooplasmic segregation breaks the radial symmetry of the egg and establishes the future anterior-posterior axis. Both phases involve the movement of a class of maternally deposited RNAs collectively known as the postplasmic/PEM (posterior end mark) RNAs (reviewed in Prodon, Yamada, Shirae-Kurabayashi, Nakamura, & Sasakura, 2007) along with mitochondria rich myoplasm and cortical endoplasmic reticulum (ER). The first phase of ooplasmic segregation is influenced by both the animal-vegetal polarization of the oocyte prior to fertilization as well as the site of sperm entry. Sperm entry triggers first a calcium wave and a subsequent contraction wave that concentrates the postplasmic/PEM RNAs and associated factors at the vegetal pole (McDougall & Sardet, 1995; Speksnijder, Sardet, & Jaffe, 1990).

The second phase of ooplasmic segregation is driven by sperm aster microtubules and cortical microfilaments (Chiba et al., 1999). It relocates the vegetal PEM components to a more equatorial position that defines what is widely referred to as the future posterior side of the embryo (Sardet, Speksnijder, & Jaffe, 1989). As in many species, however, this new axis of organization orthogonal to the AV axis is not perfectly congruent with any Cartesian axis of the hatched larva. The cortical ER, postplasmic/PEM RNAs, and myoplasm are inherited by the left and right posterior vegetal blastomeres at the eight-cell stage, and aggregate to form a structure called the centrosome attracting body (CAB) (Hibino et al., 1998). The CAB is continuously partitioned into the posterior-most vegetal cells during the subsequent rounds of division and is responsible for the unequal cleavages characteristic of these lineages (Patalano et al., 2006). Much of the anterior-posterior patterning of the embryo is thought to result directly or indirectly from the asymmetric inheritance of the CAB and the PEM RNAs. This includes a Zic family transcription factor variously known as Macho1 or Zic-r.a that has been identified as one of the key components of Conklin's myoplasm (Nishida & Sawada, 2001).

2.3 Origin of germ layers

As seen in diverse metazoans (Logan, Miller, Ferkowicz, & Mcclay, 1999; Miyawaki, Yamamoto, Saito, & Saito, 2003; Onai, 2019; Wikramanayake et al., 2003), maternal nuclear β-catenin plays a key role in ascidian germ layer specification. β-Catenin signaling is required for endodermal cell fate in the most vegetal cells as well as mesodermal cell fates in the more marginal vegetal cells. In *Ciona* and other ascidians, this involves two successive β-catenin dependent fate specification events. Nuclear β-catenin is observed in all the mesendodermal precursors at the 16-cell stage and is required for mesendodermal fate (Imai, Takada, Satoh, & Satou, 2000). In the next cell cycle, β-catenin becomes restricted to only the most vegetal daughters and is required for the segregation of endoderm from marginal fates including mesoderm and some neural ectoderm (Hudson, Kawai, Negishi, & Yasuo, 2013). In the first specification event, nuclear localization of β-catenin in the vegetal cells is antagonistic to the ubiquitous expression of the maternal transcription factor Gata.a. Gata.a promotes the transcription of ectodermal specific genes in the animal cells (Rothbacher, Bertrand, Lamy, & Lemaire, 2007), while the nuclear localization of β-catenin both prevents the expression of ectodermal genes and promotes expression of mesendodermal genes in the vegetal cells (Imai, Hudson, Oda-Ishii, Yasuo, & Satou, 2016). The second specification event in the following cell cycle also involves antagonism between vegetal β-catenin signaling and more broadly expressed Gata.a (Imai et al., 2016), where β-catenin promotes endodermal fate. The precise mechanisms linking preexisting oocyte AV polarity and ooplasmic segregation to β-catenin stabilization and nuclear entry remain unclear. Several downstream targets of β-catenin have been identified, including vegetally-expressed FGFs that are involved in a broad range of cell fate decisions at the margin of the embryo (Hudson, Sirour, & Yasuo, 2016; Imai, 2003).

3. Mechanisms of gastrulation

Most ascidians initiate gastrulation after the seventh cell cycle. For *Ciona* this is ~5 h after fertilization at 18 °C (Hotta et al., 2007). The embryo rapidly flattens along the animal-vegetal axis and invaginates to form a cup-shaped early gastrula (Fig. 2, left). The blastopore constricts over ~2 h to bring not just the endoderm but also the notochord, muscle and mesenchyme precursors into the interior of the embryo (Fig. 2, right).

3.1 Endoderm-intrinsic forces in gastrulation

The most detailed mechanistic description of forces driving the morphogenetic events of early tunicate gastrulation used a combination of fixed and live imaging in four different ascidian species (*Ciona intestinalis* type A [now known as *Ciona robusta*], *Ciona savignyi*, *Phallusia mammilata*, and

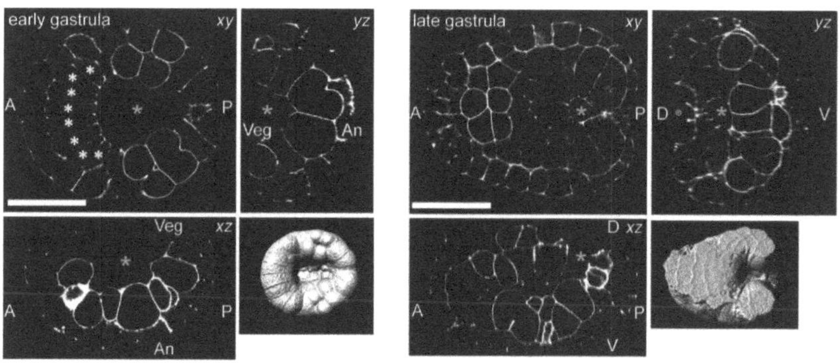

Fig. 2 Ciona gastrulation. Confocal images of gastrulating Ciona robusta embryos at the stages indicated. The panels on the left are derived from confocal stacks of a phalloidin-stained embryo early in gastrulation, with orthogonal xy, xz and yz sections shown. The surface-rendered view of the stack is from the vegetal side. The green asterisk marks the open blastopore/archenteron. Yellow asterisks mark notochord precursor cells. The panels on the right show an older embryo near the end of gastrulation/beginning of neurulation. The red dot indicates the neural groove just starting to form on the posterior dorsal side. Scale bars = 50 μm.

Boltenia villosa), all of which show essentially superimposable cleavage patterns and cell-shape changes during gastrulation (Sherrard, Robin, Lemaire, & Munro, 2010). This study generated a comprehensive model of invagination through observation, experimental manipulation and computer simulation. One surprising finding was that force-generating myosin II dynamics intrinsic to the endoderm cells is sufficient to account for invagination. Equally surprising was the finding that the concomitant cell-shape changes such as epiboly outside the endoderm not only do not contribute to invagination but actually oppose it. This is unlike *Xenopus* (Keller, Davidson, & Shook, 2003), where ectodermal epiboly is thought to help drive endodermal internalization.

Endoderm invagination is the product of two distinct and sequential steps, both driven by myosin II. In the first step the apical (vegetal-most) surfaces of the endoderm cells contract between the late 64-cell and early 112-cell stages (Apical Constriction, Fig. 3). This step is characterized by the accumulation of activated dynamic myosin II complexes at the contracting apical surfaces, as seen from immunostaining for ser19-phosphorylated myosin II regulatory light chain (1P-myosin; Sherrard et al., 2010). This apical constriction is regulated by the well-described RhoA/Rho-kinase pathway (Beane, Gross, & McClay, 2006; van Eekelen, Runtuwene, Overvoorde, & den Hertog, 2010) and can be blocked by either dominant negative RhoA or pharmacological inhibition of Rho-kinase.

The second step involves embryonic morphology shifting from a flattened disk to a cup-like shape (Collared Rounding, Fig. 3). This involves the endoderm cells contracting in the apical-basal dimension to become shorter, while at the same time expanding their basal surfaces contacting the overlying ectoderm. This change in contractility is observed as a redistribution of 1P-myosin staining to the basal and lateral surfaces. Interestingly,

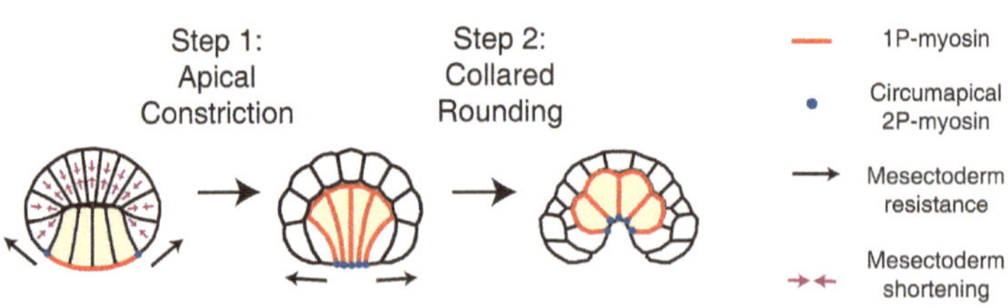

Fig. 3 Two-step model for endoderm invagination in ascidian development. *Figure from Sherrard, K., Robin, F., Lemaire, P., & Munro, E. (2010). Sequential activation of apical and basolateral contractility drives ascidian endoderm invagination.* Current Biology, *20, 1499–1510. doi:https://doi.org/10.1016/j.cub.2010.06.075 (web archive link).*

this second step is not RhoA dependent. The mechanism that regulates the transition from step 1 to step 2 and that directs the assembly of activated myosin II in the apico-basal axis remains to be determined, but a recent preprint suggests that Nodal and Ephrin signaling may be important (Fiuza, Negishi, Rouan, Yasuo, & Lemaire, 2018).

Importantly, during step 2 the apical surfaces remain constricted, although no further constriction is observed. Immunostaining shows that at the apical surfaces the more dynamic 1P-myosin is replaced with the more stable 2P-myosin which is phosphorylated on both ser19 and thr18 of the regulatory light chain. Pharmacological inhibition of RhoA at later steps of gastrulation causes the apical constrictions to relax, indicating a persistent requirement for this pathway. Collared rounding can occur even if the cell division from 64 to 112 cells is inhibited, indicating that this mechanism is robust.

Endoderm invagination in ascidian gastrulation involves changes in cell shape but no changes in cell-cell contacts. This is unlike many metazoans, where involution, ingression, intercalation, migration and other large scale tissue rearrangements lead to major changes in cell-cell neighbor relationships. Even in sea urchins, which have a relatively simple gastrulation, the mesenchyme cells delaminate and migrate, and the invaginating endoderm cells also undergo radial intercalation (Ettensohn, 1985, 1999). We speculate that the morphological simplicity of tunicate gastrulation reflects both the small size of the embryo and its very rapid development.

The Sherrard study dived deep into the molecular and biophysical mechanisms of early ascidian gastrulation as the embryo changes from ball-shaped to cup-shaped. Later aspects of gastrulation have not been studied in any detail. It is unknown whether notochord, muscle and mesenchyme are participating more actively later in gastrulation, and the mechanisms driving blastopore closure have not been investigated.

4. Gastrulation in other tunicates

As detailed in Section 1, tunicates are a diverse group of animals with dramatically different larval and adult morphologies. Given this diversity, it is reasonable to assume that the cellular and molecular mechanisms of gastrulation may also vary. Embryos of the ascidian *Halocynthia roretzi*, for example, are much larger than *Ciona* embryos, and have minor differences in early cleavage patterns. *Halocynthia* gastrulation begins at the 110-cell stage (Nishida, 1986) instead of the 112-cell stage as seen in *Ciona*. By contrast, the larvacean *Oikopleura dioica*, begins gastrulation at the 32-cell stage <2 h after fertilization. *Oikopleura* has exceptionally rapid development, with the tadpole larvae hatching in as little as 3 h following fertilization (Fujii, Nishio, & Nishida, 2008; Nishida, 2008). Gastrulation in *O. dioica* progresses in two phases. In the first phase, the vegetal-most eight cells invaginate, with the lateral four cells proceeding the central four endoderm cells. In the second phase, lateral and posterior cells invaginate. Interestingly, no change to the overall spherical shape of the embryos is evident during gastrulation. Moreover, because of the small number of invaginating cells in the gastrula, there is no obvious blastopore or archenteron. Cellular mechanisms driving invagination in *O. dioica* are not known, but it has been speculated that myosin II-driven apical constriction is important (Fujii et al., 2008).

4.1 Colonial tunicates

A subset of ascidian species are colonial and grow by repeated rounds of asexual reproduction (budding) that eventually give rise to a colony of genetically identical individuals, called zooids, arranged within a common tunic. Each zooid can be sexually mature, but colonial reproduction has several notable differences when compared to solitary species, and these have hampered embryological studies. First, all colonial ascidians are brooders, and fertilization and development occur inside the adult zooids, often with maternal contributions, such as a brood pouch, which can be difficult to replicate in vitro. Development is also significantly slower, on the order of days to a week or more. Finally, the eggs and developing embryos are opaque, hindering high-resolution studies of cell movement. Thus little is known about gastrulation in these species, and there are no recently published studies.

More germane to this review is the potential role of a gastrulation-like process during asexual development. During the budding process, colonial ascidians are regenerating all tissues and organs from a population of pluripotent stem cells. An important question is whether asexual development is controlled by novel genes and gene regulatory networks, or if embryonic pathways and networks are redeployed (Tiozzo, Brown, & De Tomaso, 2008). Interestingly, there are many colonial species scattered throughout the class Ascidiacea, including different monophyletic orders, suggesting that coloniality has arisen independently multiple times. Comparison of these species reveals a diversity of budding modes, including the source of the new body, and the timing of various developmental landmarks

(Tiozzo et al., 2008). However, in all species studied to date, asexually developing zooids go through an initial blastula-like stage. This is followed by invaginations and evaginations of the epithelium to form various tissues and organs. While in a sense this resembles gastrulation, these movements do not result in segregation of cells into presumptive germ layers, as they do in embryogenesis. The only species in which asexual budding has been studied at a molecular level is *Botryllus schlosseri* (Manni et al., 2019). In an elegant study by Ricci, Cabrera, Lotito, and Tiozzo (2016), it was found that there is regionalization of germ layer specific transcription factors in the blastula-like stage and during these epithelial folding events, but they do not correspond to the regions that are folding (i.e., it does not appear that the movement is correlated to germ layer specification). In this study, it was found that germ layer restricted transcription factors, such as Otx (ectoderm), Fox-A1 (endoderm) and Gsc (mesoderm) are expressed in distinct regions of the blastula-like structure and could be followed through organogenesis. Importantly, the expression of each of these germ layer markers corresponded to the source of mature organs during embryogenesis. For example, the endoderm marker Fox-A1 was initially expressed in a region of the blastula-like vesicle that eventually became the gut, and results were equivalent for ectodermal and mesodermal markers.

Given that coloniality has arisen multiple times, it would hardly be feasible that completely novel mechanisms for regeneration of every tissue and organ also evolved multiple times, and a more parsimonious explanation is that this is due to the ability to coopt embryonic pathways at different times during regeneration. Interestingly though, this does not appear to include a clear recapitulation of embryonic gastrulation.

4.2 The thaliaceans: Salps, doliods and pyrosomes

The thaliaceans include three orders of pelagic tunicates: the Pyrosomida, Salpida, and Doliolida. The thaliaceans are more closely related to the ascidians than to the appendicularians (see Section 1), and a conserved anatomy is evident between these two groups, particularly the large pharyngeal basket used for filter feeding in both adult ascidians and thaliaceans. Although the anatomy and reproductive cycles of thaliaceans have been described in reports dating back to the 19th century (see review Piette & Lemaire, 2015), much remains unknown about this group of animals. Thaliaceans are more difficult to collect and culture than ascidians, and they do not readily lend themselves to modern experimental methods. Despite their similarities to ascidians, the thaliaceans have diverged extensively and have evolved complex reproductive strategies involving alternating sexual and asexual generations (Holland, 2016; Piette & Lemaire, 2015). Like some ascidian species, salps and pyrosomes can form large colonies comprised of individual zooids produced by asexual budding. While asexual budding and coloniality appear to have evolved independently several times in the ascidians (Brown & Swalla, 2012), the evolutionary history of thaliacean budding is less clear and it is uncertain whether it arose before or after the split with ascidians (see discussion in Piette & Lemaire, 2015).

Embryonic development of salps appears to be particularly unusual. The salp lifecycle starts with the sexually produced embryo that goes on to form a single 'oozooid.' The adult oozoid then buds asexually to produce connected chains of 'blastozooids' known as 'aggregates.' Each blastozooid has germ cells that produce first eggs and later sperm. Depending on the species, only a single egg may be produced per blastozooid. Fertilized eggs attach to the adult body wall and develop viviparously. The salp zygote is called the 'blastophore' and its development remains poorly described, particularly with modern molecular tools. A 1960s-era description of blastophore development in *Salpa fusiformis* shows a developing embryo largely unrecognizable to those familiar with ascidian or larvacean embryos (Sutton, 1960; Fig. 4). It lacks an obvious notochord or neural tube, and does not produce a tadpole larva. Remarkably, it does not obviously gastrulate. Sutton identified a mass of blastophore cells that he terms 'neuro-chordal,' but these fail to form an identifiable notochord or neural plate and molecular markers of cell identity have not been examined in these cells. The only morphogenic event that might be conflated with gastrulation is the formation of the neuro-chordal invagination (N-C.I. in Fig. 4). Sutton described the neuro-chordal invagination as the result of infolding caused by the differential growth of adjacent epithelia, but it is unclear whether it bears any relationship to gastrulation or neurulation. Finally, although description of embryonic development of doliolids and pyrosomes is still incomplete, neither are viviparous, and both reproduce sexually as well as asexually by budding (Holland, 2016; Lemaire, 2011; Lemaire & Piette, 2015). Although neither is thought to show the extreme derivation of development observed in salps, various species within both groups show high, but variable, degrees of degeneration of the tadpole larva morphology. For example, *Doliolum* larvae are reported to make a rudimentary notochord flanked by tail muscle, but no obvious nervous system (Holland, 2016). How these altered larval morphologies may be reflected in earlier gastrula morphologies is currently unknown.

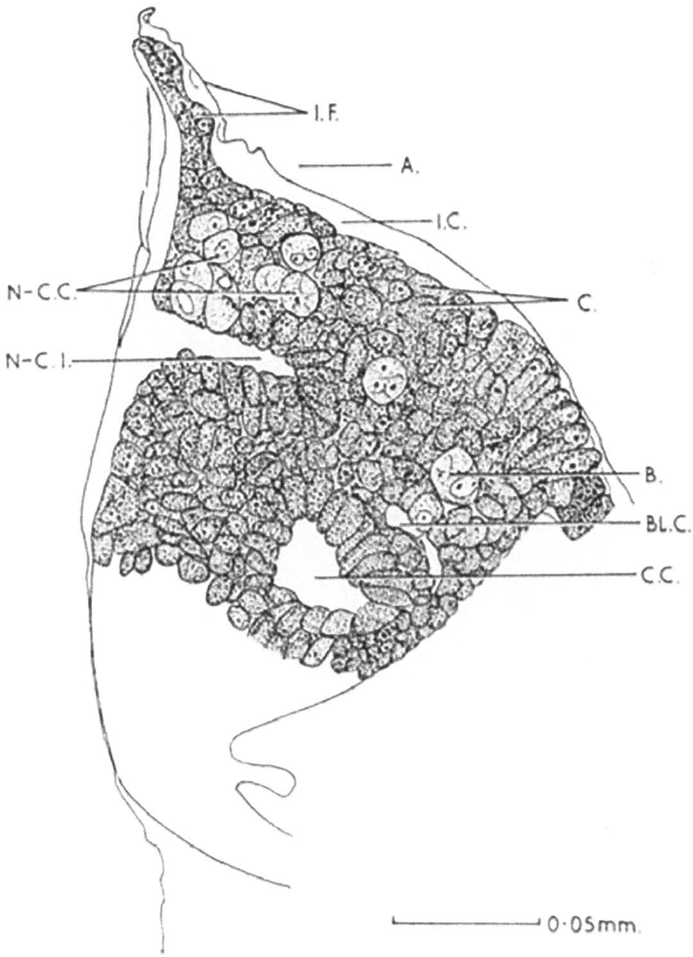

Fig. 4 Blastophore of *Salpa fusiformis*. The precise developmental age of the blastophore is not known. Abbreviations: I.F., Incubation fold; A., atrium; I.C., incubation cavity; C., calymnocyte; B., blastomere; C.C., central cavity; N-C.I., neuro-chordal invagination; N-C.C. Neuro-chordal cells. *Image from Sutton, M. F. (1960). The sexual development of Salpa fusiformis (Cuvier): Part I.* Journal of Embryology and Experimental Morphology, *8, 268–290.*

5. Peri-gastrulation events

5.1 Notochord development

As in vertebrate embryos, early ascidian notochord morphogenesis overlaps with gastrulation. The notochord is divided into two separate developmental lineages. The primary notochord is derived from anterior vegetal (A-line) blastomeres while the secondary notochord is derived from posterior vegetal (B-line) blastomeres (Nishida, 1987). Primary notochord fate is induced at the 64-cell stage and secondary notochord at the 112-cell stage via distinct but overlapping mechanisms (Hudson & Yasuo, 2006; Picco, Hudson, & Yasuo, 2007; Yasuo & Hudson, 2007). At the onset of gastrulation, the four primary notochord cells have divided once to form an arc of eight cells that are not yet contiguous with the two secondary notochord precursors (Fig. 2, yellow asterisks). The two final notochord cell divisions take place during gastrulation. The primary notochord cells are effectively on the anterior lip of the blastopore during the second-last division, and invert during gastrulation so that their previously vegetal surfaces become animal. The final primary notochord division takes place while the blastopore is closing, and is immediately followed by the onset of mediolateral intercalation behaviors that cause the notochord to converge and extend into a single-file column (Munro & Odell, 2002a, 2002b). This is in striking contrast to organisms such as *Drosophila* where the cell cycle pauses in cells that are actively gastrulating (Grosshans & Wieschaus, 2000).

Perhaps the most surprising conclusion from Sherrard et al. (2010) is the apparent absence of a contribution from the axial mesoderm (notochord and tail muscle precursors) to the forces that drive invagination during ascidian gastrulation—at least up through the formation of the cup-shaped gastrula. Although it is possible that the developing axial mesoderm could play a role in later steps, such as blastopore closure, observations of mutant *Ciona*, as well as tailless variants of the *Molgulid* species of ascidians, suggest otherwise. As detailed above, ascidian notochord morphogenesis initiates at gastrulation, although extensive convergent extension and elongation of the ascidian notochord is observed most prominently in the neurula embryo (i.e., following the completion of gastrulation) (Smith, 2018). Consistent with the absence of a role for axial mesoderm in invagination, *Ciona* embryos homozygous for the mutation *chobi* have a severe defect in notochord specification due to a mutation in the gene *brachyury* (Chiba, Jiang, Satoh, & Smith, 2009), but nevertheless successfully complete gastrulation and properly close the blastopore. In fact, presumptive notochord cells in *chobi* are transfated to endoderm, further evidencing the robustness of the endoderm invagination mechanisms to perturbations (Sherrard et al., 2010), in this case to increased cell number. Likewise, a number of species of ascidians have been described that differ from the conventional body plan in having no tail, and reduced numbers of notochord and tail muscle cells—the best described of these being *Molgula occulta* (Swalla & Jeffery, 1990;

Swalla, Makabe, Satoh, & Jeffery, 1993). Nevertheless, these tailless species gastrulate and close their blastopores. In *M. occulta*, as well as the congeneric tailless species *Molgula tectiformi*, the notochord lineage is present in the blastula embryo, as apparent by the expression of *brachyury*, but these cells fail to follow the full mitotic sequence observed in the tailed *Molgula occulta*, nor do they undergo convergent extension (Takada et al., 2002). Finally, *Ciona* mutant *aimless* carries a null mutation in the gene encoding the essential planar cell polarity (PCP) component Prickle (Jiang, Munro, & Smith, 2005). In the *Ciona* embryo *prickle* expression is limited to the notochord, and homozygous *aimless* embryos have defects in the intercalation of notochord cells, although markers of notochord differentiation are expressed normally. Despite these defects in notochord morphogenesis, *aimless* embryos gastrulate with no obvious defects, although they have greatly shortened anterior/posterior axes. Thus, the PCP pathway is not required for the basic steps of involution and blastopore closure in *Ciona*. Whether this differs significantly from the role of the PCP pathway in vertebrate gastrulation is unclear. *Xenopus* embryos in which Prickle was knocked down with morpholinos have been described as having severe defects in mesoderm involution (Takeuchi et al., 2003). However, PCP loss of function results in zebrafish appear to be more similar to *Ciona*, with a loss of mediolateral intercalation in axial mesoderm and a resulting shortened axes, but complete involution (Jessen et al., 2002; Veeman, Slusarski, Kaykas, Louie, & Moon, 2003).

After the notochord has intercalated into a single file column of 40 cells, it undergoes considerable further morphogenesis, including cell shape changes and concomitant protein localization (and relocalization) to generate constricted apical domains at the interfaces between the stacked cells (see reviews Denker & Jiang, 2012; Smith, 2018). Each notochord cell develops separate anterior and posterior apical domains (Denker, Bocina, & Jiang, 2013). Extracellular lumens then form between the cells, centered at the apical domains. As these lumens expand they fuse to form a continuous inflated, hollow lumen running the length of the notochord and surrounded by the now epithelium-like notochord cells. Moreover, unlike many vertebrates, there is no fate distinction between separate notochord sheath and core cell types. At the same time that the apical domains are forming there is a repolarization of the PCP machinery. Early in notochord development PCP activity drives mediolateral intercalation (Jiang et al., 2005). However, at the completion of intercalation the polarity of PCP activity is repositioned to the anterior/posterior axis (Newman-Smith, Kourakis, Reeves, Veeman, & Smith, 2015). The most obvious manifestation of the

PCP-dependent anterior/posterior polarity of the notochord is the posterior localization of the nucleus, and the anterior localization of the PCP components Prickle and Strabismus, as well as myosin II (reviewed in Smith, 2018).

5.2 Neural induction and neurulation

In vertebrates, induction of the central nervous system initiates late in the blastula stage and continues through gastrulation. The vertebrate neural-inducing molecules are products of the dorsal lip of the blastopore, a tissue which is variously termed the organizer in amphibians, the shield in teleosts, Hensen's node in chicks and the node in mammals. These inducers act on neighboring ectoderm to initiate the formation of the neural plate. A number of secreted factors have neural-inducing activity, including FGF, and BMP inhibitors such as Noggin and Chordin (see review Stern, 2005). The evolutionary similarities with the vertebrates are most evident in the morphology and morphogenesis of the ascidian larval nervous system (see review Hudson, 2016). While there are some parallels between ascidian and vertebrate neural induction, significant departures are also evident.

The *Ciona* neural plate has a remarkably simple structure with a grid-like arrangement of six rows and eight columns of cells that have distinct row and column identities (Hudson et al., 2007). The cells forming the plate come from 3 of the 4 quadrants of the 8-cell embryo and are induced in quite different ways. These include the *a*-lineage which gives rise to the anterior neural plate and goes on to make the anterior brain, palps and presumptive mouth (Veeman, Newman-Smith, El-Nachef, & Smith, 2010), the *A*-lineage which gives rise to posterior brain and the posterior nerve cord, and the *b*-lineage which gives rise to the dorsal-most row of cells in the posterior brain and posterior nerve chord (neural lineages reviewed in Hudson, 2016). Neural induction in *Ciona* starts at the 32-cell stage with the specification of the *a*-lineage (Bertrand, Hudson, Caillol, Popovici, & Lemaire, 2003) via vegetally expressed FGF9/16/20 (i.e., the ortholog of vertebrate FGFs 9, 16 and 20). During *a*-lineage neural specification, not only does contact between the inducer cells and the induced ectodermal cells appear to be important, but the area of contact appears to be essential in distinguishing between induced cells and neighboring non-induced cells (Tassy et al., 2006). Neural induction in the *b*-lineage is not well characterized, and the induction of *A*-lineage neural cells is very different than the *a*-lineage. The *A*-lineage of the neural plate becomes fate restricted at the 64-cell stage as the result of asymmetric cell divisions in which *suppression* of FGF signaling is required to suppress notochord fate (Yasuo & Hudson, 2007). While the *Ciona* embryo does express some secreted BMP inhibitors, none have been found to share the critical role in early neural induction seen in vertebrate embryos.

Ciona neurulation is stereotypically chordate, with a neural plate that forms a distinctive midline groove and rolls up to form a hollow, dorsal neural tube (Conklin, 1905a; Nicol & Meinertzhagen, 1988b; Satoh, 1978). The onset of neurulation is continuous with the end of gastrulation, with the neural folds first appearing at the posterior of the embryo on either side of the closing blastopore (Nicol & Meinertzhagen, 1988b). It is unclear whether this is simply a consequence of rapid early development or whether blastopore closure might be mechanistically important for early neurulation. Neural tube closure progresses from posterior to anterior (Nishida, 1986). A transient chordoneural hinge can be seen in phalloidin-stained confocal stacks connecting the forming neural tube lumen to the remnant of the archenteron via the nearly-closed blastopore.

Neurulation first involves the furrowing and formation of a hinge point in the midline of the neural plate around which neural cells will bend as the midline cells constrict apically (Nicol & Meinertzhagen, 1988b). Convergence and extension movements narrow the mediolateral axis and lengthen the anterior-posterior axis of the neural plate as it rolls up (Nicol & Meinertzhagen, 1988b). Previously lateral and now dorsal cells meet at the midline and 'zipper' together to form a closed neural tube. The posterior to anterior zippering involves contraction of the apical domain of bilateral pairs of neural cells. These contractions are sequential and are punctuated by apical junction exchange and release from one pair of midline cells to another in a posterior to anterior direction (Hashimoto, Robin, Sherrard, & Munro, 2015).

6. Conclusion

The presence of a number of traits in the tunicates, such as simple and stereotyped lineages, the extreme transparency of some ascidian species such as *Phallusia*, making them ideal for live imaging (Robin et al., 2011), their evolutionary affinity to vertebrates, and the development of a suite of genomic and imaging methods (Stolfi & Christiaen, 2012; Veeman & Reeves, 2015), has allowed them to make ongoing contributions to our understanding of gastrulation. Many questions remain, some specific to tunicate gastrulation, and others to chordate gastrulation in general. The apparent differences in the relative importance of endoderm involution in ascidian gastrulation in comparison to vertebrate models in which mesoderm and ectoderm are thought to play more prominent roles, raises the question of whether this is a derived feature of the ascidians. Has a complex gastrulation involving coordinated morphogenetic processes in different germ layers been lost in the ascidians, perhaps as a consequence of embryo simplification, or was a simple gastrulation driven predominantly by endoderm invagination a feature of the last common ancestor of the chordates? We have emphasized in this review the diversity of tunicate morphologies and apparent developmental strategies, although our detailed knowledge is limited to just a few species. Renewed investigation of lesser-studied tunicate clades, such as the salps, with modern methods may reveal new insights into the diversity of chordate developmental strategies, including those for gastrulation.

Acknowledgments

We thank Hiroki Nishida for his helpful discussion of *Oikopleura* gastrulation. This work support by NIH grant HD038701 to W.C.S., and NIH grant HD085909 and NSF grant IOS 1456555 to M.V.

References

Beane, W. S., Gross, J. M., & McClay, D. R. (2006). RhoA regulates initiation of invagination, but not convergent extension, during sea urchin gastrulation. *Developmental Biology*, *292*, 213–225. https://doi.org/10.1016/j.ydbio.2005.12.031.

Bertrand, V., Hudson, C., Caillol, D., Popovici, C., & Lemaire, P. (2003). Neural tissue in ascidian embryos is induced by FGF9/16/20, acting via a combination of maternal GATA and Ets transcription factors. *Cell*, *115*, 615–627.

Bourlat, S. J., Rota-Stabelli, O., Lanfear, R., & Telford, M. J. (2009). The mitochondrial genome structure of Xenoturbella bocki (phylum Xenoturbellida) is ancestral within the deuterostomes. *BMC Evolutionary Biology*, *9*, 107. https://doi.org/10.1186/1471-2148-9-107.

Brown, F. D., & Swalla, B. J. (2012). Evolution and development of budding by stem cells: Ascidian coloniality as a case study. *Developmental Biology*, *369*, 151–162. https://doi.org/10.1016/j.ydbio.2012.05.038.

Brusca, R. C., & Brusca, G. J. (1990). *Invertebrates*. Sunderland, MA: Sinauer.

Carlson, M., Reeves, W., & Veeman, M. (2015). Stochasticity and stereotypy in the Ciona notochord. *Developmental Biology*, *397*, 248–256. https://doi.org/10.1016/j.ydbio.2014.11.016.

Chabry, L. (1887). Contribution a l'embryologie normale et teratologique des Ascidie simples. *Journal of Anatomy and Physiology*, *23*, 167–319.

Chiba, S., Jiang, D., Satoh, N., & Smith, W. C. (2009). Brachyury null mutant-induced defects in juvenile ascidian endodermal organs. *Development*, *136*, 35–39. https://doi.org/10.1242/dev.030981.

Chiba, S., Miki, Y., Ashida, K., Wada, M. R., Tanaka, K. J., Shibata, Y., et al. (1999). Interactions between cytoskeletal components during myoplasm rearrangement in ascidian eggs. *Development, Growth & Differentiation, 41*, 265–272.

Conklin. (1905a). The organization and cell-lineage of the ascidian egg. *Journal of the Academy of Natural Sciences of Philadelphia, 13*, 1.

Conklin. (1905b). Organ-forming stubstances in the eggs of ascidians. *Biology Bulletin, VIII*, 205–230.

Conklin. (1905c). Mosaic development in ascidian eggs. *The Journal of Experimental Zoology, 2*, 145–223.

Darwin, C. (1871). The descent of man, and selection in relation to sex. *Vol 1*. John Murray. https://doi.org/10.1037/12293-000.

Delsuc, F., Brinkmann, H., Chourrout, D., & Philippe, H. (2006). Tunicates and not cephalochordates are the closest living relatives of vertebrates. *Nature, 439*, 965–968. https://doi.org/10.1038/nature04336.

Delsuc, F., Philippe, H., Tsagkogeorga, G., Simion, P., Tilak, M. K., Turon, X., et al. (2018). A phylogenomic framework and timescale for comparative studies of tunicates. *BMC Biology*. https://doi.org/10.1186/s12915-018-0499-2.

Denker, E., Bocina, I., & Jiang, D. (2013). Tubulogenesis in a simple cell cord requires the formation of bi-apical cells through two discrete Par domains. *Development, 140*, 2985–2996.

Denker, E., & Jiang, D. (2012). Ciona intestinalis notochord as a new model to investigate the cellular and molecular mechanisms of tubulogenesis. *Seminars in Cell & Developmental Biology, 23*, 308–319.

Dumollard, R., Hebras, C., Besnardeau, L., & McDougall, A. (2013). Beta-catenin patterns the cell cycle during maternal-to-zygotic transition in urochordate embryos. *Developmental Biology, 384*, 331–342. https://doi.org/10.1016/j.ydbio.2013.10.007.

Dumollard, R., Minc, N., Salez, G., Aicha, S. B., Bekkouche, F., Hebras, C., et al. (2017). The invariant cleavage pattern displayed by ascidian embryos depends on spindle positioning along the cell's longest axis in the apical plane and relies on asynchronous cell divisions. *eLife, 6*, 1–23. https://doi.org/10.7554/eLife.19290.

Ettensohn, C. A. (1985). Gastrulation in the sea urchin embryo is accompanied by the rearrangement of invaginating epithelial cells. *Developmental Biology, 112*, 383–390.

Ettensohn, C. A. (1999). Cell movements in the sea urchin embryo. *Current Opinion in Genetics & Development, 9*, 461–465. https://doi.org/10.1016/S0959-437X(99)80070-7.

Fischer, J. L. (1992). The embryological œuvre of Laurent Chabry. *Roux's Archives of Developmental Biology, 201*, 125–127. https://doi.org/10.1007/BF00188709.

Fiuza, U.-M., Negishi, T., Rouan, A., Yasuo, H., & Lemaire, P. (2018). Nodal and Eph signalling relay drives the transition between apical constriction and apico-basal shortening during ascidian endoderm invagination. *bioRxiv*. https://doi.org/10.1101/418988.

Fujii, S., Nishio, T., & Nishida, H. (2008). Cleavage pattern, gastrulation, and neurulation in the appendicularian, Oikopleura dioica. *Development Genes and Evolution, 218*, 69–79. https://doi.org/10.1007/s00427-008-0205-4.

Großhans, J., & Weischaus, E. (2000). A genetic link between morphogenesis and cell division during formation of the ventral furrow in Drosophila. *Cell, 101*, 523–531.

Haeckel, E. (1874). *Anthropogenie oder Entwicklungsgeschichte des Menschen*. Leipzig: Engelmann.

Hashimoto, H., Robin, F. B., Sherrard, K. M., & Munro, E. M. (2015). Sequential contraction and exchange of apical junctions drives zippering and neural tube closure in a simple chordate. *Developmental Cell, 32*, 241–255. https://doi.org/10.1016/j.devcel.2014.12.017.

Hibino, T., Nishikata, T., & Nishida, H. (1998). Centrosome-attracting body: A novel structure closely related to unequal cleavages in the ascidian embryo. *Development, Growth & Differentiation, 40*, 85–95. https://doi.org/10.1046/j.1440.

Holland, L. Z. (2016). Tunicates. *Current Biology, 26*, R146–R152. https://doi.org/10.1016/j.cub.2015.12.024.

Hotta, K., Mitsuhara, K., Takahashi, H., Inaba, K., Oka, K., Gojobori, T., et al. (2007). A web-based interactive developmental table for the ascidian *Ciona intestinalis*, including 3D real-image embryo reconstructions: I. From fertilized egg to hatching larva. *Developmental Dynamics, 236*, 1790–1805. https://doi.org/10.1002/dvdy.21188.

Hudson, C. (2016). The central nervous system of ascidian larvae. *Wiley Interdisciplinary Reviews: Developmental Biology, 5*, 538–561. https://doi.org/10.1002/wdev.239.

Hudson, C., Kawai, N., Negishi, T., & Yasuo, H. (2013). β-Catenin-driven binary fate specification segregates germ layers in ascidian embryos. *Current Biology, 23*, 491–495. https://doi.org/10.1016/j.cub.2013.02.005.

Hudson, C., Lotito, S., & Yasuo, H. (2007). Sequential and combinatorial inputs from nodal, Delta2/Notch and FGF/MEK/ERK signalling pathways establish a grid-like organisation of distinct cell identities in the ascidian neural plate. *Development, 134*, 3527–3537. https://doi.org/10.1242/dev.002352.

Hudson, C., Sirour, C., & Yasuo, H. (2016). Co-expression of Foxa.a, Foxd and Fgf9/16/20 defines a transient mesendoderm regulatory state in ascidian embryos. *eLife, 5*, 1–17. https://doi.org/10.7554/eLife.14692.

Hudson, C., & Yasuo, H. (2006). A signalling relay involving Nodal and Delta ligands acts during secondary notochord induction in Ciona embryos. *Development, 133*, 2855–2864. https://doi.org/10.1242/dev.02466.

Imai, K. S. (2003). Isolation and characterization of beta-catenin downstream genes in early embryos of the ascidian Ciona savignyi. *Differentiation, 71*, 346–360. https://doi.org/10.1046/j.1432-0436.2003.7106001.x.

Imai, K. S., Hudson, C., Oda-Ishii, I., Yasuo, H., & Satou, Y. (2016). Antagonism between β-catenin and Gata.a sequentially segregates the germ layers of ascidian embryos. *Development, 143*, 4167–4172. https://doi.org/10.1242/dev.141481.

Imai, K., Takada, N., Satoh, N., & Satou, Y. (2000). (Beta)-Catenin mediates the specification of endoderm cells in ascidian embryos. *Development, 127*, 3009–3020.

Iseto, T., & Nishida, H. (1999). Ultrastructural studies on the centrosome-attracting body: Electron-dense matrix and its role in unequal cleavages in ascidian embryos. *Development, Growth & Differentiation, 41*, 601–609. https://doi.org/10.1046/j.1440-169x.1999.00457.x.

Jessen, J. R., Topczewski, J., Bingham, S., Sepich, D. S., Marlow, F., Chandrasekhar, A., et al. (2002). Zebrafish trilobite identifies new roles for strabismus in gastrulation and neuronal movements. *Nature Cell Biology, 4*, 610–615. https://doi.org/10.1038/ncb828.

Jiang, D., Munro, E. M., & Smith, W. C. (2005). Ascidian prickle regulates both mediolateral and anterior-posterior cell polarity of notochord cells. *Current Biology, 15*, 79–85.

Keller, R., Davidson, L. A., & Shook, D. R. (2003). How we are shaped: The biomechanics of gastrulation. *Differentiation, 71*, 171–205. https://doi.org/10.1046/j.1432-0436.2003.710301.x.

Kocot, K. M., Tassia, M. G., Halanych, K. M., & Swalla, B. J. (2018). Phylogenomics offers resolution of major tunicate relationships. *Molecular Phylogenetics and Evolution, 121*, 166–173. https://doi.org/10.1016/j.ympev.2018.01.005.

Kowalevsky, A. (1866). Entwicklungsgeschichte der einfachen Ascidien. In *Mémoires l'Académie Impériale des Sci. St. Petersbourg. VIIe Série.* Imperial Academy at St. Petersburg.

Lemaire, P. (2011). Evolutionary crossroads in developmental biology: The tunicates. *Development, 138*, 2143–2152. https://doi.org/10.1242/dev.048975.

Lemaire, P., & Piette, J. (2015). Tunicates: Exploring the sea shores and roaming the open ocean. A tribute to Thomas Huxley. *Open Biology*, *5*, 150053. https://doi.org/10.1098/rsob.150053.

Logan, C. Y., Miller, J. R., Ferkowicz, M. J., & Mcclay, D. R. (1999). Nuclear β-catenin is required to specify vegetal cell fates in the sea urchin embryo. *Development*, *126*, 345–357.

Manni, L., Anselmi, C., Cima, F., Gasparini, F., Voskoboynik, A., Martini, M., et al. (2019). Sixty years of experimental studies on the blastogenesis of the colonial tunicate Botryllus schlosseri. *Developmental Biology*, *448*, 293–308. https://doi.org/10.1016/j.ydbio.2018.09.009.

McDougall, A., & Sardet, C. (1995). Function and characteristics of repetitive calcium waves associated with meiosis. *Current Biology*, *5*, 318–328.

Miyawaki, K., Yamamoto, M., Saito, K., & Saito, S. (2003). Nuclear localization of β -catenin in vegetal pole cells during early embryogenesis of the starfish Asterina pectinifera. *Development, Growth, and Differentiation*, *45*(2), 121–128, Japanese Society for Developmental Biologists.

Munro, E. M., & Odell, G. (2002a). Morphogenetic pattern formation during ascidian notochord formation is regulative and highly robust. *Development*, *129*, 1–12.

Munro, E. M., & Odell, G. M. (2002b). Polarized basolateral cell motility underlies invagination and convergent extension of the ascidian notochord. *Development*, *129*, 13–24.

Newman-Smith, E., Kourakis, M. J., Reeves, W., Veeman, M., & Smith, W. C. (2015). Reciprocal and dynamic polarization of planar cell polarity core components and myosin. *eLife*, *4*, e05361.

Nicol, D., & Meinertzhagen, I. A. (1988a). Development of the central nervous system of the larva of the ascidian, Ciona intestinalis L. I. The early lineages of the neural plate. *Developmental Biology*, *130*, 721–736.

Nicol, D., & Meinertzhagen, I. A. (1988b). Development of the central nervous system of the larva of the ascidian, Ciona intestinalis L. II. Neural plate morphogenesis and cell lineages during neurulation. *Developmental Biology*, *130*, 737–766.

Nishida, H. (1986). Cell-division pattern during gastrulation of the ascidian, Halocynthia-Roretzi. *Development, Growth & Differentiation*, *28*, 191–201.

Nishida, H. (1987). Cell lineage analysis in ascidian embryos by intracellular injection of a tracer enzyme. III. Up to the tissue restricted stage. *Developmental Biology*, *121*, 526–541. https://doi.org/10.1126/science.725606.

Nishida, H. (2008). Development of the appendicularian Oikopleura dioica: Culture, genome, and cell lineages. *Development, Growth & Differentiation*, *50*(Suppl. 1), S239–S256. https://doi.org/10.1111/j.1440-169X.2008.01035.x.

Nishida, H., & Satoh, N. (1983). Cell lineage analysis in ascidian embryos by intracellular injection of a tracer enzyme. I. Up to the eight-cell stage. *Developmental Biology*, *99*, 382–394.

Nishida, H., & Satoh, N. (1985). Cell lineage analysis in ascidian embryos by intracellular injection of a tracer enzyme. II. The 16- and 32-cell stages. *Developmental Biology*, *110*, 440–454.

Nishida, H., & Sawada, K. (2001). Macho-1 encodes a localized mRNA in ascidian eggs that specifies muscle fate during embryogenesis. *Nature*, *409*, 724–729. https://doi.org/10.1038/35055568.

Nishikata, T., Hibino, T., & Nishida, H. (1999). The centrosome-attracting body, microtubule system, and posterior egg cytoplasm are involved in positioning of cleavage planes in the ascidian embryo. *Developmental Biology*, *209*, 72–85. https://doi.org/10.1006/dbio.1999.9244.

Nishikata, T., Mita-Miyazawa, I., Deno, T., & Satoh, N. (1987). Muscle cell differentiation in ascidian embryos analysed with a tissue-specific monoclonal antibody. *Development*, *99*, 163–171.

Okado, H., & Takahashi, K. (1988). A simple "neural induction" model with two interacting cleavage-arrested ascidian blastomeres. *Proceedings of the National Academy of Sciences of the United States of America*, *85*, 6197–6201. https://doi.org/10.1073/pnas.85.16.6197.

Onai, T. (2019). Canonical Wnt/β-catenin and notch signaling regulate animal/vegetal axial patterning in the cephalochordate amphioxus. *Evolution & Development*, *21*, 31–43. https://doi.org/10.1111/ede.12273.

Patalano, S., Prulière, G., Prodon, F., Paix, A., Dru, P., Sardet, C., et al. (2006). The aPKC-PAR-6-PAR-3 cell polarity complex localizes to the centrosome attracting body, a macroscopic cortical structure responsible for asymmetric divisions in the early ascidian embryo. *Journal of Cell Science*, *119*, 1592–1603. https://doi.org/10.1242/jcs.02873.

Picco, V., Hudson, C., & Yasuo, H. (2007). Ephrin-Eph signalling drives the asymmetric division of notochord/neural precursors in Ciona embryos. *Development*, *134*, 1491–1497. https://doi.org/10.1242/dev.003939.

Piette, J., & Lemaire, P. (2015). Thaliaceans, the neglected pelagic relatives of ascidians: A developmental and evolutionary enigma. *The Quarterly Review of Biology*, *90*, 117–145.

Prodon, F., Yamada, L., Shirae-Kurabayashi, M., Nakamura, Y., & Sasakura, Y. (2007). Postplasmic/PEM RNAs: A class of localized maternal mRNAs with multiple roles in cell polarity and development in ascidian embryos. *Developmental Dynamics*, *236*, 1698–1715. https://doi.org/10.1002/dvdy.21109.

Racioppi, C., Kamal, A. K., Razy-Krajka, F., Gambardella, G., Zanetti, L., Di Bernardo, D., et al. (2014). Fibroblast growth factor signalling controls nervous system patterning and pigment cell formation in Ciona intestinalis. *Nature Communications*, *5*, 1–17. https://doi.org/10.1038/ncomms5830.

Ricci, L., Cabrera, F., Lotito, S., & Tiozzo, S. (2016). Redeployment of germ layers related TFs shows regionalized expression during two non-embryonic developments. *Developmental Biology*, *416*, 235–248. https://doi.org/10.1016/j.ydbio.2016.05.016.

Robin, F. B., Dauga, D., Tassy, O., Sobral, D., Daian, F., & Lemaire, P. (2011). Time-lapse imaging of live Phallusia embryos for creating 3D digital replicas. *Cold Spring Harbor Protocols*, *2011*, 1244–1246. pdb.prot065847-pdb.prot065847. https://doi.org/10.1101/pdb.prot065847.

Rothbacher, U., Bertrand, V., Lamy, C., & Lemaire, P. (2007). A combinatorial code of maternal GATA, Ets and -catenin-TCF transcription factors specifies and patterns the early ascidian ectoderm. *Development*, *134*, 4023–4032. https://doi.org/10.1242/dev.010850.

Sardet, C., Speksnijder, J., & Jaffe, L. (1989). Fertilization and ooplasmic movements in the ascidian egg. *Development*, *249*, 237–249.

Sardet, C., Speksnijder, J., Terasaki, M., Chang, P., Biologie, U., De Marine, C., et al. (1992). Polarity of the ascidian egg cortex before fertilization. *Development*, *115*, 221–237.

Satoh, N. (1978). Cellular morphology and architecture during early morphogenesis of the ascidian egg: An SEM study. *The Biological Bulletin*, *155*, 608–614. https://doi.org/10.2307/1540794.

Satoh, N., Rokhsar, D., & Nishikawa, T. (2014). Chordate evolution and the three-phylum system. *Proceedings of the Royal Society B: Biological Sciences*, *281*. https://doi.org/10.1098/rspb.2014.1729.

Sherrard, K., Robin, F., Lemaire, P., & Munro, E. (2010). Sequential activation of apical and basolateral contractility drives ascidian endoderm invagination. *Current Biology*, *20*, 1499–1510. https://doi.org/10.1016/j.cub.2010.06.075.

Smith, W. C. (2018). *Cellular processes of notochord formation*. Singapore: Springer, pp. 165–177. https://doi.org/10.1007/978-981-10-7545-2_15.

Speksnijder, J. E., Sardet, C., & Jaffe, L. F. (1990). The activation wave of calcium in the ascidian egg and its role in ooplasmic segregation. *The Journal of Cell Biology*, *110*, 1589–1598. https://doi.org/10.1083/jcb.110.5.1589.

Stern, C. D. (2005). Neural induction: Old problem, new findings, yet more questions. *Development*, *132*, 2007–2021. https://doi.org/10.1242/dev.01794.

Stolfi, A., & Christiaen, L. (2012). Genetic and genomic toolbox of the chordate Ciona intestinalis. *Genetics*, *192*, 55–66. https://doi.org/10.1534/genetics.112.140590.

Stolfi, A., & Levine, M. (2011). Neuronal subtype specification in the spinal cord of a protovertebrate. *Development*, *138*, 995–1004. https://doi.org/10.1242/dev.061507.

Stolfi, A., Ryan, K., Meinertzhagen, I. A., & Christiaen, L. (2015). Migratory neuronal progenitors arise from the neural plate borders in tunicates. *Nature*, *527*, 371–374. https://doi.org/10.1038/nature15758.

Stolfi, A., Wagner, E., Taliaferro, J. M., Chou, S., & Levine, M. (2011). Neural tube patterning by Ephrin, FGF and Notch signaling relays. *Development*, *138*, 5429–5439. https://doi.org/10.1242/dev.072108.

Sutton, M. F. (1960). The sexual development of Salpa fusiformis (Cuvier): Part I. *Journal of Embryology and Experimental Morphology*, *8*, 268–290.

Swalla, B. J., & Jeffery, W. R. (1990). Interspecific hybridization between an anural and urodele ascidian: Differential expression of urodele features suggests multiple mechanisms control anural development. *Developmental Biology*, *142*, 319–334.

Swalla, B. J., Makabe, K. W., Satoh, N., & Jeffery, W. R. (1993). Novel genes expressed differentially in ascidians with alternate modes of development. *Development*, *119*, 307–318.

Takada, N., York, J., Davis, J. M., Schumpert, B., Yasuo, H., Satoh, N., et al. (2002). Brachyury expression in tailless Molgulid ascidian embryos. *Evolution & Development*, *4*, 205–211.

Takeuchi, M., Nakabayashi, J., Sakaguchi, T., Yamamoto, T. S., Takahashi, H., Takeda, H., et al. (2003). The prickle-related gene in vertebrates is essential for gastrulation cell movements. *Current Biology*, *13*, 674–679.

Tassy, O., Daian, F., Hudson, C., Bertrand, V., & Lemaire, P. (2006). A quantitative approach to the study of cell shapes and interactions during early chordate embryogenesis. *Current Biology*, *16*, 345–358. https://doi.org/10.1016/j.cub.2005.12.044.

Tiozzo, S., Brown, F. W., & De Tomaso, A. W. (2008). Regeneration and stem cells in ascidians. In T. Bosch (Ed.), *Stem cells: From hydra to man* (pp. 95–112). Springer.

Tokuoka, M., Imai, K. S., Satou, Y., & Satoh, N. (2004). Three distinct lineages of mesenchymal cells in Ciona intestinalis embryos demonstrated by specific gene expression. *Developmental Biology*, *274*, 211–224. https://doi.org/10.1016/j.ydbio.2004.07.007.

van Eekelen, M., Runtuwene, V., Overvoorde, J., & den Hertog, J. (2010). RPTPalpha and PTPepsilon signaling via Fyn/Yes and RhoA is essential for zebrafish convergence and extension cell movements during gastrulation. *Developmental Biology*, *340*, 626–639. https://doi.org/10.1016/j.ydbio.2010.02.026.

Veeman, M. T., Newman-Smith, E., El-Nachef, D., & Smith, W. C. (2010). The ascidian mouth opening is derived from the anterior neuropore: Reassessing the mouth/neural tube relationship in chordate evolution. *Developmental Biology*, *344*, 138–149. https://doi.org/10.1016/j.ydbio.2010.04.028.

Veeman, M., & Reeves, W. (2015). Quantitative and in toto imaging in ascidians: Working toward an image-centric systems biology of chordate morphogenesis. *Genesis*, *53*, 143–159. https://doi.org/10.1002/dvg.22828.

Veeman, M. T., Slusarski, D. C., Kaykas, A., Louie, S. H., & Moon, R. T. (2003). Zebrafish prickle, a modulator of noncanonical Wnt/Fz signaling, regulates gastrulation movements. *Current Biology*, *13*, 680–685.

Wang, W., Niu, X., Stuart, T., Jullian, E., Mauck, W. M., Kelly, R. G., et al. (2019). A single-cell transcriptional roadmap for cardiopharyngeal fate diversification. *Nature Cell Biology*, *21*, 674–686. https://doi.org/10.1038/s41556-019-0336-z.

Whittaker, J. R. (1973). Segregation during ascidian embryogenesis of egg cytoplasmic information for tissue-specific enzyme development. *Proceedings of the National Academy of Sciences of the United States of America*, *70*, 2096–2100. https://doi.org/10.1073/pnas.70.7.2096.

Wikramanayake, A. H., Hong, M., Lee, P. N., Pang, K., Byrum, C. A., Bince, J. M., et al. (2003). An ancient role for nuclear beta-catenin in the evolution of axial polarity and germ layer segregation. *Nature*, *426*, 446–450. https://doi.org/10.1038/nature02113.

WoRMS. (2019). *WoRMS—World register of marine species [WWW document]*. Available from http//www.marinespecies.org. URL http://www.marinespecies.org.

Yasuo, H., & Hudson, C. (2007). FGF8/17/18 functions together with FGF9/16/20 during formation of the notochord in Ciona embryos. *Developmental Biology*, *302*, 92–103. https://doi.org/10.1016/j.ydbio.2006.08.075.

CHAPTER NINE

Mesoderm and endoderm internalization in the *Xenopus* gastrula

Rudolf Winklbauer*
Department of Cell and Systems Biology, University of Toronto, Toronto, ON, Canada
*Corresponding author: e-mail address: r.winklbauer@utoronto.ca

Contents

Abstract

Mesoderm and endoderm internalization in the *Xenopus* embryo are based on a number of region-specific morphogenetic processes that co-act in the vegetal half of the gastrula. In the multilayered wall surrounding the blastocoel, the apical layer engages in bottle cell formation and associated invagination and involution movements, and in cell intercalation in the plane of the layer. Of these epithelial-type processes, only bottle cell formation has been analyzed mechanistically. In the deep layers of the blastocoel wall, cell-on-cell migration drives the internalization of mesoderm by various forms of involution and of the endodermal cell mass by vegetal rotation. In the mesoderm, cells migrate in a mesenchymal mode with the aid of locomotory protrusions, whereas cells of the vegetal cell mass resemble free bottle cells that engage in ingression-type amoeboid migration. Cells rearrange by differential migration leading to parallel or

Current Topics in Developmental Biology, Volume 136
ISSN 0070-2153
https://doi.org/10.1016/bs.ctdb.2019.09.002

orthogonal forms of intercalation and respective types of convergent extension. The interaction of the various apical and deep layer processes gives rise to dorsal multilayer invagination, ventrolateral internal involution, peak involution and orthogonal convergent extension of the dorsal posterior mesoderm, vegetal rotation, and blastopore constriction. It is speculated how these multilayer gastrulation movements could be derived from mechanisms in invertebrate single-epithelium gastrulae.

1. Introduction

Non-vertebrate animal embryos typically pass through a stage where they consist of a hollow sphere bounded by a single-layered cellular wall, usually an epithelium. The prospective germ layers map onto the surface of this blastula, and to attain their definitive positions, gastrulation must take place. Depending on the organism, delamination or ingression of cells from the outer layer, invagination and involution of parts of the layer, or any combination of these processes internalizes the mesoderm and endoderm, assisted by cell rearrangements in the plane of the epithelial layer (Gustafson & Wolpert, 1963; Pilot & Lecuit, 2005; Siewing, 1969).

While the closest relatives of vertebrates, cephalochordates and urochordates, possess a single-layered blastula wall, an order-of-magnitude increase in vertebrate egg size is accompanied by the wall becoming multilayered (Fig. 1) (Shook & Keller, 2008). This must affect gastrulation. In the following, mesoderm and endoderm internalization are described for the anuran amphibian, *Xenopus laevis*, where an ancestral mode of vertebrate multilayer gastrulation has been conserved (Collazo, Bolker, & Keller, 1994; Shook & Keller, 2008). Processes are compared to that in other species, to see how they could be derived from epithelial gastrulation as an adaptation to the multilayered structure of the vertebrate embryo.

2. Outline of *Xenopus* gastrulation

During early cleavage, the large blastomeres of the *Xenopus* embryo form a single layer around an incipient blastocoel, but the embryo subsequently becomes multilayered (Fig. 1B). Periclinal divisions generate deep cell layers beneath a superficial layer that shows apical epithelial specializations such as tight junctions, adherens junctions and desmosomes (Fig. 1C) (Chalmers, Strauss, & Papalopulu, 2003; Müller & Hausen, 1995). Cells in superficial and deep layers are tightly packed, express C-cadherin and

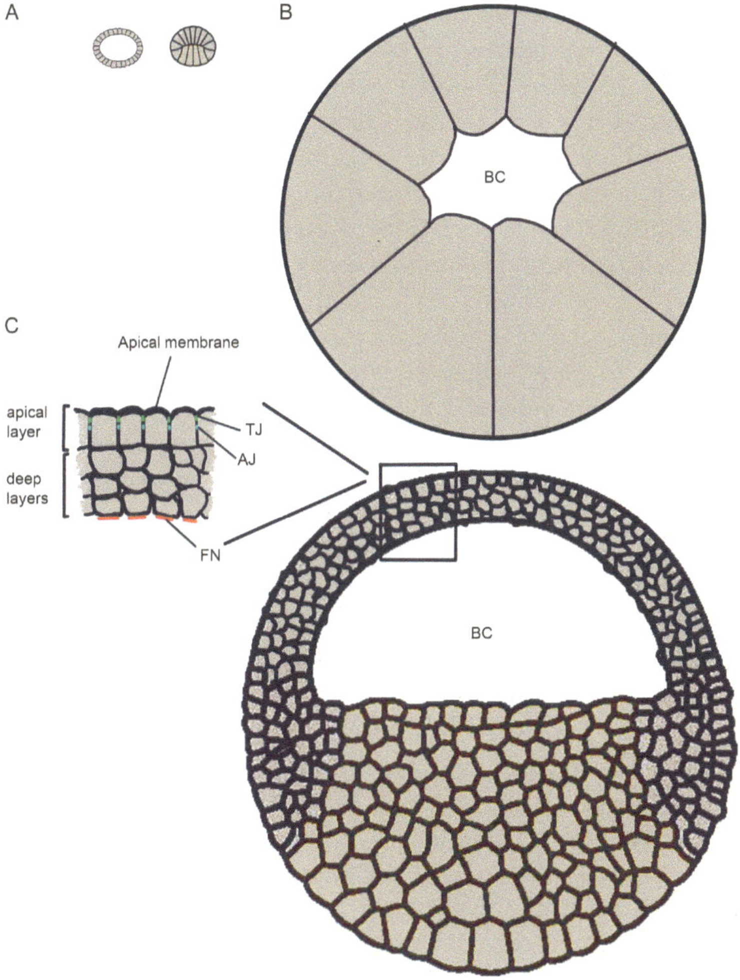

Fig. 1 Size and structure of chordate blastulae. (A) *Branchiostoma* (left) and *Ciona* (right) blastula stage embryos and (B) *Xenopus* cleavage stage (top) and late blastula (bottom) drawn to scale. The *Xenopus* blastula diameter is about 1.3 mm. (C) Multilayered structure of blastula wall. TJ, tight junctions; AJ, adherens junctions; FN, fibronectin fibril extracellular matrix. BC, blastocoel.

epithelial intermediate filaments, and are not separated by a basal lamina, suggesting that the blastocoel wall is a vertebrate-typical multilayered epithelium, with an apical layer on the surface (Shook & Keller, 2003; Winklbauer, 2009).

During cleavage, the delamination of deep layers leads to a partial germ layer segregation. Whereas in non-vertebrate bilaterian blastulae, mesoderm and endoderm reside side by side in a single layer (Fig. 2A), most mesoderm is covered by a non-mesodermal layer in *Xenopus* (Fig. 2B) (Keller, 1975; Shook, Majer, & Keller, 2004). Yet germ layers are still part of a single, continuous blastocoel wall that forms a thin blastocoel roof (BCR) animally, a thick vegetal cell mass opposite, and a marginal zone in between. These regions roughly correspond to prospective ectoderm, endoderm, and mesoderm (Fig. 2Bb′) (Dale & Slack, 1987; Keller, 1975, 1976; Moody, 1987).

In epithelial gastrulation, invagination by the inward bending of the epithelial layer is a well-studied internalization movement (Pearl, Li, & Green, 2017). Counter-bending is an integral part of invagination that is required at the transition to non-bending regions (Fig. 2Aa′). Often, it propagates wave-like such that more and more cells are sequentially internalized by the process of involution. In the *Xenopus* embryo, gastrulation still appears as a folding of the multilayered blastocoel wall that internalizes the vegetal cell mass and the marginal zone and brings the germ layers to their definitive positions (Fig. 2B; see also Fig. 4 C and D). The deep part of the marginal zone, the prospective mesoderm, becomes apposed to the ectodermal BCR, whereas apical suprablastoporal endoderm lines the archenteron. The floor of the archenteron is provided by the vegetal cell mass (Fig. 2Bb″ and b″″) (Ewald, Peyrot, Tyszka, Fraser, & Wallingford, 2004; Ibrahim & Winklbauer, 2001; Keller, 1975, 1976; Nieuwkoop & Florschütz, 1950; Winklbauer & Schürfeld, 1999). Internalization is based on region-specific autonomous processes in the lower half of the gastrula, which include bottle cell formation and invagination, convergence, involution and extension of mesoderm, and deep bottle cell migration leading to vegetal rotation (Fig. 2B).

3. Bottle cell formation

Apical constriction of cells is common in epithelial morphogenesis. In *Xenopus*, apically constricted cells with a slender sub-apical neck and a bulbous cell body form dorsally at the onset of gastrulation in a narrow zone from small marginal and large vegetal apical cells. With a 2-h delay, bottle cells also appear laterally and ventrally, completing a ring-shaped zone of apical constriction (Figs. 2Bb′, b″ and 3A, B) (Hardin & Keller, 1988; Keller, 1978, 1981; Keller & Schoenwolf, 1977; Kurth & Hausen, 2000; Lee & Harland, 2007; Nieuwkoop & Florschütz, 1950).

In isolation, the apical domain of bottle cells contracts autonomously, while cell body elongation and neck formation require interaction with

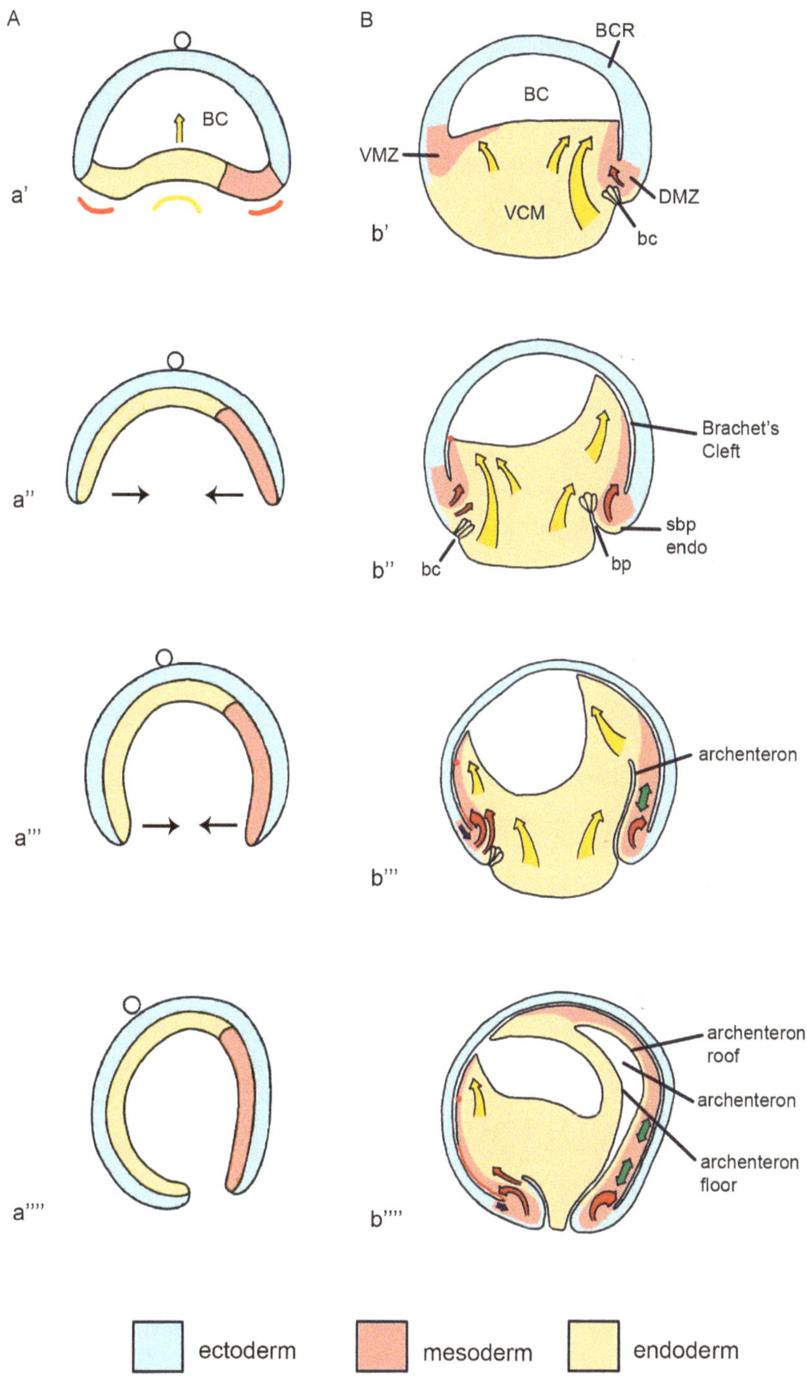

Fig. 2 See figure legend on next page.

adjacent cells (Hardin & Keller, 1988). Apically, an F-actin-rich cell cortex extends into numerous surface protrusions that form a reticulate meshwork. Together with the dense packing of large vesicles below the apex (Fig. 3B) (Kurth & Hausen, 2000; Lee & Harland, 2007), the dynamic actin cytoskeleton implicates macropinocytosis in apical constriction (Wen & Winklbauer, 2017). Consistent with such a mechanism, inhibition of endocytosis by dominant-negative dynamin or Rab5 impedes constriction (Lee & Harland, 2010), while inhibition of myosin II in the apical cortex has only weak effects (Lee & Harland, 2007; Pfister, Shook, Chang, Keller, & Skoglund, 2016; Skoglund, Rolo, Chen, Gumbiner, & Keller, 2008). Moreover, the RhoGEF, Plekhg5, which has been implicated in vesicle trafficking (Lüningschrör et al., 2017), is necessary and sufficient for bottle cell formation (Popov, Ray, Skoglund, Keller, & Chang, 2018).

Bottle cell constriction is not directly linked to invagination. In *Xenopus*, ventral bottle cells bulge inward basally, but the outer surface remains flat as adjacent cells expand and bend apically to compensate for the constriction (Fig. 3A). Dorsally, bottle cell formation leads to a surface indentation due to columnar cells that form a straight surface between the constricted and the apically expanded cells (intermediate, dark and light gray cells, respectively, in Fig. 3A) (Hausen & Riebesell, 1991; Lee & Harland, 2007, 2010; Nieuwkoop & Florschütz, 1950). The columnar cells could be generated by coordinated shear movements of previously expanded cells that reduce their apical surfaces while transmitting apical expansion and counter-bending to the next row of cells (Fig. 3A). If the process continued, involution of the apical layer would ensue through a propagated change in the orientation of the surface (Keller, 1978). A movement of epithelial cells that corresponds to this behavior has been described as telescoping, an invagination in the absence of apical constriction that involves a characteristic apical overlapping of rhomboid cells (light gray cells in Fig. 3) (Li, Economou, & Green, 2019; Pearl et al., 2017). Although its detailed cellular mechanism is

Fig. 2 Gastrula stages. (A) *Branchiostoma* as an example of epithelial gastrulation by invagination. Yellow curve, arrow: inward bending; red curves: counter-bending. Arrows, blastopore constriction. a′, initial; a″, early; a‴ middle; a⁗ late gastrula. (B) *Xenopus* multilayer gastrulation. b′, initial; b″, early; b‴ middle; b⁗ late gastrula. Yellow arrows, vegetal rotation movements; red arrows, deep cell involution; green double-arrows, dorsal convergent extension. BC, blastocoel; BCR, blastocoel roof; DMZ, dorsal marginal zone; VMZ, ventral marginal zone; VCM, vegetal cell mass; sbp endo, suprablastoporal endoderm; bc, bottle cells; bp, blastopore. *Panel (A): After Holland, L. Z., & Onai, T. (2012). Early development of cephalochordates (amphioxus).* Wiley Interdisciplinary Reviews: Developmental Biology, 1, *167–183.*

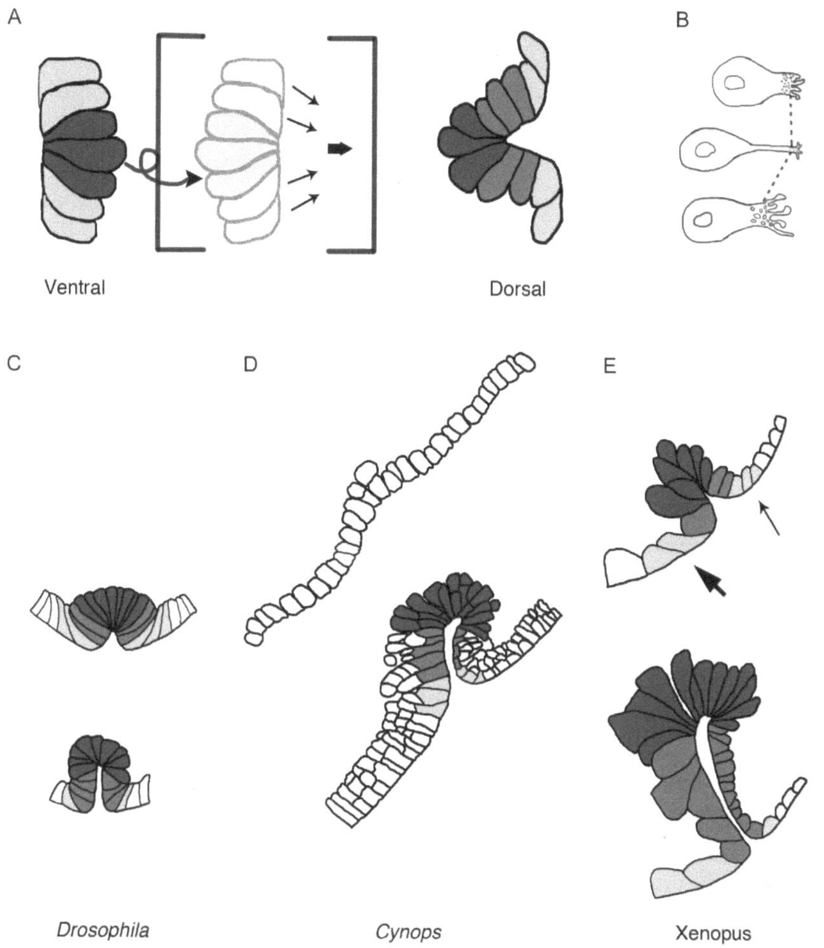

Fig. 3 Epithelial invagination and involution. (A) *Xenopus* bottle cell formation in apical layer. Ventral bottle cells without invagination (left). Hypothetical mechanism of invagination by telescoping (arrows) (center, bracket) leading to dorsal cell array (right). (B) Bottle cell structure and migration (steps from top to bottom) toward the left, sites of apical connections indicated by dashed line. (C) Stages of *Drosophila* mesoderm invagination. (D) Invagination of *Cynops* apical layer explant. (E) *Xenopus* apical layer, dorsal side. Dark gray, bottle cells; medium gray, cylindrical cells after telescoping; light gray, rhomboid, apically overlapping telescoping cells. Light arrow, suprablastoporal apical layer; bold arrow, apical layer of vegetal cell mass. *Panel (A): After Hausen, P., & Riebesell, M. (1991). The early development of* Xenopus laevis. *An atlas of the histology. Berlin-Heidelberg-New York: Springer. Panel (B): After Baker, P. C. (1965). Fine structure and morphogenetic movements in the gastrula of the treefrog, Hyla regilla.* The Journal of Cell Biology, 24, *95–116. Panel (C): After Sweeton, D., Parks, S., Costa, M., & Wieschaus, E. (1991). Gastrulation in Drosophila: The formation of the ventral furrow and posterior midgut invaginations.* Development, 112, *775–789. Panel (D): After Komazaki, S. (1993). Movement of an epithelial layer isolated from early embryos of the newt, Cynops* pyrrhogaster. *II. Formation of a blastoporal groove and archenteron in a superficial epithelial layer isolated from initial gastrula.* Development, Growth & Differentiation, 35, *471–478.*

not yet clear, the movement seems to be the missing link in many explanations of invagination and involution. The different effects of dorsal and ventral bottle cell constriction suggest that telescoping is central to invagination.

An intriguing feature of bottle cell invagination is its dependence on the size of the constricting field. When bottle cells are ectopically induced in *Xenopus* ectoderm, small patches form a deep, narrow invagination; larger ones a shallow groove; and still larger patches a flat plate of columnar cells (Kurth & Hausen, 2000). Also, *Drosophila* mesoderm invagination involves constriction of a long but narrow array of epithelial cells. If the mesoderm region is experimentally widened, a flat plate develops, and bending occurs only in stripes on either side (Heer et al., 2017). Apparently, for bottle cells to bend an epithelial layer, the constricting region has to be sufficiently narrow.

4. Dorsal multilayer invagination

4.1 Apical layer processes

The immediate morphogenetic effect of bottle cell formation in the *Xenopus* gastrula is minute (Fig. 3A), but subsequently, dorsal bottle cells partake in a major process, multilayer tissue invagination. In vertebrate late embryos, multilayered epithelia are common, and their invagination involves specialized mechanisms such as cell migration in parallel to the surface of the epithelium (Pearl et al., 2017). In the *Xenopus* gastrula, the apical layer retains behavior known from other single-layered epithelia (Fig. 3C–E). As the bottle cells move to the interior, they remain connected to the apical layer, which follows inward to line the slit-like archenteron (Fig. 3E) (Damm & Winklbauer, 2011; Hausen & Riebesell, 1991; Keller, 1978; Keller & Schoenwolf, 1977; Nieuwkoop & Florschütz, 1950). In the process, a mesodermal blastopore lip forms (Hardin & Keller, 1988), over which the apical layer rolls inward. Suprablastoporal involution is matched by a mirror-symmetric movement of the apical layer that covers the vegetal cell mass (Fig. 3E). This "vegetal involution" gradually internalizes the vegetal surface as it progresses centripetally (Figs. 2B and 6F) (Bauer, Huang, & Moody, 1994; Ewald et al., 2004; Keller, 1978; Nieuwkoop & Florschütz, 1950). To further deepen the archenteron, internalized bottle cells gradually re-spread and become squamous. In fact, the archenteron is severely truncated when bottle cells are extirpated after formation, even as involution continues (Hardin & Keller, 1988; Keller, 1981).

Formation of a slit-like invagination that terminates in a "bouquet" of bottle cells can occur in the absence of deep cells in a purely epithelial context. An example is *Drosophila* mesoderm invagination (Sweeton et al., 1991). At an early stage, bottle-shaped and columnar cells form a shallow indentation, which is surrounded by apically overlapping cells characteristic of telescoping (dark, intermediate and light gray cells in Fig. 3C). Later, an increase in the number of columnar cells, and their widening at the expense of height, lengthens the indentation into a slit (Fig. 3C). Likewise, an explanted apical layer of the urodele amphibian *Cynops* invaginates, with bottle cells at the tip of an archenteron slit, a sharp bend at the suprablastoporal involution zone, and a shallow bend at the site of vegetal involution, although cell shapes are not as distinct due to explants becoming pseudostratified, or multilayered through cell divisions (Fig. 3D) (Komazaki, 1993). Changes in the apical layer of *Xenopus* are strikingly similar to those in the epithelial examples (Fig. 3E) (Damm & Winklbauer, 2011; Hausen & Riebesell, 1991; Keller & Schoenwolf, 1977; Nieuwkoop & Florschütz, 1950), suggesting a combined invagination-involution process in this layer by epithelium-intrinsic mechanisms, namely, the apical constriction of bottle cells and the telescoping of adjacent cells.

However, in amphibians, bottle cells and invaginating apical layers have to penetrate into a coherent cell mass. They could do so by pushing in, but apically connected bottle cells could also crawl collectively into the deep layers by amoeboid motility of the cell body. Consistent with such a mechanism, bottle cells induced in *Xenopus* ectoderm form an archenteron-like pocket in an ectopic context (Kurth & Hausen, 2000; Lustig, Kroll, Sun, & Kirschner, 1996), and Holtfreter (1944) found that a patch of bottle cells can migrate into explanted endoderm to form an invagination. From an analysis of the ultrastructure of anuran bottle cells, Baker (1965) inferred that their neck region is stretched by the advancing cell body and shortened again by active contraction, which leads to the movement of the bottle cell bouquet (Fig. 3B). Multilayer invagination can thus combine autonomous apical layer cell shape changes with bottle cell migration.

4.2 Deep cell movements

The deep cells, which the bottle cells encounter, are migratory themselves (Fig. 4). As described below, vegetal endoderm cells are bottle shaped themselves, and move away from the apical surface toward the blastocoel floor, but turn toward the periphery above the level of the apical bottle cells.

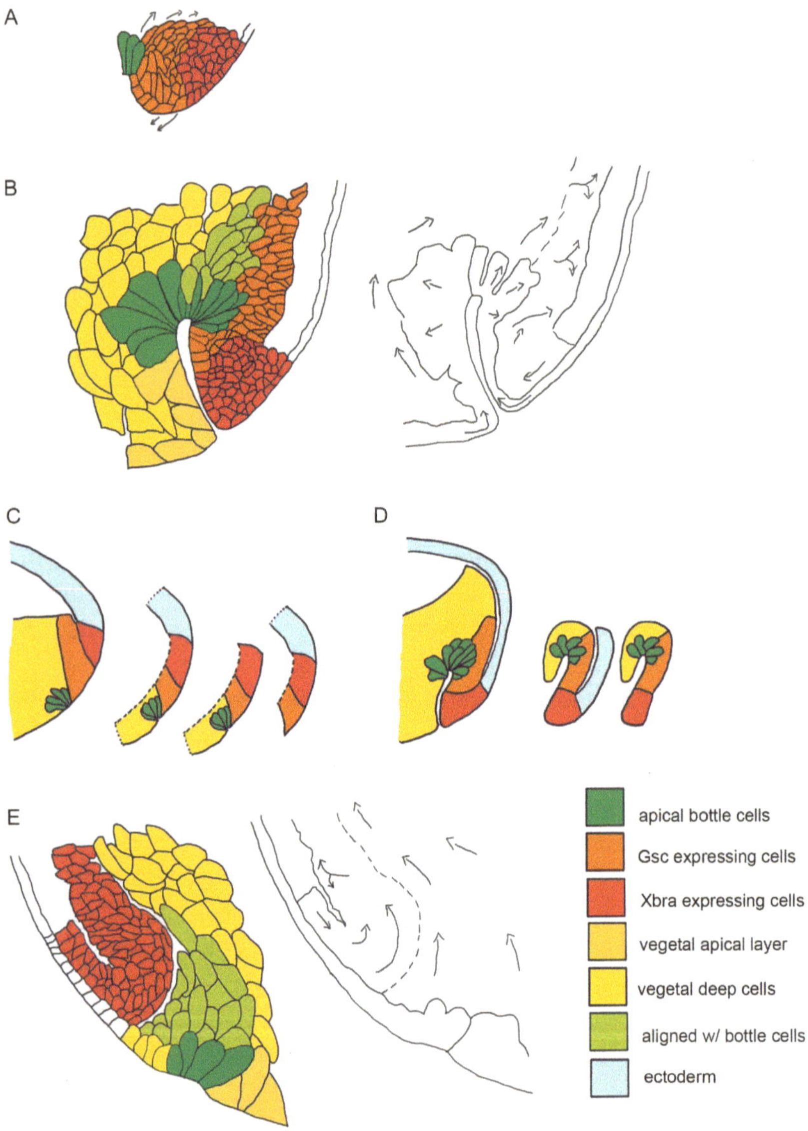

Fig. 4 Dorsal multilayer invagination and ventral internal involution. (A) Apical layer involution (arrows, telescoping cells) and deep cell migration (arrows) in Gsc domain of early gastrula. (B) Mid-gastrula stage after multilayer invagination, cell shapes (left) and directions of movements (arrows, right). (C and D) Transition from initial (C) to middle gastrula stage (D) by folding of multilayered blastocoel wall. Folding is reproduced in explants of different regions, and inverts a vegetal to animal sequence of mesoderm tissues (C) to what will later become the anterior to posterior axis (D). (E) Internal involution of mesoderm and endoderm on ventral side of gastrula, cell shapes (left) and directions of movements (arrows, right).

More vegetally, deep vegetal cells thus move perpendicular to the bottle cells radiating from the tip of the archenteron, but more animally, deep and apical bottle cells align and interdigitate (light and dark green cells, respectively, in Fig. 4B) (Keller & Schoenwolf, 1977; Regen & Steinhardt, 1986), and move together on arc-shaped trajectories (Hardin & Keller, 1988). In *Cynops* (Komazaki, 1993) and in *Pleurodeles* (Boucaut et al., 1996) similar deep bottle-shaped cells lead the apical bottle cells.

The deep blastopore lip also moves. Its lower part is marked by the transcription factor Goosecoid (Gsc) and internalizes in the early gastrula (Fig. 4A and E) (Cho, Blumberg, Steinbeisser, & De Robertis, 1991; Vodicka & Gerhart, 1995). The upper part expresses the transcription factor Brachyury (Xbra) (Ibrahim & Winklbauer, 2001; Wilkinson, Bhatt, & Herrmann, 1990) and involutes later. The Gsc deep cells are inversely bottle shaped with locomotory protrusion at their pointed leading ends, and they migrate in a mesenchymal mode over each other and around the Xbra domain, evacuating the lower blastopore lip (Figs. 4A, B and 7) (Damm & Winklbauer, 2011; Evren et al., 2014; Huang & Winklbauer, 2018). Once inside, they are redirected and attracted by the PDGF-A-secreting BCR (Damm & Winklbauer, 2011).

In the Gsc domain, deep cells and apical layer normally move together, but their translocation can be separated experimentally. When the kinase Pak1 is inhibited, apical layer involution is arrested, but deep cells internalize normally (Evren et al., 2014). Conversely, when the small GTPase Rnd1 is knocked down, apical layers involute despite the deep Gsc cells being arrested (Ogata et al., 2007). Altogether, bottle cell migration, supra- and sub-blastoporal apical layer invagination-involution, and deep vegetal and lip cell migration act coordinately to internalize cells during multilayer invagination in the first half of gastrulation, but the interplay of these multiple processes and their relative contributions have yet to be analyzed.

4.3 Convergent extension by parallel intercalation

During involution, cells move vegetally to the site of internalization at the blastopore, which itself shifts vegetally, and as cells traverse the spherical embryo surface from higher to lower latitudes, they converge. Further convergence occurs when cells internalize and move from a wide pre-involution perimeter closer to the animal-vegetal axis (Evren et al., 2014). In apical and deep layers, the mode of convergence in the Gsc domain is the same. In the embryo, apical cells closer to the blastopore move faster than cells further away (Keller, 1978, 1981), and a similar velocity gradient is seen in deep cells in explants (Evren et al., 2014). When a fast cell separates from a slower one behind, lateral cells fill the gap, and the tissue lengthens and narrows. For the deep layer, this form of convergent extension had been described as oriented radial intercalation (Wilson & Keller, 1991). In addition to the animal-vegetal velocity gradient, faster movement of cells near the surface of the lip, compared to deeper cells, rotates the Gsc domain inward (Evren et al., 2014) (Fig. 4B).

Convergent extension in the Gsc domain differs from canonical convergent extension in the Xbra-expressing dorsal mesoderm: in the Xbra domain, cells translocate orthogonal to the axis of tissue extension (see below), whereas in the Gsc domain, they migrate in parallel to this axis (Fig. 5E). A further example of parallel convergent extension is the elongation of the archenteron in the sea urchin gastrula (Ettensohn, 1985; Hardin, 1989; Hardin & Cheng, 1986). Here, cells move animally, in parallel to the long axis of the extending archenteron, to intercalate circumferentially. Although of epithelial type, the cells extend lamellipodia on their basal side

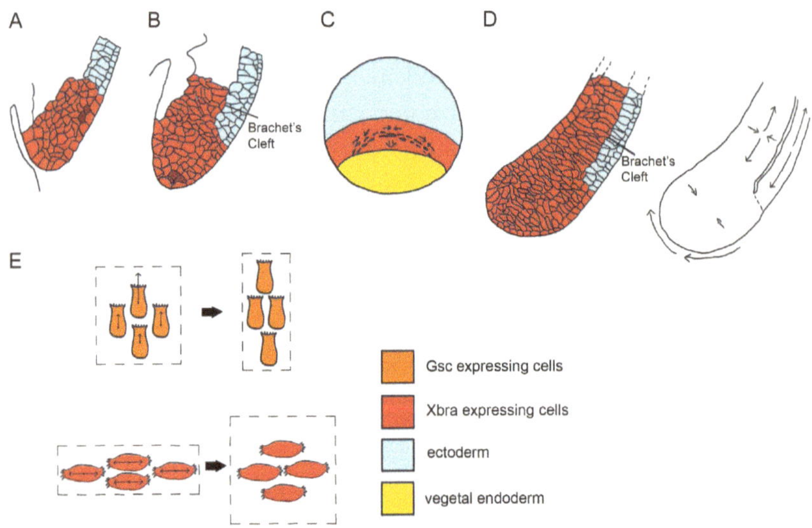

Fig. 5 Peak involution and convergent extension of the dorsal Xbra domain. (A) Early gastrula before, (B) mid-gastrula after peak involution. (C) Mediolaterally interdigitating cells (dark red in A and B) form vegetal alignment zone in the Xbra domain. (D) Convergent extension in the dorsal Xbra domain, cell shapes (left) and directions of movements (arrows, right). Only radial, not mediolateral elongation of cells (as shown for vegetal alignment zone in (C)) is seen in sagittal sections. (E) Parallel intercalation (top) and orthogonal intercalation (bottom) both lead to a convergence of cells and extension of the tissue.

in the direction of tissue elongation, stretch along this axis, and rearrange as they become isometric again (Hardin, 1989). If apical and basal parts of the cells would segregate by division, the basally derived deep cells might use protrusions to migrate underneath the apical layer in a mesenchymal mode (see Fig. 7E), reproducing the layout of the *Xenopus* Gsc domain.

5. Peak involution of the dorsal Xbra domain

After the near internalization of the Gsc domain, the Xbra domain still occupies the outer part of the blastopore lip (Fig. 4A). During a short burst of internalization where the involution rate peaks (Winklbauer & Schürfeld, 1999), the sectional profile of the dorsal lip becomes pointed, with the Xbra domain forming the apex, and such bending of the domain is preserved in involuting explants (Figs. 4A–D and 5A, B) (Evren et al., 2014). The cellular basis of peak involution is not fully understood. A contributing factor is suggested by the vegetal alignment zone: deep cells elongate mediolaterally in a narrow stripe within the Xbra region, forming an arc across the dorsal mid-line (Fig. 5A–C) (Lane & Keller, 1997; Shih & Keller, 1992), and as it shortens by mediolateral cell intercalation, it should move the outer part of the dorsal lip vegetally. Peak involution is a distinct phase of lip internalization as confirmed by its molecular control. Although Xbra is expressed throughout early development, for its essential role in peak involution its expression is specifically controlled by the Eph receptor EphA4 and the Pak1 kinase downstream of it (Evren et al., 2014).

Peak involution concludes a folding process in which dorsally, an animal-vegetal sequence of regions—ectoderm, Xbra-expressing chordamesoderm, Gsc-expressing prechordal mesoderm, and vegetal endoderm—is transformed into the anterior-posterior sequence of the internalized mesoderm regions and the proper inside-out positioning of germ layers (Fig. 4C and D). Folding is autonomous, as is obvious from explants. Like the single-layer marginal zone explant from *Cynops* gastrulae, the *Xenopus* multilayered version folds back on itself. Moreover, when further dissected, its parts reproduce in isolation both the bending at the site of bottle cell formation, and counter-bending at the blastopore lip (Fig. 4C and D) (Evren et al., 2014).

6. Ventral internal involution

Ventrally, internalization is already under way in the deep layers when bottle cells form, and continues while they remain on the surface

(Ibrahim & Winklbauer, 2001; Nieuwkoop & Florschütz, 1950). Like on the dorsal side, the large cells of the vegetal mass deep to the apical bottle cells are also bottle-shaped and oriented inward-upward (Fig. 4E). However, this cell type also occupies the space between bottle cells and the Xbra domain in the ventral upper blastopore lip, taking the place held dorsally by the Gsc domain. The cells migrate around the Xbra domain, but instead of approaching the ectoderm, they move deep into the endoderm (Fig. 4E) (Ibrahim & Winklbauer, 2001). Vogt (1929) observed a similar endodermal fate of the ventral lip in other anurans, but not in urodeles.

Xbra is expressed in the upper ventral lip except in its apical layer (Fig. 4E) (Ibrahim & Winklbauer, 2001; Smith, Price, Green, Weigel, & Herrmann, 1991). Intriguingly, cells on the internalized side of the ventral Xbra domain are shaped like the cells of the dorsal Gsc domain, with pointed, protrusion-bearing leading edges and rounded read ends. They initially move on curved trajectories in parallel to the endodermal lower lip cells, but diverge to join the internalized mesoderm above the blastopore lip (Fig. 4E). Labeled cell patches become drawn out by the late gastrula stage into lines of single cells surrounding the lower, vegetal end of Brachet's cleft (Ibrahim & Winklbauer, 2001). This stretching of cell patches suggests parallel convergent extension in both lower and upper lip regions. Like in the Gsc domain, convergence occurs as cells move from wide pre-involution perimeters closer to the animal-vegetal axis, and as the blastopore constricts to eventually close. Convergence is only partly translated into an extension of the tissue; in part it also thickens the ventral lip (Keller & Danilchik, 1988). A further difference to the dorsal side is the expression of Xbra in the lower part of the BCR (Fig. 4E). The region is fated to become mesoderm, but initially behaves BCR-like with respect to cell sorting and fibronectin matrix formation. The behavior changes as the BCR mesoderm moves into the lip to replace cells moving interiorly (Ibrahim & Winklbauer, 2001).

Initially, ventral deep cells involute independently of the apical layer, but eventually invagination occurs (Ewald et al., 2004; Ibrahim & Winklbauer, 2001; Korvin-Pavlovskaya & Cherdantsev, 2016; Nieuwkoop & Florschütz, 1950). Ventral invagination is a highly variable trait in primitive vertebrates. In the urodeles *Cynops* and *Pleurodeles*, bottle cells invaginate late (Imoh, 1988; Shi et al., 1990), like in anurans, while in lung fish, ventral bottle cells appear in the late gastrula, but do not invaginate, i.e., no blastopore forms on the ventrolateral side (Kerr, 1902). In lampreys, but also in the urodele *Spelerpes*, not even bottle cells form ventrally, although internal involution occurs (Glaesner, 1909; Goodale, 1911; Hatta, 1907). Deep cell independency from the apical layer culminates in the teleost gastrula where invagination has been replaced at all dorso-ventral levels by forms of internal involution termed single-cell or synchronized ingression, respectively (Cooper & Virta, 2007). These examples show how transition to a multilayered gastrula provides opportunity for evolutionary divergence.

7. Orthogonal convergent extension of the dorsal Xbra domain

Brachet's cleft is first established dorsally to separate Gsc-expressing prechordal mesoderm from the BCR. During peak involution, the cleft extends to also separate newly internalized Xbra-expressing chordamesoderm from prospective neural BCR (Fig. 5A and B) (Damm & Winklbauer, 2011). This is made possible by changes in tissue separation behavior in regions of the blastopore lip (Wacker, Grimm, Joos, & Winklbauer, 2000). After peak involution, the internalized part of the dorsal Xbra domain converges and extends in parallel with the attached neural ectoderm of the BCR (Fig. 5B and D). In both tissues, movement is driven by mediolateral cell intercalation (Elul, Koehl, & Keller, 1997; Keller & Danilchik, 1988; Keller, Shih, & Sater, 1992; Shih & Keller, 1992; Wallingford & Harland, 2001).

Below the vegetal end of Brachet's cleft, dorsal lip cells have to converge as well to constrict and close the blastopore (Fig. 2B), and in explants, mediolateral intercalation is seen to spread, adding converging cells to the lip (Keller, Cooper, Danilchik, Tibbetts, & Wilson, 1989; Shih & Keller, 1992). Convergence in the lip is partly translated into a thickening of the lip, and partly into a slow post-peak involution that moves cells to the interior (Winklbauer & Schürfeld, 1999), thus keeping the height of the lip constant (Fig. 5D). This requires the continuous formation of new sections of Brachet's cleft (Fig. 5B and D) by changes in tissue separation behavior (Wacker et al., 2000), and faster extension of the ectoderm compared to the mesoderm. Convergent extension is indeed more pronounced in the neuroectodermal part of explants than in adjacent chordamesoderm (Keller & Danilchik, 1988). If redirected inwardly at the apex of the lip, increased expansion near the surface would generate a bending moment and drive involution (Fig. 5B and D).

Dorsal convergent extension moves the involuting blastopore lip vegetally along the surface of the embryo while at the same time, the site

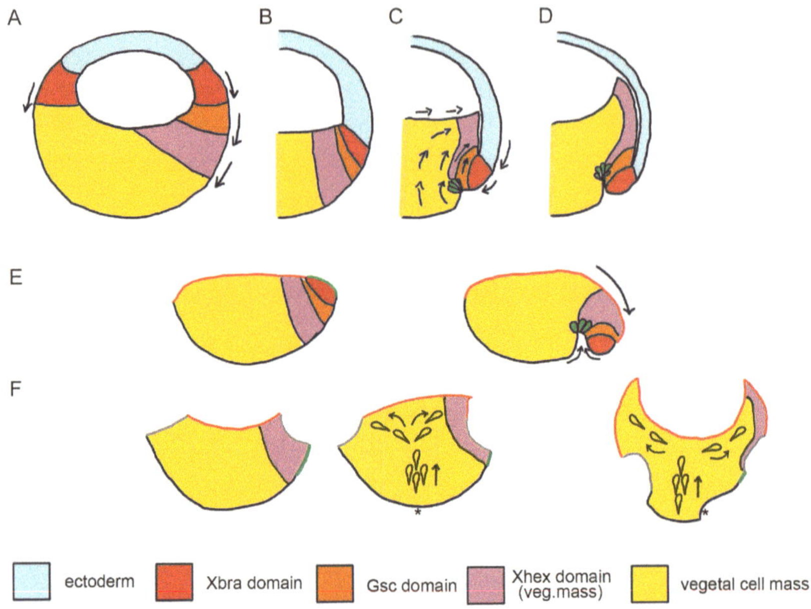

Fig. 6 Internalization of vegetal cell mass. (A and B) Pre-gastrulation expansion of BCR and contraction of vegetal mass shift the marginal zone vegetally. (C) Vegetal rotation of dorsal periphery. (D) Formation of leading edge mesendoderm moving animally on BCR. (E) Explanted mid-sagittal slice of vegetal cell mass performs vegetal rotation in vitro. (F) Visualization of lipid-rich vegetal cell mass and its deformation. Red, blastocoel floor; green, bottle cell region; black, vegetal apical surface. Orientation of bottle-shaped cells and direction of movement are indicated. Asterisk, vegetal pole is internalized early by vegetal involution such that the blastopore eventually closes over the former ventral side.

of vegetal involution regresses centripetally beyond the former vegetal pole, such that the blastopore eventually closes on the former ventral side (Keller, 1975) (see Fig. 6F, asterisk). When convergent extension is attenuated with an inhibitory construct of the scaffold protein Disheveled, blastopore constriction ceases during mid-gastrula stages; unexpectedly, vegetal involution is also arrested. Under these conditions, only anterior archenteron forms (Ewald et al., 2004). By contrast, interfering with EphA4 or Pak1 function, or bottle cell excision, truncates the archenteron anteriorly (Evren et al., 2014). Thus, the anterior archenteron is formed by invagination and bottle cell re-spreading and the posterior one by convergent extension.

In contrast to parallel intercalation in the Gsc domain, Xbra cells elongate and migrate orthogonally to the axis of tissue elongation and wedge themselves between anteroposterior neighbors (Keller, 2002; Keller et al., 2000).

At the same time, they also stretch radially (Damm & Winklbauer, 2011; Keller et al., 1989; Keller & Schoenwolf, 1977), probably to limit thickening and instead promote lengthening of the tissue (Fig. 5D). In this example of orthogonal convergent extension (Fig. 5E), medial and lateral lamellipodia of a bipolar cell move over the surface of adjacent cells, stretch the cell (Fig. 7G) and establish new contacts. Contraction of the actomyosin cortex between contacts shortens cells again (Pfister et al., 2016). Notably, the proposed mechanism combines elements of cell-on-cell migration by lamellipodia protrusion with those of junction remodeling where mutually attached cells contract their contact zones.

Emphasizing the second of these elements, late- or post-gastrula dorsal mesoderm intercalation has been described as classical junction remodeling, involving the shrinkage of anteroposterior and the expansion of new mediolateral cell contacts (Fig. 7G) (Shindo & Wallingford, 2014). However, oriented protrusions form also at post-gastrula stages, and the mechanisms at both stages may be very similar. The co-occurrence of junction remodeling (Pilot & Lecuit, 2005) and lamellipodial migration (Sun et al., 2017) within individual cells characterizes also the orthogonal intercalation in the epithelial context of *Drosophila* germ band extension, and this combination of processes could generally underlie mesenchymal cell-on-cell migration. Conceptually, a transition from epithelial to deep cell orthogonal intercalation could simply be achieved by periclinal cell division (Fig. 7G).

8. Internalization of the vegetal cell mass by ingression-type deep bottle cell migration

Before gastrulation, the vegetal cell mass takes part in a slow expansion of the animal and contraction of the vegetal blastula wall (Bauer et al., 1994; Boternebrood & Narraway, 1990; Jones, Broadbent, Thomas, Smith, & Beddington, 1999; Keller, 1978; Papan, Boulat, Velan, Fraser, & Jacobs, 2007a, 2007b; Winklbauer & Damm, 2012), which is reflected in the positioning of marker genes (Fig. 6A and B). At the onset of gastrulation, the Xbra and Gsc domains suddenly shift vegetally together with the rim of the BCR (Fig. 6C) (Papan et al., 2007a; Winklbauer & Schürfeld, 1999). The floor of the blastocoel accelerates it movement toward the periphery, bending downward at the BCR to transiently form a trench and to eventually establish Brachet's cleft by applying itself to the BCR. The blastocoel floor is replenished by upward moving cells from the deep vegetal mass. Together, movements form a vortex at the vegetal periphery, which

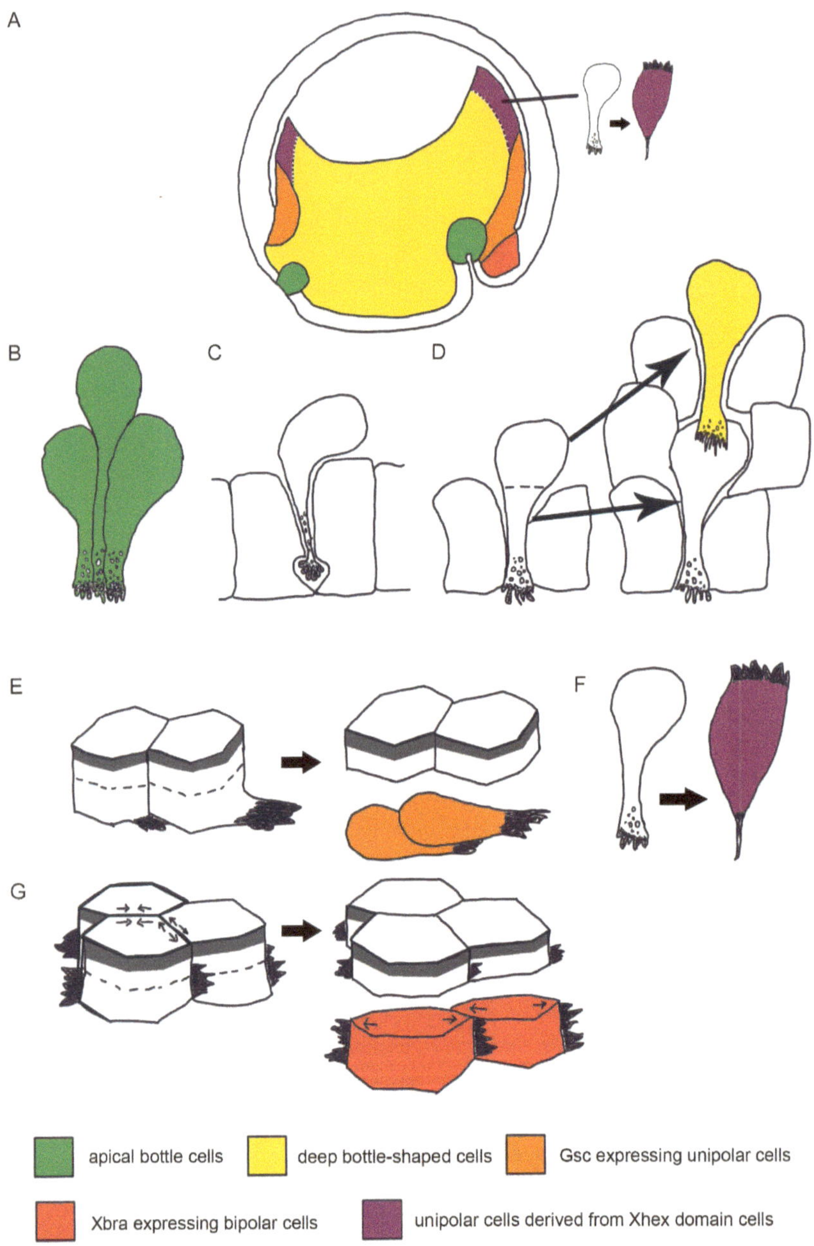

Fig. 7 Hypothetical derivation of deep cell motility modes from epithelial morphogenetic processes. (A) Distribution of cell types in gastrula. (B) Apically connected bottle cells. (C) Bottle cell ingression. (D) Derivation of deep migrating bottle cell from surface bottle cell. (E) Derivation of unipolar mesenchymal deep cells from basal part of epithelial cells engaged in parallel intercalation. (F) Transition from deep migrating bottle cell to unipolar mesenchymal cell of leading edge mesendoderm. (G) Derivation of bipolar intercalating cells from basal part of epithelial cells engaged in orthogonal intercalation. Dark serrate structures, lamellipodia; gray sub-apical zones, apical junction belt. *Panel (C): After Katow, H., and Solursh, M. (1980). Ultrastructure of primary mesenchyme cell ingression in the sea urchin* Lytechinus pictus. The Journal of Experimental Zoology, 213*, 231–246.*

suggested the term vegetal rotation (Fig. 6C) (Winklbauer & Schürfeld, 1999). A similar establishment of Brachet's cleft is suggested by respective gastrula shape changes in anurans like *Rana* (Morgan, 1905), in urodele amphibians such as *Cynops* (Yanagi et al., 2015) and *Pleurodeles* (Shi, Delarue, Darribère, Riou, & Boucaut, 1987), and in lampreys, the most primitive vertebrates (Glaesner, 1909; Hatta, 1907). Thus, vegetal rotation is most likely a conserved feature of primitive vertebrate gastrulation.

In *Xenopus*, vegetal rotation is preserved in explanted slices of the vegetal half of the gastrula (Fig. 6E). Further dissection of explants revealed that the main driving force is generated within the vegetal cell mass while the mesoderm is rotated passively, initiating its involution (Winklbauer & Schürfeld, 1999). Involution becomes an active process as Gsc cells migrate inward. At this time, the downward rotation of blastopore lip and BCR has ceased, and the continued centrifugal movement of the blastocoel floor is instead diverted upward, leading to the formation of a mesendodermal leading edge (purple region in Fig. 6D). This process requires interaction with the BCR, not for mechanical support but for signaling (Winklbauer & Schürfeld, 1999). PDGF-A from the BCR guides the radial intercalation of prechordal mesoderm cells (Damm & Winklbauer, 2011) and directs leading edge cells animally (Nagel, Tahinci, Symes, & Winklbauer, 2004; Nagel & Winklbauer, 2018), which together with ephrin/Eph receptor and PAPC-dependent mesoderm cell repulsion by the BCR control the spreading of the mesoderm layer (Luu et al., 2015; Rohani, Canty, Luu, Fagotto, & Winklbauer, 2011; Rohani, Winklbauer, & Fagotto, 2014).

Vegetal movements are not restricted to the dorsal periphery. By magnetic resonance imaging, the distortion of the vegetal cell mass can be followed in live embryos (Fig. 6F) (Papan et al., 2007b). It shows a shrinkage of the lower vegetal part and a massive expansion of the blastocoel floor and the ectoderm-apposed surfaces derived from it. The shape changes are due to cell rearrangements by differential migration. In the lower vegetal mass, velocity increases from vegetal to animal. In an instance of parallel intercalation (Fig. 5E), faster cells migrate away from slower cells behind, and resultant gaps are filled by lateral cells, thus narrowing the tissue and transporting cells toward the blastocoel floor. Here, cells reorient and migrate to the periphery, turning the animal-vegetal elongation of the vegetal mass into an expansion of the blastocoel floor (Fig. 6F) (Moosmann et al., 2013; Papan et al., 2007a, 2007b; Wen & Winklbauer, 2017; Winklbauer & Schürfeld, 1999).

Vegetal cells strikingly resemble invaginating and ingressing bottle cells in various vertebrates and invertebrates (Fig. 7A–D) (Wen & Winklbauer, 2017). These bottle cells have a small apical surface with a cortical layer thrown into extensive folds and accumulations of variously sized vesicles—the hallmarks of macropinocytosis—and a bulbous, protrusionless cell body (Fig. 7B and C). Deep vegetal cells are similarly shaped and move in amoeboid fashion without locomotory protrusions (Fig. 7D) (Wen & Winklbauer, 2017). Cells glide by a rearward flow of the cell surface (Kubota, 1981), but also progress through cyclic shape changes as they elongate, widen their front end, round up, and elongate again. Strikingly, the retracting tail of the cells is highly protrusive and accumulates vesicles by macropinocytosis, an endocytic mechanism, which in these cells depends on the Eph receptor ligand, ephrinB1 (Wen & Winklbauer, 2017).

Amoeboid migration does not require specific substrate adhesion (Paluch, Aspalter, & Sixt, 2016), and newt vegetal cells can translocate on agar (Kubota, 1981). *Xenopus* vegetal cells are separated by wide interstitial spaces, but surprisingly, the cell adhesion molecule C-cadherin is essential for their migration. Cadherin-dependent cell contacts are small and rare, and often mediated by lateral filopodia. They regulate α-catenin localization at the cell cortex, and require ephrinB1 and the engulfment adaptor Gulp1 for their detachment by trogocytosis (Gong et al., 2019; Wen & Winklbauer, 2017). The cells also interact with small puncta of cell surface fibronectin (Winklbauer, 1998), and inhibition of this supposed signaling interaction attenuates vegetal rotation (Wen & Winklbauer, 2017) and internalization of the vegetal mass, which becomes pinched off by the constricting blastopore (Winklbauer & Keller, 1996). Apparently, blastopore closure is not sufficient for the internalization of the vegetal mass.

Macropinocytosis at the trailing edge of vegetal deep cells indicates a more than superficial resemblance to bottle cells. Generally, a basic bottle cell type could be subdivided into apical bottle cells, which give the name to the type and can engage in bottle cell migration while linked apically (Fig. 7B); ingressing bottle cells, which leave the apical layer by migrating individually (Fig. 7C); and deep bottle-shaped cells, which are not generated by ingression, but originate in the deep layers, like in the vegetal cell mass (Fig. 7C). The similar translocation mode of ingressing and deep bottle cells was termed ingression-type migration (Wen & Winklbauer, 2017). As in ingressing bottle cells in other systems, a transition to a mesenchymal type occurs when some of the deep bottle cells begin to extend lamellipodia as they form the leading edge of the internalized vegetal mass (Fig. 7A and F) (Huang & Winklbauer, 2018).

Conceptually, deep bottle cells can be derived from apical bottle cells by divisions perpendicular to the apical-basal axis (Fig. 7D), although this would require to implement a macropinocytotic tail in the putative absence of apical landmarks. At any rate, migration would take the form of an invagination in the apical bottle cells and of ingression-type migration in the deep cells. Both cell modifications could use the same orienting cues and substratum interactions, retaining coordinated movements in a multilayer context as seen, for example, during dorsal invagination (light and dark green cells in Fig. 4B).

9. Blastopore closure

The end of gastrulation is operationally defined by the closure of the blastopore below the vegetal cell mass (Fig. 2Bb′′′′). Interestingly, blastopore constriction is resumed in Disheveled-inhibited, convergent extension-attenuated embryos in the late gastrula (Ewald et al., 2004), and delayed blastopore closure occurs also in embryos ventralized by UV treatment (Mise & Wakahara, 1994), indicating that dorsal convergent extension is not absolutely necessary. Directly measuring the constriction force in marginal zone explants confirmed that dorsal, lateral and ventral sections all generate such force (Shook, Kasprowicz, Davidson, & Keller, 2018).

The cellular mechanism of blastopore constriction is not clear, but cells becoming oriented to encircle the closing blastopore toward the end of gastrulation (Hausen & Riebesell, 1991) suggests that circumblastoporal cell intercalation and convergence may contribute. Molecularly, expression of a construct that turns Xbra into a transcriptional repressor arrested gastrulation at a small blastopore stage, consistent with a role of Xbra in the final steps of blastopore closure (Conlon, Sedgwick, Weston, & Smith, 1996). Convergence not only constricts, but also thickens the blastopore lip to form a circumblastoporal collar. The independence of convergent extension and blastopore constriction is relevant evolutionarily. Although gastrula-stage convergent extension is common in amphibians (Wallingford, Fraser, & Harland, 2002), in some anurans with large, slowly developing embryos, the mesodermal circumblastoporal collar is formed first before the body axis extends after blastopore closure (del Pino et al., 2007).

10. Conclusions

Epithelial sheet bending and cell intercalation in the plane of the sheet are basic mechanisms of invertebrate epithelial gastrulation, which seem to be retained in the *Xenopus* gastrula. However, only bottle cell formation has been studied mechanistically, whereas other processes are inferred from cell shape changes and comparisons with other systems. Surprisingly little is known about the mechanisms of epithelial in-plane cell rearrangements in the *Xenopus* gastrula apical layer, despite the accessibility of the processes in an otherwise opaque embryo. Cell migration processes are better understood. Differential cell-on-cell migration in the mesenchymal or amoeboid mode, respectively, underlies deep cell rearrangement by parallel intercalation in mesoderm and vegetal cell mass, whereas orthogonal intercalation of bipolar cells drives dorsal convergent extension and perhaps blastopore closure.

All these processes are pre-programmed in astonishing detail, to render movements highly autonomous and diminish the need for passive adaptations of tissue parts to overall deformations. Consequences of this design feature are a redundancy of mechanisms, and a modularity that allows gastrulation movements to be variable in their timing and rates (Ewald et al., 2004; Winklbauer & Schürfeld, 1999). Experimentally, the onset of movements can be varied to an extreme extent by applying temperature or oxygen gradients to the embryo, such that gastrulation starts ventrally instead of dorsally, completely reversing the normal sequence but without compromising the normal completion of gastrulation (Black, 1989; Black, Hawk, & Larkin, 1996).

The analysis of *Xenopus* gastrulation yields concepts that may in the future help to derive more stringently multilayer from single-layer morphogenesis. Cooperation of apical and basolateral mechanisms in cells is an important feature of epithelial morphogenesis (e.g., Gustafson & Wolpert, 1963; Munro & Odell, 2002; Sun et al., 2017; Williams-Masson, Heid, Lavin, & Hardin, 1998), and the segregation of apical and basal domains into separate cells (Fig. 7E and G), or the duplication of apical-basal polarity properties in deep daughter cells (Fig. 7D) are two possible scenarios to derive multilayer mechanisms. With both, epithelial and deep cell movements would naturally be coordinated just as motility is in the different parts of individual epithelial cells, and minimal changes would be required for the transition to vertebrate gastrulation. Once established, apical and deep cell populations could evolve new properties independently, to adapt gastrulation to novel conditions such as a further increased egg size in amniotes.

Acknowledgments

I thank Olivia Luu for preparing the figures, Martina Nagel for checking references, Debanjan Barua for suggestions to improve the manuscript, and Erich Damm for making countless SEM micrographs available for examination. Research of R.W. is funded by the Canadian Institutes of Health Research (PJT-15614) and the Natural Sciences and Engineering Research Council of Canada (RGPIN-2017-06667).

References

Baker, P. C. (1965). Fine structure and morphogenetic movements in the gastrula of the treefrog, Hyla regilla. *The Journal of Cell Biology*, *24*, 95–116.

Bauer, D. V., Huang, S., & Moody, S. A. (1994). The cleavage stage origin of Spemann's organizer: Analysis of the movements of blastomere clones before and during gastrulation. *Development*, *120*, 1179–1189.

Black, S. D. (1989). Experimental reversal of the normal dorsal-ventral timing of blastopore formation does not reverse axis polarity in Xenopus laevis embryos. *Developmental Biology*, *134*, 376–381.

Black, S. D., Hawk, S. N., & Larkin, K. (1996). Restricting oxygen supply to the prospective dorsal side does not reverse axis polarity in embryos of Xenopus laevis. *Development Genes and Evolution*, *206*, 147–152.

Boternebrood, E. C., & Narraway, J. M. (1990). Epiboly connected with cleavage in morula and early blastula stages of Xenopus laevis, a study using time-lapse photography. *Roux's Archives of Developmental Biology*, *198*, 303–307.

Boucaut, J. C., Clavilier, L., Darribere, T., Delarue, M., Riou, J.-F., & Shi, D.-L. (1996). What mechanisms drive cell migration and cell interactions in Pleurodeles? *The International Journal of Developmental Biology*, *40*, 675–683.

Chalmers, A. D., Strauss, B., & Papalopulu, N. (2003). Oriented cell divisions asymmetrically segregate aPKC and generate cell fate diversity in the early Xenopus embryo. *Development*, *130*, 2657–2668.

Cho, K., Blumberg, B., Steinbeisser, H., & De Robertis, E. (1991). Molecular nature of Spemann's organizer: The role of the Xenopus homeobox gene goosecoid. *Cell*, *67*, 1111–1120.

Collazo, A., Bolker, J. A., & Keller, R. (1994). A phylogenetic perspective on teleost gastrulation. *The American Naturalist*, *144*, 133–152.

Conlon, F. L., Sedgwick, S. G., Weston, K. M., & Smith, J. C. (1996). Inhibition of Xbra transcription activation causes defects in mesodermal patterning and reveals autoregulation of Xbra in dorsal mesoderm. *Development*, *122*, 2427–2435.

Cooper, M. S., & Virta, V. C. (2007). Evolution of gastrulation in the ray-finned (actinopterygian) fishes. *Journal of Experimental Zoology. Part B, Molecular and Developmental Evolution*, *308*, 591–608.

Dale, L., & Slack, J. M. W. (1987). Fate map for the 32-cell stage of Xenopus laevis. *Development*, *99*, 527–551.

Damm, E. W., & Winklbauer, R. (2011). PDGF-A controls mesoderm cell orientation and radial intercalation during *Xenopus* gastrulation. *Development*, *138*, 565–575.

del Pino, E. M., Venegas-Ferrin, M., Romero-Carjaval, A., Montenegro-Larrea, P., Saenz-Ponce, N., Moya, I. M., et al. (2007). A comparative analysis of frog early development. *Proceedings of the National Academy of Sciences of the United States of America, 104*, 11882–11888.

Elul, T., Koehl, M., & Keller, R. (1997). Cellular mechanism underlying neural convergent extension in Xenopus laevis embryos. *Developmental Biology, 191*, 243–258.

Ettensohn, C. A. (1985). Gastrulation in the sea urchin embryo is accompanied by the rearrangement of invaginating epithelial cells. *Developmental Biology, 112*, 383–390.

Evren, S., Wen, J. W. H., Luu, O., Damm, E. W., Nagel, N., & Winklbauer, R. (2014). EphA4-dependent Brachyury expression is required for dorsal mesoderm involution in the Xenopus gastrula. *Development, 141*, 3649–3661.

Ewald, A. J., Peyrot, S. M., Tyszka, J. M., Fraser, S. E., & Wallingford, J. B. (2004). Regional requirements for Dishevelled signaling during Xenopus gastrulation: Separable effects on blastopore closure, mesendoderm internalization and archenteron formation. *Development, 131*, 6195–6209.

Glaesner, L. (1909). Studien zur Entwicklungsgeschichte von Petromyzon fluviatilis. I. Furchung und gastrulation. *Zoologische Jahrbücher, 29*, 139–190.

Gong, J., Gaitanos, T. N., Luu, O., Huang, Y., Gaitanos, L., Lindner, J., et al. (2019). Gulp1 controls Eph/ephrin trogocytosis and is important for cell rearrangements during development. *The Journal of Cell Biology, 218*, 3455–3471. jcb.201901032.

Goodale, H. D. (1911). The early development of Spelerpes bilineatus (Green). *The American Journal of Anatomy, 12*, 173–247.

Gustafson, T., & Wolpert, L. (1963). The cellular basis of morphogenesis and sea urchin development. *International Review of Cytology, 15*, 139–214.

Hardin, J. (1989). Local shifts in position and polarized motility drive cell rearrangement during sea urchin gastrulation. *Developmental Biology, 136*, 430–445.

Hardin, J., & Cheng, L. Y. (1986). The mechanisms and mechanics of archenteron elongation during sea urchin gastrulation. *Developmental Biology, 115*, 490–501.

Hardin, J., & Keller, R. (1988). The behavior and function of bottle cells during gastrulation of Xenopus laevis. *Development, 103*, 211–230.

Hatta, S. (1907). On the gastrulation of Petromyzon. *Journal of College Science, Imperial University Tokyo, 21*, 1–44.

Hausen, P., & Riebesell, M. (1991). *The early development of* Xenopus laevis. *An atlas of the histology*. Berlin-Heidelberg-New York: Springer.

Heer, N. C., Miller, P. W., Chanet, S., Stoop, N., Dunkel, J., & Martin, A. C. (2017). Actomyosin-based tissue folding requires a multicellular myosin gradient. *Development, 144*, 1876–1886.

Holtfreter, J. (1944). A study of the mechanics of gastrulation; part II. *The Journal of Experimental Zoology, 95*, 171–212.

Huang, Y., & Winklbauer, R. (2018). Cell migration in the Xenopus gastrula. *Wiley Interdisciplinary Reviews: Developmental Biology*, 7, e325.

Ibrahim, H., & Winklbauer, R. (2001). Mechanisms of mesendoderm internalization in the Xenopus gastrula: Lessons from the ventral side. *Developmental Biology, 240*, 108–122.

Imoh, H. (1988). Formation of germ layers and roles of the dorsal lip of the blastopore in normally developing embryos of the newt Cynops pyrrhogaster. *The Journal of Experimental Zoology, 246*, 258–270.

Jones, M. C., Broadbent, J., Thomas, P. Q., Smith, J. C., & Beddington, R. S. P. (1999). An anterior signalling centre in *Xenopus* revealed by the homeobox gene Xhex. *Current Biology, 9*, 946–954.

Keller, R. E. (1975). Vital dye mapping of the gastrula and neurula of Xenopus laevis. I. Prospective areas and morphogenetic movements of the superficial layer. *Developmental biology, 42*, 222–241.

Keller, R. E. (1976). Vital dye mapping of the gastrula and neurula of Xenopus laevis. II. Prospective areas and morphogenetic movements of the deep layer. *Developmental biology*, *51*, 118–137.

Keller, R. E. (1978). Time-lapse cinemicrographic analysis of superficial cell behavior during and prior to gastrulation in Xenopus laevis. *Journal of Morphology*, *157*, 223–248.

Keller, R. (1981). An experimental analysis of the role of bottle cells and the deep marginal zone in gastrulation of Xenopus laevis. *The Journal of Experimental Zoology*, *216*, 81–101.

Keller, R. (2002). Shaping the vertebrate body plan by polarized embryonic cell movements. *Science*, *298*, 1950–1954.

Keller, R., & Danilchik, M. (1988). Regional expression, pattern and timing of convergence and extension during gastrulation of Xenopus laevis. *Development*, *103*, 193–209.

Keller, R. E., & Schoenwolf, G. C. (1977). An SEM study of cellular morphology, contact, and arrangement, as related to gastrulation in Xenopus laevis. *Wilhelm Roux's Archives of Developmental Biology*, *182*, 165–186.

Keller, R., Cooper, M. S., Danilchik, M., Tibbetts, P., & Wilson, P. A. (1989). Cell intercalation during notochord development in Xenopus laevis. *The Journal of Experimental Zoology*, *251*, 134–154.

Keller, R., Shih, J., & Sater, A. (1992). The cellular basis of the convergence and extension of the Xenopus neural plate. *Developmental Dynamics*, *193*, 199–217.

Keller, R., Davidson, R., Edlund, A., Elul, T., Ezin, M., Shook, D., et al. (2000). Mechanisms of convergence and extension by cell intercalation. *Philosophical Transactions of the Royal Society of London B*, *355*, 897–922.

Kerr, J. G. (1902). The development of Lepidosiren paradoxa. Part II. *The Quarterly Journal of Microscopical Science*, *45*, 1–40.

Komazaki, S. (1993). Movement of an epithelial layer isolated from early embryos of the newt, Cynops pyrrhogaster. II. Formation of a blastoporal groove and archenteron in a superficial epithelial layer isolated from initial gastrula. *Development, Growth & Differentiation*, *35*, 471–478.

Korvin-Pavlovskaya, E. G., & Cherdantsev, V. G. (2016). Geometry of movement of the outer surface of the embryo during Xenopus gastrulation. *Russian Journal of Developmental Biology*, *47*, 223–237.

Kubota, H. Y. (1981). Creeping locomotion of the endodermal cells dissociated from gastrulae of the Japanese newt, Cynops pyrrhogaster. *Experimental Cell Research*, *133*, 137–148.

Kurth, T., & Hausen, P. (2000). Bottle cell formation in relation to mesodermal patterning in the Xenopus embryo. *Mechanisms of Development*, *97*, 117–131.

Lane, M. C., & Keller, R. (1997). Microtubule disruption reveals that Spemann's organizer is subdivided into two domains by the vegetal alignment zone. *Development*, *124*, 895–906.

Lee, J.-Y., & Harland, R. M. (2007). Actomyosin contractility and microtubules drive apical constriction in Xenopus bottle cells. *Developmental Biology*, *311*, 40–52.

Lee, J.-Y., & Harland, R. M. (2010). Endocytosis is required for efficient apical constriction during Xenopus gastrulation. *Current Biology*, *20*, 253–258.

Li, J., Economou, A. D., & Green, J. B. A. (2019). Epithelial invagination by vertical telescoping. *bioRxiv*, 1–16. https://doi.org/10.1101/515981.

Lüningschrör, P., Binotti, B., Dombert, B., Heimann, P., Perez-Lara, A., Slotta, C., et al. (2017). Plekhg5-regulated autophagy of synaptic vesicles reveals a pathogenic mechanism in motoneuron disease. *Nature Communications*, *8*, 678.

Lustig, K. D., Kroll, K. L., Sun, E. E., & Kirschner, M. W. (1996). Expression cloning of a Xenopus T-related gene (Xombi) involved in mesodermal patterning and blastopore lip formation. *Development*, *122*, 4001–4012.

Luu, O., Damm, E. W., Parent, S. E., Barua, D., Smith, T. H., Wen, J. W., et al. (2015). PAPC mediates self/non-self-distinction during Snail1-dependent tissue separation. *The Journal of Cell Biology, 208*, 839–856.

Mise, N., & Wakahara, M. (1994). Dorsoventral polarization and formation of dorsal axial structures in Xenopus laevis: Analyses using UV irradiation of the full-grown oocyte and after fertilization. *The International Journal of Developmental Biology, 38*, 447–453.

Moody, S. A. (1987). Fates of the blastomeres of the 32-cell-stage Xenopus embryo. *Developmental Biology, 122*, 300–319.

Moosmann, J., Ershov, A., Altapova, V., Baumbach, T., Prasad, M. S., LaBonne, C., et al. (2013). X-ray phase-contrast in vivo microtomography probes new aspects of Xenopus gastrulation. *Nature, 497*, 374–377.

Morgan, T. H. (1905). The relation between normal and abnormal development of the embryo of the frog: X. A re-examination of the early stages of normal development from the point of view of the results of abnormal development. *Archiv für Entwicklungsmechanik der Organismen, 19*, 588–614.

Müller, H.-A. J., & Hausen, P. (1995). Epithelial cell polarity in early Xenopus development. *Developmental Dynamics, 202*, 405–420.

Munro, E. M., & Odell, G. M. (2002). Polarized basolateral cell motility underlies invagination and convergent extension of the ascidian notochord. *Development, 129*, 13–24.

Nagel, M., & Winklbauer, R. (2018). PDGF-A suppresses contact inhibition during directional collective cell migration. *Development, 145*, 1–15. pii: dev162651.

Nagel, M., Tahinci, E., Symes, K., & Winklbauer, R. (2004). Guidance of mesoderm cell migration in the Xenopus gastrula requires PDGF signaling. *Development, 131*, 2727–2736.

Nieuwkoop, P. D., & Florschütz, P. A. (1950). Quelques caracteres speciaux de la gastrulation et de la neurulation de l'oeuf de Xenopus laevis, Daud. Et de quelques autres Anoures. 1ere partie. Etude descriptive. *Archives de Biologie, 61*, 113–150.

Ogata, S., Morokuma, J., Hayata, T., Kolle, G., Niehrs, C., Ueno, N., et al. (2007). TGF-beta signaling-mediated morphogenesis: Modulation of cell adhesion via cadherin endocytosis. *Genes & Development, 21*, 1817–1831.

Paluch, E. K., Aspalter, I. M., & Sixt, M. (2016). Focal adhesion-independent cell migration. *Annual Review of Cell and Developmental Biology, 32*, 469–490.

Papan, C., Boulat, B., Velan, S. S., Fraser, S. E., & Jacobs, R. E. (2007a). Formation of the dorsal marginal zone in *Xenopus laevis* analyzed by time-lapse microscopic magnetic resonance imaging. *Developmental Biology, 305*, 161–171.

Papan, C., Boulat, B., Velan, S. S., Fraser, S. E., & Jacobs, R. E. (2007b). Two-dimensional and three-dimensional time-lapse microscopic magnetic resonance imaging of Xenopus gastrulation movements using intrinsic tissue-specific contrast. *Developmental Dynamics, 236*, 494–501.

Pearl, E. J., Li, J., & Green, J. B. A. (2017). Cellular systems for epithelial invagination. *Philosophical Transactions of the Royal Society B, 372*, 20150526.

Pfister, K., Shook, D. R., Chang, C., Keller, R., & Skoglund, P. (2016). Molecular model for force production and transmission during vertebrate gastrulation. *Development, 143*, 215–727.

Pilot, F., & Lecuit, T. (2005). Compartmentalized morphogenesis in epithelia: From cell to tissue shape. *Developmental Dynamics, 232*, 685–694.

Popov, I. K., Ray, H. J., Skoglund, P., Keller, R., & Chang, C. (2018). The RhoGEF protein Plekhg5 regulates apical constriction of bottle cells during gastrulation. *Development, 145*, 168922.

Regen, C. M., & Steinhardt, R. A. (1986). Global properties of the Xenopus blastula are mediated by a high-resistance epithelial seal. *Developmental Biology, 113*, 147–154.

Rohani, N., Canty, L., Luu, O., Fagotto, F., & Winklbauer, R. (2011). EphrinB/EphB-signaling controls embryonic germ layer separation by contact-induced cell detachment. *PLoS Biology*, *9*, e1000597.

Rohani, N., Winklbauer, R., & Fagotto, F. (2014). Ephrin-Eph signaling in embryonic tissue separation. *Cell Adhesion & Migration*, *8*, 308–326.

Shi, D. L., Delarue, M., Darribère, T., Riou, J. F., & Boucaut, J. C. (1987). Experimental analysis of the extension of the dorsal marginal zone in Pleurodeles waltl gastrulae. *Development*, *100*, 147–161.

Shi, D.-L., Beetschen, J.-C., Delarue, M., Riou, J.-F., Daguzan, C., & Boucaut, J.-C. (1990). Lithium induces dorsal-type migration of mesodermal cells in the entire marginal zone of urodele amphibian embryos. *Roux's Archives of Developmental Biology*, *199*, 1–13.

Shih, J., & Keller, R. (1992). Patterns of cell motility in the organizer and dorsal mesoderm of Xenopus laevis. *Development*, *116*, 915–930.

Shindo, A., & Wallingford, J. B. (2014). PCP and septins compartmentalize cortical actomyosin to direct collective cell movement. *Science*, *343*, 649–652.

Shook, D., & Keller, R. (2003). Mechanisms, mechanics and function of epithelial-mesenchymal transitions in early development. *Mechanisms of Development*, *120*, 1351–1383.

Shook, D., & Keller, R. (2008). Epithelial type, ingression, blastopore architecture and the evolution of chordate mesoderm morphogenesis. *The Journal of Experimental Zoology*, *310B*, 85–110.

Shook, D. R., Majer, C., & Keller, R. (2004). Patterns and morphogenesis of presumptive superficial mesoderm in two closely related species, Xenopus laevis and Xenopus tropicalis. *Developmental Biology*.

Shook, D. R., Kasprowicz, E. M., Davidson, L. A., & Keller, R. (2018). Large, long range tensile forces drive convergence during Xenopus blastopore closure and body axis elongation. *eLife*, 7, e26944.

Siewing, R. (1969). *Lehrbuch der vergleichenden Entwicklungsgeschichte der Tiere*. Berlin: Parey Verlag.

Skoglund, P., Rolo, A., Chen, X., Gumbiner, B. M., & Keller, R. (2008). Convergence and extension at gastrulation require a myosin IIB-dependent cortical actin network. *Development*, *135*, 2435–2445.

Smith, J. C., Price, B. M. J., Green, J. B. A., Weigel, D., & Herrmann, B. G. (1991). Expression of a Xenopus homolog of Brachyury (T) is an immediate-early response to mesoderm induction. *Cell*, *67*, 79–87.

Sun, Z., Amourda, C., Shagirov, M., Hara, Y., Saunders, T. E., & Toyama, Y. (2017). Basolateral protrusion and apical contraction cooperatively drive Drosophila germ-band extension. *Nature Cell Biology*, *19*, 375–383.

Sweeton, D., Parks, S., Costa, M., & Wieschaus, E. (1991). Gastrulation in Drosophila: The formation of the ventral furrow and posterior midgut invaginations. *Development*, *112*, 775–789.

Vodicka, M. A., & Gerhart, J. C. (1995). Blastomere derivation and domains of gene expression in the Spemann Organizer of Xenopus laevis. *Development*, *121*, 3505–3518.

Vogt, W. (1929). Gestaltungsanalyse am Amphibiankeim mit ortlicher Vitalfarbung. II Teil. Gastrulation und Mesodermbildung bei Urodelen und Anuren. *Wilhelm Roux' Archiv für Entwicklungsmechanik der Organismen*, *120*, 384–706.

Wacker, S., Grimm, K., Joos, T., & Winklbauer, R. (2000). Development and control of tissue separation at gastrulation in Xenopus. *Developmental Biology*, *224*, 428–439.

Wallingford, J. B., & Harland, R. M. (2001). Xenopus Dishevelled signaling regulates both neural and mesodermal convegent extension: Parallel forces elongating the body axis. *Development*, *128*, 2581–2592.

Wallingford, J. B., Fraser, S. E., & Harland, R. M. (2002). Convergent extension: The molecular control of polarized cell movement during embryonic development. *Developmental cell, 2*, 695–706.

Wen, J. W. H., & Winklbauer, R. (2017). Ingression-type cell migration drives vegetal endoderm internalization in the *Xenopus* gastrula. *eLife, 6*, e27190.

Wilkinson, D., Bhatt, S., & Herrmann, B. (1990). Expression pattern of the mouse T gene and its role in mesoderm formation. *Nature, 343*, 657–659.

Williams-Masson, E. M., Heid, P. J., Lavin, C. A., & Hardin, J. (1998). The cellular mechanism of epithelial rearrangement during morphogenesis of the Caenorhabditis elegans dorsal hypodermis. *Developmental Biology, 204*, 263–276.

Wilson, P., & Keller, R. (1991). Cell rearrangement during gastrulation of Xenopus: Direct observation of cultured explants. *Development, 112*, 289–300.

Winklbauer, R. (1998). Conditions for fibronectin fibril formation in the early Xenopus embryo. *Developmental Dynamics: An official Publication of the American Association of Anatomists, 212*, 335–345.

Winklbauer, R. (2009). Cell adhesion in amphibian gastrulation. *International Review of Cell and Molecular Biology, 278*, 215–275.

Winklbauer, R., & Damm, E. (2012). Internalizing the vegetal cell mass before and during amphibian gastrulation: Vegetal rotation and related movements. *WIREs Developmental Biology, 1*, 301–306.

Winklbauer, R., & Keller, R. E. (1996). Fibronectin, mesoderm migration and gastrulation in Xenopus. *Developmental Biology, 177*, 413–426.

Winklbauer, R., & Schürfeld, M. (1999). Vegetal rotation, a new gastrulation movement involved in the internalization of the mesoderm and endoderm in Xenopus. *Development, 126*, 3703–3713.

Yanagi, T., Ito, K., Nishihara, A., Minamino, R., Mori, S., Sumida, M., et al. (2015). The Spemann organizer meets the anterior-most neuroectoderm at the equator of early gastrulae in amphibian species. *Development, Growth & Differentiation, 57*, 218–231.

CHAPTER TEN

Convergent extension in the amphibian, *Xenopus laevis*

Ray Keller*, Ann Sutherland
Department of Biology, University of Virginia, Charlottesville, VA, United States
*Corresponding author: e-mail address: rek3k@virginia.edu

Contents

Current Topics in Developmental Biology, Volume 136
ISSN 0070-2153
https://doi.org/10.1016/bs.ctdb.2019.11.013

Abstract

This review is a comprehensive analysis of the cell biology and biomechanics of Convergent Extension in *Xenopus*.

1. Introduction

The *Xenopus* embryo is an excellent system for analysis of Convergent Extension (CE). Its size and mechanical properties are unmatched for making grafts, recombinants and explants of tissues, which allow high resolution imaging, testing physiological and mechanical interactions of cells and tissues, distinguishing active movements from passive ones dependent on or modulated by mechanical linkages to other tissues, and for revealing emergent properties of the "Mechanome" that shapes the embryo (Lang, 2008). *Xenopus* displays variants of CE involving an epithelial layer and one to several layers of deep (inner) mesenchymal layers, a tissue organization that presents a broad range of mechanisms for analysis.

CE is the convergence (narrowing) of a tissue coupled with its extension (elongation) in an orthogonal axis (Vogt, 1929). CE is a kinematic description of tissue deformation (strain) without specifying the force (stress) producing it, as those may be internally generated forces ("active" CE), external forces from mechanically linked tissues ("passive" CE), or some combination of the two. CE is a tissue-level description, although in use, the term often assumes or implies a particular cellular mechanism, namely, cell intercalation. However, CE may be driven or modulated by cell shape change (Condic, Fristrom, & Fristrom, 1991), oriented cell division *without* growth (Bissen & Weisblat, 1989), cell division with growth (Williams & Solnica-Krezel, 2020), or some combination thereof. Convergence can also be coupled differently (Section 3), or completely uncoupled from extension (Solnica-Krezel & Sepich, 2012). In early *Xenopus* embryos cell division is not accompanied by growth and the total cellular volume remains nearly constant (Tuft, 1962), the exception being the swelling of notochordal cells due to expansion of intracellular vacuoles after neurulation, which can contribute to extension (Adams, Keller, & Koehl, 1990).

2. Active, force-producing CE occurs in presumptive notochordal and somitic mesoderm and in presumptive hindbrain-spinal cord during gastrulation and neurulation

Vital dye mapping (Keller, 1975, 1976), as well as tracing cells with fluorescent dextrans (Keller, Shih, & Sater, 1992; Keller & Tibbetts, 1989; Lane, Davidson, & Sheets, 2004) and with time-lapse imaging (Keller, 1978; Keller, Cooper, Danilchik, Tibbetts, & Wilson, 1989; Wilson & Keller, 1991; Wilson, Oster, & Keller, 1989) show that the presumptive notochordal, somitic, and mesendodermal tissues of the Involuting Marginal Zone (IMZ) and the presumptive hindbrain and spinal cord tissues in the Non-Involuting Marginal zone (NIMZ) of *Xenopus* undergo CE during gastrulation and neurulation. CE of these tissues closes the blastopore and elongates the future body axis in one stroke (Fig. 1A–D), and shapes the future vertebral column, the hindbrain and the spinal cord (Keller, 1991) (Fig. 1D and E). The vegetal endoderm and ventral mesoderm undergo CE in early tailbud stages, which straightens and further elongates the embryo (Drawbridge & Steinberg, 2000; Larkin & Danilchik, 1999) (Fig.1D and E). Tracing cells labeled with first generation fluorescent dextrans (Gimlich & Braun, 1985) showed that CE of the presumptive notochordal and somitic mesoderm (Keller & Tibbetts, 1989) and hindbrain/spinal cord (Keller, Shih, & Sater, 1992) occurs by mediolateral (across the emerging body axis) intercalation of cells to form a narrower, longer array in both the multilayered deep, mesenchymal region and in the superficial, epithelial layer of these tissues (Fig. 1B–D). Time-lapse imaging showed that the epithelial cells, despite being connected by a circumferential, apical junctional complex, intercalate to form a narrower, longer array during CE (Keller, 1978). This concept of epithelial cell rearrangement was pioneered by work on *Drosophila* imaginal disc eversion by Fristrom and others (Fristrom, 1976, 1982).

"Sandwich explants," adapted from Schechtman (1942), showed that presumptive notochordal/somitic mesoderm and spinal cord/hindbrain of the IMZ/NIMZ undergo CE while unattached to a substrate and independent of other force-generating processes in the embryo (Keller & Danilchik, 1988) (Fig. 1F, G–H). In explants, mesendodermal and neural regions undergo CE in series, in opposite directions, and of opposite polarity, as indicated by patterning of cell behavior and anterior-posterior (A-P)-specific

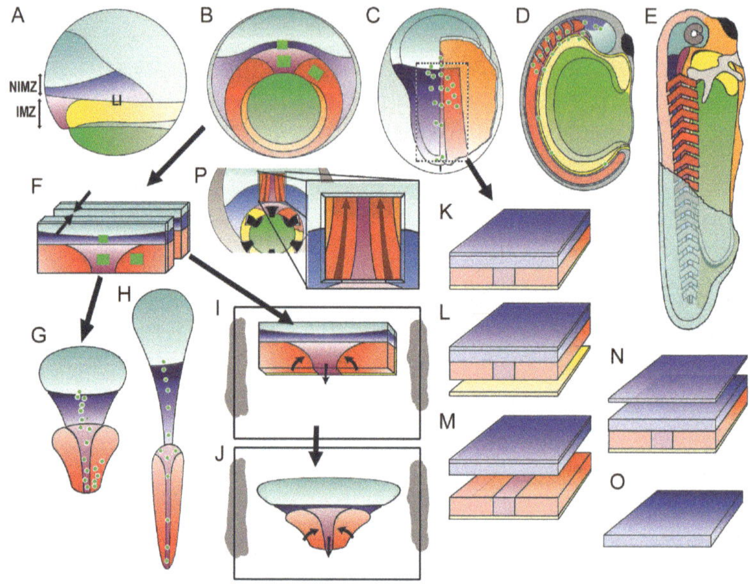

Fig. 1 Diagrams show the *Xenopus* early gastrula fate map in lateral (A) and vegetal (B) views, the early neurula in dorsal view, right side cutaway to show postinvolution mesoderm (C), the closed neural tube stage in sagittal view (D), and the early tadpole in a cutaway view (E). Ectodermal fates are: epidermis (dull blue), forebrain (bright blue), and hindbrain/spinal cord (dark blue). Mesodermal fates are: notochord (magenta), somitic mesoderm (red), head, heart and lateroventral mesoderm (orange). Endodermal fates are: archenteron roof suprablastoporal endoderm (yellow, removed to show underlying mesoderm in B and explants (F–J), archenteron roof/bottle cell endoderm (light green), and vegetal endoderm (dark green). The future anterior-posterior (A-P) axes are shaded from A (dark) to P (light). Also indicated are the IMZ (Involuting Marginal Zone), the NIMZ (Non-Involuting Marginal Zone), and the LI (Limit of Involution). Construction of experimental preparations are shown for the "Keller Sandwich" (F–H), the Open-faced explant as made (F–I) and undergoing convergent extension (J), the "dorsal isolate" or "Wilson explant (C-K), with endodermal epithelium removed to reveal deep notochordal and somitic mesoderm (L), with neural and mesodermal separated to reveal their apposing surfaces (M), with epithelial neural removed to reveal deep neural with underlying mesoderm (N), and isolated neural deep layer (O), and the Davidson "windowed" embryo to reveal deep notochordal and somitic mesoderm (P). The square green patches and green dots are schematic representations of the graded separation of labeled patches of cells during convergent extension.

markers (Doniach, Phillips, & Gerhart, 1992; Poznanski & Keller, 1997), rather than in parallel, the former beneath the latter, as *in vivo* (*cf.* F–H to B–D, Fig. 1). The "sandwiches" provide a covering epithelial layer, which prevents an abnormal tissue thickening, driven by deep cell surface tension (Section 16; Shook, Wen, et al., 2018), and it shields the deep cells from the culture media which at the time were inadequate for vigorous deep cell motility. These results suggested that these tissues are stiff and can push (generate a compressive force). Uniaxial, compressive stress relaxation tests with the "Histowiggler," an early computer-controlled mechanical measuring device, showed that these tissues stiffen by a factor of 3–4 in the axis of extension but not in the axis of convergence, at the onset of CE (Moore, Keller, & Koehl, 1995), and they can also push with a force of about a micro-Newton (Moore, 1994). These studies paved the way for deeper analyses of the mechanical properties underlying CE (Zhou, Kim, & Davidson, 2009), and raised questions of whether the deep mesenchymal layers, the epithelial layer, or both, generate the forces driving CE (Section 15).

3. CE is driven by both radial intercalation (RI) and mediolateral intercalation (MI) of cells

To image cell intercalation, "open-faced" explants (made from half the sandwich preparation, thus leaving the deep cells exposed, Fig. 1F–I) were lightly compressed between two coverslips, one forming the imaging interface and the other supported on high vacuum silicone grease (Fig. 1F, I and J) (Shih & Keller, 1992a, 1992b; Wilson, 1991; Wilson & Keller, 1991), a restraint that permits CE but limits thickening due to the tissue surface tension mentioned above (Shook, Wen, et al., 2018). Too much pressure increases friction, which retards CE, in extreme cases resulting in tissue differentiation *in situ* without normal movement (see Davidson, Keller, & DeSimone, 2004a). All explants with exposed deep cells were cultured in "Danilchik's Solution" (Keller, Danilchik, Gimlich, & Shih, 1985), which mimics the ionic composition of the blastocoel fluid (Gillespie, 1983), supports vigorous deep cell motility, and contains 0.1% BSA to prevent attachment to the coverslip, thus allowing CE (Fig. 1I and J). These explants are made from early to mid-gastrulation and imaged shortly thereafter (Shih & Keller, 1992a, 1992b; Wilson & Keller, 1991), and for later gastrula (Stages 11–13) and neurula stages (Stages 14–18), "Dorsal Isolates" (Zhou et al., 2009) or "Wilson Explants" (Keller et al., 1989; Wilson, 1991; Wilson et al., 1989) were made by excising the neural plate with underlying mesendoderm (Fig. 1C–K) and removing the mesendodermal epithelial layer, which allowed imaging the dorsal notochordal and somitic mesoderm (see Fig. 1L).

Low-angle epi-illumination of deep mesodermal tissue in the gastrula (open-faced, Fig. 1I and J) (Wilson & Keller, 1991) and neurula stages

(Fig. 1L) (Wilson, 1991; Wilson et al., 1989) showed directly that CE occurs by intercalation of cells along the axis perpendicular to the plane of the tissue (designated as "radial intercalation," RI), thereby producing tissue thinning (Fig. 2A). This intercalation was biased to occur between anterior-posterior neighbors, thus contributing to the A-P extension component of CE (Fig. 2B). Deep mesodermal cells also intercalate between one another across the emerging body axis, designated as "mediolateral intercalation" (MI) to form a M-L narrower and A-P longer array (Keller et al., 1989; Shih & Keller, 1992a, 1992b; Wilson et al., 1989; Wilson & Keller, 1991), as predicted by the tracing of contiguous, labeled patches of cells above (Fig. 2C).

The evidence argues that MI and RI act orthogonally, MI producing mediolateral tension and RI producing radial tension, and thus tissue compression in these axes. First, the structure of the IMZ reveals cells and their protrusions in both planar and radial orientations, which likely generate compressive forces along these axes (Fig. 2D–G, Section 4). The relative magnitude of each determines the specific patterns of extension in the A-P axis and thickening or thinning in the D-V axis for a given amount of convergence in the M-L axis, much like elongating a loaf of bread dough involves forces from side-to-side and top-to-bottom (Fig. 2H). The amount of thickening and extension resulting from a given amount of convergence is a tissue-specific property, with the notochord extending more and thinning less compared to the somitic mesoderm which extends less and thickens more (Keller et al., 1989; Wilson et al., 1989) (Fig. 2H). Second, embryos and explants undergoing CE show cells extending protrusions, and putatively exerting traction and generating tension both mediolaterally and radially (magenta and green, respectively, Figs. 2D–G) (Keller, 1984; Keller et al., 1989; Keller & Schoenwolf, 1977). Third, blocking radial and mediolateral forces independently would be revealing, but both are affected by the same pathways in frogs (Goto, Davidson, Asashima, & Keller, 2005; Ossipova, Chu, et al., 2014). However, in the mouse embryo blocking MI, and therefore the convergence force, by knocking out Ptk7, revealed a vigorous radial force previously unsuspected because the somitic mesoderm normally thickens; this force dramatically thins and widens presumptive somitic tissue beyond what is normal (Yen et al., 2009) (Fig. 2H). This suggests that RI and MI, and their resulting forces, act orthogonally to shape axial and paraxial tissues as a function of their relative strengths. Also, failure to display a morphogenic movement, such as thinning, may not indicate absence or failure of the underlying force-generating mechanism, as it

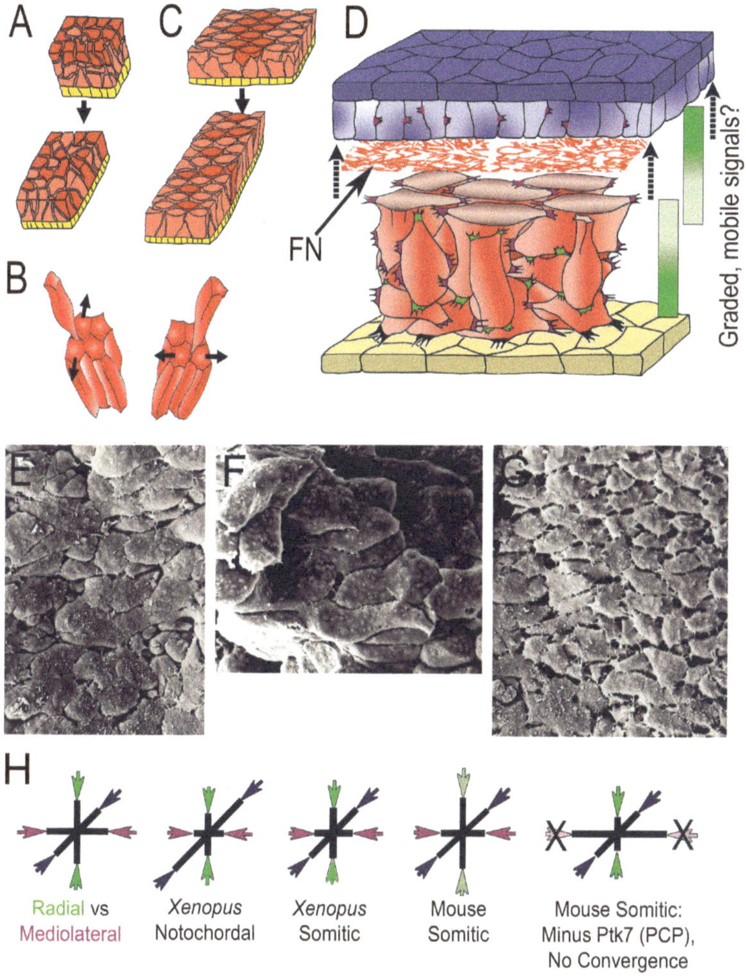

Fig. 2 Diagrams illustrate intercalation of deep mesodermal cells along the radial axis of the embryo (normal to the surface epithelium (yellow)), designated Radial Intercalation (RI) (A), and their intercalation mediolaterally (across the future body axis) designated mediolateral intercalation (MI), together, resulting in convergent extension (CE). In this case, RI is biased to occur between anterior and posterior neighbors (B, left) rather than between medial and lateral neighbors (B, right). The context and details of RI and MI during early CE are illustrated in D, showing the overlying presumptive neural tissue (epithelial layer, dark blue; deep layer light blue), a network of fibronectin fibrils (FN) between the basal layer of the presumptive neural tissue and outer layer of the deep mesodermal cells (red), which show mediolaterally polarized protrusive activity, reflective of intercellular traction and development of mediolateral tensile stress (magenta), and radially polarized protrusive activity acting orthogonally (green). At later stages FN is also assembled at the interface of the deep mesoderm and the overlying epithelial *(Continued)*

may exist but is losing out to larger, opposing forces in the overall morphogenic system. The embryo is shaped by both regional and global, long-range mechanical forces (Shook, Kasprowicz, et al., 2018) that comprise the embryonic Mechanome, and further analysis of the mechanical relationship between RI and MI, and their regulation, may go a long way in understanding the specificity of tissue shaping.

4. Mechanisms underlying the RI component of CE

The mechanism of RI during CE is less understood than that in the animal cap (AC) where the deep, non-epithelial cells elongate and partially interdigitate along the radial axis (normal to the surface) of the embryo, and then intercalate between one another and contact with the underside of the epithelium to form a single layer of greater area (Keller, 1980). The behavior of excised tissue suggests a model in which the inner layer of deep cells extend protrusions between the intermediate layers to the contact the underside of the epithelium. There they are "captured" by the boundary, and then contract, and wedge between their neighbors, thereby generating a single layer and a compressive, spreading force that produces epiboly (Keller, 1980). The normal, planarly oriented cell divisions that maintain a minimum number of layers are also essential, and both this orientation and RI are dependent on Integrin $\alpha5\beta1$-mediated assembly of, and interaction with fibronectin (FN) fibrils on the inner surface of the AC (blastocoel roof) (Fig. 2D) (Marsden & DeSimone, 2001). A pathway involving cadherin adhesion, Rac and Pak signaling, the actomyosin cytoskeleton, and the resulting mechanical tension are necessary for FN assembly (Dzamba, Jakab, Marsden, Schwartz, & DeSimone, 2009; Marsden & DeSimone, 2003). Planar assembly of FN fibrils is likely essential for RI during CE as well; inhibition of several PCP genes results in abnormal assembly

Fig. 2—Cont'd roof of the gastrocoel/archenteron (yellow). Integrin-FN signaling is essential for RI, as are mobile, chemotactic, signaling molecules (C3, PDGFA, Szabo et al., 2016; Damm & Winklbauer, 2011) in two other tissues in early *Xenopus* embryos, raising the possibility in this tissue. Scanning electron micrographs of the late gastrula show the notochord in dorsal (E) and transverse (F) view, and the somitic mesoderm in dorsal view (G). The Tri-axial Force Model of tissue shaping (H) show the postulated relationship between radial (green), and mediolateral (magenta) forces overcoming the resistance to extension in the third, A-P axis (blue) to varying, tissue-specific degrees in the *Xenopus* notochord and somitic mesoderm. Experimental support for tri-axial tissue molding forces comes from dramatic flattening and widening of mouse somitic mesoderm after loss of convergence forces.

of FN fibrils throughout the radial axis of the mesoderm rather than only at its planar interfaces, and RI and MI are retarded (Goto et al., 2005). A radial, short range chemotactic mechanism is involved in RI-driven extension of the mesodermal prechordal plate (Damm & Winklbauer, 2011) and in the RI-driven epiboly of the AC (Szabo et al., 2016). The first involves a mobile isoform of PDGFA and the second involves C3 and its receptor, C3A. The transient or permanent presence of an epithelium, depending on the circumstances, is necessary for CE in the IMZ (Wilson, 1991; Wilson et al., 1989; Wilson & Keller, 1991). This epithelium could be a source of a radially polarizing signal driving the radial component of CE; alternatively, it could lower tissue surface tension, which also contributes to normal CE (Ninomiya & Winklbauer, 2008). Computational models of RI (Szabo et al., 2016; also see Longo et al., 2004) suggest "boundary capture" on the underside of the epithelium or the FN layer as an essential element in stabilizing RI. New computational methods are applicable to RI and MI together (Section 10), an advance which may guide experimental investigations of RI-MI interactions (Fig. 2).

5. The mechanism and function of mediolateral intercalation behavior (MIB) in mesodermal CE

Time-lapse imaging of open-faced explants (Fig. 1F, I and J) and dorsal isolates lacking endoderm (Fig. 1L) showed that MI of deep mesodermal cells occurs by a characteristic "mediolateral intercalation behavior" (MIB), in which initially randomly oriented protrusive activity becomes mediolaterally oriented and bipolar (Keller et al., 1989; Shih & Keller, 1992a, 1992b; Wilson, 1991; Wilson et al., 1989; Wilson & Keller, 1991) (Fig. 3). Filo-lamelliform protrusions extend mediolaterally between neighboring cells in a bipolar manner, attach to neighboring cells *via* C-cadherin adhesions, and contract in the mediolateral axis, thereby pulling themselves between one another in repeated cycles (see Section 6). MIB appears to be a typical cell-substrate motility, with the exception that cells' "substrate" is one another and the bipolar cells exert roughly balanced traction in both directions (Keller et al., 2000; Keller, Shih, & Domingo, 1992; Section 8) (Fig. 3). Expression of MIB is progressive, beginning anteriorly and laterally and progressing posteriorly and medially (Section 12). The notochordal-pre-somitic boundary forms as a function of Eph/Ephrin signaling (Fagotto, Rohani, Touret, & Li, 2013; Reintsch, Habring-Mueller, Wang, Schohl, & Fagotto, 2005; Section 17), beginning anteriorly

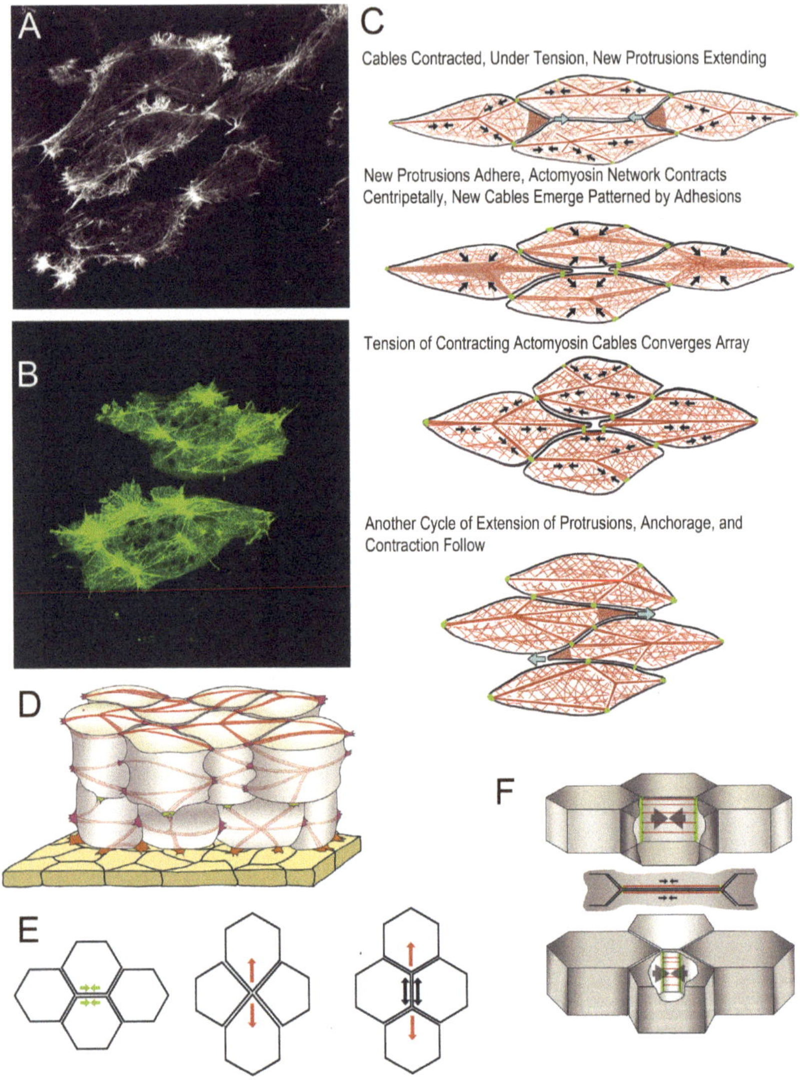

Fig. 3 Scanning laser confocal micrographs from time-lapse movies show examples of the Node and Cable Network (NCN) actin cytoskeleton with Moesin-AB-GFP (A, B). Diagrams show phases of Mediolateral Intercalation Behavior (MIB) in the Cell-on-Cell Traction Model of intercalation (C). Black arrows indicate contraction of F-actin network/F-actin cables (red), green indicates C-cadherin adhesions, and blue arrows indicate extension of protrusions. Anterior sides of cells are black, the posterior gray. A semi-transparent diagram shows surface cables and lateral cables of the cortical NCN (red) anchored to mediolateral (magenta) and radial (green) tractive protrusions, and protrusions connected to the underside of the epithelial endoderm of the archenteron roof (yellow). An illustration of the Epithelial Junctional Remodeling mode of epithelial cell *(Continued)*

progressing posteriorly, and thereafter bipolar cells contacting the boundary cease protrusive activity at the boundary end and transition into a monopolar mode with inwardly directed protrusive activity (Shih & Keller, 1992a, 1992b). As, or shortly after it forms, fibronectin (Rosario & DeSimone, 2009) and fibrillin (Skoglund, Dzamba, Coffman, Harris, & Keller, 2006) are assembled into fibrils in the boundary; these are essential for notochord cell intercalation and stabilization of the boundary (Skoglund & Keller, 2007). After the boundary forms, the notochordal and somitic modes of RI, MI, and MIB progressively differentiate as new cell behaviors add to or replace MIB, such as columnarization and thickening in the pre-somitic tissue (Keller et al., 1989; Wilson et al., 1989), and modified polarity, intercalation, and vacuolation to form a fiber-wound hydrostat in the notochord (Keller et al., 1989). These and other late modifications of the basic mechanism of MIB remain poorly understood (see Section 18).

Many studies in *Xenopus* and other species show that the MIB and the resulting CE of the axial/paraxial mesoderm are dependent on the "non-canonical Wnt"/vertebrate planar cell polarity (PCP) and a number of downstream effector proteins (Butler & Wallingford, 2017; Darken et al., 2002; Djiane, Riou, Umbhauer, Boucaut, & Shi, 2000; Glickman, Kimmel, Jones, & Adams, 2003; Goto & Keller, 2002; Habas, Dawid, & He, 2003; Habas, Kato, & He, 2001; Heisenberg et al., 2000; Jessen et al., 2002; Shendo, 2018; Sokol, 1996; Tada & Smith, 2000; Wallingford et al., 2000; Williams, Yen, Lu, & Sutherland, 2014; Yen et al., 2009). The common failure of MIB and CE in the wake of PCP

Fig. 3—Cont'd intercalation (E) shows the initial active junction-shortening contraction (green arrow left), the formation of the tetrad of cells meeting at a vertex, and the initiation of "resolution" with a centripetal contraction of the actomyosin cytoskeleton of the same cells (red arrows, middle), and the completed intercalation (right) showing the active contraction of resolution (red arrows) and the passive response of the intercalated cells (black arrows). A diagram shows the features of the junction shortening step of the epithelial-type intercalation behavior of notochord cells (F). Septins (green) form barriers that confine the active myosin to the actomyosin cables (red) of the cortical regions of the two apposed cell boundaries (in lateral view, top, bottom and vertical view, middle); contraction of actomyosin cables shortens the boundaries (arrows) until the vertices meet (bottom). *Panel (E) Based on reference Collinet, C., Rauzi, M., Lenne, P.-F., & Lecuit, T. (2015). Local and tissue-scale forces drive oriented junction growth during tissue extension.* Nature Cell Biology, 17, *1247. Panel (F) Based on reference Huebner, R.J., & Wallingford, J.B. (2019). Coming to consensus: A unifying model emerges for convergent extension.* Developmental Cell, 46, *389–396.*

perturbations suggests that, in fact, MIB is a major driving force for mesodermal CE in *Xenopus* and other chordates as well, although there is much variation in MIB itself and in its deployment across species. In *Xenopus* the PCP pathway is necessary but not sufficient for polarization during CE. Polarization requires integrin α5β1-fibronectin signaling, as blocking integrin function alone results in expansion of protrusive activity all around the cell, and super-activation of integrins results in suppression of all protrusive activity (Davidson, Marsden, Keller, & DeSimone, 2006).

6. The node and cable network (NCN) and iterated actomyosin contraction is the "power stroke" of MIB

The polarized extension of protrusions and their adherence to medial and lateral neighbors sets up the "power stroke" of MIB, which is the cell shortening and resulting traction on the adherent, neighboring cells that pulls the cells between one another and transmits tensile force across the mediolateral aspect of the tissue. This force is generated by a contractile, mediolaterally oriented actomyosin cytoskeleton organized as a Node and Cable Network (NCN) (Davidson et al., 2006; Skoglund, Rolo, Chen, Gumbiner, & Keller, 2008) (Fig. 3A and B). The NCN consists of actin cables connected to one another at internal nodes to form a network, and to neighboring cells by C-cadherin adhesions (see below), thereby forming a contractile system spanning many cells (Pfister, Shook, Chang, Keller, & Skoglund, 2016; Skoglund et al., 2008) (Fig. 3C). The contractile activity of the NCN, as indicated by the major displacements of the internal nodes, the C-cadherin adhesions, and the shortening of the cables, is biased toward being parallel to the mediolateral axis of the tissue (Fig. 3C). These contractions are dependent on myosin heavy chain IIB (MHC-IIB) and regulation by the Myosin Regulatory Light Chain (MRLC) in a dose dependent fashion (Pfister et al., 2016; Skoglund et al., 2008). Knock down of Myosin IIB (Skoglund et al., 2008) or MRLC activity (Pfister et al., 2016) resulted in decreases in the measured mediolateral convergence force and failure of axial/paraxial elongation and blastopore closure.

Time-lapse imaging of the dynamics of fluorescently labeled C-cadherin and the actin cytoskeleton shows a coordinate development of cell-cell adhesions and contraction of the actin cytoskeleton during intercalation. This suggests that C-cadherin-mediated Cell-on-Cell (CCT)traction is a major mechanism for transmitting the force generated by the actomyosin cytoskeleton across the mediolateral aspect of the tissue (Pfister et al., 2016).

Fluorescently labeled C-cadherin appears as puncta throughout cell perimeter and as larger plaques at lamellipodia. As the F-actin of the lamellipodia flows toward the center of the bipolar cell, it coalesces into a "proto-nodes" and the C-cadherin plaques elongate parallel to and colocalize with the distal actin cytoskeleton, as the actin cytoskeleton matures into the definitive nodes and connecting cables. Neighboring cells expressing red or green fluorescent C-cadherin show colocalization in adhesions. Measuring the displacement of proto-nodes and adhesion sites in neighboring cells is consistent with the idea that the contractile actin cytoskeleton of individual cells is linked into a transcellular mechanical unit by C-cadherin adhesions (Pfister et al., 2016). These adhesions are used to exert traction on other cells and at the same time to serve as contractile (active) substrates for other cells (Fig. 3C). The NCN constitutes a global system of dynamic contractile elements, constantly remodeled with formation of new adhesions and loss of old ones as a function of new protrusive activity, contact and adhesion during rapid intercalation, thereby transmitting tensile force across the entire tissue (Yen et al., 2009) (Fig. 3C and D). These findings are consistent with pioneering work showing the importance of regulated activity of C-cadherin in CE (Brieher & Gumbiner, 1994).

Integrating the punctuated actin dynamics described by Kim and Davidson (Kim & Davidson, 2011) and actin dynamics of the NCN described by Pfister et al. (2016) argues for a model in which the iterated contractile, condensation of an F-actin network constantly remodels the NCN to reflect the turnover dynamics of C-cadherin adhesions, which in turn are pre-patterned by the polarization of protrusive activity. Defined, quantified, single, iterated cycles of actin concentration in the mid-cell region, as measured by moe-GFP intensity, were strongly and inversely correlated with cell area (Kim & Davidson, 2011). These "punctuated F-actin contractions" occur in several cell types but in the dorsal mesoderm the contractions were more frequent, moved larger distances, and they were aligned with the long axis of the cell, as the cells became progressively more elongated during CE in the second half of gastrulation. Inhibitor and FRAP studies showed that actin polymerization and F-actin contraction were involved in the punctuated F-actin dynamics. Overexpression of the Frizzled receptor, Xfz7 resulted in more persistent and long-lived contractions whereas the expression of a dominant negative Disheveled, Xdd1 randomized the orientation of the contractions of mesodermal cells, thereby linking the contractions to the non-canonical Wnt/PCP pathway. Each iteration of polarized protrusive activity results in "new grips," new C-cadherin adhesions to other bipolar cells, and these nascent C-cadherin plaques are enlarged and reinforced with the retrograde/centripetal movement and contraction of the actomyosin network (Pfister et al., 2016), thereby serving as anchorages for the contraction-driven compaction of actomyosin network into the cables of NCN (Skoglund et al., 2008). In short, each iterated contraction "scans" for and integrates the retrograde flow of actin from C-cadherin anchorages into the existing NCN. Analysis of the change in lengths of cables connecting a series of nodes and adhesion sites across several cells revealed combinations of contractions (shortening), stasis (no change in length), and relaxation (lengthening) of these cables over time (Pfister et al., 2016). These observations suggest ongoing variation in contractile force of these elements across many cells, with the overall tension of the intracellular-transcellular system being a function of the weakest link. The contractile array is not static but constantly remodeled on a time scale of seconds, and the tensile paths across the multicellular array are constantly remodeled as geometry of the NCN reflects the integration of newly made adhesions and the loss of old ones, always self-aligned mediolaterally by the mediolateral polarization of the protrusive activity and protrusion-associated C-cadherin adhesions (Pfister et al., 2016) (Fig. 3C).

7. Cell-on-cell traction rather than cell on matrix traction generates most of the tissue-level, tensile forces driving cell intercalation

The intercalating deep mesodermal cells also make adhesions with and exert traction on a flexible, fibrillar, fibronectin network assembled between the inner surface of the ectoderm and the outer surface of the mesoderm, a tissue boundary designated as Brachet's Cleft (Fig. 2D) (Davidson, Keller, & DeSimone, 2004b). In fact, $\alpha5\beta1$integrin-mediated adhesion to FN, and tension generated by the actomyosin cytoskeleton are both necessary to assemble this matrix, as observed directly by time-lapse imaging of fluorescently labeled fibrils in living tissue (Davidson et al., 2004b), suggesting the hypothesis that cells could use flexible FN fibrils as tensile tethers to pull themselves between one another. However, FN-$\alpha5\beta1$integrin signaling is essential for expression of MIB and for CE (Davidson et al., 2006), making it necessary to test the *mechanical function of fibronectin fibrils independent from fibronectin's molecular, signaling function.* Rosario and others (Rosario, Dzamba, Weber, Davidson, & DeSimone, 2009) ectopically expressed the 70kd multimerization domain of fibronectin, which acts as dominant inhibitor of fibronectin fibrillogenesis, thereby producing embryos that lack fibrils but have molecular fibronectin available for the integrin signaling essential for cell polarity. These embryos, and explants made from them, show near normal CE, arguing that the cells exert enough tension on the fibronectin to support multimerization and fibrillogenesis but not enough to play a large role in the generating the tensile forces driving CE. Thus, the bulk of the tractional forces are likely due to cell-on-cell traction *via* dynamic C-cadherin linkages of the contractile actomyosin cytoskeleton of individual cells. However, cell-matrix, cell-cell interactions, and the mechanical properties of both are linked in multiple ways and levels that are essential for CE. Cadherin adhesion, tissue tension and non-canonical Wnt signaling regulate fibronectin matrix organization (Dzamba et al., 2009), and in turn, integrin-ECM interactions regulate the Cadherin-dependent cell adhesion that is essential for CE (Goto et al., 2005). These results suggest that the early deep mesodermal cell intercalation is driven by a Cell-on-Cell Traction (CCT) rather than cell-on-matrix traction.

8. Balanced traction, regulation of contraction, and the logic of cell intercalation

Evidence from an abnormal mode of neural cell intercalation argues that the bipolar, CCT mode of intercalation requires balanced traction in the medial and lateral directions to produce CE. Cells of the deep neural tissue normally show monopolar protrusive activity biased toward the midline. When cultured without underlying mesoderm (and thus lacking a midline), they display statistically bipolar protrusive activity when averaged over time. However, this reflects a repeated switch of their major protrusion, and their displacement, from lateral to medial and back again. The cells just exchange places, the pattern of intercalation becomes chaotic, and CE is greatly reduced (Elul & Keller, 2000; Elul, Koehl, & Keller, 1997). This suggests that balanced traction is essential for an effective bipolar mode and raises the question of how the bipolar deep mesodermal cells maintain this balance. In addition, it indicates that *frequency of neighbor exchanges* does not assure nor is it a measure of CE. CE seems to be first an issue of generating large-scale tensile force across the span of the tissue, with cell intercalation being an essential follow up (barring any permanent cell shape change).

Also, how is the contractile activity and the relative strength of the linked actomyosin cytoskeletons of the aligned series of cells spanning the tissue regulated? Myosin IIB knockdown results in failure of blastopore closure and CE in a dose dependent manner (Pfister et al., 2016; Skoglund et al., 2008), as well as a reduction in convergence force (Pfister et al., 2016). Moderate knockdowns of Myosin IIB, result in larger and more rapid shortening and lengthening of the internodal segments of the NCN cables and movements of the nodes, suggesting that IIB may be crosslinking actin in the cables and thus enabling them to carry more tensile stress (Skoglund et al., 2008), which may be a specialized function of IIB (Kelley et al., 1996; Rochlin, Itoh, Adelstein, & Bridgman, 1995). How are the contractile force, tensile properties, and stiffness regulated across the convergence dimension such that local lesions do not occur? The mean stress (force) produced by CE is 5.0 ± 1.6 Pascal (Pa) but CE is mechanically adaptive and can produce threefold more force in a stiffer environment (Zhou, Pal, Maiti, & Davidson, 2015), and in fact the tissue stiffness increases 8–10-fold during CE (Zhou et al., 2009). The evidence suggests that force production should match mechanical resistance to produce a consistent rate of CE. This implies mechanical feedback and the process appears to be adaptive with mechanical accommodation of force production and stiffness (Zhou et al., 2015). Normally, mesodermal MIB/CE operates in an arc anchored at each end, as a homogeneous process that rarely fails within its domain (Yen et al., 2009). However, when artificially put in series with the weaker process of CT (see Shook, Wen, et al., 2018) in explants, CE/CT combination as a whole does not accommodate, the weaker tissue stretches and produces a plateau in force generation (Shook, Kasprowicz, et al., 2018).

9. An epithelial, junction remodeling model of notochordal cell intercalation

A mechanism of cell intercalation similar to the Epithelial Junction Remodeling (EJR) mode of the epithelial germband of *Drosophila* (Fig. 3E) has been described for *Xenopus* notochord (Fig. 3F) (Kim et al., 2010; Shindo & Wallingford, 2014). Accumulation of pulsed actin, a specific MRLC, Myl9, a higher concentration of phosphorylated myosin (pMyoII), and increased tension are all found along the elongated mediolateral junctions compared to non-mediolateral junctions (those off axis 20 degrees or greater) (Fig. 3F). These parameters are correlated with junction shortening, which brings the vertices of the cells into contact, followed by "resolution" by expansion of a transverse junction, thus resulting in intercalation (Fig. 3F). The myosin activity is confined to a broad area of the apposing, mediolaterally oriented, shortening junctions by Sept7 located at cell vertices of mediolateral and non-mediolateral junctions. When Sept7 is knocked down, the high concentration of actin at these vertices is lost, and the higher pMyoII at mediolateral junctions, compared to non-mediolateral junctions, is averaged out. Sept7 is localized at the vertices in a PCP-dependent manner where it forms a barrier to retain the pMyoII, and thus the high myosin activity and greater tension is localized to mediolateral junctions, compared to the non-mediolateral junctions (Shindo & Wallingford, 2014).

As a general model for deep mesenchymal cell intercalation, EJR-based model seemed in conflict with the CCT, and an effort was made to find a consensus of the two (Huebner & Wallingford, 2019). However, on deeper examination there is no conflict as the two models address two different stages and modes of cell intercalation. The CCT model emerged from making open-faced IMZ explants of the early gastrula (Stage 10–10.5) and imaging them immediately (Shih & Keller, 1992a, 1992b; Wilson & Keller, 1991) or making dorsal isolates of the same tissue at mid to late gastrula (Stage 11–11.5 and 12–14) and imaging immediately or soon after (Keller et al., 1989; Skoglund et al., 2008; Wilson et al., 1989), with all preparations cultured on non-adhesive BSA-coated substrates. These methods capture the early and mid CE period in which the mesoderm narrows rapidly, the notochord narrowing rapidly from 30 to 40 cells or more in width outside the blastoporal lip to just over 4 or 5 in width at stage 12.5 and further narrowing to just over two, on average by Stage 14 *in vivo* (Keller, 1984; Keller et al., 1989) (Fig. 4A and B). It narrows less than that in open-faced explants, the degree of intercalation depending on the amount of compression with a restraining coverslip (Shih & Keller, 1992a, 1992b). The cells are participating in both RI and MI at the same time, and thus lamellipodia, the resulting, tractive cell adhesions, and cell intercalations are orthogonal to one another (Figs. 2D–G and 3D) and therefore present a complex morphology and arrangement of cells, dynamics of protrusive activity and exchange of focal junctions that are incompatible with the EJR Model.

However, this version of the EJR model is in many respects compatible to events at a later stage and constitutes a major advance in understanding a heretofore ignored problem, the transition from the largely-monopolar inwardly directed motility of the midneurula stage notochord to the late stage neurula/tailbud fiber-wound hydrostat (Fig. 4A and B). EJR model arises from explants made at early to mid-gastrula, cultured for a "half a day" (presumably ~12 h) at 15–16C, on fibronectin coated surfaces before

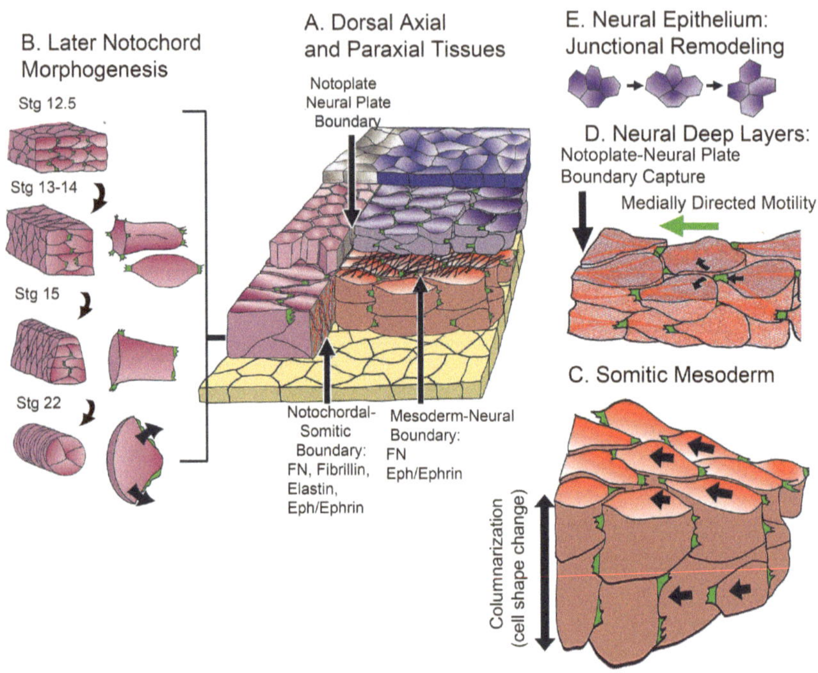

Fig. 4 A diagram (A) shows relationships in the early neurula between notochord (magenta), the deep layer notoplate of the neural plate above it (magenta), the epithelial layer of the notoplate (gray), the somitic mesoderm (red), the deep layer of the neural plate (blue-gray), the epithelial layer of the neural plate (blue) and the endodermal gastrocoel (archenteron) roof (yellow). Diagrams (B) show differentiation of new patterns of MIB at Nieuwkoop-Faber developmental stages (magenta): the transition from a mix of bipolar protrusive activity of internal cells and monopolar, inwardly directed protrusive activity of boundary cells (Stages12.5–14) and the transition from monopolar inwardly directed protrusive activity to circumferential spreading, forming pizza-slice shaped cells, and rounding of the notochord at the onset of vacuolation (Stages 15–22) (Keller et al., 1989). A diagram (C) shows convergence of the pre-somitic mesoderm by the bipolar mode of intercalation in the now double-layered somitic mesoderm, which is later accompanied by columnarization of the medial somitic mesoderm, the second component of somitic convergence. A diagram (D) shows the medially directed monopolar protrusive activity and crawling mode of intercalation in the deep neural layer, using actomyosin cables similar to those seen in the notochordal and somitic mesoderm. A diagram (E) illustrates the Epithelial Junctional Remodeling (EJR) mode of cell intercalation characteristic of the outer neural epithelial layer (Butler & Wallingford, 2018).

imaging, and at this temperature the stage control embryos would be nearing Nieuwkoop-Faber Stage15–16 (Huebner & Wallingford, 2019; Shindo & Wallingford, 2014). At these stages, the notochord *in vivo* has narrowed to two side-by-side rows of interdigitated cells, which then proceed to the final stage of spreading their outside "apical aspects" circumferentially, in the transverse plane of the notochord to form an array of cells shaped like "pizza slices," which are overlapped to form a newly rounded, fiber-wound notochord (Adams et al., 1990; Keller et al., 1989) (Fig. 4B). Scanning Electron Microscopy (SEM) suggest that the dominant protrusive activity at this stage occurs on a broad front comprising the long diagonals of the "slices" as these edges inter-leaf (Keller et al., 1989) (Fig. 4B). However, in explants cultured from the early bipolar stage onward on rigid FN-coated surfaces, the "pizza slice intercalation" is unlikely to occur, as offering these bipolar cells an external, undeformable substratum would likely result in zero net force on the cells, conditions that greatly retard CE (Davidson et al., 2004a). This would account for the very wide notochords and lack of NSBs in these studies. However, making "Wilson explants" (Wilson et al., 1989) or "dorsal isolates" (Zhou et al., 2009) with the endodermal layer removed (Fig. 1C–L) at Stage 16 and culturing them immediately on non-adhesive substrates in Danilchik's solution could allow imaging of the normal "pizza-slice phase" of terminal notochord cell intercalation. This would also avoid possible unwanted effects of long-term culture on a very stiff substrate of coverslip-bound fibronectin, as substrate stiffness can affect cell differentiation (Discher, Janmey, & Wang, 2005; Engler, Sen, Sweeney, & Discher, 2006), junction formation (Liua et al., 2010) and cell motile behavior *in vivo* (Barriga, Franze, Charras, & Mayor, 2018).

10. Comparison of the mesenchymal, cell-on-cell traction (CCT) model and the epithelial junction remodeling (EJR) model of cell intercalation

The CCT and EJR models share the concept of an actively contracting cytoskeleton, but they differ in its organization. In the case of CCT, contractile cables in one cell are linked by transient, focal C-cadherin adhesions to cables of other cells, and the cables have two roles; first, they can shorten and exert traction on another cell, and second, they can serve as "substrate cables," maintaining their length or contracting but in either case, serving as "substrate" for the traction of another cell (Fig. 3C). The cables are constantly reorganized as function of newly formed protrusions, the resulting formation of new adhesions, and their engagement with the iterative actomyosin contractions (Kim & Davidson, 2011; Pfister et al., 2016). In the CCT model, many different cycles of protrusion, adhesion, and actomyosin contraction occur to bring about any one intercalation event, whereas in the EJR model, actomyosin dynamics are associated with iterated contractions of long continuous apical junctions, often spanning multiple cells in the case of "rossettes, in the pre-"resolution" stage (Blankenship, Backovic, Sanny, Weitz, & Zallen, 2006) (Fig. 3E). Both models share the concept of "active (contractive) substrates," in the case of the CCT, internal cables connecting any two or more tractive protrusions (as the plasma membrane is a liquid and offers no resistance), and in the case of the EJR the contractile elements are associated with a long region of two plasma membranes bound continuously by a junctional complex, the elements of which must be recycled (Levayer & Lecuit, 2013) or passed through vertices (Fristrom, 1982).

The concepts of mediolateral and non-mediolateral junctions do not apply in the same way to cells in CCT and in the EJR models, nor is tension associated with them in the same way. Movies of CCT show cells to be generally fusiform but oscillating from rotund to elongate shapes, generally aligned mediolaterally, and shearing past one another, sometimes as individuals and sometimes as multicellular chains (Pfister et al., 2016; Shih & Keller, 1992a, 1992b). As they move by one another their apposed sides give the illusion of a "shortening" junction whereas in reality it is a shear zone. In contrast, the EJR model deals with the continuous "weld" of the apical junctional complex, rather than the C-Cadherin "spot welds" of deep mesodermal cells. As a consequence, the EJR model defines higher tension, shortening mediolateral apical junctions in contrast to lower tension, lengthening non-mediolateral apical junctions. In the CCT model there is no such distinction, as generation of tension is not associated with the necessity of removing the components of a circumferential-apical junctional complex from cell boundaries that are shortening and adding them to others that are expanding. In the CCT, turnover of "spot-weld" C-Cadherin junctions serve to allow protrusions to "get a new" grip, a new anchorage for the next contraction of the NCN.

Finally, it should be noted that basolateral protrusive activity and CCT is likely used by the basolateral aspect of epithelial cells whereas the EJR is not used in mesenchymal cells, which lack the defining apical junctional complex. Basolateral protrusions are found in intercalating dorsal epithelial cells of the nematode embryo (Williams-Masson, Heid, Lavin, & Hardin, 1998), the neural plate epithelial cells of the mouse (Williams et al., 2014), and in the *Drosophila* germband (Sun et al., 2017). In all three cases, the intercalation of the deep part of the cell leads the intercalation of the apical part more frequently than the reverse, suggesting that CCT of the basolateral sides may be the dominant force-generating machine.

11. Computational models of CE by MIB mediated CCT

Several computational models of MIB have been created, and two capture many of the essential features (Belmonte, Swat, & Glazier, 2016; Brodland, 2006). In the first, a finite element model, a lamellipodium extends between two cells, filling the entire interface between them, contacts a fourth cell, and then shortens, pulling the extending and the contacted cell toward one another. This step invokes a doubling of the interfacial tension along the protrusion, as it now has double the number of interfaces and thus twice the interfacial tension. In principle, the single, large extension seems a proxy for tension generated by the many, smaller extensions of lamellipodia between two cells, their adherence to one or both, or to the fourth cell, and there by anchoring the iterated cell shortening seen in *ex vivo movies* (Kim & Davidson, 2011). Using values for tissue viscosity and tension that are reasonable from the literature, the model results in cell intercalation and CE. If the bounding tissues offer no resistance, the cells are nearly isodiametric at the end, whereas if resistance is offered, the cells elongate in the axis of convergence and CE fails; if intermediate resistance is offered, the cells show intermediate cell elongation and CE, which fits with cell shapes seen in *Xenopus* explants undergoing CE with increasing resistance (Shih & Keller, 1992b).

The second is a Cellular Potts/Glazier-Graner Hogeweg model, which explores a number of relevant processes and is closely integrated with experimental data (Belmonte et al., 2016). This work discusses a number of models, and the one most relevant here, the Anisotropic Filipodial-Tension Model, models the number, angular distribution, range, frequency of formation and breakage, strength of cell-cell (filipodial) connections. It examines parameters of planar polarization such as misaligned and passive cells (lack filopodia but can be contacted by filopodia) and refractory cells (cannot be contacted). The model shows the inverse relationship of tissue surface tension and filipodial tension, a result confirmed by experimental suppression of CE by increasing surface tension (Ninomiya & Winklbauer, 2008), and it explores the predicted effects of filopodial lifetime, range, number of interactions, and angular range. The investigators also evaluate cell contact mediated behaviors, the effects of degrees of misaligned polarization, and the potential effect of mechanical feedback on failed CE under conditions of high misaligned polarization. 3D versions of the model make it useful for analysis of bipolar and monopolar mechanisms of intercalation and orthogonal processes, such as radial intercalation in the presence of mediolateral intercalation. These models best capture the important parameters gleaned from observed behavior of early deep mesodermal cell intercalation (Pfister et al., 2016; Shih & Keller, 1992b; Skoglund et al., 2008) and offer an excellent guide for further experimental exploration, including convergence on two axes and extension on the third, and two converging and one extending axis (see Sections 3 and 4).

12. Large scale patterning of MIB is essential for Normal CE function

Large scale patterning of MIB is essential for mesodermal CE to function in both closing the blastopore in the second half of gastrulation, and in elongating the body axis through neurulation (Keller et al., 2000; Keller, Shih, & Domingo, 1992; Shih & Keller, 1992b). The large-scale patterning in the embryo (Fig. 5A) was described by imaging large open-faced explants comprising most or all of the IMZ (Fig. 5B–D) and mapping the pattern back on to the gastrula. Time-lapse imaging revealed a progressive A-P expression of MIB across the late involuting IMZ. MIB/CE begins at Stage 10.5 2h after the onset of blastopore formation in the presumptive anterior somitic mesoderm on both sides of the midline adjacent to the vegetal endoderm, and it proceeds medially toward the midline where the two lateral components meet to form the arc-shaped vegetal alignment zone (VAZ; Fig. 5C and E) (Lane & Keller, 1997; Shih & Keller, 1992b). As this arc grows in thickness, the notochordal-somitic boundary (NSB) forms within it and progresses the full length of the A-P axis of the body plan to separate the notochordal and somitic mesoderm (Fig. 5C and F). After formation of the NSB, MIB proceeds from A to P in the lateral regions of the presumptive notochordal and presumptive somitic mesoderm (yellow arrows, VE, Fig. 5C–F). From this lateral origin, next to the vegetal endoderm in the case of the somitic mesoderm, it proceeds toward the notochord, and in the case of the notochordal mesoderm, it proceeds toward the midline (Fig. 5D–G and H). Mapped on to the IMZ of the gastrula (Fig. 5A), the A-P progression takes the form of an arc-like pattern that progresses posteriorly in a continuous manner (yellow arrows). However, it can be better visualized as discrete hoops anchored on both ends at the vegetal

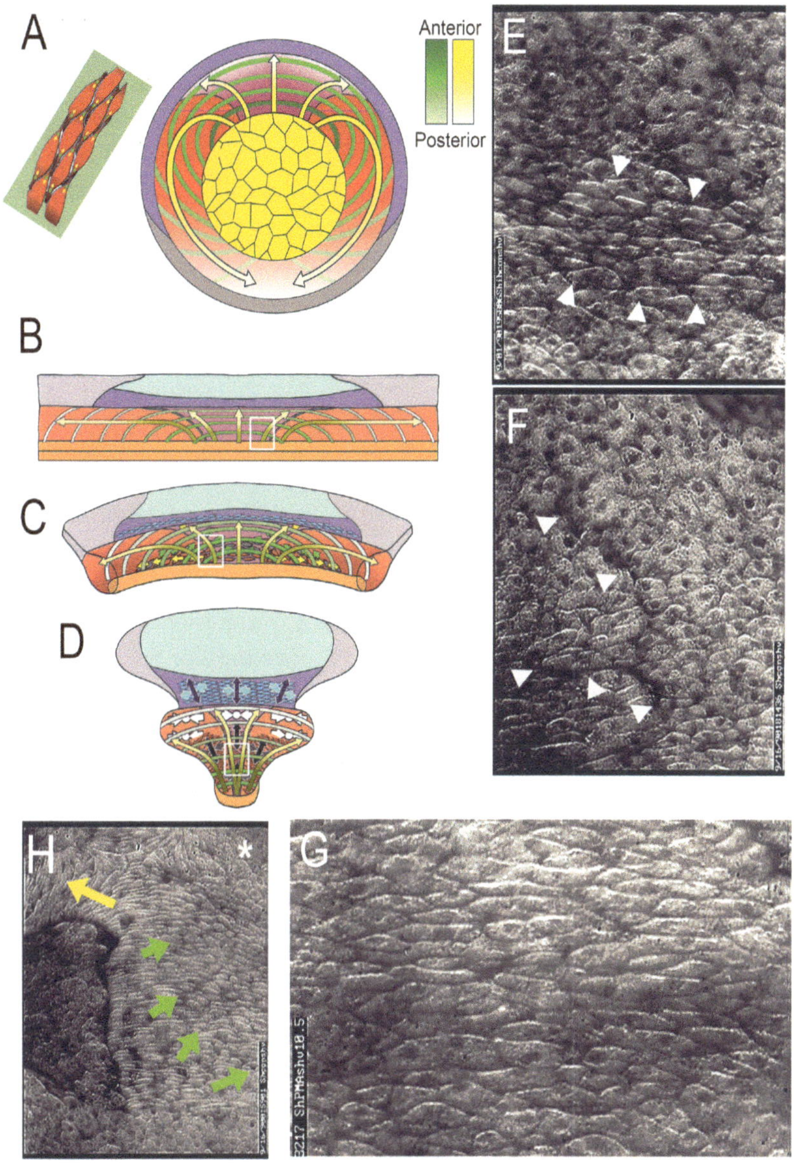

Fig. 5 The spatiotemporal pattern of expression of mediolateral intercalation behavior (MIB) is a continuous progression from presumptive anterior (yellow arrows, dark ends) to presumptive posterior (yellow arrows, light ends) mapped on to the IMZ (A). It occurs in an arc-like pattern, here illustrated as discrete hoops progressing from early and anterior (dark green) to late and posterior (light green); each hoop is anchored at the vegetal endoderm (yellow), and represents an arc of intercalating deep mesodermal cells

(Continued)

endoderm and arcing across the dorsal IMZ from the presumptive A (darker hoops) to P (lighter hoops). The origin and anchorage of the arc of MIB at the vegetal endoderm in the region of bottle cell formation is particularly interesting. The animal edge of the vegetal endoderm is the source of Nodal signaling (see Luxardi, Marchal, Thomé, & Kodjabachian, 2010). It is the site of strong mechanical anchorage of the mesoderm to the endodermal bottle cells and thus the vegetal endoderm. It is the transition from the lateral border of the CE-expressing (pre-somitic) mesoderm to the boundary of the directed (collective) migration-expressing head, heart, and lateroventral mesoderm at the leading edge of the mesodermal mantle (Keller et al., 2000; Keller, Shih, & Domingo, 1992; Keller & Shook, 2004). Thus, it is a largely uninvestigated nexus of signaling and mechanical importance.

The advancing front of MIB/CE takes as MIB begins at the point of involution and progresses posteriorly from this initiation as the IMZ rolls over the lip. The posteriorward progression probably occurs as a result of a combination of forces. First the straightening effect of the stiffening IMZ tissue (Moore et al., 1995) undergoing progressive MIB/CE from the inside out might tend to pull the IMZ over the lip. Second, it might be pushed from behind by Convergent Thickening (CT), an isotropic convergence and thickening of the preinvolution IMZ driven by transient increase in tissue surface tension (Shook, Wen, et al., 2018). Third, the *tissue*

Fig. 5—Cont'd (upper left (A); The arcs are pushed toward the blastoporal lip by preinvolution CT (Convergent Thickening) and as they involute the arcs shorten in A-P order with the expression of MIB, thereby pulling the archenteron roof and dorsal, somitic and notochordal mesoderm over the vegetal endoderm and closing the blastopore. This pattern was determined by mapping the expression pattern of MIB in large open-faced explants on to the IMZ (B) where A-P progressive arc-shortening results in CE (C-D). Images from low angle epi-illuminated open-faced explants show key features: the early onset of MIB in the vegetal alignment zone at Stage 10.5 (arrow heads in E, and white box in B); the initial, focal formation and subsequent straightening of the notochordal-somitic mesodermal boundary through the intercalating cells of the VAZ shortly thereafter (arrow heads in F, white box in C), a large, elongated, aligned and intercalating population of cells in the early central notochordal field (G, and white box in D). and a late neurula/early tailbud explant in which CE was retarded with excessive coverslip pressure, which results in partial progression of MIB and the subsequent vacuolation of the notochordal cells (G). Note the progression of vacuolation from the lateral boundary of the notochord posteriorly (yellow arrow) and medially from this lateral origin (green arrows), and undifferentiated, blebbing, rounded cells at the posterior medial notochordal field (asterisk). See references Shih and Keller (1992a, 1992b).

separation behavior (Wacker, Grimm, Winklbauer, Joos, & Winklbauer, 2000) must operate to support the posterior lengthening of the Cleft of Brachet, itself a complex process of Eph/Ephrin function downstream of many regulatory inputs (Medina, Swain, Kuerner, & Steinbeisser, 2004; Olivia Luu et al., 2015; Rohani, Canty, Luu, Fagotto, & Winklbauer, 2011; Winklbauer, Medina, Swan, & Steinbeisser, 2001). The mediolateral convergence of IMZ tissue in this arc-like, laterally anchored pattern results in narrowing and lengthening of the gastrocoel/archenteron roof, elongating the dorsal body axis, and closing the blastopore in one stroke. Note that the MIB-driven shortening of the arcs pulls the left and right vegetal margins (where they were attached to the vegetal endoderm *in vivo*) in an arc back toward the midline, thus producing CE. *In vivo*, however, the ends of the arcs are anchored in the vegetal endoderm and thus their shortening by MIB *pulls the dorsal tissues vegetally, over the vegetal endoderm* until the blastopore is closed. Note also that the degree of CE increases toward the posterior that posterior parts of the presumptive A-P axes in the notochord and somitic mesoderm (the arrow heads) are brought together as the Limit of Involution converges and closes the blastopore. Lastly, the arcs of convergence are truncated and not continuous with those in the notochord in the lateral and ventral regions IMZ (Fig. 5A). This is because the notochord shears posteriorly with respect to the somitic mesoderm, which shows less extension and is lengthened, in part, by addition of cells streaming around both sides of the posteriorly moving blastoporal region (Keller et al., 1989; Wilson, 1991; Wilson et al., 1989); this posterior shearing acts as a spline or "zipper" that pulls the two sides of the blastopore together. In the giant explant, the posteriorly moving "front" of activation MIB/CE is at the concave, neck-down point of the explant (Fig. 5D), and in the embryo, it lies just inside the blastoporal lip and proceeds posteriorly into the freshly involuting tissue. Again, CT occurs in the preinvolution IMZ, and in the case of the early involuting mesoderm, it transitions into animally-directed migration of head, heart and lateroventral mesoderm (not shown in Fig. 5A), and in the case of the late involuting mesoderm, CT transitions into CE of notochordal and somitic mesoderm at the point of involution (Shook, Wen, et al., 2018).

Understanding the A-P patterning of MIB is a fundamental problem. The participating cells do not appear to be strictly predetermined to express MIB at a particular time and place, as labeled notochordal cells from various A-P and M-L regions of the early gastrula scattered randomly by transplantation into unlabeled, host IMZs, adopt MIB in the normal A-P and L-M progressions, although with some delay, suggesting a response to some endogenous, ongoing signaling process (Domingo & Keller, 1995). The extent and progression of CE in *Xenopus* is dependent on the activity of Nodal and its inhibitor, Lefty (Branford & Yost, 2002). Luxardi and others (Luxardi et al., 2010) present evidence that distinct *Xenopus* Nodal Related (XNRs) signals act sequentially *in vivo* in two phases. In the early phase, XNR5 and XNR6, acting together, induce mesendodermal tissue differentiation, and in the late phase XNR1 and XNR2 regulate morphogenic movements of gastrulation, including CE, by regulating effector genes *papc, has2 and pdgfrα, independent of the Wnt/PCP pathway.* Interestingly, they found that the Wnt/PCP pathway was not regulated by the Nodal pathway, and they propose a parallel regulation in which Nodals activate cell movement effectors and Wnt/PCP polarizes the activities of these effectors.

Graded Nodal signaling along the A-P axis, and the resulting differences in A-P positional identity appear to be necessary for MIB (Ninomiya, Elinson, & Winklbauer, 2004); the nature of this positional effect is unknown. Both the L-M and A-P progressions of MIB may require positive feedback. In large explants that are compressed and not allowed to move significantly, expression of MIB fails to reach presumptive posterior midline of the notochordal field nor does it reach the presumptive posterior somitic tissue (Shih & Keller, 1992b), suggesting that CE potentiates its own progress, perhaps by bringing tissues in the range of the signaling source. Additionally, there appears to be an A-P/early-late discontinuity in the mechanism of MIB regulation, as there is an early critical period before the formation of the VAZ (Stage 10.5) in which microtubules (Lane & Keller, 1997) and the activity of a RhoGef (Kwan & Kirschner, 2005) are essential for further progress of MIB, but not thereafter.

The geometry and progressive shortening of the hoop pattern of MIB is essential for proper involution and blastopore closure. Inhibition of Lefty, a Nodal feedback inhibitor, results in excessive rate and extent of expression of CE, beyond its usual range, and causes defects in involution and blastopore closure (Branford & Yost, 2002). Lithium treatment of the blastula results in activation of the Wnt pathway, expression of "Spemann Organizer" properties, and expression of CE uniformly throughout the entire IMZ, which often results in failure of involution and extension of a proboscis of converging/extending tissue (Kao & Elinson, 1988). This likely occurs because the asymmetric, vegetally anchored, hoop-like pattern of progressively expressed MIB in the normal embryo is transformed into a symmetric, concentric pattern, unanchored to the vegetal endoderm, and perhaps showing little or no progressivity of MBI expression. Rapid and symmetric expression of CE would likely result in an outward extension of tissue too stiff to bend over the lip.

The alignment of the elongated fusiform deep mesodermal cells expressing MIB may be related to differences between their A-P surfaces. The PCP protein Prickle (Pk) is enriched on the anterior surface and another, Dishevelled (Dvl) is enriched on the posterior surface (Shendo, 2018). The A-P surfaces appear to be important for cell alignment and intercalation; apposition of two notochords in dorsal explants in parallel orientation results in rapid loss of the fibronectin matrix between them and the cells intercalate to form one, longer notochord. In contrast, juxtaposing two notochords in anti-parallel configuration results in loss of matrix and a visible boundary, but cells from opposite sides will not intercalate or mix across the boundary of apposition (Flournoy, 2009). The difference is that in the parallel orientation, A-surfaces face P-surfaces and in the anti-parallel case, A-surfaces face A-surfaces and P-surfaces face P-surfaces. At the ends of the recombinants, the cells reorient, as if they could read A-P positional values. On one end, they turn their A-surfaces facing the A end and their P-surfaces facing the P end, and on the other end they turn the opposite way, but in both cases their A-surfaces now facing P-surfaces of cells the other half, and they intercalate and form axes transverse to the originals (Flournoy, 2009). These behaviors suggest that anterior and posterior surfaces must face the opposite surfaces of other cells for intercalation to occur. If and how this is related to A-P differences, including the localization of Prickle and Dvl, in frog and other species, is not clear. Also, how do offsets of A-P positional identity regulate expression of MIB (Ninomiya et al., 2004)? Cells in neurula stage explants show a PCP-dependent oscillatory actomyosin contraction that is out of phase in sequential A-P rows of cells (Shindo, Inoue, Kinoshita, & Wallingford, 2019). The function of this phenomenon, and whether it occurs in the early phase of deep mesodermal intercalation in the last half of gastrulation-early neurulation is not known.

13. Patterning of mesodermal CE relative to other landmarks and presumptive tissues

A common misunderstanding is the perception that the tissue around the lateral sides of the blastopore constitutes the "dorsoventral axis." In fact, it is the physical A-P axis of the somitic mesoderm, and the dorsalization event that occurs along this axis is actually the recruitment of late involuting cells from the ventral side of the blastopore into becoming "dorsal" somitic mesoderm at progressively more posterior levels (Lane et al., 2004; Lane & Keller, 1997; Lane & Sheets, 2000, 2006). The future lateral (or "ventral") aspect of the somitic mesoderm is at the ends of the arcs of convergence next to the vegetal endoderm and the medial (or "dorsal") aspect of the somitic mesoderm is next to the presumptive notochordal boundary (Fig. 5A). The "ventral tissues," the head, heart and lateroventral mesoderm, lie vegetally in the IMZ, involute early (first) and progressively, in that order, whereas the presumptive "dorsal" mesoderm is in the animal part of the IMZ, involutes late, and forms the notochordal and somitic tissues (Keller & Shook, 2004). The early involuting mesoderm at the ventral side of the vegetal endoderm escapes dorsalization as it involutes early, and the late involuting, presumptive posterior somitic tissue lying at the "ventral" side of the blastopore of the mid-gastrula is subject to dorsalizing signals and moves around both sides of the closing blastopore, shears along the side of the notochord, and joins the posterior end of the segmental plate as somitic mesoderm (Keller et al., 1989; Wilson et al., 1989). In the "giant," sandwich explant, CE proceeds to completion and most of the far ventral (lateral in the explant) regions are swept up into the somitic mesodermal files but in open-faced giant explants, friction retards movement, CE does not go to completion, and only about two-thirds of the somitic mesoderm forms (Shih & Keller, 1992b). It is popular to use the original fate maps (Keller, 1975, 1976), perhaps because they better fit simple models of movement and patterning, but the revisions cited above are important changes and better guides going forward.

14. Neural cell intercalation and CE

The neural plate of *Xenopus* undergoes CE in parallel with the underlying mesoderm, matching its behavior quite closely (Keller, 1975, 1976, 1991). In urodele amphibians, the extension of the neural tissue occurs in large measure by mechanical coupling (attachment) to the underlying, elongating notochord (Jacobson & Gordon, 1976), but in *Xenopus* presumptive neural tissue in the sandwich explant shows robust CE without underlying mesoderm (Keller & Danilchik, 1988) (Fig. 1F–H). Neural CE is induced by planar signals emanating from the posterior mesoderm (the Organizer) (Doniach et al., 1992; Keller, Shih, Sater, & Moreno, 1992; Poznanski & Keller, 1997), and the extending neural tissue is patterned in proper A-P order as shown by neural markers (Distaless, Otx2, Krox20, HoxB1, HoxB9)

(Doniach et al., 1992; Poznanski & Keller, 1997). Without contact with the underlying notochord the notoplate/floorplate does not form by molecular markers or by morphology (Keller, Shih, & Sater, 1992; Poznanski & Keller, 1997). Patches of fluorescently labeled cells grafted into the these extending neural structures *in vivo,* or in explants, are dispersed along the A-P axis with increasing separation with distance posteriorly, indicating progressively greater CE and mediolateral cell intercalation, as in the mesoderm (Keller & Tibbetts, 1989). Although the deep mesoderm expresses MIB and CE in open-faced, gastrula explants, the deep neural region does neither in early gastrula, open-faced explants, whereas early neurula deep cell explants do (see below). The gastrula stage deep cells may require mechanical assistance from an actively extending epithelial layer (Butler & Wallingford, 2018) or physiological support.

As in the case of mesodermal CE (Section 5), neural CE in *Xenopus* is dependent on the PCP pathway as is neural tube closure (Wallingford & Harland, 2001, 2002). Grafts of labeled, normal animal cap into the presumptive neural region of Stage 11 gastrulae results in its induction to form neural tissue and it undergoes CE, neural tube closure, and normal body axis elongation. However, similar grafts expressing XDsh-D fail to show CE, fail to close the neural tube, and the body axis has an extreme dorsally concave flexure, as expected if the neural component failed in extension coupled to the extending mesoderm. Disrupting *Xenopus* Dishevelled (XDsh) with Xdd-1 (affecting canonical and non-canonical Wnt signaling) and XDsh-2 (non-canonical only) results in failure of neural tube closure in half or more of the embryos. Targeting inhibition of XDsh to the medial and lateral parts of the presumptive neural plate showed that is needed for CE but not neural fold formation nor for their medial movement. XDsh inhibition prevented CE of both the midline region and the lateral regions of the neural plate followed XDsh-2 inhibition, but neural tube closure was blocked only by the midline perturbation. The conclusion was that neural tube closure failed because the folds were too far apart due to failure of CE to sufficiently narrow the midline tissue (Goto & Keller, 2002; Wallingford & Harland, 2001, 2002). On second thought, a broader neural plate, with more cells across its mediolateral extent, would not necessarily compromise meeting of the folds, since, as with barrel staves, the more units (cells or staves) there are, the less angle of "wedging" required by each for closing the circle. The PCP pathway also compromises events on the radial axis (Ossipova, Chu, et al., 2014) and apical constriction and cell wedging (Ossipova, Kim, et al., 2014).

The epithelial layer of the *Xenopus* neural plate shows cell intercalation by the ERJ mechanism, through the shortening of mediolaterally oriented junctions and subsequent expansion of non-mediolaterally oriented junctions (Butler & Wallingford, 2018) (Fig. 4A and E). The PCP proteins Prickle2 and Vangl2 were found to be enriched in shortening junctions and depleted in the expanding junctions, corrected for concentration due solely to shortening. FRAP studies revealed that the stable fraction of both proteins was higher in shortening junctions and in proportion to the rate of shortening. Imaging a GFP fusion to the myosin regulatory light chain Myl9 showed enrichment of myosin at the shortening junctions, an enrichment that was lost with expression of inhibitory Xdd1 and P2-ΔPΔL (Wallingford & Harland, 2002).

The deep cells of the neural plate also converge on the midline as the plate narrows, and their behavior can be observed in neural plate explants with the overlying neural epithelium removed (Elul et al., 1997; Elul & Keller, 2000) (Fig. 1C–N). The notoplate/floorplate differentiates over the underlying notochord in these explants, and with the notoplate present, deep neural cells display a monopolar, medially-biased protrusive activity toward the midline notoplate. When they reach the notoplate-neural plate boundary, they stop and the cells behind them intercalate between their now-stalled medial neighbors and thus contribute to CE of the neural plate (Fig. 4A and D) (Elul & Keller, 2000). Cells contacting the notoplate boundary rarely, if ever, leave, supporting a "boundary capture" mechanism as proposed by Jacobson and others (Jacobson & Moury, 1995). This ensures maximum efficiency of intercalation, that is, nearly all the intercalations are productive in producing CE (Shih & Keller, 1992b). Regulation of this behavior may involve two signals. Removing the midline notochord/notoplate region of a neural plate and abutting an ectopic notochord/notoplate the other side results in the cells near the ectopic notochord expressing the normal monopolarity and orienting toward the ectopic notochord. In contrast, cells farther away express monopolarity but orient toward the "ghost" midline that had been removed, suggesting that a persistent midline signal orients any monopolarized cell in the field. In contrast, monopolarization itself requires a short-lived signal from the midline (Ezin, Skoglund, & Keller, 2003). Testing the notochord and notoplate independently suggests that the notoplate by itself can polarize cells nearby but not farther away, whereas the combination of notochord and notoplate does both (Ezin, Skoglund, & Keller, 2006).

The isolated, explanted deep cell layer of neural plate (Fig. 1O) shows relatively little CE compared to deep neural layer isolated with underlying mesoderm. Live imaging revealed vigorous, bipolar protrusive activity when averaged over time, but at any given time they were monopolar and indecisive, constantly alternating between medially and laterally directed protrusion, displacement, and intercalation, exchanging places with one another rather than intercalating to produce CE (Elul & Keller, 2000) These observations underscore, first, that the frequency of neighbor changes, even if oriented, does not necessarily result in CE. CE instead depends more on generating mediolaterally oriented tensile force, which compresses the tissue mediolaterally, and thus brings about convergence, rather than a simple positional exchange. Second, if the bipolar mode is used, traction must be approximately balanced at the two ends at any one time; if it is unbalanced, the cell will simply make canceling excursions first one way and then the other. Finally, the unbalanced bipolar mode may be an atavistic behavior, a remnant of a former bipolar mode, elements of which are used for the current monopolar mode and uncovered in its absence. Morphogenic mechanisms are not *de novo*, complete, unitary machines but composites of new elements added to the evolutionarily preceding mechanisms, which have since been modified to their new role in evolutionary time.

15. Forces and mechanics of the progressive expression of mesodermal MIB

CE of the notochordal, somitic and the posterior neural tissues close the blastopore and elongate the dorsal side of the embryo in pre-tailbud stages by producing a tensile convergence force and the compressive extension force to accomplish this transformation. "Giant Sandwich Explants" of the entire IMZ/NIMZ mounted in a tensile force measuring device measures the total tensile convergence force of both mesodermal and neural CE (Shook, Kasprowicz, et al., 2018). Convergence force rises at the onset of gastrulation to 1.5–2.0 μN, when it plateaus through the end of gastrulation, and then rises to over 4.0 μN by the tailbud stage. The convergence force generated before Stage 10.5, and thus before the onset of CE, is generated by CT, an active, force-producing process acting in the preinvolution IMZ, as described above (Shook, Wen, et al., 2018). At Stage 10.5 (early mid-gastrula), MIB/CE begins but contributes increased force progressively as the process recruits more cells. The force plateaus temporarily as less stiff lateral tissues undergoing CT, connected to the stronger, stiffer midline region of CE/MIB, begin to undergo strain (stretch). This interpretation is confirmed by the fact that the plateau is lost in older explants having undergone the CT to CE transition, or in dorsal IMZ explants, which lack the intervening CT expressing region (Shook, Kasprowicz, et al., 2018), thereby directly connecting the CE/MIB region to the force measuring machine. *In vivo*, CE occurs postinvolution and is directly connected to the vegetal endodermal boundary region, without an intervening region of CT, which occurs preinvolution *in vivo* (Shook, Wen, et al., 2018). It is important that CT and CE occur in parallel, the first outside and the second inside the blastoporal lip, rather than serially, as in the explant, as any break in the arc-pattern of MIB/CE results in catastrophic failure of gastrulation (Keller, 1981; Schechtman, 1942). Direct connection of MIB/CE in older or dorsal explants probably approximates the *convergence force at the blastoporal lip* but underestimates the *total force along the dorsal axis* generated *in vivo*, however, as both neural and mesodermal tissues extend away from the axis of measurement; thus they pull on the measuring platters at an increasingly steep angle and thus less efficiently. The same is true *in vivo* with regard to the blastoporal lip, and thus these measurements may approximate the force at the blastoporal lip. Force of blastopore closure was measured at 0.5 μN with a dual cantilever device, and the structural stiffness of the tissue increased by 1.5-fold (Feroze, Shawky, von Dassow, & Davidson, 2015; also see this work for maps of cellular strain). In summary, the IMZ/NIMZ generates constant tensile convergence force for many hours. The MIB/CE component of this force increases as more cells undergo MIB in parallel, and it is transmitted to the ventral ends of the explant with increasing efficiency as MIB progresses toward the ends (Shook, Kasprowicz, et al., 2018) Accordingly, morpholino knockdown of Myosin IIB (Skoglund et al., 2008) or reduction of MRLC activity (Pfister et al., 2016) results in proportional failure of blastopore closure (Skoglund et al., 2008), altered behavior of the NCN-actin dynamics (Pfister et al., 2016; Skoglund et al., 2008), and generation of less convergence force (Shook, Kasprowicz, et al., 2018).

Early experiments showed that the sandwich explant can push (Moore, 1994) and that the tissues undergoing CE stiffen from about 3 Pa to 10 Pa from the early to mid-gastrula (Moore et al., 1995). Recent work showed that dorsal isolates, consisting of the complete neural and axial/paraxial mesodermal/endodermal components, stiffen from less than 20 Pa to over 80 Pa from gastrulation to the end of neurulation (Stage 22) (Zhou et al., 2009).

This is a key factor in the morphogenesis of this period, as tissue must stiffen in order to push, and the stiffening is more dependent on the cortical actomyosin cytoskeleton than on matrix (fibronectin fibril) assembly in this period (Zhou et al., 2009). These investigations also employed the superposition principle of analysis of composite materials, paired with microsurgical construction of explants of different compositions (normal explants, ones with no notochords, with two notochords, etc.), to estimate the relative contributions of the different tissues to embryonic stiffness (Zhou et al., 2009). The results showed that midline tissues are stiffer than lateral tissues, but of the midline notochordal and paraxial tissue, stiffness was the same with and without notochord, and medial paraxial mesoderm adjacent to the notochord was the stiffest tissue. Explants lacking neural tissue did not differ significantly from the whole dorsal isolate, but ones lacking endodermal tissue were stiffer, meaning that the endoderm is the weakest tissue (Zhou et al., 2009). Using the composite analysis, these investigators concluded that compared to the dorsal isolate as a whole, the endoderm is the least stiff by 10-fold, the neural and notochordal tissues are similar, and similar to the dorsal isolate as whole, and the paraxial mesoderm is the stiffest at near twofold that of the isolate as a whole. Interestingly, they note that this tissue undergoes major changes in cellular architecture, notably the thickening and formation of the "buttresses" against the underside of the neural plate, due to a columnarization component to somitic convergence (Keller et al., 1989; Wilson et al., 1989). These buttresses have been proposed to aid folding of the neural plate (Schroeder, 1970). However, cell size and tissue architecture seem to make less contribution to the bulk mechanical properties of these embryonic tissues than the cortical actin cytoskeleton (Shawky, Balakrishnan, Stuckenholz, & Davidson, 2018), and this stiffness, and thus the capacity of the medial paraxial mesoderm to generate an extension force, may be related to the accumulation of pMLC, a regulator of the actomyosin cytoskeleton at the apposed regions of the pre-myocoel surfaces of this tissue (Zhou et al., 2009).

These results suggest that the paraxial mesoderm may be the major contributor to CE forces driving both the convergence and the extension in this period of neural tube closure, a suggestion supported by the fact that isolated pre-somitic mesoderm will show CE whereas isolated notochord does not (Keller, 1978; Wilson, 1991). Other studies using a gel-based force sensor show that the mean stress (force) produced by CE is 5.0 ± 1.6 Pa, and again, the notochord is not necessary. By increasing the gel stiffness, they found that dorsal isolates accommodate to the mechanical environment by increasing stress production in stiffer gels (Zhou et al., 2015). In stiff gels, the notochord was curved and mushroomed at the rear, suggesting involvement force generation but again tests of this parameter with and without notochords revealed no difference.

16. Evaluating the contributions of the epithelial and deep layers to forces driving CE

This issue is complicated by the fact that the two layers may have both signaling and mechanical interactions. Dorsal sandwich explants (Fig. 1B, F and G), in which the epithelial layer is replaced with animal cap epithelium, a region that does not undergo CE, nevertheless show CE, and so do controls having the native epithelium removed and replaced, but neither do so as well as intact explants (Keller & Danilchik, 1988). This argues that the deep, mesenchymal layer can drive the CE of itself and that of an attached epithelium, which does not normally undergo CE in its native position but does not rule out force generated by the native epithelium of the IMZ when operating *with* its deep layer. The contribution of the epithelium alone, independent of the deep region, could not be evaluated because of the instability of epithelial sheets cultured alone and without adhesion to substrates. CE of early dorsal explants fails on removal of the epithelium early but not later (Wilson, 1991), suggesting a transient, early effect, perhaps the radial specification of matrix deposition to particular interfaces (Goto et al., 2005). In other situations, ongoing contact with an epithelium, either endodermal or ectodermal, is essential for radial intercalation and CE of the somitic mesoderm (Wilson et al., 1989; Wilson, 1991). The epithelial contribution could perhaps be assayed by a composite analysis of mechanical properties (see Zhou et al., 2009).

Contributions of the epithelial layer may not be due solely to the EJR mechanism of cell intercalation. The epithelial sheets of the nematode (Williams-Masson et al., 1998), the mouse (Williams et al., 2014), and the *Drosophila* germband (Sun et al., 2017) have basolateral protrusive activity, and in all these situations, neighbor exchange, the "resolution" step of the basolateral end of the cells, occurs shortly or substantially before that of the apical region, suggesting that basolateral protrusive activity and cell traction may play the major role in driving epithelial cell intercalation (see below). Epithelial tissues of the *Xenopus* gastrula may use basolateral protrusive activity and cell traction, as the mesodermal, endodermal and ectodermal epithelia of the early embryo have basolateral protrusions (Keller, 1980; Keller et al., 1989; Keller & Schoenwolf, 1977).

This raises the question of how much tissue-level force the EJR can produce. In the *Drosophila* germband, the initial contraction of the shortening boundary that brings cells to a common vertex (green arrows Fig. 3E, left) is followed by a medial (centripetal) actomyosin contraction in the same cells that elongates the transverse boundary (red arrows, Fig. 3E, middle and left), presumably passively stretching it (black arrows, Fig. 3E, right), and thus accomplishing "resolution" and elongating the array of cells in the orthogonal axis (Collinet et al., 2015). This process normally produces elongated arrays of cells that are aligned in the axis of germband extension, and thus they contribute to the extension, However, if the posterior midgut invagination is prevented, the germband does not elongate properly and the local elongated arrays become misaligned and compressed (Collinet et al., 2015). These results are a welcomed illumination of the mysterious "resolution" step, and they suggest that in this case the EJR functions to remodel apical junctions, but externally generated force, in this case tension produced by the posterior midgut invagination, is required for elongation or CE. This work also suggests that the basolateral protrusions of these particular cells (Sun et al., 2017) do not generate sufficient force to bring about tissue-level CE. The broader question is whether the EJR mechanism functions primarily in remodeling junctions, a necessary component of CE by cell intercalation (Fristrom, 1982), but the major stress driving tissue strain during CE is generated by basolateral CCT.

17. The role of tissue boundary formation, Eph/Ephrin signaling, and tissue surface tension in CE

For CE to occur properly, the boundary properties on all sides of the tissue must be regulated to maintain physical integrity and to manage tissue surface tension (Ninomiya & Winklbauer, 2008). As the IMZ involutes, its deep cells come into contact with the inner surface of the blastocoel roof, and they display *tissue separation behavior* (*TSB*), meaning that they will not reintegrate with the cell layers they just left. They thus maintain a clear separation boundary, called the Cleft of Brachet between the outer neural ectoderm/epidermis and the inner mesoderm (Wacker et al., 2000). TSB is assayed by placing pre and post involution tissues on the inner surface of the blastocoel roof; the preinvolution ones will reintegrate into the roof, the postinvolution ones will remain separate (Wacker et al., 2000).

TSB involves non-canonical Wnt signaling through the Wnt receptor, XFizzled-7, paraxial protocadherin (PAPC), and the ankyrin repeat domain protein 5 (xANR5), and finally, activation of RhoA and Rho kinase (Medina et al., 2004; Olivia Luu et al., 2015; Rohani et al., 2011; Winklbauer et al., 2001). Ephrin B ligands and EphB receptors signal across the tissue boundary in both directions, resulting in a repeated cycle of transient attraction, attachment, followed by repulsion and detachment. This simultaneously maintains a dynamic tissue contact but permits the shearing between the layers that is essential for mesendoderm migration on the roof of the blastocoel, as well as any differential rates of CE in axial/paraxial mesoderm and neural tissue.

The immediate response to removing pre- and postinvolution tissue from the embryo, as in making sandwich or open-faced explants, is a tissue surface tension driven reduction in its newly exposed deep cell surface area. This results in thickening, concave curling at its deep surface, and eventually rounding up and thereby retarding CE. On the other, outer side of the deep mesoderm, the overlying endomesodermal epithelial sheet also has the effect of lowering the tissue surface tension of the underlying deep mesoderm. Lowering tissue surface tension appears necessary to allow CE, a process that involves an increase in surface area. Reducing the epithelial properties by co-expression of lethal giant larvae (Lgl) with atypical protein kinase C (aPKC) results in rounding and decreased CE of tissue aggregates by raising their surface tension, and regulating epithelial contact control fine modeling of deep mesenchymal tissue shape (Ninomiya & Winklbauer, 2008).

Open-faced explants having the native epithelium on one side but lacking the surface tension reducing effect of tissue separation on the deep side, will thicken immediately on separation and must be restrained from doing so by two non-adhesive glass coverslips (Shih & Keller, 1992a, 1992b). This is a limitation of open-faced explant system. Culturing open-faced preparations on FN-coated, rigid substrates such as glass or plastic will usually prevent thickening but that comes with a number of problems and uncertainties. It also promotes unregulated and abnormal migration and spreading of inherently migratory cells, probably because of the extreme stiffness of and avidity of the glass or plastic for binding FN. The same properties likely account for the fact that it greatly retards or stops CE of tissues such as somitic mesoderm and notochord, which employ the bipolar intercalation mode, a mode that would be expected to fail to produce intercalation on an undeformable, adhesive substrate. The extreme stiffness compared to the endogenous flexible fibrillar array, which does thicken and perhaps stiffens over time (Davidson et al., 2004b), could result in a number of developmental changes (Barriga et al., 2018; Discher et al., 2005; Engler et al., 2006; Liua et al., 2010). The second convergence movement in gastrulation, CE, involves an increase in surface area and requires a lowering of surface tension, and the first convergence movement in gastrulation, CT, involves decrease in surface area and is driven, at least in part, by an increase in tissue surface tension due to loss of affinity of the deep mesenchymal IMZ for the overlying mesendodermal epithelium (Shook, Wen, et al., 2018).

18. Late elongation and straightening of the body plan is driven by notochord straightening and elongation, and elongation of the vegetal endoderm

In the early neurula stage, mediolateral intercalation brings the notochord from 4 to 6 rows of cells down to an array consisting of two rows of boundary contacting lateral cells and rare internal, bipolar cells by midneurula (Stages 15 and 16) (Fig. 4B). From there, the cells begin to spread laterally, become thinner in their anterior-posterior dimension to form the "pizza-slice" stack of notochordal cells (Fig. 4B), which then take up water in vacuoles, thereby pressurizing a "hydrostat" (Adams et al., 1990). Meanwhile the surrounding matrix thickens and differentiates in composition and structure (Buisson et al., 2014; Trapani, Bonaldo, & Corallo, 2017). The notochord then straightens and elongates as a function of the pattern of its fiber-wound matrix (Adams et al., 1990; Koehl, Quillin, & Pell, 2000) and thereby contributes to the lifting, straightening and elongation of the tail (Fig. 1D and E). Mechanical models illuminate the relationship between the fiber-winding angle, hydrostatic pressure, and change in shape of the notochord (Koehl et al., 2000). The epithelioid mode of notochord intercalation with its broad panel of septin-confined, actomyosin activity (Kim et al., 2010; Shindo & Wallingford, 2014) (Fig. 3F) fits well with the transition to the circumferential spreading that forms the "pizza-slice" cells, and highlights the importance of further work in this period, which involves further differentiation of matrix, including assembly and distribution of fibronectin, fibrillin, elastin, and laminin around the notochord (Skoglund et al., 2006; Skoglund & Keller, 2007; Trapani et al., 2017). In addition, the complex "adhesome" consisting of Dystroglycan (Dg), laminin, and myosin IIA functions in adhesion, actin organization, polarity, and finally, vacuolation in this period (Buisson et al., 2014), a key step in forming the pressurized fiber-wound hydrostat (Koehl et al., 2000).

19. Late endoderm elongation

Notochord straightening is accompanied by elongation of the vegetal endoderm on the ventral side (Fig. 1D and E), a process that has been attributed to the lateral plate mesoderm in *Xenopus* (Larkin & Danilchik, 1999) and to the endoderm in urodeles (Drawbridge & Steinberg, 2000). Explants and cell labeling experiments on the urodele (tailed) amphibian, *Ambystoma mexicanum*, shows that the vegetal endoderm actively elongates, a process requiring signals from the blastopore region (Drawbridge & Steinberg, 2000). Fate mapping of *Xenopus* gut formation shows that the original archenteron is nearly lost due to closure and the endoderm forms the gut tube through subsequent outward radial intercalation of cells toward a "boundary capture" at the surrounding basal lamina (Chalmers & Slack, 2000, 1998). Similar to the cell intercalation seen at the mesoderm and the neural tissue, this intercalation is Rho/ROCK/Myosin II and PCP (Vangl2) dependent, and it forms the definitive cavity and elongates the gut in one stroke (Chung, Nascone-Yoder, Grover, Drysdale, & Wallingford, 2010; Dush & Nascone-Yoder, 2013, 2019; Reed et al., 2009). This represents yet another example of the nearly omnipresent process of CE in early development.

20. CE as a large-scale mechanical patterning mechanism

CE strains (stretches) tissues that otherwise do not show strain and thus shapes the embryo, but this strain is also used in patterning aspects of the embryo, how many is yet to be determined. The presumptive epidermis shows a pattern of strain toward the closing blastopore in the late gastrula and early neurula (Keller, 1975), and later in the tailbud stage, the epidermal multiciliated cells generate fluid flow along this pattern (Chien, Keller, Kintner, & Shook, 2015). This flow pattern that can be redirected by new micro-aspiration-induced strain, which regulates the polarity of the multiciliated cells in a stage specific manner and involving microtubule alignment and differential turnover of cell adhesion and PCP proteins (Chien et al., 2015). More striking, there is two-dimensional gradient of endogenous CE-generated strain in the posterior roof of the gastrocoel, the left-right organizer (LRO), which regulates the cell polarity, and the apical position, and length of the cilia responsible for LRO function in a strain-dependent manner (Chien, Srinivasan, Keller, & Kintner, 2018). In both the epidermis and the LRO, the PCP pathway is necessary for strain-mediated patterning, but the essential, large-scale coordinator is the pattern of mechanical strain. The key parameter in these experiments is cellular strain, and the tissue strain applied by large-scale CE movements to other, passive tissues is translated into cellular strain by the cells' resistance to passive intercalation. As the tissue is strained, cells that slide by one another easily would passively rearrange and would show little or no cellular strain, whereas those that have higher resistance to passive rearrangement, will show cellular strain increasing in proportion to their resistance [see Blanchard et al. (2009) for an excellent discussion]. One could imagine that cells could differentiate resistance to rearrangement at specific developmental periods, to specific degrees, and in specific places, and thus produce a temporally and regionally specific cellular strain in response to a given tissue strain, and respond to this strain level with regulatory changes.

21. Outlook

Many advances have and are being made on understanding CE in *Xenopus* early development, the most significant ones being in resolving the cell biology and biomechanics of key underlying processes. Several problems stand out as challenging but important ones. The signaling processes that underlie the spatiotemporal progressions of MIB are key to understanding how the activities of large populations of cells are coordinated to produce the elegant sweep of MIB/CE across the embryonic landscape. Second, we know too little about the generation and magnitude of the forces that are contributed by the various tissues involved, particularly how local and global forces and mechanical properties are tuned to one another. Third, it seems likely that the forces generated by the progression of force-generating cellular processes and properties feedback, positively or negatively, to regulate their own progression and spatial patterning, and this needs to be understood. Fourth, it is important to understand the regulation of the radial and mediolateral cell behaviors and integration of the forces they generate during CE. Finally, it is important to understand that the specificity of morphogenic movements emerges from contextual information (mechanical and physiological) at many levels. Local force-cell behaviors can result in many different outcomes depending on their spatial pattern, their mechanical connectivity, their geometries, and their anchor points in the larger system. The old adage that the whole is greater than the sum of its parts has turned out to be so true.

References

Adams, D. S., Keller, R., & Koehl, M. A. (1990). The mechanics of notochord elongation, straightening, and stiffening in the embryo of Xenopus laevis. *Development*, *110*, 115–130.

Barriga, E., Franze, K., Charras, G., & Mayor, R. (2018). Tissue stiffening coordinates morphogenesis by triggering collective cell migration *in vivo*. *Nature*, *554*, 523–527. https://doi.org/10.1038/nature25742, Epub 2018 Feb 14.

Belmonte, J. M., Swat, M. H., & Glazier, J. A. (2016). Filopodial-tension model of convergent-extension of tissues. *PLoS Computational Biology*, *12*(6), e1004952. https://doi.org/10.1371/journal.pcbi.1004952.

Bissen, S. T., & Weisblat, D. A. (1989). The durations and compositions of cell cycles in embryos of the leech, *Helobdella triserialis*. *Development*, *106*, 105–118.

Blanchard, G. B., Kabla, A. J., Schultz, N. L., Butler, L. C., Sanson, B., Gorfinkiel, N., et al. (2009). Tissue tectonics: Morphogenetic strain rates, cell shape change and intercalation. *Nature Methods*, *6*, 458–464.

Blankenship, J. T., Backovic, S. T., Sanny, J. S. P., Weitz, O., & Zallen, J. A. (2006). Multicellular rosette formation links planar cell polarity to tissue morphogenesis. *Developmental Cell*, *11*, 459–470.

Branford, W. W., & Yost, H. J. (2002). Lefty-dependent inhibition of nodal- and Wnt-responsive organizer gene expression is essential for normal gastrulation. *Current Biology*, *12*, 2136–2214.

Brieher, W., & Gumbiner, B. (1994). Regulation of C-cadherin function during activin induced morphogenesis of *Xenopus* animal caps. *The Journal of Cell Biology*, *126*, 519–527.

Brodland, G. W. (2006). Do lamellipodia have the mechanical capacity to drive convergent extension. *The International Journal of Developmental Biology*, *50*, 151–155.

Buisson, N., Sirour, C., Moreau, N., Denker, E., Bouffant, R., Goullancourt, A., et al. (2014). An adhesome comprising laminin, dystroglycan and myosin IIA is required during notochord development in Xenopus laevis. *Development*, *141*, 4569–4579.

Butler, M. T., & Wallingford, J. B. (2017). Planar cell polarity in development and disease. *Nature Reviews. Molecular Cell Biology*, *18*, 375–388. https://doi.org/10.1038/nrm.2017.11.

Butler, M. T., & Wallingford, J. B. (2018). Spatial and temporal analysis of PCP protein dynamics during neural tube closure. *eLife*, *7*, e36456. https://doi.org/10.7554/eLife.36456.

Chalmers, A. D., & Slack, J. M. (1998). Development of the gut in *Xenopus laevis*. *Developmental Dynamics*, *212*, 509–521.

Chalmers, A. D., & Slack, J. M. (2000). The *Xenopus* tadpole gut: Fate maps and morphogenetic movements. *Development*, *127*, 1457–1466.

Chien, Y. H., Keller, R., Kintner, C., & Shook, D. R. (2015). Mechanical strain determines the axis of planar polarity in ciliated epithelia. *Current Biology*, *25*, 2774–2784.

Chien, Y.-H., Srinivasan, S., Keller, R., & Kintner, C. (2018). Mechanical strain determines cilia, length, motility and planar position in the left-right organizer. *Developmental Cell*, *45*, 316–330.e4. https://doi.org/10.1016/j.devcel.2018.04.007.

Chung, M.-I., Nascone-Yoder, N. M., Grover, S. A., Drysdale, T. A., & Wallingford, J. B. (2010). Direct activation of Shroom3 transcription by Pitx proteins drives epithelial morphogenesis in the developing gut. *Development*, *137*, 1339–1349.

Collinet, C., Rauzi, M., Lenne, P.-F., & Lecuit, T. (2015). Local and tissue-scale forces drive oriented junction growth during tissue extension. *Nature Cell Biology*, *17*, 1247.

Condic, M. L., Fristrom, D., & Fristrom, J. W. (1991). Apical cell shape changes during *Drosophila* imaginal leg disc elongation: A novel morphogenetic mechanism. *Development*, *111*, 23–33.

Damm, E. W., & Winklbauer, R. (2011). PDGF-A controls mesoderm cell orientation and radial intercalation during *Xenopus* gastrulation. *Development*, *138*, 565–575.

Darken, R. S., Scola, A. M., Rakeman, A. S., Das, G., Mlodzik, M., & Wilson, P. A. (2002). The planar polarity gene strabismus regulates convergent extension movements in *Xenopus*. *The EMBO Journal*, *21*, 976–985.

Davidson, L. A., Keller, R., & DeSimone, D. W. (2004a). Patterning and tissue movements in a novel explant of the marginal zone of *Xenopus laevis*. *Gene Expression Patterns*, *4*, 457–466.

Davidson, L., Keller, R., & DeSimone, D. (2004b). Assembly and remodeling of the fibrillar fibronectin extracellular matrix during gastrulation and neurulation in Xenopus laevis. *Developmental Dynamics*, *231*, 888–895.

Davidson, L., Marsden, M., Keller, R., & DeSimone, D. (2006). Integrin $\alpha5\beta1$ and fibronectin regulate polarized protrusions required for *Xenopus* convergence and extension. *Current Biology*, *16*, 833–844.

Discher, D., Janmey, P., & Wang, Y.-L. (2005). Tissue cells feel and respond to the stiffness of their substrate. *Science*, *310*, 1139–1143. https://doi.org/10.1126/science.1116995.

Djiane, A., Riou, J., Umbhauer, M., Boucaut, J., & Shi, D. (2000). Role of frizzled 7 in the regulation of convergent extension movements during gastrulation in *Xenopus laevis*. *Development*, *127*, 3091–3100.

Domingo, C., & Keller, R. (1995). Induction of notochord cell intercalation behavior and differentiation by progressive signals in the gastrula of *Xenopus laevis*. *Development*, *121*, 3311–3321.

Doniach, T., Phillips, C. R., & Gerhart, J. C. (1992). Planar induction of anteroposterior pattern in the developing central nervous system of Xenopus laevis. *Science*, *257*, 542–545.

Drawbridge, J., & Steinberg, M. S. (2000). Elongation of axolotl tailbud embryos requires GPI-linked proteins and organizer-induced, active, ventral trunk endoderm cell rearrangements. *Developmental Biology*, *223*, 27–37.

Dush, M. K., & Nascone-Yoder, N. M. (2013). Jun N-terminal kinase maintains tissue integrity during cell rearrangement in the gut. *Development*, *140*, 1457–1466.

Dush, M. K., & Nascone-Yoder, N. M. (2019). Vangl2 coordinates cell rearrangements during gut elongation. *Developmental Dynamics*, *248*, 569–582.

Dzamba, B. J., Jakab, K. R., Marsden, M., Schwartz, M. A., & DeSimone, D. W. (2009). Cadherin adhesion, tissue tension, and noncanonical Wnt signaling regulate fibronectin matrix organization. *Developmental Cell*, *16*, 421–432.

Elul, T., & Keller, R. (2000). Monopolar protrusive activity: A new morphogenic cell behavior in the neural plate dependent on vertical Interactions with the mesoderm in Xenopus. *Developmental Biology*, *224*, 3–19.

Elul, T., Koehl, M., & Keller, R. (1997). Cellular mechanism underlying neural convergent extension in *Xenopus laevis* embryos. *Developmental Biology*, *191*, 243–258.

Engler, A. J., Sen, S., Sweeney, H. L., & Discher, D. E. (2006). Matrix elasticity directs stem cell lineage specification. *Cell*, *126*, 677–689.

Ezin, A. M., Skoglund, P., & Keller, R. (2003). The midline (notochord and notoplate) patterns the cell motility underlying convergence and extension of the *Xenopus* neural plate. *Developmental Biology*, *256*, 100–113.

Ezin, A. M., Skoglund, P., & Keller, R. (2006). Presumptive floor-plate (notoplate) induces behaviors associated with convergent extension in medial but not lateral neural plate cells of *Xenopus*. *Developmental Biology*, *300*, 670–686.

Fagotto, F., Rohani, N., Touret, A.-S., & Li, R. (2013). A molecular base for cell sorting at embryonic boundaries: Contact Inhibition of cadherin adhesion by Ephrin/Eph-dependent contractility. *Developmental Cell, 27*, 1–16.
Feroze, R., Shawky, J. H., von Dassow, M., & Davidson, L. A. (2015). Mechanics of blastopore closure during amphibian gastrulation. *Developmental Biology, 398*, 5767.
Flournoy, C. (2009). Anterior-posterior tissue positional identity and anterior-posterior cell polarity maintain the axis of cell intercalation during Xenopus convergent extension (Ph.D. thesis). Charlottesville: University of Virginia.
Fristrom, D. (1976). The mechanism of evagination of the imaginal discs of Drosophila melanogaster. III. Evidence for cell rearrangement. *Developmental Biology, 54*, 163–171.
Fristrom, D. (1982). Septate junctions in imaginal disks of Drosophila: A model for the redistribution of septa during cell rearrangement. *The Journal of Cell Biology, 94*, 77–87.
Gillespie, J. I. (1983). The distribution of small ions during the early development of *Xenopus laevis* and *Ambystoma mexicanum* embryos. *The Journal of Physiology, 344*, 359–377.
Gimlich, R., & Braun, J. (1985). Improved fluorescent compounds for tracing cell lineage. *Developmental Biology, 109*, 509–514.
Glickman, N. S., Kimmel, C. B., Jones, M. A., & Adams, R. J. (2003). Shaping the zebrafish notochord. *Development, 130*, 873–887.
Goto, T., Davidson, L. A., Asashima, M., & Keller, R. (2005). Planar cell polarity genes regulate polarized extracellular matrix deposition during frog gastrulation. *Current Biology, 15*, 787–793.
Goto, T., & Keller, R. (2002). The planar cell polarity gene strabismus regulates convergence and extension and neural fold closure in *Xenopus*. *Developmental Biology, 247*, 165–181.
Habas, R., Dawid, I. B., & He, X. (2003). Coactivation of Rac and Rho by Wnt/Frizzled signaling is required for vertebrate gastrulation. *Genes & Development, 17*, 295–309.
Habas, R., Kato, Y., & He, X. (2001). Wnt/Frizzled activation of Rho regulates vertebrate gastrulation and requires a novel Formin homology protein Daam1. *Cell, 107*, 843–854.
Heisenberg, C.-P., Tada, M., Rauch, G.-J., Saude, L., Concha, M., Geisler, R., et al. (2000). Siberblick/Wnt11 mediates convergent extension movements during zebrafish gastrulation. *Nature, 405*, 76–81.
Huebner, R. J., & Wallingford, J. B. (2019). Coming to consensus: A unifying model emerges for convergent extension. *Developmental Cell, 46*, 389–396.
Jacobson, A. G., & Gordon, R. (1976). Changes in the shape of the developing vertebrate nervous system analyzed experimentally, mathematically, and by computer simulation. *The Journal of Experimental Zoology, 197*, 191–246.
Jacobson, A. G., & Moury, J. D. (1995). Tissue boundaries and cell behavior during neurulation. *Developmental Biology, 171*, 98–110.
Jessen, J. R., Topczewski, J., Bingham, S., Sepich, D. S., Marlow, F., Chandrasekhar, A., et al. (2002). Zebrafish trilobite identifies new roles for Strabismus in gastrulation and neuronal movements. *Nature Cell Biology, 4*, 610–615.
Kao, K. R., & Elinson, R. P. (1988). The entire mesodermal mantle behaves as Spemann's organizer in dorsoanterior enhanced embryos. *Developmental Biology, 127*, 64–77.
Keller, R. E. (1975). Vital dye mapping of the gastrula and neurula of *Xenopus laevis*. I. Prospective areas and morphogenetic movements of the superficial layer. *Developmental Biology, 42*, 222–241.
Keller, R. E. (1976). Vital dye mapping of the gastrula and neurula of Xenopus laevis. II. Prospective areas and morphogenetic movements of the deep layer. *Developmental Biology, 51*, 118–137.
Keller, R. E. (1978). Time-lapse cinemicrographic analysis of superficial cell behavior during and prior to gastrulation in *Xenopus laevis*. *Journal of Morphology, 157*, 223–248.

Keller, R. E. (1980). The cellular basis of epiboly: An SEM study of deep-cell rearrangement during gastrulation in *Xenopus* laevis. *Journal of Embryology and Experimental Morphology, 60*, 201.

Keller, R. E. (1981). An experimental analysis of the role of bottle cells and the deep marginal zone in gastrulation of *Xenopus laevis*. *The Journal of Experimental Zoology, 216*, 81–101.

Keller, R. (1984). The cellular basis of gastrulation in *Xenopus laevis*: Active, postinvolution convergence and extension by mediolateral interdigitation. *American Zoologist, 24*, 589–603.

Keller, R. (1991). Early embryonic development of *Xenopus laevis*. Chapter 5. In B. Kay & B. Peng (Eds.), *Methods in cell biology, Vol. 36, Xenopus laevis: Practical uses in cell and molecular biology* (pp. 61–113). Academic Press, Inc.

Keller, R. E., Cooper, M., Danilchik, M., Tibbetts, P., & Wilson, P. (1989). Cell intercalation during notochord development in *Xenopus laevis*. *The Journal of Experimental Zoology, 251*, 134–154.

Keller, R., & Danilchik, M. (1988). Regional expression, pattern and timing of convergence and extension during gastrulation of *Xenopus laevis*. *Development, 103*, 193–209.

Keller, R., Danilchik, M., Gimlich, R., & Shih, J. (1985). The function and mechanism of convergent extension during gastrulation of *Xenopus laevis*. *Journal of Embryology and Experimental Morphology, 89*(Suppl), 185–209.

Keller, R., Davidson, R., Edlund, A., Elul, T., Ezin, M., Shook, D., et al. (2000). Mechanisms of convergence and extension by cell intercalation. *Philosophical Transactions of the Royal Society of London B, 355*, 897–922.

Keller, R. E., & Schoenwolf, G. C. (1977). An SEM study of cellular morphology, contact, and arrangement, as related to gastrulation in *Xenopus laevis*. *Wilhelm Roux's archives of developmental biology, 182*, 165–186.

Keller, R., Shih, J., & Domingo, C. (1992). The patterning and functioning of protrusive activity during convergence and extension of the *Xenopus* organizer. *Development Supplement, 1992*, 81–91.

Keller, R., Shih, J., & Sater, A. (1992). The cellular basis of the convergence and extension of the *Xenopus* neural plate. *Developmental Dynamics, 193*, 199–217.

Keller, R., Shih, J., Sater, A., & Moreno, C. (1992). Planar induction of convergence and extension of the neural plate by the organizer of Xenopus. *Developmental Dynamics, 193*, 218–234.

Keller, R., & Shook, D. (2004). Gastrulation in amphibians. In C. Stern (Ed.), *Gastrulation*. (pp. 171–204). NY: Cold Spring Harbor Laboratory Press.

Keller, R. E., & Tibbetts, P. (1989). Mediolateral cell intercalation in the dorsal axial mesoderm of *Xenopus laevis*. *Developmental Biology, 131*, 539–549.

Kelley, C., Sellers, J., Gard, D., Bui, D., Adelstein, R., & Baines, I. (1996). Xenopus nonmuscle myosin heavy chain Isoforms have different subcellular localizations and enzymatic activities. *The Journal of Cell Biology, 134*, 675–687.

Kim, H. Y., & Davidson, L. A. (2011). Punctuated actin contractions during convergent extension and their permissive regulation by the non-canonical Wnt-signaling pathway. *Journal of Cell Science, 124*, 635–646.

Kim, S. K., Shindo, A., Park, T. J., Oh, E., Ghosh, S., Gray, R., et al. (2010). Planar cell polarity acts through septins to control collective cell movement and ciliogenesis. *Science, 329*, 1337–1340.

Koehl, M. A. R., Quillin, K. J., & Pell, C. A. (2000). Mechanical design of fiber-wound hydraulic skeletons: The stiffening and straightening of embryonic notochords. *American Zoologist, 40*, 28–41.

Kwan, K. M., & Kirschner, M. W. (2005). A microtubule-binding Rho-GEF controls cell morphology during convergent extension of *Xenopus laevis*. *Development, 132*, 4599–4610.

Lane, M. C., Davidson, L., & Sheets, M. (2004). BMP antagonism by Spemann's organizer regulates rostral-caudal fate of mesoderm. *Developmental Biology*, *275*, 356–374.
Lane, M. C., & Keller, R. (1997). Microtubule disruption reveals that Spemann's organizer is subdivided into two domains by the vegetal alignment zone. *Development*, *124*, 895–906.
Lane, M. C., & Sheets, M. D. (2000). Designation of the anterior/posterior axis in pregastrula *Xenopus laevis*. *Developmental Biology*, *225*, 37–58.
Lane, M. C., & Sheets, M. D. (2006). Heading in a new direction: Implications of the revised fate map for understanding Xenopus laevis development. *Developmental Biology*, *296*, 12–28.
Lang, M. (2008). Lighting up the mechanome. In *Frontiers of engineering: Reports on leading-edge engineering from the 2007 symposium* (pp. 39–47). Washington, DC: The National Academies Press. https://doi.org/10.17226/12027.
Larkin, K., & Danilchik, M. (1999). Ventral cell rearrangements contribute to the anterior-posterior lengthening between neurula and tailbud stages in Xenopus laevis. *Developmental Biology*, *216*, 550–560.
Levayer, R., & Lecuit, T. (2013). Oscillation and polarity of E-Cadherin asymmetries control actomyosin flow patterns during morphogenesis. *Developmental Cell*, *26*, 162–175.
Liua, Z., Tanb, J. L., Cohena, D. M., Yanga, M. T., Sniadeckia, N. J., Ruiza, S. A., et al. (2010). Mechanical tugging force regulates the size of cell–cell junctions. *PNAS*, *107*, 9944–9949.
Longo, D., Peirce, S. M., Skalak, T., Davidson, L., Marsden, M., Dzamba, B., et al. (2004). Multicellular computer simulation of morphogenesis: Blastocoel roof thinning and matrix assembly in *Xenopus laevis*. *Developmental Biology*, *271*, 210–222.
Luxardi, G., Marchal, L., Thomé, V., & Kodjabachian, L. (2010). Distinct *Xenopus* Nodal ligands sequentially induce mesendoderm and control gastrulation movements in parallel to the Wnt/PCP pathway. *Development*, *137*, 417–426.
Marsden, M., & DeSimone, D. (2001). Regulation of cell polarity, radial intercalation and epiboly in Xenopus: Novel roles for integrin and fibronectin. *Development*, *128*, 3635–3647.
Marsden, M., & DeSimone, D. (2003). Integrin-ECM interactions regulate cadherin-dependent cell adhesion and are required for convergent extension in *Xenopus*. *Current Biology*, *13*, 1182–1191.
Medina, A., Swain, R., Kuerner, K.-M., & Steinbeisser, H. (2004). Xenopus paraxial protocadherin has signaling functions and is involved in tissue separation. *The EMBO Journal*, *23*, 3249–3258.
Moore, S. W. (1994). A fiber optic system for measuring dynamic mechanical properties of embryonic tissues. *IEEE Transactions on Biomedical Engineering*, *41*, 45–50.
Moore, S. W., Keller, R. E., & Koehl, M. A. R. (1995). The dorsal involuting marginal zone stiffens anisotropically during its convergent extension in the gastrula of Xenopus laevis. *Development*, *121*, 3131–3140.
Ninomiya, H., Elinson, R. P., & Winklbauer, R. (2004). Antero-posterior tissue polarity links mesoderm convergent extension to axial patterning. *Nature*, *430*, 364–367.
Ninomiya, H., & Winklbauer, R. (2008). Epithelial coating controls mesenchymal shape change through tissue-positioning effects and reduction of surface-minimizing tension. *Nature Cell Biology*, *10*, 61–69.
Olivia Luu, O., Erich, W., Damm, E. W., Parent, S. E., Barua, D., Smith, T. H. L., et al. (2015). PAPC mediates self/non–self-distinction during Snail1-dependent tissue separation. *The Journal of Cell Biology*, *208*, 839–856. www.jcb.org/cgi/doi/10.1083/jcb.201409026.

Ossipova, O., Chu, C.-W., Filatre, J., Brott, B. K., Itoh, K., & Sokol, S. Y. (2014). The involvement of PCP proteins in radial cell intercalations during Xenopus embryonic development. *Developmental Biology, 408*, 316–327.

Ossipova, O., Kim, K., Lake, B. B., Itoh, K., Ioannou, K., & Sokol, S. Y. (2014). Role of Rab11 in planar cell polarity and apical constriction during vertebrate neural tube closure. *Nature Communications, 5*, 3734. https://doi.org/10.1038/ncomms4734. www.nature.com/naturecommunication.

Pfister, K., Shook, D., Chang, C., Keller, R., & Skoglund, P. (2016). Molecular model for force production and transmission during vertebrate gastrulation. *Development, 143*, 715–727.

Poznanski, A., & Keller, R. (1997). The role of planar and early vertical signaling in patterning the expression of Hoxb-1 in Xenopus. *Developmental Biology, 184*, 351–366.

Reed, R. A., Womble, M. A., Dush, M. K., Tull, R. R., Bloom, S. K., Morckel, A. R., et al. (2009). Morphogenesis of the primitive gut tube is generated by Rho/ROCK/Myosin II–mediated endoderm rearrangements. *Developmental Dynamics, 238*, 3111–3125.

Reintsch, W., Habring-Mueller, A., Wang, R., Schohl, A., & Fagotto, F. (2005). β-catenin controls cell sorting at the notochord-somite boundary independently of cadherin-mediated adhesion. *The Journal of Cell Biology, 170*, 675–686.

Rochlin, M. W., Itoh, K., Adelstein, R. S., & Bridgman, P. C. (1995). Localization of myosin II A and B isoforms in cultured neurons. *Journal of Cell Science, 108*, 3661–3670.

Rohani, N., Canty, L., Luu, O., Fagotto, F., & Winklbauer, R. (2011). EphrinB/EphB signaling controls embryonic germ layer separation by contact-induced cell detachment. *PLoS Biology, 9*(3). e1000597. https://doi.org/10.1371/journal.pbio.

Rosario, T., & DeSimone, D. W. (2009). The extracellular matrix in development and morphogenesis: A dynamic view. *Developmental Biology, 341*, 126–140.

Rosario, T., Dzamba, B., Weber, G., Davidson, L., & DeSimone, D. (2009). The physical state of fibronectin matrix differentially regulates morphogenetic movements in vivo. *Developmental Biology, 327*, 386–398.

Schechtman, A. M. (1942). The mechanism of amphibian gastrulation. I. Gastrulation-promoting interactions between various region of an anuran egg (Hyla regilla). *University of California Publications in Zoology, 51*, 1–39.

Schroeder, T. E. (1970). Neurulation in *Xenopus laevis*. An analysis and model based on light and electron microscopy. *Journal of Embryology and Experimental Morphology, 23*, 427–462.

Shawky, J. H., Balakrishnan, U. L., Stuckenholz, C., & Davidson, L. A. (2018). Multiscale analysis of architecture, cell size and the cell cortex reveals cortical F-actin density and composition are major contributors to mechanical properties during convergent extension. *Development, 145*, dev161281. https://doi.org/10.1242/dev.161281.

Shendo, A. (2018). Models of convergent extension during morphogenesis. *WIREs Developmental Biology, 7*, e293. https://doi.org/10.1002/wdev.293.

Shih, J., & Keller, R. (1992a). Cell motility driving mediolateral intercalation in explants of *Xenopus laevis*. *Development, 116*, 901–914.

Shih, J., & Keller, R. (1992b). Patterns of cell motility in the organizer and dorsal mesoderm of *Xenopus*. *Development, 116*, 915–930.

Shindo, A., Inoue, Y., Kinoshita, M., & Wallingford, J. B. (2019). PCP-dependent transcellular regulation of actomyosin oscillation facilitates convergent extension of vertebrate tissue. *Developmental Biology, 446*, 159–167. https://doi.org/10.1016/j.ydbio.2018.12.017.

Shindo, A., & Wallingford, J. B. (2014). PCP and septins compartmentalize cortical actomyosin to direct collective cell movement. *Science, 343*, 649–652.

Shook, D. R., Kasprowicz, E. M., Davidson, L. A., & Keller, R. (2018). Large long range tensile forces drive convergence, blastopore closure, and body axis elongation. *eLife*, *7*, e26944. pii: e26944. https://doi.org/10.7554/eLife.26944.

Shook, D. R., Wen, J., Rolo, A., Francica, B., Dobins, D., Skoglund, P., et al. (2018). Characterization of convergent thickening, a major convergence force producing morphogenic movement in amphibians. *bioRxiv*, 270892. https://doi.org/10.1101/270892.

Skoglund, P., Dzamba, B., Coffman, C. R., Harris, W., & Keller, R. (2006). *Xenopus* fibrillin is expressed in the organizer and is the earliest component of matrix at the developing notochord-somite boundary. *Developmental Dynamics*, *235*, 1974–1983.

Skoglund, P., & Keller, R. (2007). *Xenopus* fibrillin is required for convergence and extension. *Developmental Biology*, *301*, 404–416.

Skoglund, P., Rolo, A., Chen, X., Gumbiner, B., & Keller, R. (2008). Convergence and extension at gastrulation requires a myosin IIB dependent cortical actin network. *Development*, *135*, 2435–2444.

Sokol, S. Y. (1996). Analysis of Dishevelled signaling pathways during *Xenopus* development. *Current Biology*, *6*, 1456–1467.

Solnica-Krezel, L., & Sepich, D. S. (2012). Gastrulation: Making and shaping germ layers. *Annal Review of Cell and Developmental Biology*, *28*, 687–717.

Sun, Z., Amourda, C., Shagirov, M., Hara, Y., Saunders, T. E., & Toyama, Y. (2017). Basolateral protrusion and apical contraction cooperatively drive Drosophila germ-band extension. *Nature Cell Biology*, *19*, 375–383. https://doi.org/10.1038/ncb3497.

Szabo, A., Cobo, I., Omara, S., McLachlan, S., Keller, R., & Mayor, R. (2016). The molecular basis of radial intercalation during tissue spreading in early development. *Developmental Cell*, *37*, 213–225.

Tada, M., & Smith, J. C. (2000). Xwnt is a target of *Xenopus* Brachyury: Regulation of gastrulation movements via Dishevelled, but not through the canonical Wnt pathway. *Development*, *127*, 2227–2238.

Trapani, V., Bonaldo, P., & Corallo, D. (2017). Role of the ECM in notochord formation, function and disease. *Journal of Cell Science*, *130*, 3203–3211. https://doi.org/10.1242/jcs.175950.

Tuft, P. H. (1962). The uptake and distribution of water in the embryo of Xenopus laevis (Daudin). *The Journal of Experimental Biology*, *39*, 1–19.

Vogt, W. (1929). Gestaltanalyse am Amphibienkein mit ortlicher Vitalfarbung. II. Teil. Gastrulation und Mesodermbildung bei Urodelen und Anuren. *Wilhelm Roux' Archiv für Entwicklungsmechanik der Organismen*, *120*, 384–706.

Wacker, S., Grimm, K., Winklbauer, R., Joos, T., & Winklbauer, R. (2000). Development and control of tissue separation at gastrulation in *Xenopus*. *Developmental Biology*, *224*, 428–439.

Wallingford, J. B., & Harland, R. M. (2001). *Xenopus* Dishevelled signaling regulates both neural and mesodermal convergent extension: Parallel forces elongating the body axis. *Development*, *128*, 2581–2592.

Wallingford, J., & Harland, R. (2002). Neural tube closure requires Dishevelled-dependent convergent extension of the midline. *Development*, *129*, 5815–5825.

Wallingford, J., Rowning, B. A., Vogeli, K. M., Rothbacher, U., Fraser, S., & Harland, R. (2000). Dishevelled controls cell polarity during *Xenopus* gastrulation. *Nature*, *405*, 81–85.

Williams, M., & Solnica-Krezel, L. (2020). Cellular and molecular mechanisms of convergence and extension in zebrafish. In L. Solnica-Krezel (Ed.), *Current topics in developmental biology*, *Vol. 136*. (pp. 377–407). Elsevier.

Williams, M., Yen, W., Lu, X., & Sutherland, A. (2014). Distinct apical and basolateral mechanisms drive planar cell polarity-dependent convergent extension of the mouse neural plate. *Developmental Cell*, *29*, 34–46.

Williams-Masson, E. M., Heid, P. J., Lavin, C. A., & Hardin, J. (1998). The cellular mechanisms of epithelial rearrangement during morphogenesis of Caenorhabditis elegans dorsal hypodermis. *Developmental Biology*, *204*, 263–276.
Wilson, P. (1991). *The development of the axial mesoderm in* Xenopus laevis. Ph.D. Thesis Charlottesville, VA: University of Virginia.
Wilson, P., & Keller, R. (1991). Cell rearrangement during gastrulation of Xenopus: Direct observation of cultured explants. *Development*, *112*, 289–300.
Wilson, P., Oster, G., & Keller, R. E. (1989). Cell rearrangement and segmentation in *Xenopus*: Direct observation of cultured explants. *Development*, *105*, 155–166.
Winklbauer, R., Medina, A., Swan, R. K., & Steinbeisser, H. (2001). Frizzled-7 signaling controls tissue separation behavior during *Xenopus* gastrulation. *Nature*, *413*, 856–860.
Yen, W. W., Williams, M., Periasamy, A., Conaway, M., Burdsal, C., Keller, R., et al. (2009). PTK7 is essential for polarized cell motility and convergent extension during mouse gastrulation. *Development*, *136*, 2039–2048.
Zhou, J., Kim, H. Y., & Davidson, L. A. (2009). Actomyosin stiffens the vertebrate embryo during crucial stages of elongation and neural tube closure. *Development*, *136*, 677–688.
Zhou, J., Pal, S., Maiti, S., & Davidson, L. A. (2015). Force production and mechanical accommodation during convergent extension. *Development*, *142*, 692–701.

CHAPTER ELEVEN

Mechanisms of zebrafish epiboly: A current view

Ashley E.E. Bruce[a,*], Carl-Philipp Heisenberg[b]
[a]Department of Cell & Systems Biology, University of Toronto, Toronto, ON, Canada
[b]Institute of Science and Technology Austria, Klosterneuburg, Austria
*Corresponding author: e-mail address: ashley.bruce@utoronto.ca

Contents

Abstract

Epiboly is a conserved gastrulation movement describing the thinning and spreading of a sheet or multi-layer of cells. The zebrafish embryo has emerged as a vital model system to address the cellular and molecular mechanisms that drive epiboly. In the zebrafish embryo, the blastoderm, consisting of a simple squamous epithelium (the enveloping layer) and an underlying mass of deep cells, as well as a yolk nuclear syncytium (the yolk syncytial layer) undergo epiboly to internalize the yolk cell during gastrulation. The major events during zebrafish epiboly are: expansion of the enveloping layer and the internal yolk syncytial layer, reduction and removal of the yolk membrane ahead of the advancing blastoderm margin and deep cell rearrangements between the enveloping layer and yolk syncytial layer to thin the blastoderm. Here, work addressing the cellular and molecular mechanisms as well as the sources of the mechanical forces that underlie these events is reviewed. The contribution of recent findings to the current model of epiboly as well as open questions and future prospects are also discussed.

Current Topics in Developmental Biology, Volume 136
ISSN 0070-2153
https://doi.org/10.1016/bs.ctdb.2019.07.001

1. Introduction and overview

Gastrulation produces diverse animal body plans via germ layer specification and coordinated changes in cell morphology and motility. Epiboly is an evolutionarily conserved gastrulation movement characterized by tissue thinning and spreading (Trinkaus & Lentz, 1967) which is essential for normal development in many vertebrate and invertebrate species. Due to the prominence of epiboly during zebrafish development, the zebrafish embryo has proven to be a particularly useful model system for studying the cellular and molecular mechanisms that drive this process. Classic work, primarily from J.P. Trinkaus and colleagues, in the related teleost *Fundulus heteroclitus*, has provided a critical experimental and conceptual foundation for the study of zebrafish epiboly. Here we discuss the current view of zebrafish epiboly, highlighting recent findings and open questions.

Epiboly is the first morphogenetic event during zebrafish development, occurring from late-blastula (4.3 hours post-fertilization, hpf) to the bud stage (10 hpf) (Fig. 1). Just prior to epiboly initiation, at sphere stage (4 hpf), the embryo consists of the blastoderm, which is a mass of about 4000 cells, that sits on top of a large multi-nucleated yolk cell (Kimmel, Ballard, Kimmel, Ullmann, & Schilling, 1995). The surface cells of the

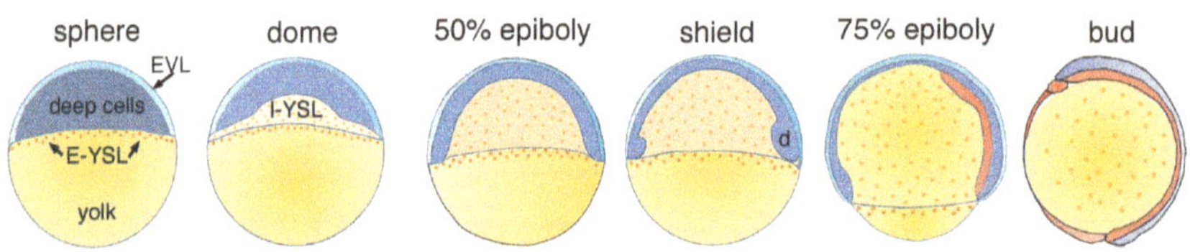

Fig. 1 Schematics of zebrafish epiboly. Lateral views of embryos at selected stages of epiboly with the animal pole at the top and vegetal pole at the bottom. Sphere stage (4 hpf): the blastoderm (blue), consisting of the enveloping layer (EVL) and deep cells, maintains a flat interface with the underlying yolk cell (yellow). Yolk syncytial nuclei are shown in brown. Dome stage (4.3 hpf): epiboly starts with yolk cell doming and deep cells rearrange to form an inverted cup over the yolk. Yolk syncytial nuclei are present in the internal yolk syncytial layer (I-YSL). During epiboly progression, the yolk syncytial layer and blastoderm move vegetally. 50% epiboly (5.3 hpf): the blastoderm has reached the equator. Shield stage (6 hpf): the shield forms marking the dorsal side (d). 75% epiboly (8 hpf): the blastoderm and yolk syncytial layer continue epiboly. Mesendoderm has internalized (red). Bud (10 hpf): epiboly and gastrulation are complete with the closure of the blastopore at the vegetal pole. EVL: enveloping layer; E-YSL: external yolk syncytial layer: I-YSL: internal yolk syncytial layer. Internalized mesendoderm in red, ectoderm in blue.

blastoderm comprise a simple squamous epithelial sheet called the enveloping layer (EVL), which is attached at its margin to the yolk cell. The EVL encloses a multi-layered deep cell mass and together they give rise to all embryonic tissues. At the surface of the animal most region of the yolk cell is the yolk syncytial layer (YSL), which contains internal and external yolk syncytial layer nuclei (YSN) (Kimmel et al., 1995; Kimmel & Law, 1985). The YSL also contains mitochondria, endoplasmic reticulum and Golgi membranes, consistent with it being a transcriptionally and metabolically active region of the yolk cell (Carvalho & Heisenberg, 2010; Kondakova & Efremov, 2014; Popgeorgiev et al., 2011; Rawson, Zhang, Kalicharan, & Jongebloed, 2000). The YSL is contiguous with the yolk cytoplasmic layer (YCL), a thin surface layer of cytoplasm that covers the remainder of the yolk surface and surrounds most of the yolk platelets (Solnica-Krezel & Driever, 1994). The yolk cell also contains a network of microtubule and actin cytoskeletal structures which is described further below (Cheng, Miller, & Webb, 2004; Solnica-Krezel & Driever, 1994; Strähle & Jesuthasan, 1993).

During epiboly, all three regions (the EVL, deep cells and the yolk syncytial layer) spread vegetally to enclose the yolk cell by the end of gastrulation (Figs. 1–3). Epiboly can be broken down into four major events: (1) expansion of the EVL epithelial sheet (2) expansion of the internal YSL (3) reduction and removal of the YCL and (4) deep cell rearrangements between the EVL and YSL (Trinkaus, 1984). Thus, zebrafish epiboly unfolds in a coordinated manner in three parallel, yet distinct cellular contexts. Studies of the same overall process in three mechanistically different tissue contexts may ultimately lead to an understanding of how different

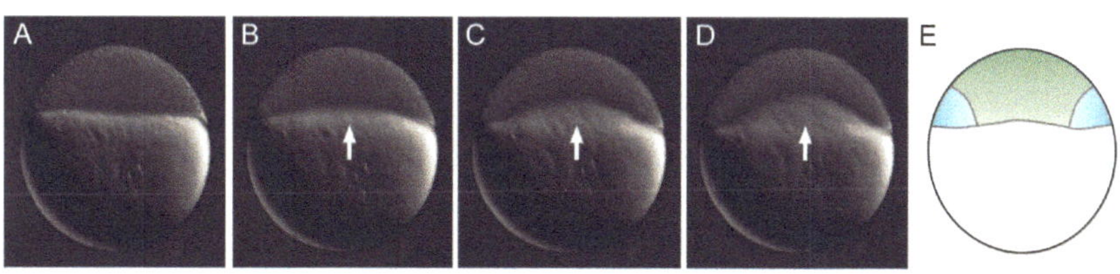

Fig. 2 (A–D) Lateral views of live embryos, with animal pole to the top, from sphere stage (4 hpf) to 30% epiboly (4.7 hpf). (A) sphere stage, yolk-blastoderm interface is flat. (B) as doming starts, the yolk begins to bulge centrally (arrow). (C) Dome stage, the central region of the yolk has domed. (D) 30% epiboly, the yolk cell continues to bulge upwards (arrow) and the blastoderm begins to spread. (E) Schematic with green shading indicating the central blastoderm and blue shading marking the marginal blastoderm. See text for details.

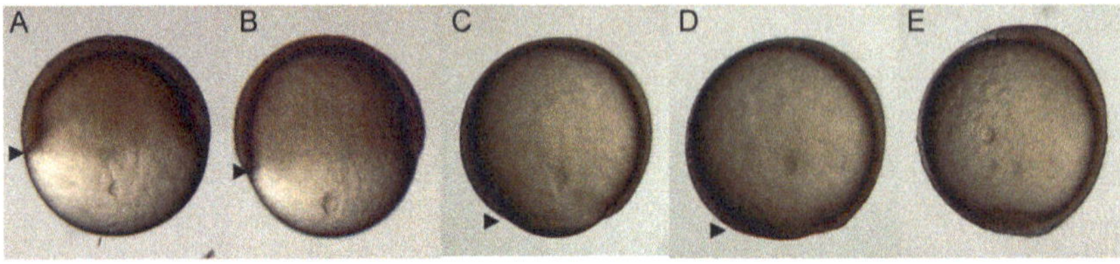

Fig. 3 Epiboly progression. (A–E) Lateral views of live embryos with animal pole to the top. (A) 50% epiboly (B) 60% epiboly (C) 80% epiboly (D) 90% epiboly (E) bud stage. Arrowhead indicates the margin.

morphogenetic mechanisms are functionally integrated at the level of the organism. The interrelatedness of epiboly in each cellular context provides experimental advantages but also challenges, which the more recent use of computational models and simulations of epiboly may help overcome.

1.1 Setting the stage for epiboly

Prior to the initiation of epiboly, a number of critical events take place. In the blastoderm, the EVL differentiates from surface cells and functions as a protective barrier. Later the EVL will generate the embryonic skin or periderm (Ballard, 1966) and ultimately contribute to the larval epidermis (Fukazawa et al., 2010). Localized expression of apical-basal polarity markers including E-cadherin and the tight junction marker ZO-1 can be detected by the 64-cell stage, although expression of epithelial differentiation markers does not commence until sphere stage (Babb et al., 2001; Fukazawa et al., 2010; Imboden, Goblet, Korn, & Vriz, 1997; Sagerström, Gammill, Veale, & Sive, 2005). Prior to EVL lineage restriction at around 3.5 hpf, surface cells divide to produce either two surface cells (S-S division), or one surface and one deep cell (S-D division). Following lineage restriction surface cells divide exclusively to generate more surface cells (S-S), which will go on to form the EVL (Warga & Kimmel, 1990).

Examination of pre-EVL development from the 128-cell (2.25 hpf) to 2k-cell (3.2 hpf) stages revealed that the shape of surface cells predicts cell division orientation (Xiong et al., 2014). Highly flattened surface cells undergo S-S divisions, while more columnar cells undergo S-D divisions. Surface cell division thus appears to conform to the long-axis rules in which cell division planes are perpendicular to the long axis of the cell (Minc & Piel, 2012). Over time, surface cells continue to flatten and eventually all surface divisions are S-S and lineage restriction is established. It has also been suggested that local mechanical forces and cell-cell interactions function to homogenize cell shapes over time, although the molecular details are not well understood (Xiong et al., 2014). Experimental manipulations demonstrated that a regulated interplay between cell surface number and cell shape confers a robustness to pre-EVL morphogenesis. For example, when surface cell number was experimentally reduced, cells spread to fill the gap resulting in an increase in S-S divisions and normal development (Xiong et al., 2014). By sphere stage, the EVL is lineage restricted and cells have begun to express epithelial differentiation genes. Interestingly, EVL lineage restriction does not reflect cell commitment, as when individual EVL cells are transplanted from sphere stage embryos into the deep cell layers they change their fate (Ho, 1992).

The maternal-zygotic transition (MZT) occurs prior to epiboly initiation and blocking zygotic transcription inhibits epiboly (Kane et al., 1996; Kane & Kimmel, 1993). MZT begins at the 10th cell cycle and is accompanied cell cycle lengthening and loss of division synchrony (Harvey et al., 2013; Kane, Warga, & Kimmel, 1992). MZT is linked to EVL lineage restriction and the lengthened interphase correlates with the beginning of random deep cell motility and pseudopod formation (Kane & Kimmel, 1993; Kane et al., 1992; Warga & Kimmel, 1990). The 9th to 10th cell division is also when the YSL forms with the collapse of marginal blastoderm cells into the yolk, with release of their nuclei and cytoplasmic contents (Kimmel & Law, 1985). This eventually leads to the remaining blastoderm cells dissociating from the yolk cell and the complete closure of the yolk cell at its animal pole by a plasma membrane forming the surface of the I-YSL. The yolk syncytial nuclei in the newly formed yolk syncytial layer typically undergo three more synchronous divisions prior to epiboly initiation (Kane & Kimmel, 1993). The external yolk syncytial nuclei (e-YSN) in the external YSL nucleate a longitudinal microtubule array that extends into the yolk cytoplasmic layer towards the vegetal pole (Fig. 4D) (Solnica-Krezel & Driever, 1994). When epiboly starts, many e-YSN spread under the blastoderm to populate the internal YSL (Kimmel et al., 1995; Kimmel & Law, 1985).

2. Epiboly initiation

Epiboly starts with doming (4.3 hpf) when the initially flat blastoderm-yolk interface changes (Fig. 2). The yolk cell domes upwards centrally, thinning the blastoderm in this region, while the interface shifts downward at the margin (Petridou, Grigolon, Salbreux, Hannezo, &

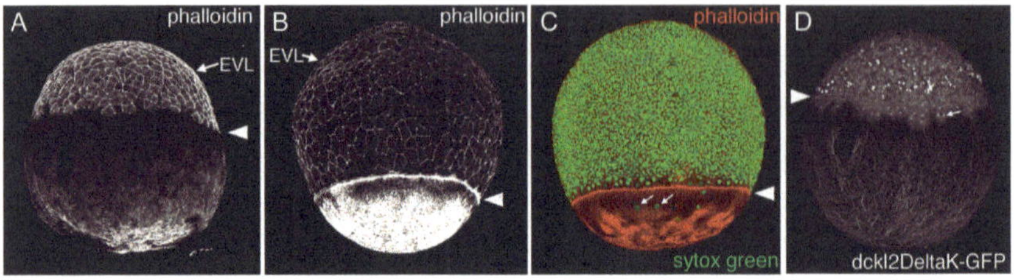

Fig. 4 Actin and microtubule cytoskeletons. (A–D) Lateral views of confocal projections, markers as indicated. Margin indicated by white arrowhead. (A) 30% epiboly stage embryo stained for phalloidin. EVL margin is uneven and actin is present diffusely in the vegetal region of the yolk. (B) Phalloidin stained embryo at 75% epiboly. EVL margin is straight and actin circumferential ring as formed in the E-YSL, vegetal actin is still present. (C) 75% epiboly stage embryos stained for phalloidin (red) and nuclei (green). The E-YSL actin ring is visible as well as migrating E-YSN (arrows). (D) Sphere stage dckl2DeltaK-GFP transgenic embryo. The longitudinal yolk microtubule network is visible, arrow indicates centrosomes of an E-YSN nucleating a branch of the network.

Heisenberg, 2019; Warga & Kimmel, 1990). Doming produces a distinct geometry that differs from a uniform bulging of the yolk (Morita et al., 2017). Four main events occur during doming: (1) the expansion and thinning of EVL cells (2) the start of deep cell radial intercalation (3) expansion of the I-YSL and (4) narrowing of E-YSL (Fig. 5A and B). Work from a number of groups has established that the initiation phase of epiboly involves cellular and molecular mechanisms that are distinct from the epiboly progression phase (Du, Draper, Mione, Moens, & Bruce, 2012; Morita et al., 2017; Petridou et al., 2019).

As doming begins, EVL cells continue to flatten by apical surface expansion and cell height reduction, with no change in cell volume (Campinho et al., 2013; Marsal, Hernández-Vega, & Martin-Blanco, 2017). Over the course of epiboly, EVL spreading results in a more than threefold increase in surface area (Marsal et al., 2017). At the same time, during the sphere to dome transition, the E-YSL becomes compressed along the animal-vegetal axis and highly folded, as first observed in *Fundulus* (Betchaku & Trinkaus, 1978; Trinkaus, 1951, 1984). The EVL is attached to the E-YSL via tight junctions in both Fundulus and zebrafish (Köppen, Fernández, Carvalho, Jacinto, & Heisenberg, 2006; Trinkaus, 1951). In *Fundulus*, severing the EVL-YSL connection results in EVL retraction, demonstrating both that the EVL sheet is under tension and that the EVL attachment to the YSL is essential for its epiboly (Trinkaus, 1951). Furthermore, removal of the *Fundulus* blastoderm prior to epiboly initiation results in

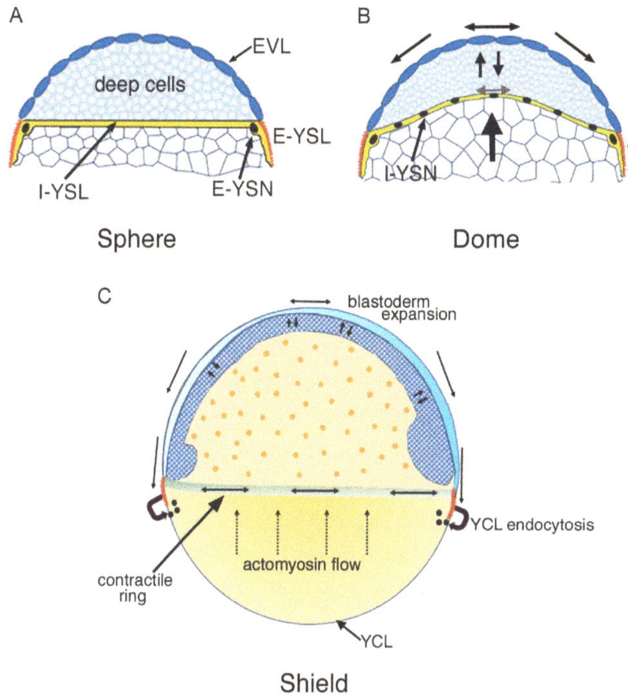

Fig. 5 Model of epiboly. Schematic cross sections of sphere (A) and dome (B) stage embryos. During doming, the I-YSL begins to expand (gray arrow) and the E-YSL narrows, leading to the upward movement of the yolk in the central blastoderm. The EVL spreads while central deep cells fluidize and then radial intercalate to thin the blastoderm. The YSL is in yellow, red marks the E-YSL surface. (C) Schematic cross section of shield stage embryo. During epiboly progression, the actomyosin ring (green) in the E-YSL contracts circumferentially, pulling the attached EVL vegetally as the I-YSL expands and deep cells radially intercalate. Actomyosin flow from vegetal to animal in the yolk generates an animal-vegetal contraction that also contributes to blastoderm vegetal movement. A region of localized endocytosis removes the YCL membrane as the YSL expands. See text for details.

precocious *E*-YSL narrowing and I-YSL expansion, indicating that YSL epiboly is autonomous and that the blastoderm normally restrains YSL epiboly (Trinkaus, 1951). These findings also show that initial EVL spreading cannot be the cause of the E-YSL narrowing (via pushing) since this process also occurs when the blastoderm is removed. Instead, it points to active contraction of the E-YSL and/or active expansion of the I-YSL leading to compression of the E-YSL. Similar to *Fundulus*, at the start of doming, the E-YSL narrows and the EVL undergoes autonomous spreading via a reduction in surface tension (Morita et al., 2017; Solnica-Krezel & Driever, 1994). What triggers the reduction in EVL surface tension is not understood, but it depends upon EVL differentiation (Morita et al., 2017), consistent with work from several groups demonstrating that EVL differentiation is required for normal epiboly (Fukazawa et al., 2010; Pei et al., 2007; Sabel et al., 2009). Thus, EVL differentiation could lead to the expression of molecules involved in regulating EVL spreading.

Deep cell behavior also changes with epiboly onset. It was recently shown that there is a ~10-fold decrease in deep cell viscosity in the central but not marginal region of the blastoderm that occurs transiently just at doming onset (Petridou et al., 2019). Computational modeling based on the shape of the domed yolk cell and force generation via active EVL expansion, indicates that a central blastoderm restricted reduction in viscosity could account for the domed blastoderm shape (Petridou et al., 2019). One experimental test for the necessity of central blastoderm fluidization for doming was to artificially increase cell adhesion and, in turn, tissue viscosity by transplanting E-cadherin coated beads into the central blastoderm. The blastoderm failed to thin in the region of the transplant, indicating that fluidization of central blastoderm deep cells is required for doming (Petridou et al., 2019).

Cell autonomy of the change in deep cell viscosity was demonstrated by explants and cell transplantation experiments (Petridou et al., 2019). The decrease in viscosity occurs via changes in cell adhesion based on the observation that the number, size and duration of deep cell contacts decreases in central but not marginal regions of the doming blastoderm (Petridou et al., 2019). The reduction in central cell-cell contacts occurs during cycles 11–13, just prior to doming. At this time, mitotic rounding occurs to a similar extent in marginal and central regions, but following mitosis central cells fail to reestablish cell contacts, resulting in their gradual loss (Petridou et al., 2019). Although there is still much to understand about the molecular control of epiboly initiation, non-canonical Wnt signaling mediated by Wnt11 has been implicated in preventing deep cell fluidization at the margin by promoting cell cohesion (Petridou et al., 2019). In *wnt11* mutant embryos, fluidization occurs throughout the blastoderm and correlates with reduced cell contacts. Wnt11 appears to prevent fluidization by promoting actomyosin contractility in marginal cells (Petridou et al., 2019).

These recent findings suggest an intriguing role for cell division in morphogenesis via the regulation of tissue viscosity (Petridou et al., 2019). Doming occurs rapidly and evenly and one reason for this may be related to the rapid change in the central blastoderm, akin to a rigidity phase transition, that results from a small change in cell cohesion. The viscosity decrease centrally also reverses rapidly and how this is regulated is unknown but it could be related to continued doming of the yolk cell and the start of deep cell radial intercalation movements. Deep cell radial intercalation, described first in *Xenopus* epiboly (Keller, 1980), was originally characterized in zebrafish epiboly by Warga and Kimmel (1990). Deep cells move along radial trajectories without intercalating into the EVL as the yolk cell domes and the blastoderm thins (Kimmel, Warga, & Schilling, 1990; Warga & Kimmel, 1990).

Deep cell radial intercalation results in widespread cell mixing that is most prominent in the central blastoderm region from dome to 30% epiboly stages (Wilson, Cretekos, & Helde, 1995). Radial intercalation was previously thought to occur unidirectionally toward the EVL as deep cells were pushed outwards by the doming yolk cell (Kimmel et al., 1995). However, more recent work showed that radial intercalation is bidirectional with nearly equal movements towards and away from the EVL (Bensch, Song, Ronneberger, & Driever, 2013). The deep cells are confined between the EVL and YSL and this impacts their morphogenesis. Morita et al. (2017) combined theory and experimentation to show that doming occurs as the result of two interrelated force-generating processes: autonomous EVL expansion coupled with deep cell radial intercalation leading to anisotropic stress generation within the blastoderm.

Deep cell morphology changes as cells start intercalating. They become more radially elongated and actin concentrates at the leading tip, oriented either towards or away from the EVL (Morita et al., 2017). Deep cell arrangement and orientation does not however show the highly ordered organization seen in *Xenopus* during epiboly (Keller, 1980) and deep cells show a distribution of orientations (Morita et al., 2017). Deep cell polarization was hypothesized to cause radial stress in the blastoderm leading to cell intercalations. This was supported by the observation that yolk granules near the blastoderm change shape and elongate radially as deep cells intercalate, suggesting that stresses in the deep cell layer are relayed to and affect the yolk granules. The role of deep cell radial intercalation during doming is to thin the blastoderm, rather than initiate spreading. This was shown by replacing 60% of deep cells at 3 hpf with a similar volume of buffer and observing relatively normal EVL spreading and yolk cell doming but reduced blastoderm thinning. Simulations of doming were consistent with blastoderm spreading being driven by EVL surface tension reduction alone when deep cells were depleted and with deep cell intercalations normally producing radial stress in the blastoderm resulting in even blastoderm thinning (Morita et al., 2017).

EVL expansion was shown to be required for doming and also to promote deep cell radial intercalation by generating radial stress in the blastoderm. Particularly during the first 50 min after doming onset, the density of deep cells decreases while deep cell motility increases which is likely important both for tissue fluidization and radial intercalation (Morita et al., 2017; Petridou et al., 2019). Complementing other approaches to understanding epiboly Hernández-Vega et al. (2017) used hydrodynamic regression, a non-invasive method to identify regions of the embryo where mechanical stress and power are high during morphogenesis at the mesoscopic level. They found that active power is located mainly in the blastoderm during epiboly initiation, consistent with the work described above (Hernández-Vega et al., 2017; Marsal et al., 2017).

How does narrowing of the E-YSL contribute to doming? It may be the initiating force, as it appears to be in *Fundulus*, with its effectiveness depending upon the concomitant changes taking place in the EVL and deep cells (Morita et al., 2017; Petridou et al., 2019; Trinkaus, 1984). Alternatively, active I-YSL expansion could be the driving force for doming and lead to E-YSL compression. Since embryo shape and volume remain constant and the yolk is incompressible, E-YSL narrowing results in animally-directed yolk granule flow at the center of the embryo leading to doming (Marsal et al., 2017). This suggest that both the EVL and yolk cell influence deep cell radial intercalation movements. In *Fundulus* and zebrafish E-YSL narrowing is completed before epiboly progression begins (Solnica-Krezel & Driever, 1994; Trinkaus, 1984).

3. Epiboly progression

The blastoderm and YSL continue to spread vegetally to cover the yolk cell during epiboly progression which extends from doming until the closure of the blastopore at 10 hpf (Fig. 4). Epiboly pauses briefly at 50% epiboly as other gastrulation movements commence (Solnica-Krezel & Driever, 1994; Warga & Kimmel, 1990). The relative positions of the deep cells, EVL and YSL shift dynamically during the progression phase but the tightly linked YSL and EVL are always positioned vegetal to the deep cell margin (Solnica-Krezel & Driever, 1994). This organization is consistent with the major force for epiboly progression coming from the yolk cell.

The yolk cell tows the attached EVL along with it as the deep cells rearrange in between the EVL and YSL. After doming, the E-YSL is thicker than the I-YSL and rich with microfilaments and microtubules, similar to *Fundulus* (Betchaku & Trinkaus, 1978; Kondakova & Efremov, 2014; Köppen et al., 2006; Trinkaus, 1984). The I-YSL expands to cover the yolk cell during epiboly as the E-YSL undergoes active contraction (Trinkaus, 1984). In *Fundulus* microvilli on the I-YSL surface diminish in size and are thought to contribute to the expanding yolk membrane (Betchaku & Trinkaus, 1978; Lentz & Trinkaus, 1967; Trinkaus, 1984). Whether this is the case in zebrafish has not been examined. Most studies have focused on the E-YSL where the actomyosin-based epiboly motors reside, which are essential for epiboly progression. Hydrodynamic regression analysis has also shown that after 50% epiboly the main mechanical power source in the embryo is localized to the E-YSL (Hernández-Vega et al., 2017).

Beginning at 50% epiboly, actin becomes enriched at the EVL and deep cell margins (Cheng et al., 2004; Zalik, Lewandowski, Kam, & Geiger, 1999). In the yolk cell, actomyosin flow from vegetal to animal starts at 40% epiboly and coalesces into a narrow circumferential band by 70–80% epiboly (Behrndt et al., 2012). Circumferential contraction of the E-YSL actomyosin network is important for epiboly progression and ultimately for closure of the blastopore once the blastoderm passes the equator of the yolk cell (Behrndt et al., 2012; Köppen et al., 2006). Using a combination of theory and experimental approaches, a second actomyosin motor was postulated to operate throughout epiboly progression, including stages prior to 50% epiboly (Behrndt et al., 2012). This flow-friction motor results from the retrograde flow of actin and myosin towards the blastoderm margin. Resistance to the flow as the result of frictional forces is thought to lead to contraction along the animal-vegetal axis which then exerts a pulling force on the EVL margin (Behrndt et al., 2012; Heisenberg & Bellaïche, 2013).

As the YSL expands, the YCL is removed by a region of localized endocytosis in the E-YSL which was first observed in *Fundulus* and proposed to contribute to epiboly movements (Betchaku & Trinkaus, 1986; Solnica-Krezel & Driever, 1994). The regulation of endocytosis in the region between the contiguous E-YSL and YCL is not understood. Although inhibition of yolk endocytosis by targeting dynamin function does not affect epiboly, inhibition of endocytosis by morpholino knockdown of Rab5ab does (Kenyon et al., 2015; Lepage, Tada, & Bruce, 2014; Marsal et al., 2017). Additional work is required to determine the extent to which YCL membrane removal contributes to the E-YSL forces that drive epiboly progression.

As an alternative to the flow-friction model, it was proposed that contraction of the E-YSL coupled with yolk membrane removal can account for epiboly progression after 50% epiboly (Hernández-Vega et al., 2017; Marsal et al., 2017). In this model, recruitment of actin and myosin by cortical animally-directed flow would lead to contraction of the E-YSL both circumferentially and along the animal-vegetal axis, exerting a pulling force both on the EVL and on the YCL. It is proposed that the EVL responds by spreading while the more rigid YCL cannot deform. The different responses of the EVL and YCL to E-YSL contraction coupled with yolk membrane removal would result in vegetal spreading of the EVL (Hernández-Vega et al., 2017; Marsal et al., 2017). Future studies will have to explore how retrograde flow of actin and myosin towards the blastoderm margin can be generated under the assumption that the actomyosin network within the E-YSL is coupled to a rigid YCL, and whether such mechanical coupling indeed exists.

3.1 EVL morphogenesis during epiboly progression

Epiboly of the EVL necessitates the regulation of tissue tension to allow spreading. Campinho et al. (2013) demonstrated that oriented cell division plays a role. Following EVL and deep cell lineage restriction, the EVL cell cycle slows in comparison to the deep cells (Kane et al., 1992). Consistent with this, EVL divisions are most numerous during early epiboly stages (Campinho et al., 2013; Kane et al., 1992). Cell divisions decrease cell volume by half and the average number of EVL cells during early epiboly stages increases from 496 at sphere stage to 1072 at 55% epiboly (Campinho et al., 2013). In addition, EVL cells preferentially divide along the animal-vegetal (AV) axis, aligned with the axis of tissue spreading and suggesting that these two processes are related (Campinho et al., 2013).

Early studies in *Fundulus* demonstrated that the epibolizing EVL is under tension along the AV axis as well as around the marginal circumference (Trinkaus, 1951, 1984). In zebrafish, laser cutting experiments showed that from 40% to 50% epiboly tension along the AV axis and marginal circumference are similar, while at mid- to late-epiboly (50–80% epiboly), tension in the EVL is greater along the AV axis than along the marginal circumference (Campinho et al., 2013). This main axis of planar tension within the EVL determines the preferential division orientation of EVL cells, and oriented EVL cell divisions again can reduce tissue tension anisotropy within the EVL. When EVL cell divisions are blocked or their preferential division orientation is lost, epiboly progression remains intact, but EVL cells display an enhanced tendency to undergo ectopic fusion within the plane of the tissue (Campinho et al., 2013).

As the EVL spreads over the yolk during epiboly, the marginal EVL first widens to reach the EVL equator and then narrows again to eventually close over the yolk cell. This involves cell shape changes and a reduction in the number of marginal EVL cells which takes place without compromising EVL integrity or barrier function (Keller & Trinkaus, 1987; Köppen et al., 2006). Marginal EVL cells in *Fundulus* become tapered at the region of contact with the E-YSL and some cells retract from the margin with the gap being closed by flanking cells. Cell rearrangements also take place in sub-marginal regions (Keller & Trinkaus, 1987).

In zebrafish, the EVL margin initially has an irregular appearance and filopodia are apparent at the leading edge of cells where they contact the E-YSL (Fig. 4A) (Köppen et al., 2006; Lachnit, Kur, & Driever, 2008; Zalik et al., 1999). With the formation of the actomyosin band in the E-YSL at mid-epiboly, the EVL margin becomes straight, filopodia are reduced, and marginal EVL cells begin to narrow mediolaterally and elongate along the animal-vegetal axis (Fig. 4B) (Cheng et al., 2004; Köppen et al., 2006). Some marginal EVL cells do not narrow at the contact point with the E-YSL and this is correlated with local levels of actin in the YSL (Köppen et al., 2006). The *TRAF2 and NCK interacting kinase a, tnika*, a member of the germinal center kinase (GCK) family, functions in the E-YSL to recruit actin and myosin to the EVL-YSL interface. Actomyosin contraction in the E-YSL then leads to local shape changes in adjacent EVL cells, and ultimately to the reduction in the EVL marginal circumference as it advances vegetally (Köppen et al., 2006). Hydrodynamic regression analysis showed that the power source in the E-YSL was impaired in *tnika* knockdown embryos, consistent with the E-YSL providing the motive force for EVL epiboly (Hernández-Vega et al., 2017).

There remains a great deal to learn about the molecular control of EVL spreading which requires changes in cell-cell contact lengths and apical-basal cell height. EVL integrity is maintained via tight junctions, adherens junctions and desmosomes and some of the adhesion molecules involved include E-cadherin, EpCAM, Celsr and the desmosomal cadherins desmocollin and desmoglein (Babb et al., 2001; Babb & Marrs, 2004; Carreira-Barbosa et al., 2009; Goonesinghe, Luan, Hurlstone, & Garrod, 2012; Slanchev et al., 2009). It has been suggested that expression of keratin may be important for EVL spreading, analogous to the role of keratins in cell spreading during wound healing (Morita et al., 2012). Tissue spreading and cell rearrangements are likely to involve modulation of the uniform actomyosin cortex present around EVL cells (West et al., 2017; Zalik et al., 1999). Increasing tissue tension by transplanting EVL cells expressing RhoA locally impairs doming (Morita et al., 2017). Myosin is implicated in EVL cell elongation and also potentially in mitotic spindle orientation (Campinho et al., 2013). Although EVL cells elongate along the AV axis, myosin is not planar polarized (Morita et al., 2017).

One evolutionarily conserved mechanism to regulate actomyosin accumulation to enable EVL spreading is via the Arf-GEF Cytohesin or Steppke identified in *Drosophila* (West et al., 2017). In both *Drosophila* dorsal closure and zebrafish EVL epiboly, Cytohesin diminishes tissue tension by reducing actomyosin accumulation at adherens junctions, thereby allowing the tissue to elongate under stress. This has been shown to operate in a negative feedback loop in *Drosophila* in which actomyosin recruits Steppke, which then antagonizes actomyosin contractility leading to tissue relaxation (West et al., 2017). Remodeling cell contacts and increasing apical surface area involves membrane trafficking (Wirtz-Peitz & Zallen, 2009). Work from *Fundulus* showed that apical membrane turnover in the EVL increases when cells are actively rearranging suggesting that membrane trafficking is linked to local mechanical tension (Fink & Cooper, 1996). Inhibition of endocytosis by interfering with dynamin function produces abnormal EVL morphologies and slows epiboly. Dynamin plays a role in the maintenance of epithelial integrity and functions by antagonizing the activity of RhoA to promote epiboly (Lepage et al., 2014).

4. Deep cell movements

Deep cell rearrangements over the course of epiboly transform the initially six to eight cell layered blastoderm into one that is two to three cell layers thick (Bensch et al., 2013; Kane & Adams, 2002). As the YSL and EVL move vegetally, deep cells move into the open space, thinning and spreading the cell mass (Bensch et al., 2013; Lachnit et al., 2008). Global analysis of deep cell movements from sphere stage through 50% epiboly revealed temporal changes in deep cell radial intercalation behavior (Bensch et al., 2013). Intercalations are more oriented along the animal-vegetal axis from 30% onward than during doming. In addition, during mid-epiboly stages, more intercalations take place in superficial cell layers, close to the

EVL, than in deeper cell layers (Bensch et al., 2013). As seen during doming, the behavior of central deep cells differs from marginal cells. Central cells exhibit higher rates of upward (toward the EVL) and downward (away from the EVL) intercalations than marginal cells (Bensch et al., 2013). Radial intercalation was also examined in maternal-zygotic *spg* (MZ*spg*) embryos which are mutant for the transcription factor Pou5f3 and exhibit a prominent deep cell epiboly delay. The number of intercalations was significantly lower in MZ*spg* embryos compared to wild type as the result of cell motility defects (Bensch et al., 2013).

The efficiency of deep cell rearrangements depends largely upon E-cadherin mediated adhesion (Babb & Marrs, 2004; Kane, McFarland, & Warga, 2005; Shimizu et al., 2005). The radial intercalation defects in MZ*spg* embryos are linked to aberrant E-cadherin trafficking (Bensch et al., 2013; Lachnit et al., 2008; Onichtchouk et al., 2010; Song et al., 2013). In wild-type embryos, E-cadherin is initially localized to the cell membrane and then begins to shift to a more vesicular pattern of localization at sphere stage (Song et al., 2013). Vesicular localization becomes more prominent at later epiboly stages. In Pou5f3 mutants, E-cadherin remains localized to the plasma membrane, which results in decreased cell cohesion, and impaired deep cell rearrangements (Bensch et al., 2013; Song et al., 2013). Pou5f3 appears to affect *E*-cadherin endosomal localization via transcriptional regulation of Epidermal growth factor (EGF). EGF signaling is proposed to regulate p120 activity, which is involved in stabilizing the cadherin complex at the membrane by blocking endocytosis (Harris & Tepass, 2010). Overall, these findings point to the importance of dynamic regulation of E-cadherin for efficient deep cell rearrangements during epiboly.

Radial intercalation has also been examined in localized regions of the blastoderm at 70–90% epiboly (Kane et al., 2005). At late-epiboly stages, superficial deep cell intercalations result in close contact and E-cadherin mediated adhesion of deep cells to the basal EVL surface. EVL-deep cell adhesion is impaired in E-cadherin mutant and morphant embryos, which results in deep cells "de-intercalating" causing a deep cell epiboly delay (Kane et al., 2005). In E-cadherin mutants, YSL and EVL epiboly is not impaired, indicating that they do not depend upon the deep cells for their movement (Kane et al., 2005). Other regulators of E-cadherin include prostaglandin E_2 (PGE_2), which plays a permissive role in cell motility during zebrafish gastrulation and the G-protein subunits of the $G\alpha 12/13$ class (Lin et al., 2009; Sumi et al., 2018).

5. E-YSN and yolk cell microtubules

The yolk cell contains an elaborate microtubule network oriented along the AV axis that extends from a subset of E-YSN towards the vegetal pole (Fig. 4D) (Solnica-Krezel & Driever, 1994). The network shortens over the course of epiboly as the YCL is removed and the YSL expands (Solnica-Krezel & Driever, 1994). Pharmacological treatments that either destabilize or stabilize microtubules in the entire embryo lead to epiboly delay (Solnica-Krezel & Driever, 1994; Strähle & Jesuthasan, 1993). Factors implicated in regulating yolk microtubule stability include the zebrafish CLIP-170 homolog, Clip1a (Hsu, Liang, Chen, & Chung, 2006; Weng et al., 2013). The steroid pregnenolone binds to Clip1a which leads to microtubule polymerization (Weng et al., 2013). Progesterone is also involved in promoting microtubule growth in the yolk cell during epiboly and it was recently proposed that progesterone and pregnenolone might function together to regulate microtubule dynamics at different phases of epiboly (Eckerle, Ringler, Lecaudey, Nitschke, & Driever, 2017). Recent work showed an intriguing role in epiboly for the large atypical cadherin Dachsous 1b (Dchs1b), which is linked to planar cell polarity (PCP) and cell adhesion in other systems (Li-Villarreal et al., 2015). Maternal-zygotic *dchs1b* mutants exhibit a range of early defects including an epiboly delay that is correlated with abnormally bundled microtubules in the yolk cell (Li-Villarreal et al., 2015). Mutant embryos also have defects in the blastoderm, therefore the exact cause of the epiboly delay is currently unclear. The role of Dchs1b in the early embryo appears to be independent of the planar cell polarity pathway, and it remains to be determined how Dachous interacts with and regulates yolk cell microtubules.

Understanding the exact function of microtubules in the yolk cell has been hampered by the lack of experimental approaches to manipulate microtubules just in the yolk cell as opposed to in the entire embryo. Early work suggested that the primary role of the yolk microtubules is in e-YSN migration during epiboly progression starting at about 60% epiboly, although the mechanism was unclear (Fig. 4C and D) (Solnica-Krezel & Driever, 1994). Recent work provided support for a cargo model for e-YSN migration in which the nuclei associate directly with microtubule motors that carry them towards the vegetal pole (Fei et al., 2019). In addition, a YSL specific *wnt8* enhancer was used to express a microtubule stabilizing factor just in the yolk cell, which led to alterations in e-YSN migration but had little overall effect on epiboly, supporting the view that main function of the yolk microtubules is in nuclear migration rather than as part of the yolk cell motor for blastoderm epiboly (Fei et al., 2019). One open question is whether there are interactions between yolk microtubules and actin that are important for epiboly. For example, in *Xenopus* oocytes microtubules have been shown to slow actomyosin-based cortical flows (Canman & Bement, 1997).

6. Conclusions and perspectives

Despite the fact the zebrafish zygote is smaller than *Fundulus* (0.7 mm versus 2 mm) and gastrulation is faster (blastopore closure at 10 hpf in zebrafish and 46 hpf in *Fundulus*) the process of epiboly in each species is remarkably similar (Armstrong & Child, 1965; Kimmel et al., 1995). In *Fundulus*, EVL epiboly is driven by its tight attachment to the expanding YSL, which is also the case in zebrafish. In *Fundulus* this was demonstrated by detaching the EVL from the YSL and showing that YSL epiboly not only continues but accelerates (Betchaku & Trinkaus, 1978; Trinkaus, 1984). Similar observations have also been made in zebrafish (unpublished observations CPH), supporting the notion of an evolutionary conserved role of the YSL in epiboly. In addition, laser ablation of a region of the actomyosin ring in the YSL retards the vegetal movement of the bordering EVL margin (Behrndt et al., 2012) and knock-down of the tight junction protein Claudin E affects the EVL-YSL boundary resulting in delayed epiboly (Siddiqui, Sheikh, Tran, & Bruce, 2010). Taken all together, movement of the EVL margin appears to be passive. However, EVL cells can actively spread following wounding (Hunter, Willoughby, Bruce, & Fernandez-Gonzalez, 2018). In addition, unlike *Fundulus*, marginal EVL cells exhibit filopodia at their leading edges up until 50% epiboly which could indicate active cell movements (Betchaku & Trinkaus, 1978; Köppen et al., 2006; Zalik et al., 1999). An argument against this is that the EVL has never been observed to overtake the E-YSL during epiboly in any experimental condition. The EVL undergoes cell rearrangements to overcome the geometric challenges of spreading over the yolk cell, suggesting that there is also an active component to EVL morphogenesis.

Epiboly does do not depend on the deep cells in either *Fundulus* or zebrafish (Kane et al., 2005; Morita et al., 2017; Trinkaus, 1984). Deep cell motility and regulated adhesion are required for normal blastoderm thinning but the YSL and EVL can undergo epiboly relatively normally even when the deep cells are severely delayed. Therefore, the deep cells do not play an instructive role in the process. In contrast, deep cell epiboly does depend on the EVL. It was recently been shown that the EVL is involved in triggering deep cell radial intercalations, but the mechanism remains to be determined (Morita et al., 2017). It has been speculated that EVL expansion provides a larger substrate and more space for deep cells to adhere to or that signaling from the EVL to the deep cells could be involved, similar to what has been observed in *Xenopus* (Szabó et al., 2016). It is currently unclear whether the EVL-deep cell interaction is adhesive prior to late-epiboly stages since deep cells adjacent to the EVL move laterally relative to it during early epiboly stages (Bensch et al., 2013).

Taken together all findings made so far on the epiboly movements in zebrafish and *Fundulus*, a working model of zebrafish epiboly emerges (Fig. 5): doming movements at the onset of epiboly begin when the EVL actively expands to reduce surface tension, the deep cells first fluidize to reduce resistance against blastoderm spreading and then undergo radial intercalations to produce anisotropic stress that promotes blastoderm spreading. Doming may involve the active expansion of the I-YSL, whereby the I-YSL provides an initial pushing force to narrow the E-YSL. The E-YSL first narrows and during epiboly progression contracts circumferentially and animal-vegetally to generate the actomyosin-based pulling forces that drive epiboly. E-YSL contraction causes membrane folding which could be the trigger for the localized zone of endocytosis at the YSL-YCL border that removes the YCL membrane as the YSL expands. By the end of epiboly, the YCL is replaced by the YSL. Normal epiboly must require coordination between the EVL, deep cells and yolk cell but how this is accomplished is not yet fully understood. However, the strength of the zebrafish system is that a wide range of experimental and theoretical approaches can be brought to bear on the problem in future studies.

Acknowledgments

AEEB thanks Morley Willoughby for helpful comments on the manuscript and lab members for the images in Fig. 4. Work in the Bruce lab is supported by the Natural Sciences Research Council of Canada.

References

Armstrong, P. B., & Child, I. B. (1965). Stages in the normal development of *Fundulus heteroclitus*. *Biological Bulletin*, *128*(2), 143–168.

Babb, S., Barnett, J., Doedens, A., Cobb, N., Liu, Q., Sorkin, B., et al. (2001). Zebrafish E-cadherin: Expression during early embryogenesis and regulation during brain development. *Developmental Dynamics*, *221*(2), 231–237. https://doi.org/10.1002/dvdy.1132.

Babb, S. G., & Marrs, J. A. (2004). E-cadherin regulates cell movements and tissue formation in early zebrafish embryos. *Developmental Dynamics, 230*(2), 263–277. https://doi.org/10.1002/dvdy.20057.

Ballard, W. W. (1966). The role of the cellular envelope in the morphogenetic movements of teleost embryos. *Journal of Experimental Zoology, 161*(2), 193–200. https://doi.org/10.1002/jez.1401610204.

Behrndt, M., Salbreux, G., Campinho, P., Hauschild, R., Oswald, F., Roensch, J., et al. (2012). Forces driving epithelial spreading in zebrafish gastrulation. *Science, 338*(6104), 257–260. https://doi.org/10.1126/science.1224143.

Bensch, R., Song, S., Ronneberger, O., & Driever, W. (2013). Non-directional radial intercalation dominates deep cell behavior during zebrafish epiboly. *Biology Open, 2*(8), 845–854. https://doi.org/10.1242/bio.20134614.

Betchaku, T., & Trinkaus, J. (1978). Contact relations, surface activity, and cortical microfilaments of marginal cells of the enveloping layer and of the yolk syncytial and yolk cytoplasmic layers of fundulus before and during epiboly. *Journal of Experimental Zoology, 206*(3), 381–426. https://doi.org/10.1002/jez.1402060310.

Betchaku, T., & Trinkaus, J. (1986). Programmed endocytosis during epiboly of fundulus-heteroclitus. *American Zoologist, 26*(1), 193–199.

Campinho, P., Behrndt, M., Ranft, J., Risler, T., Minc, N., & Heisenberg, C.-P. (2013). Tension-oriented cell divisions limit anisotropic tissue tension in epithelial spreading during zebrafish epiboly. *Nature Cell Biology, 15*(12), 1405–1414. https://doi.org/10.1038/ncb2869.

Canman, J., & Bement, W. (1997). Microtubules suppress actomyosin-based cortical flow in *Xenopus oocytes. Journal of Cell Science, 110*(Pt. 16), 1907–1917.

Carreira-Barbosa, F., Kajita, M., Morel, V., Wada, H., Okamoto, H., Arias, A., et al. (2009). Flamingo regulates epiboly and convergence/extension movements through cell cohesive and signalling functions during zebrafish gastrulation. *Development, 136*(5), 877, 383–392. https://doi.org/10.1242/dev.034751.

Carvalho, L., & Heisenberg, C.-P. (2010). The yolk syncytial layer in early zebrafish development. *Trends in Cell Biology, 20*(10), 586–592. https://doi.org/10.1016/j.tcb.2010.06.009.

Cheng, J. C., Miller, A. L., & Webb, S. E. (2004). Organization and function of microfilaments during late epiboly in zebrafish embryos. *Developmental Dynamics, 231*(2), 313–323. https://doi.org/10.1002/dvdy.20144.

Du, S., Draper, B. W., Mione, M., Moens, C. B., & Bruce, A. (2012). Differential regulation of epiboly initiation and progression by zebrafish Eomesodermin A. *Developmental Biology, 362*(1), 11–23. https://doi.org/10.1016/j.ydbio.2011.10.036.

Eckerle, S., Ringler, M., Lecaudey, V., Nitschke, R., & Driever, W. (2017). Progesterone modulates microtubule dynamics and epiboly progression during zebrafish gastrulation. *Developmental Biology, 434*(2), 249–266. https://doi.org/10.1016/j.ydbio.2017.12.016.

Fei, Z., Bae, K., Parent, S. E., Wan, H., Goodwin, K., Theisen, U., et al. (2019). A cargo model of yolk syncytial nuclear migration during zebrafish epiboly. *Development, 146*(1) dev.169664. https://doi.org/10.1242/dev.169664.

Fink, R. D., & Cooper, M. S. (1996). Apical membrane turnover is accelerated near cell–cell contacts in an embryonic epithelium. *Developmental Biology, 174*(2), 180–189. https://doi.org/10.1006/dbio.1996.0064.

Fukazawa, C., Santiago, C., Park, K., Deery, W. J., de la Canny, S., Holterhoff, C. K., et al. (2010). poky/chuk/ikk1 is required for differentiation of the zebrafish embryonic epidermis. *Developmental Biology, 346*(2), 272–283. https://doi.org/10.1016/j.ydbio.2010.07.037.

Goonesinghe, A., Luan, X.-M., Hurlstone, A., & Garrod, D. (2012). Desmosomal cadherins in zebrafish epiboly and gastrulation. *BMC Developmental Biology, 12*(1), 1. https://doi.org/10.1186/1471-213x-12-1.

Harris, T. J., & Tepass, U. (2010). Adherens junctions: From molecules to morphogenesis. *Nature Reviews. Molecular Cell Biology, 11*(7), 502–514. https://doi.org/10.1038/nrm2927.

Harvey, S. A., Sealy, I., Kettleborough, R., Fenyes, F., White, R., Stemple, D., et al. (2013). Identification of the zebrafish maternal and paternal transcriptomes. *Development, 140*(13), 2703–2710. https://doi.org/10.1242/dev.095091.

Heisenberg, C.-P., & Bellaïche, Y. (2013). Forces in tissue morphogenesis and patterning. *Cell, 153*(5), 948–962. https://doi.org/10.1016/j.cell.2013.05.008.

Hernández-Vega, A., Marsal, M., Pouille, P., Tosi, S., Colombelli, J., Luque, T., et al. (2017). Polarized cortical tension drives zebrafish epiboly movements. *The EMBO Journal, 36*(1), 25–41. https://doi.org/10.15252/embj.201694264.

Ho, R. K. (1992). Cell movements and cell fate during zebrafish gastrulation. *Development Supplement*, 65–73.

Hsu, H.-J., Liang, M.-R., Chen, C.-T., & Chung, B.-C. (2006). Pregnenolone stabilizes microtubules and promotes zebrafish embryonic cell movement. *Nature, 439*(7075), 480–483. https://doi.org/10.1038/nature04436.

Hunter, M. V., Willoughby, P., Bruce, A. E., & Fernandez-Gonzalez, R. (2018). Oxidative stress orchestrates cell polarity to promote embryonic wound healing. *Developmental Cell, 47*(3), 377–387.e4. https://doi.org/10.1016/j.devcel.2018.10.013.

Imboden, M., Goblet, C., Korn, H., & Vriz, S. (1997). Cytokeratin 8 is a suitable epidermal marker during zebrafish development. *Comptes Rendus de l'Académie Des Sciences, Series III, Sciences de La Vie, 320*(9), 689–700. https://doi.org/10.1016/s0764-4469(97)84816-0.

Kane, D., & Adams, R. (2002). Life at the edge: Epiboly and involution in the zebrafish. *Results and Problems in Cell Differentiation, 40*, 117–135.

Kane, D., Hammerschmidt, M., Mullins, M., Maischein, H., Brand, M., van Eeden, F., et al. (1996). The zebrafish epiboly mutants. *Development, 123*(47), 55.

Kane, D., & Kimmel, C. (1993). The zebrafish midblastula transition. *Development, 119*(2), 447–456.

Kane, D. A., McFarland, K. N., & Warga, R. M. (2005). Mutations in half baked/E-cadherin block cell behaviors that are necessary for teleost epiboly. *Development, 132*(5), 1105–1116. https://doi.org/10.1242/dev.01668.

Kane, D., Warga, R., & Kimmel, C. (1992). Mitotic domains in the early embryo of the zebrafish. *Nature, 360*(6406), 735–737.

Keller, R. (1980). The cellular basis of epiboly: An SEM study of deep-cell rearrangement during gastrulation in *Xenopus laevis*. *Journal of Embryology and Experimental Morphology, 60*(201), 234.

Keller, R. E., & Trinkaus, J. P. (1987). Rearrangement of enveloping layer cells without disruption of the epithelial permeability barrier as a factor in Fundulus epiboly. *Developmental Biology, 120*(1), 12–24. https://doi.org/10.1016/0012-1606(87)90099-6.

Kenyon, E. J., Campos, I., Bull, J. C., Williams, H. P., Stemple, D. L., & Clark, M. D. (2015). Zebrafish Rab5 proteins and a role for Rab5ab in nodal signalling. *Developmental Biology, 397*(2), 212–224. https://doi.org/10.1016/j.ydbio.2014.11.007.

Kimmel, C., Ballard, W., Kimmel, S., Ullmann, B., & Schilling, T. (1995). Stages of embryonic development of the zebrafish. *Developmental Dynamics, 203*(3), 253–310. https://doi.org/10.1002/aja.1002030302.

Kimmel, C. B., & Law, R. D. (1985). Cell lineage of zebrafish blastomeres II. Formation of the yolk syncytial layer. *Developmental Biology, 108*(1), 86–93. https://doi.org/10.1016/0012-1606(85)90011-9.

Kimmel, C., Warga, R., & Schilling, T. (1990). Origin and organization of the zebrafish fate map. *Development, 108*(4), 581–594.

Kondakova, E. A., & Efremov, V. I. (2014). Morphofunctional transformations of the yolk syncytial layer during zebrafish development. *Journal of Morphology*, *275*(2), 206–216. https://doi.org/10.1002/jmor.20209.

Köppen, M., Fernández, B., Carvalho, L., Jacinto, A., & Heisenberg, C.-P. (2006). Coordinated cell-shape changes control epithelial movement in zebrafish and drosophila. *Development*, *133*(14), 2671–2681. https://doi.org/10.1242/dev.02439.

Lachnit, M., Kur, E., & Driever, W. (2008). Alterations of the cytoskeleton in all three embryonic lineages contribute to the epiboly defect of Pou5f1/Oct4 deficient MZspg zebrafish embryos. *Developmental Biology*, *315*(1), 1–17. https://doi.org/10.1016/j.ydbio.2007.10.008.

Lentz, T., & Trinkaus, J. (1967). A fine structural study of cytodifferentiation during cleavage, blastula, and gastrula stages of *Fundulus heteroclitus*. *Journal of Cell Biology*, *32*(1), 121–138.

Lepage, S. E., Tada, M., & Bruce, A. E. (2014). Zebrafish dynamin is required for maintenance of enveloping layer integrity and the progression of epiboly. *Developmental Biology*, *385*(1), 52–66. https://doi.org/10.1016/j.ydbio.2013.10.015.

Lin, F., Chen, S., Sepich, D. S., Panizzi, J., Clendenon, S. G., Marrs, J. A., et al. (2009). Gα12/13 regulate epiboly by inhibiting E-cadherin activity and modulating the actin cytoskeleton. *The Journal of Cell Biology*, *184*(6), 909–921. https://doi.org/10.1083/jcb.200805148.

Li-Villarreal, N., Forbes, M. M., Loza, A. J., Chen, J., Ma, T., Helde, K., et al. (2015). Dachsous1b cadherin regulates actin and microtubule cytoskeleton during early zebrafish embryogenesis. *Development*, *142*(15), 2704–2718. https://doi.org/10.1242/dev.119800.

Marsal, M., Hernández-Vega, A., & Martin-Blanco, E. (2017). Contractility, differential tension and membrane removal lead zebrafish epiboly biomechanics. *Cell Cycle*, *16*(14), 1328–1335. https://doi.org/10.1080/15384101.2017.1327489.

Minc, N., & Piel, M. (2012). Predicting division plane position and orientation. *Trends in Cell Biology*, *22*(4), 193–200. https://doi.org/10.1016/j.tcb.2012.01.003.

Morita, H., Grigolon, S., Bock, M., Krens, G. S., Salbreux, G., & Heisenberg, C.-P. (2017). The physical basis of coordinated tissue spreading in zebrafish gastrulation. *Developmental Cell*, *40*(4), 354–366.e4. https://doi.org/10.1016/j.devcel.2017.01.010.

Morita, H., Kajiura-Kobayashi, H., Takagi, C., Yamamoto, T. S., Nonaka, S., & Ueno, N. (2012). Cell movements of the deep layer of non-neural ectoderm underlie complete neural tube closure in Xenopus. *Development*, *139*(8), 1417–1426. https://doi.org/10.1242/dev.073239.

Onichtchouk, D., Geier, F., Polok, B., Messerschmidt, D. M., Mössner, R., Wendik, B., et al. (2010). Zebrafish Pou5f1-dependent transcriptional networks in temporal control of early development. *Molecular Systems Biology*, *6*(1), 1–18. https://doi.org/10.1038/msb.2010.9.

Pei, W., Noushmehr, H., Costa, J., Ouspenskaia, M. V., Elkahloun, A. G., & Feldman, B. (2007). An early requirement for maternal FoxH1 during zebrafish gastrulation. *Developmental Biology*, *310*(1), 10–22. https://doi.org/10.1016/j.ydbio.2007.07.011.

Petridou, N. I., Grigolon, S., Salbreux, G., Hannezo, E., & Heisenberg, C.-P. (2019). Fluidization-mediated tissue spreading by mitotic cell rounding and non-canonical Wnt signalling. *Nature Cell Biology*, *21*(2), 169–178. https://doi.org/10.1038/s41556-018-0247-4.

Popgeorgiev, N., Bonneau, B., Ferri, K. F., Prudent, J., Thibaut, J., & Gillet, G. (2011). The apoptotic regulator Nrz controls cytoskeletal dynamics via the regulation of Ca2+ trafficking in the Zebrafish Blastula. *Developmental Cell*, *20*(5), 663–676. https://doi.org/10.1016/j.devcel.2011.03.016.

Rawson, D., Zhang, T., Kalicharan, D., & Jongebloed, W. (2000). Field emission scanning electron microscopy and transmission electron microscopy studies of the chorion, plasma membrane and syncytial layers of the gastrula-stage embryo of the zebrafish *Brachydanio rerio*: A consideration of the structural and functional relationships with respect to cryoprotectant penetration. *Aquaculture Research*, *31*(3), 325–336. https://doi.org/10.1046/j.1365-2109.2000.00401.x.

Sabel, J. L., d'Alençon, C., O'Brien, E. K., Otterloo, E., Lutz, K., Cuykendall, T. N., et al. (2009). Maternal interferon regulatory factor 6 is required for the differentiation of primary superficial epithelia in Danio and Xenopus embryos. *Developmental Biology*, *325*(1), 249–262. https://doi.org/10.1016/j.ydbio.2008.10.031.

Sagerström, C. G., Gammill, L. S., Veale, R., & Sive, H. (2005). Specification of the enveloping layer and lack of autoneuralization in zebrafish embryonic explants. *Developmental Dynamics*, *232*(1), 85–97. https://doi.org/10.1002/dvdy.20198.

Shimizu, T., Yabe, T., Muraoka, O., Yonemura, S., Aramaki, S., Hatta, K., et al. (2005). E-cadherin is required for gastrulation cell movements in zebrafish. *Mechanisms of Development*, *122*(6), 747–763. https://doi.org/10.1016/j.mod.2005.03.008.

Siddiqui, M., Sheikh, H., Tran, C., & Bruce, A. E. (2010). The tight junction component Claudin E is required for zebrafish epiboly. *Developmental Dynamics*, *239*(2), 715–722. https://doi.org/10.1002/dvdy.22172.

Slanchev, K., Carney, T. J., Stemmler, M. P., Koschorz, B., Amsterdam, A., Schwarz, H., et al. (2009). The epithelial cell adhesion molecule EpCAM is required for epithelial morphogenesis and integrity during zebrafish epiboly and skin development. *PLoS Genetics*, *5*(7) e1000563 https://doi.org/10.1371/journal.pgen.1000563.

Solnica-Krezel, L., & Driever, W. (1994). Microtubule arrays of the zebrafish yolk cell: Organization and function during epiboly. *Development*, *120*(9), 2443–2455.

Song, S., Eckerle, S., Onichtchouk, D., Marrs, J. A., Nitschke, R., & Driever, W. (2013). Pou5f1-dependent EGF expression controls E-cadherin endocytosis, cell adhesion, and zebrafish epiboly movements. *Developmental Cell*, *24*(5), 486–501. https://doi.org/10.1016/j.devcel.2013.01.016.

Strähle, U., & Jesuthasan, S. (1993). Ultraviolet irradiation impairs epiboly in zebrafish embryos: Evidence for a microtubule-dependent mechanism of epiboly. *Development*, *119*(3), 909–919.

Sumi, A., Hayes, P., D'Angelo, A., Colombelli, J., Salbreux, G., Dierkes, K., & Solon, J. (2018). adherens junction length during tissue contraction is controlled by the mechanosensitive activity of actomyosin and junctional recycling. *Developmental Cell*, *47*(4), 453–463.e3, Pubmed ID: 30458138. https://doi.org/10.1016/j.devcel.2018.10.025.

Szabó, A., Cobo, I., Omara, S., McLachlan, S., Keller, R., & Mayor, R. (2016). The molecular basis of radial intercalation during tissue spreading in early development. *Developmental Cell*, *37*(3), 213–225. https://doi.org/10.1016/j.devcel.2016.04.008.

Trinkaus, J. (1951). A study of the mechanism of epiboly in the egg of *Fundulus heteroclitus*. *Journal of Experimental Zoology*, *118*(2), 269–319. https://doi.org/10.1002/jez.1401180204.

Trinkaus, J. (1984). Mechanism of *Fundulus* epiboly—A current view. *American Zoologist*, *24*(3), 673–688. https://doi.org/10.1093/icb/24.3.673.

Trinkaus, J., & Lentz, T. (1967). Surface specializations of *Fundulus* cells and their relation to cell movements during gastrulation. *Journal of Cell Biology*, *32*(1), 139–153.

Warga, R., & Kimmel, C. (1990). Cell movements during epiboly and gastrulation in zebrafish. *Development*, *108*(4), 569–580.

Weng, J.-H., Liang, M.-R., Chen, C.-H., Tong, S.-K., Huang, T.-C., Lee, S.-P., et al. (2013). Pregnenolone activates CLIP-170 to promote microtubule growth and cell migration. *Nature Chemical Biology*, *9*(10), 636–642. https://doi.org/10.1038/nchembio.1321.

West, J. J., Zulueta-Coarasa, T., Maier, J. A., Lee, D. M., Bruce, A., Fernandez-Gonzalez, R., et al. (2017). An actomyosin-Arf-GEF negative feedback loop for tissue elongation under stress. *Current Biology*, *27*(15), 2260–2270.e5. https://doi.org/10.1016/j.cub.2017.06.038.

Wilson, E., Cretekos, C., & Helde, K. (1995). Cell mixing during early epiboly in the zebrafish embryo. *Developmental Genetics*, *17*(1), 6–15. https://doi.org/10.1002/dvg.1020170103.

Wirtz-Peitz, F., & Zallen, J. A. (2009). Junctional trafficking and epithelial morphogenesis. *Current Opinion in Genetics & Development*, *19*(4), 350–356. https://doi.org/10.1016/j.gde.2009.04.011.

Xiong, F., Ma, W., Hiscock, T. W., Mosaliganti, K. R., Tentner, A. R., Brakke, K. A., et al. (2014). Interplay of cell shape and division orientation promotes robust morphogenesis of developing epithelia. *Cell*, *159*(2), 415–427. https://doi.org/10.1016/j.cell.2014.09.007.

Zalik, S. E., Lewandowski, E., Kam, Z., & Geiger, B. (1999). Cell adhesion and the actin cytoskeleton of the enveloping layer in the zebrafish embryo during epiboly. *Biochemistry and Cell Biology*, *77*(6), 527–542. https://doi.org/10.1139/o99-058.

Further reading

Cha, Y. I., Kim, S.-H., Sepich, D., Buchanan, G. F., Solnica-Krezel, L., & DuBois, R. N. (2006). Cyclooxygenase-1-derived PGE2 promotes cell motility via the G-protein-coupled EP4 receptor during vertebrate gastrulation. *Genes & Development*, *20*(1), 77–86. https://doi.org/10.1101/gad.1374506.

Speirs, C. K., Jernigan, K. K., Kim, S.-H., Cha, Y. I., Lin, F., Sepich, D. S., et al. (2010). Prostaglandin Gbetagamma signaling stimulates gastrulation movements by limiting cell adhesion through Snai1a stabilization. *Development*, *137*(8), 1327–1337. https://doi.org/10.1242/dev.045971.

CHAPTER TWELVE

Zebrafish gastrulation: Putting fate in motion

Diana Pinheiro, Carl-Philipp Heisenberg*
Institute of Science and Technology Austria, Klosterneuburg, Austria
*Corresponding author: e-mail address: heisenberg@ist.ac.at

Contents

Abstract

Gastrulation entails specification and formation of three embryonic germ layers—ectoderm, mesoderm and endoderm—thereby establishing the basis for the future body plan. In zebrafish embryos, germ layer specification occurs during blastula and early gastrula stages (Ho & Kimmel, 1993), a period when the main morphogenetic movements underlying gastrulation are initiated. Hence, the signals driving progenitor cell fate specification, such as Nodal ligands from the TGF-β family, also play key roles in regulating germ layer progenitor cell segregation (Carmany-Rampey & Schier, 2001; David & Rosa, 2001; Feldman et al., 2000; Gritsman et al., 1999; Keller et al., 2008). In this review, we summarize and discuss the main signaling pathways involved in germ layer progenitor cell fate specification and segregation, specifically focusing on recent advances in understanding the interplay between mesoderm and endoderm specification and the internalization movements at the onset of zebrafish gastrulation.

Gastrulation is the process whereby a seemingly uniform mass of cells is transformed into a multilayered embryo composed of distinct germ layers. In triploblastic organisms, such as zebrafish, the gastrula consists of three germ layers: (i) ectoderm, which gives rise to epidermis and nervous system;

Current Topics in Developmental Biology, Volume 136
ISSN 0070-2153
https://doi.org/10.1016/bs.ctdb.2019.10.009

(ii) mesoderm, which will produce bone, muscle, connective tissue, urogenital and circulatory systems; and (iii) endoderm, which gives rise to the gastrointestinal tract and associated structures. Gastrulation requires both the activity of key signaling pathways to induce germ layer specification and patterning, as well as large scale cell movements to segregate germ layer progenitors into distinct and highly reproducible positions along the future body axes. Although the precise patterning and morphogenetic events occurring during gastrulation vary across species, the signaling modules controlling them are not only remarkably conserved (e.g. Nodal, Wnt), but also play critical roles in other key developmental processes, such as left-right asymmetry, organ morphogenesis and tissue homeostasis (reviewed in Nusse & Clevers, 2017; Pauklin & Vallier, 2015; Shen, 2007).

In zebrafish, embryo morphogenesis is initiated with the onset of epiboly movements, which spread the blastoderm, consisting of a superficial enveloping layer (EVL) and the deep cells (DCs), over the yolk cell (for review, see chapter "Mechanisms of zebrafish epiboly: A current view" by Bruce and Heisenberg in the same issue). Gastrulation starts at mid-epiboly (soon after 5 h post-fertilization, hpf) with internalization of mesoderm and endoderm (mesendoderm) progenitors at the germ ring margin below the non-internalizing ectoderm, and continues for several hours until the blastoderm completely engulfs the yolk cell (10 hpf) (Kimmel, Ballard, Kimmel, Ullman, & Schilling, 1995). Since mesendoderm internalization starts earlier and is more prominent on the dorsal side of the embryo (Keller, Schmidt, Wittbrodt, & Stelzer, 2008; Kimmel et al., 1995), the dorsal-ventral (DV) axis becomes morphologically distinguishable at this stage. Following internalization, mesendoderm progenitors migrate away from the margin and further differentiate into mesoderm or endoderm (Chng, Ho, Tian, & Reversade, 2013; Dumortier, Martin, Meyer, Rosa, & David, 2012; Keller et al., 2008; Montero et al., 2005; Montero, Kilian, Chan, Bayliss, & Heisenberg, 2003; Myers, Sepich, & Solnica-Krezel, 2002; Norris et al., 2017; Pauli et al., 2014; Pézeron et al., 2008; Sepich, Calmelet, Kiskowski, & Solnica-Krezel, 2005; Warga & Nusslein-Volhard, 1999). From 70% epiboly onwards, DCs initiate convergence and extension movements, thereby establishing the anterior-posterior body axis (Fig. 1A) (Glickman, Kimmel, Jones, & Adams, 2003; Keller et al., 2008; Kimmel et al., 1995; Myers et al., 2002; Sepich et al., 2005).

In this review, we will summarize recent advances in understanding the key signaling pathways underlying germ layer specification along with the collective cell behaviors shaping the zebrafish gastrula. Since mesendoderm

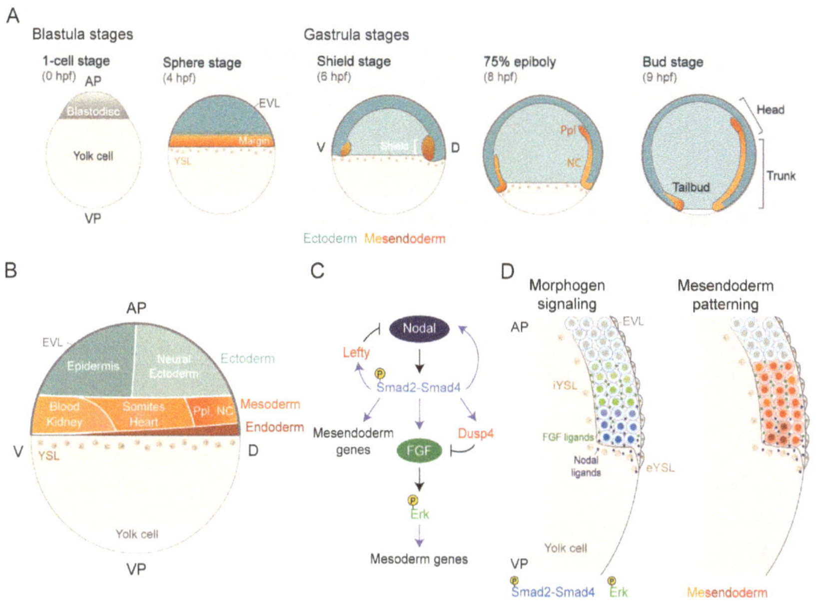

Fig. 1 Mesendoderm induction and patterning in zebrafish embryos. (A) Schematic representation of zebrafish development. Developmental stages are indicated at the top and the approximate time point is shown below in hours post-fertilization (hpf). Embryos are oriented with the animal pole to the top and the dorsal side to the right. Upon fertilization, the blastodisc positioned on top of the yolk cell initiates several rounds of rapid and metasynchronous cell cleavages to generate the blastula embryo. Mid-blastula transition (MBT) starts at 512-cell stage and marks the onset of zygotic transcription. Soon after, the blastoderm marginal cells collapse into the yolk to form the yolk syncytial layer (not shown). Around 4 hpf, the sphere stage embryo, consisting of a superficial enveloping layer and the prospective germ layer progenitors underneath, initiates morphogenesis. Gastrulation starts after mid-epiboly with specification and internalization of mesendoderm progenitors (color-coded in orange and red) at the germ ring margin under the prospective ectoderm cells (colored in light blue). By the end of gastrulation, epiboly movements have spread the blastoderm over the entire yolk cell and the future body axes have been established (Kimmel et al., 1995). (B) Zebrafish fate map at the onset of gastrulation. Ectoderm progenitors are depicted in blue and are found in more animal positions, while mesoderm and endoderm progenitors are colored in orange and red, respectively, and are found in more marginal regions of the blastoderm. Both mesodermal and ectodermal cell fates are arranged along the dorsal-ventral axis and color-coded in shades of orange and blue. Although tissue territories are outlined here, these do not intend to designate precise tissue boundaries, as cell fates are often intermingled in the early zebrafish gastrula (Kimmel, Warga, & Schilling, 1990). (C) Schematic diagram depicting the main signaling pathways controlling mesendoderm cell fate specification in zebrafish embryos. Black and blue arrows indicate protein-protein interactions and transcriptional regulations, respectively. Binding of Nodal ligands to its receptors initiates a cascade of events ultimately resulting in phosphorylation of Smad2 and its subsequent association with

(Continued)

cell fate is plastic and dependent on positional information until the early gastrula stage (Ho & Kimmel, 1993), we discuss potential mechanisms coupling cell commitment and tissue morphogenesis during zebrafish gastrulation, a key open question in developmental biology.

1. The fundamentals of germ layer specification and patterning

In zebrafish, as in frogs, formation of the dorsal organizer (also known as shield, Glossary) and germ layer specification are associated with the establishment of the DV axis. Soon after fertilization, DV patterning is initiated by microtubule-dependent transport of dorsal determinants encoding components of the canonical Wnt signaling pathway deposited in the egg, for instance Huluwa, into marginal blastomeres at the future dorsal side of the blastula (for review see chapter "Setting up for gastrulation in zebrafish" by Marlow in the same issue). The activity of these dorsal determinants along with the antagonizing functions of Sfrp1a and Frzb in the remaining blastoderm result in localized activation of canonical Wnt signaling at the dorsal organizer (Gore et al., 2005; Langdon & Mullins, 2011; Lu, Thisse, & Thisse, 2011; Schneider, Steinbeisser, Warga, & Hausen, 1996; Tran et al., 2012;

Fig. 1—Cont'd Smad4. Activated Smad2-Smad4 complexes then accumulate in the nucleus, where they cooperate with other transcription factors (not shown) to promote mesendoderm specification by inducing transcription of both Nodal and FGF ligands, as well as inducing genes directly implicated in mesendodermal cell commitment (e.g. Gsc, Sox32, not shown). Nodal activity is regulated extracellularly by Lefty1/2 antagonists and members of the DAN family (not shown). FGF signaling is established at a distance from the germ ring margin, due to the short-range activity of its inhibitor Dusp4. Upon FGF activation, phosphorylated Erk (P-Erk) results in expression of genes implicated in acquisition of a mesodermal fate (e.g. Noto, Ntl, not shown). (D) Schematic representation of the Nodal and FGF signaling gradients and resulting mesendoderm induction and patterning within the germ ring margin at 50% epiboly. Embryo cross section is shown with the animal pole to the top. The respective size of the Nodal and FGF signaling domains is shown proportionally and not in exact number of cell rows. While Nodal signaling, represented by the nuclear accumulation of P-Smad2, is active within the first 5 cell rows from the blastoderm margin, FGF signaling, denoted here by nuclear accumulation of P-Erk, is active up to about 10 cell tiers from the YSL. Abbreviations: AP, Animal pole; VP, Vegetal pole; D, Dorsal side; V, Ventral side; EVL, Enveloping layer; YSL, Yolk syncytial layer (iYSL refers to internal YSL, eYSL designates the external YSL); Ppl, Prechordal plate; NC, Notochord. *Panel (B): Adapted from Kimelman, D. (2006) Mesoderm induction: From caps to chips.* Nature Reviews. Genetics 7, *360–372; Schier, A.F., & Talbot, W.S. (2005). Molecular genetics of axis formation in zebrafish.* Annual Review of Genetics 39, *561–613.*

Yan et al., 2018). As a result, β-catenin (β-cat) translocates into the nucleus of dorsal blastomeres, where it associates with members of the Lef family to activate expression of dorsal-specific genes (e.g. Bozozok) required to counteract the action of key ventralizing factors, namely members of the Bone morphogenetic protein family (BMPs) and the transcriptional regulators vox/vent/ved, and initiate mesendoderm specification dorsally (Leung et al., 2003; Melby, Beach, Mullins, & Kimelman, 2000; Ryu et al., 2001; Schulte-Merker, Lee, McMahon, & Hammerschmidt, 1997; Shimizu et al., 2000, 2002; Stachel, Grunwald, & Myers, 1993; Yamanaka et al., 1998). Indeed, expression of the Nodal-related protein 1 (Ndr1, also known as Squint, Sqt) is first induced by β-cat, and subsequent activation of the Nodal signaling pathway is critical for mesendoderm specification and patterning (Feldman et al., 1998; Gritsman et al., 1999; Long, Ahmad, & Rebagliati, 2003; Rebagliati, Toyama, Haffter, & Dawid, 2002; Shimizu et al., 2000). Nodal signaling also contributes to the establishment of the dorsal organizer by inducing BMP antagonists, such as Chordin, at the dorsal margin (Schulte-Merker et al., 1997). As a result, canonical Wnt signaling, and consequently Nodal signaling, is high at the dorsal side of the embryo, while BMP signaling is active in the ventrolateral domain. The antagonistic interplay between these signaling pathways ultimately establishes the DV axis and initiates mesendoderm specification dorsally. Consistent with a central role for Nodal and BMP signals in patterning the DV axis, ectopically generating opposing gradients of these signals in the zebrafish embryo is sufficient to induce an entire secondary axis (Xu, Houssin, Ferri-Lagneau, Thise, & Thisse, 2014).

1.1 Nodal signaling and mesendoderm patterning

Nodal/TGF-β ligands were identified in organisms ranging from hydra to human and shown to play an essential role in early axis patterning in all deuterostomes and, more recently also in the cnidarian *Hydra magnipapillata* (Duboc, Röttinger, Besnardeau, & Lepage, 2004; Flowers, Courteau, Poustka, Weng, & Venuti, 2004; Morokuma, Ueno, Kawanishi, Saiga, & Nishida, 2002; Shen, 2007; Watanabe et al., 2014). In zebrafish, only Ndr1 (Sqt) and Ndr2 (also known as Cyclops, Cyc) are necessary for germ layer specification (Feldman et al., 1998; Gritsman et al., 1999; Long et al., 2003; Rebagliati et al., 2002). An additional TGF-β ligand, termed Growth differentiation factor 3 (Gdf3, also known as Dvr1 or Vg1), was recently shown to be maternally provided and essential for Nodal signaling and mesendoderm specification in zebrafish embryos (Bisgrove, Su, & Yost, 2017;

Montague & Schier, 2017; Pelliccia, Jindal, & Burdine, 2017). Since Ndr1 and 2 can form heterodimers with Gdf3, it was proposed that Gdf3 and the Nodal ligands may function, at least to some extent, as cofactors to elicit fast and robust activation of Nodal signaling during early embryonic development (Bisgrove et al., 2017; Montague & Schier, 2017; Pelliccia et al., 2017). Yet, when and where pre-existing Gdf3 associates with secreted Nodal ligands, and whether Gdf3 indeed elicits Nodal signaling or rather acts to increase the potency and/or range of Nodal, as proposed in other systems (Shen, 2007), still requires further investigation.

In zebrafish, Nodal ligands signal through a triple receptor complex composed of type I and II Activin receptors and a coreceptor known as One-Eyed-Pinhead (Oep). Receptor complex activation results in phosphorylation of the intracellular signal transducers Smad2/3, and their subsequent association with Smad4. Consistent with previous observations in frog and mouse embryos, Smad2 appears to be the main transducer of Nodal signaling during early zebrafish development (Dubrulle et al., 2015; Howell, Mohun, & Hill, 2001; Robertson, 2014). Accordingly, maternal-zygotic Smad2 mutants lack both endoderm and mesoderm, a phenotype identical to that of Nodal loss-of-function (Dubrulle et al., 2015; Feldman et al., 1998; Gritsman et al., 1999). Upon Smad4 binding, phosphorylated Smad2/4 complexes are then retained in the nucleus and associate with additional transcription factors (e.g. Fork Head Box H1 and Mixer) to regulate target gene expression (reviewed in Massagué, 2012; Hill, 2018). Among those target genes are the Nodal ligands themselves along with their secreted inhibitors Lefty1/2, as well as key genes involved in mesoderm (e.g. Goosecoid (Gsc), No tail (Ntl)—also known as Brachyury and T) and endoderm (Sox32—also known as Casanova, Cas) specification (Fig. 1C) (Bennett et al., 2007; Bisgrove, Yost, & Essner, 1999; Dickmeis et al., 2001; Meno et al., 1999; Pogoda, Solnica-Krezel, Driever, & Meyer, 2000). While recent studies advanced our understanding of the molecular mechanisms involved in the establishment of the Nodal signaling domain *in vivo* and its ability of induce cell fate in a dose-dependent manner, key open questions in this field remain under active research.

1.1.1 Setting up the Nodal signaling domain

Nodal signaling in zebrafish is initiated dorsally by maternal activation of Wnt/β-catenin signaling (Dougan, Warga, Kane, Schier, & Talbot, 2003; Harvey & Smith, 2009; Shimizu et al., 2000; Van et al., 2015). Prior to the start of gastrulation, however, nuclear localization of phosphorylated-Smad2 (P-Smad2), the main downstream effector of Nodal signaling during zebrafish gastrulation, can be detected along the entire DV axis of the blastoderm margin and extends ~5–6 cell rows towards the animal pole, with the highest nuclear accumulation in the most marginal region of the blastoderm (not shown, Fig. 1D) (Dougan et al., 2003; Dubrulle et al., 2015; Harvey & Smith, 2009; Van et al., 2015). The molecular and biophysical mechanisms underlying this apparent spreading of the Nodal signaling domain are still under intense investigation (reviewed in Hill, 2018; Rogers & Müller, 2019). Since Nodal signaling can trigger the expression of additional Nodal ligands and its own antagonists Lefty1/2, it was proposed that the Nodal signaling domain is established through a reaction-diffusion patterning system (Bennett et al., 2007; Bisgrove et al., 1999; Chen & Schier, 2001, 2002; Dickmeis et al., 2001; Meno et al., 1999; Müller et al., 2012). This theoretical framework, formulated by Alan Turing, suggests that chemicals with differential diffusivity and able to cross-react, e.g. as an activator-inhibitor pair, are sufficient to elicit pattern formation during development (Glossary, reviewed in Kondo & Miura, 2010). Consistent with this, classical experiments showed that ectopic Nodal, secreted from a localized source, activates target gene expression in surrounding tissues, while concomitant expression of the Nodal antagonist Lefty limits the range of Nodal signaling activity (Chen & Schier, 2001, 2002; Jing, Zhou, Wang, & Chen, 2006; Tian, Andree, Jones, & Sampath, 2008). Subsequent work using photo-convertible versions of Ndr1/2 and Lefty1/2 showed that Ndr2 displays lower effective diffusion than Ndr1, while Lefty1/2 diffuse faster than either Nodal ligand (Müller et al., 2012). Recent *in vitro* work further supports the notion that a synthetic circuit driven by the different diffusion ranges of the Nodal/Lefty pair is sufficient to elicit pattern formation in HEK293 cells (Sekine, Shibata, & Ebisuya, 2018). While this is a plausible model, several assumptions remain to be experimentally tested. For instance, diffusivity of the endogenous Nodal/Lefty pair is still unknown. Moreover, Nodal protein stability, in addition to its diffusivity, has long been proposed to correlate with its signaling range (Jing et al., 2006; Tian et al., 2008; Wang, Wang, Wohland, & Sampath, 2016), indicating that both properties need to be considered when analyzing Nodal and Lefty activities in a reaction-diffusion system.

Evidence for a relay-based mechanism, where Nodal ligands effectively expand the Nodal signaling domain by activating their own expression in neighboring cells (Glossary), has also emerged. Indeed, Ndr1/2 ligand secretion by the yolk syncytial layer (YSL), a multinucleate syncytium formed at the yolk cell surface between 1000-cell and high-stage, was shown to promote Nodal signaling activation, and subsequent expression of additional Nodal ligands, in the overlying blastoderm, thereby expanding the domain of active Nodal signaling to the ventrolateral side of zebrafish embryos (Hong, Jang, Brown, McBride, & Feldman, 2011; Van et al., 2015; Veil et al., 2017; Xiang et al., 2007; Xu et al., 2012). As Nodal ligands and their antagonists Lefty1/2 are coexpressed in cells with active Nodal signaling, spreading of the Nodal signaling domain during early embryonic development requires an additional mechanism (Van et al., 2015). Such mechanism is thought to rely on expression of the microRNA miR-430, which transiently delays the onset of Nodal-induced Lefty1/2 production. Once the delay in Lefty translation is overcome, these antagonists then restrict relay-mediated spreading of the Nodal signaling domain to ~5–6 cell rows from the blastoderm margin towards the animal pole (Fig. 1D) (Van et al., 2015). The postulated joint action of miR-430 and the relay-based mechanism predict that the final size of the Nodal signaling domain is set by the duration of Lefty1/2 translational repression, which effectively controls the extent of Nodal ligand accumulation within the blastoderm margin. This model further posits that the marginal cells located closest to the YSL are the first cells to activate Nodal expression and signaling and, therefore signal for the longest period. Consistent with this, nuclear accumulation of P-Smad2 displays an AV gradient (Fig. 1D) (Van et al., 2015). Yet, Lefty1/2 double mutants were recently shown to display only a slight expansion of the Nodal signaling domain (Rogers et al., 2017), raising questions as to the importance of limiting Lefty translation for spatiotemporally controlling Nodal activation, and further suggesting that additional Nodal antagonists may be required to control the AV size of its signaling domain.

Interestingly, more recent observations, showing that FGF signaling can activate some Nodal target genes beyond the Nodal expression domain (Chen & Schier, 2001; Rodaway et al., 1999; Van et al., 2015; van Boxtel, Economou, Heliot, & Hill, 2017), suggest that the Nodal signaling domain might in fact be smaller than previously anticipated from the expression pattern of its downstream targets. This further suggests that FGF signaling is able to induce Nodal target gene expression at larger distances from the blastoderm margin (6–10 cell rows, Fig. 1D), thereby pointing at a potential secondary relay-mechanism, whereby spreading of Nodal signaling is achieved by the induction of a secondary signaling molecule with a longer signaling range (Glossary). Consistent with this, Nodal activation was found to induce the expression of Fgf8/3 ligands, which can then signal over larger distances to induce target gene expression, for instance Ntl, up to 10 cell rows away from the germ ring margin (Fig. 1C and D) (Bennett et al., 2007; Mathieu et al., 2004; Rodaway et al., 1999; Scholpp & Brand, 2004; Van et al., 2015; van Boxtel et al., 2017). Concomitantly, Nodal signaling is also thought to limit short-range FGF signaling by inducing the expression of the FGF inhibitor Dusp4 (van Boxtel et al., 2017). This short-range inhibition might contribute to the segregation of mesendoderm precursors into mesoderm and endoderm lineages, since active FGF signaling specifically induces mesoderm specification (Fig. 1 C and D) (Mathieu et al., 2004; Rodaway et al., 1999; Van et al., 2015; van Boxtel et al., 2017).

Collectively, auto-regulation, differential diffusivities of the Nodal/Lefty pair and an FGF-dependent secondary relay mechanism are likely to cooperate to establish a gradient of Nodal signaling along the DV axis of the blastoderm margin and induce the expression of critical Nodal targets beyond the Nodal expression domain. Further work is needed to understand the relative contribution of these mechanisms for the endogenous distribution of Nodal signaling domain and robust germ layer patterning. Indeed, recent studies showed that, although Lefty diffusivity is not essential for the initial establishment of the Nodal signaling domain, it is required to ensure correct scaling of the size of the Nodal signaling domain with embryo size (Almuedo-Castillo et al., 2018). Expressing membrane-tethered versions of Nodal and FGF ligands, as recently employed for investigating *Drosophila* wing disc patterning (Alexandre, Baena-Lopez, & Vincent, 2014), and using optogenetic tools to obtain high spatiotemporal control over the activation status of these signaling pathways (Sako et al., 2016) might help in dissecting the specific contributions of Nodal and FGF signaling in mesendoderm cell fate specification and patterning at the onset of zebrafish gastrulation.

1.1.2 Dose-dependent responses to Nodal signaling

Nodal ligands were long proposed to function as morphogens (Glossary) and be able to generate dose-dependent responses within developing tissues (see French-Flag Model, Glossary). Consistent with this, low levels of Nodal signaling are sufficient to induce the expression of Nodal target genes expressed in the notochord, such as Ntl or Floating head (Flh), whereas the expression of target genes in the endoderm and anterior axial mesendoderm (prechordal plate, ppl), such as Sox32 and Gsc, requires considerably higher

Nodal activity (Gritsman, Talbot, & Schier, 2000). Yet, how dose-dependent responses to Nodal signaling during embryonic development integrate the level and duration of Nodal signaling is still unclear. Supporting the notion that signaling duration is also critical for mesendoderm patterning, endoderm and ppl specification were shown to require sustained Nodal activation, whereas transient activation is sufficient for notochord (and non-axial mesoderm) induction (Fig. 1B) (Aoki et al., 2002; Dougan et al., 2003; Dubrulle et al., 2015; Gritsman et al., 2000; Hagos & Dougan, 2007) Moreover, manipulating the duration of Nodal signaling directly, using photoactivatable versions of the Nodal receptors, showed that extending the duration of Nodal signaling results in ppl specification at the expense of endoderm cells (Sako et al., 2016).

In addition to the distribution of Nodal ligands and the duration of active Nodal signaling, the kinetics of target gene induction was recently proposed to affect the response of target genes and set their spatial range of expression upon Nodal activation (Dubrulle et al., 2015). This kinetic view of morphogen signaling integrates target gene-specific properties likely to modify their transcriptional activation, e.g. affinity for active Smad2, maximum transcription rate and competence to respond to active Smad2. Here, the onset of induction and rate of transcription appear to be the best predictors for the spatial range of target gene expression, with earlier and higher rates of transcription correlating with a wider expression range (Dubrulle et al., 2015). Given that FGF signaling was recently shown to expand the expression range of some Nodal targets, such as Ntl, beyond the Nodal expression domain (van Boxtel et al., 2017), it will be interesting to evaluate to what extent the spatial range of Nodal targets is set by the kinetics of both Nodal and FGF target gene induction.

1.2 Ectoderm specification: The *default* cell fate?

While mesendoderm progenitors are induced close to the germ ring margin upon Nodal signaling activation, the prospective ectoderm originates from more animal positions within the zebrafish gastrula (Fig. 1A and B). Early ectoderm specification was proposed to represent a sort of default developmental state arising in the absence or upon reduced Nodal signaling activity (Ho & Kimmel, 1993; Kiecker, Bates, & Bell, 2016). Consistent with this, mesoderm repression results in ectoderm expansion, whereas mesoderm expansion comes at the expense of the prospective ectoderm (Feldman et al., 2002; Feldman, Dougan, Schier, & Talbot, 2000;

Gritsman et al., 1999; Londin, Mentzer, & Sirotkin, 2007; Thisse, Wright, & Thisse, 2000). Ectoderm patterning into different neural territories, in contrast, requires the activity of additional factors, namely BMP, FGF, Retinoid acid and Wnt signals (reviewed in Schier & Talbot, 2005). Although all cells in the embryo are initially competent to respond to Nodal signaling, prospective ectoderm progenitors lose this competence over time (Ho & Kimmel, 1993). Progressive downregulation of Oep during gastrulation was recently implicated in reducing the competence of blastoderm cells to respond to Nodal signals (Vopalensky, Pralow, & Vastenhouw, 2018; Zhang, Talbot, & Schier, 1998). Yet, additional mechanisms are likely to be required to restrict mesendodermal competence, since Oep overexpression does not result in a dramatic developmental phenotype (Zhang et al., 1998). Churchill, a zinc finger transcription factor previously implicated in limiting the transcriptional response to Nodal signaling (Londin et al., 2007), might also be involved in reducing mesendodermal competence during gastrulation, although genetic evidence is still lacking.

2. Shaping the zebrafish gastrula: Cell and tissue morphogenesis

Gastrulation involves a complex set of highly coordinated cell movements that cooperate to set up the future body axes (Fig. 2A). Four main morphogenetic movements cooperate to shape the developing zebrafish embryo: (i) epiboly, which distributes the blastoderm over the yolk cell; (ii) internalization (also called emboly), which positions the mesoderm and endoderm progenitors underneath the prospective ectoderm; (iii) convergence movements, which narrow the three germ layers dorsally and mediolaterally; and finally, (iv) extension movements, which elongate the embryo from head to tail (Fig. 2A). Remarkably, these 'morphogenetic modules' are consistently observed among Metazoa, although the underlying cell behaviors are not always conserved (reviewed in Solnica-Krezel & Sepich, 2012). Since the interplay between germ layer cell fate specification and morphogenesis is particularly apparent at the onset of zebrafish gastrulation, we focus on mesendoderm internalization and its subsequent animal pole-directed migration.

2.1 Mesendoderm internalization movements

Soon after 50% epiboly, the first mesendoderm progenitor cells are specified at the blastoderm margin, leading to the formation of a transient

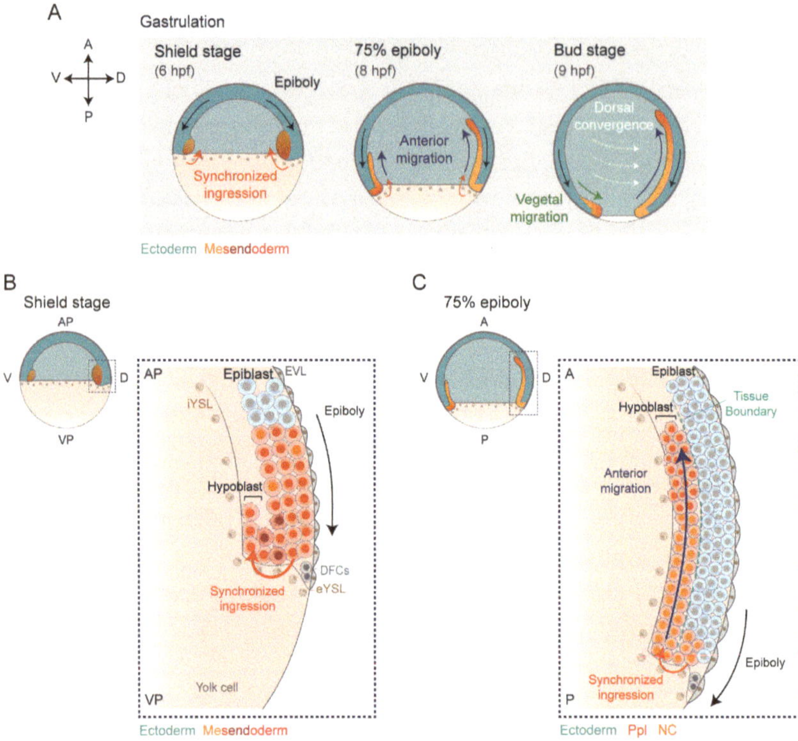

Fig. 2 Gastrulation movements in zebrafish embryos. (A) Schematic representation of the main morphogenetic movements during zebrafish gastrulation. Developmental stages and the approximate time point, shown in hours post-fertilization (hpf), are indicated at the top. Cross section with animal/anterior up and the dorsal side to the right. Gastrulation starts after mid-epiboly with internalization (orange arrows) of mesendoderm progenitors (color-coded in orange and red) at the germ ring margin underneath the prospective ectoderm (colored in light blue). Following internalization, mesendoderm cells migrate towards the animal pole (blue arrows). While in the dorsal region, mesendoderm progenitors migrate towards the anterior for several hours, ventral-derived mesoderm progenitors eventually change direction and rather move vegetally (green arrows). From 70% epiboly onwards, ectoderm and mesendoderm progenitors in the lateral region initiate convergence (white arrows) and extension movements, thereby forming the anterior-posterior body axis. (B) Mesendoderm progenitor cell internalization at shield stage. Embryo cross section shown with animal pole up and the dorsal side to the right. The lower panel is a magnified representation of the boxed area above. Mesoderm and endoderm progenitors specified at the germ ring margin ingress in a coordinated manner beneath the epiblast. The onset of internalization movements gives rise to the formation of two cell layers—the hypoblast, composed of post-involution mesendoderm progenitors and the epiblast, which consists of ectoderm progenitors and pre-involution mesendoderm progenitors. (C) Anterior migration of dorsal axial mesendoderm tissue at 75% epiboly. Embryo cross section shown with anterior up and the dorsal side to the right. The lower panel is a magnified representation of the boxed area above and highlights anterior migration of the dorsal axial mesendoderm and the characteristic separation of the germ layers at the epiblast/hypoblast boundary during gastrulation. Abbreviations: AP, Animal pole; VP, Vegetal pole; A, Anterior axis; P, Posterior axis; D, Dorsal side; V, Ventral side; EVL, Enveloping layer; YSL, Yolk syncytial layer (iYSL refers to internal YSL, eYSL designates the external YSL); DFCs, Dorsal Forerunner cells; Ppl, Prechordal plate; NC, Notochord.

thickening of the margin termed 'germ ring' (Keller et al., 2008; Montero et al., 2005; Warga & Kimmel, 1990). Mesendoderm progenitors begin to internalize at the dorsal side of the germ ring, with cells moving over the margin and underneath the non-internalizing ectoderm progenitors (Keller et al., 2008; Warga & Kimmel, 1990). The internalization movements subsequently spread along the entire germ ring margin (Fig. 2B) (Keller et al., 2008; Warga & Kimmel, 1990). Dorsal mesendoderm internalization becomes apparent by the formation of the embryonic 'shield' (Fig. 1A), a structure homologous to the blastopore lip in frogs or the Hensen node in chick (Solnica-Krezel & Sepich, 2012). The internalization movements ultimately result in the formation of two cell layers: the outer epiblast, composed of future ectoderm progenitors and pre-internalization mesendoderm progenitors, and the inner hypoblast, consisting of internalized mesendoderm progenitors (Fig. 2B and C) (Keller et al., 2008; Montero et al., 2005; Warga & Kimmel, 1990). Once internalized, mesendodermal progenitors first migrate away from the germ ring margin and eventually lateral-derived progenitor cells engage in dorsal-directed convergence movements to form the body axis (Fig. 2A) (Dumortier et al., 2012; Montero et al., 2003, 2005; Norris et al., 2017; Pauli et al., 2014; Pézeron et al., 2008; Sepich et al., 2005). In contrast to *Xenopus* embryos, mesendoderm internalization in zebrafish appears to be mostly independent of subsequent convergence and extension movements, as in zebrafish mutants with strong defects in these morphogenetic movements, mesendoderm internalization appears largely unaffected (Solnica-Krezel et al., 1996).

Although the relative position of the germ layers is evolutionarily conserved, the cellular behaviors underlying mesendoderm internalization differ between species. Broadly speaking, differences arise depending on the tissue type where mesendoderm progenitors are initially specified—epithelia in ascidians, worms, flies, chick and mouse or mesenchyme in frogs and fish—and whether mesendoderm cells internalize as individual cells, a movement called 'ingression,' or as a coherent cell mass in a movement commonly termed 'invagination' or 'involution' (reviewed in Solnica-Krezel & Sepich, 2012). Zebrafish mesendoderm exhibits mixed characteristics—although progenitors are adherent and internalization is highly coordinated among neighboring cells, single mesendoderm cells can also internalize autonomously (Babb & Marrs, 2004; Carmany-Rampey & Schier, 2001; D'Amico & Cooper, 1997, 2001; David & Rosa, 2001; Keller et al., 2008; Montero et al., 2005; Shih & Fraser, 1995). This led to the proposal that mesendoderm progenitors undergo 'synchronized cell ingression' (Adams & Kimmel, 2004).

The molecular and cellular mechanisms underlying mesendoderm internalization remain poorly understood, and only a small number of genes has so far been implicated in specifically regulating this morphogenetic process in zebrafish. In particular, it is still unclear (i) how mesendodermal cell fate specification and internalization are related to one another; (ii) how mesendoderm ingression is initiated and synchronized among coingressing cells; (iii) what cue(s) control mesendoderm cell polarization; and finally, (iii) which molecular and cellular mechanisms control boundary formation at the epiblast-hypoblast interface. In the next sections, we will address these questions in detail and discuss the state-of-the-art in the field.

2.1.1 Dual role of Nodal signals in mesendoderm specification and internalization

Blocking Nodal signaling prevents both mesendoderm cell fate specification and internalization, whereas ectopic Nodal activation in naïve cells is sufficient to drive endoderm specification and internalization, regardless of cell positioning within the zebrafish embryo (Carmany-Rampey & Schier, 2001; David & Rosa, 2001; Feldman et al., 2000; Gritsman et al., 1999; Keller et al., 2008). These findings support the notion that Nodal signals, and not the germ ring margin as such, are critical for cell internalization, and further suggest that cell fate acquisition and cell internalization may be intrinsically linked by Nodal signaling. Yet, Nodal signaling may have, at least partially, separable and independent functions in regulating mesendoderm cell fate specification and internalization, since inducing endoderm cell fate specification downstream of the Nodal receptors does not seem to be sufficient to trigger cell internalization in transplantation assays (Liu, Woo, & Weiner, 2018). Consistent with this, the Nodal coreceptor Oep was suggested to play additional roles in regulating cell motility (Warga & Kane, 2003), although the signaling pathway(s) involved remain unclear.

Since inhibiting Nodal signaling at shield stage, soon after the start of gastrulation, has little impact on both mesendoderm specification and internalization (Hagos & Dougan, 2007), Nodal signals are expected to act primarily before the onset of gastrulation, up until 50% epiboly. One possibility to explain this observation is that mesendoderm cell internalization, similarly to mesendoderm specification, is triggered above a certain threshold of Nodal signaling level and/or duration, which can only be achieved by early Nodal signaling activation. Consistent with this, mesendoderm internalization is always initiated within the most marginal region of the germ ring, where Nodal signaling activity is highest (Dubrulle et al., 2015; Keller et al., 2008; Montero et al., 2005; Harvey & Smith, 2009; Van et al., 2015; Warga & Kimmel, 1990), and dorsal marginal cells are both the first cells to activate Nodal signaling and undergo internalization (Keller et al., 2008; Shimizu et al., 2000; Warga & Kimmel, 1990). Recently generated optogenetic versions of the Nodal receptors (Sako et al., 2016) will be ideal tools to dissect the link between Nodal signaling dose and the onset and dynamics of the mesendoderm internalization movements.

2.1.2 Directed cell migration, a key mechanism to ensure germ layer segregation

Inspired by the Differential Adhesion Hypothesis proposed by Steinberg (Glossary), it was suggested that germ layer segregation during zebrafish and frog gastrulation is triggered by a cell sorting process, dependent on differential tissue surface tension (TST, determined by cell-cell adhesion and cortical tension, see Differential Tension Hypothesis in the Glossary) among the three germ layer progenitors (Krieg et al., 2008; Maître et al., 2012; Schötz et al., 2008; Steinberg, 1962). Subsequent studies, however, have shown that although differential TST is sufficient to drive sorting of germ layer progenitors *in vitro*, it does not seem to be critical *in vivo* (Krens et al., 2017; Ninomiya et al., 2012). Instead, germ layer segregation might be triggered by directed migration of mesendoderm progenitors (Fig. 2B) (Giger & David, 2017; Krens et al., 2017). Consistent with this, both endogenous ppl progenitor cells, and induced endoderm cells display extensive protrusive activity during internalization, and blocking the activity of the GTPase Rac or Arp2/3, two key factors involved in protrusion formation and cell migration, impairs mesendoderm cell internalization (Giger & David, 2017; Krens et al., 2017; Liu et al., 2018).

Although little is known about the molecular cue(s) and signaling mechanisms controlling mesendoderm cell motility and polarization, Nodal signaling appears—at least to some extent—to be involved in regulating both processes. Nodal activity induces expression of the Rac1-exchange factor Prex1, and Prex1 knockdown affects both endoderm internalization and migration (Woo, Housley, Weiner, & Stainier, 2012). A number of additional cytoskeletal-remodeling factors (e.g. Ezrin, Fascin) are Nodal targets, yet their role in mesendoderm internalization remains to be systematically tested (Bennett et al., 2007; Dickmeis et al., 2001; Diz-Muñoz et al., 2010; Link et al., 2006; Liu et al., 2011). Nodal signaling also promotes cell-cell contact formation both directly by inducing expression of the adhesion receptors E- and N-cadherin (E- or N-cad) and regulating E-cad endocytosis/recycling, and indirectly by upregulating cellular contractility (Barone, Lang, Krens, Calin, & Heisenberg, 2017; Giger & David, 2017; Krieg et al., 2008; Maître et al., 2012; Ogata et al., 2007; Pinheiro & Bellaïche, 2018), pointing at the possibility that Nodal may act in cell internalization by controlling cell-cell contact formation. While in cell transplantation experiments, E-cad downregulation in transplanted endodermal donor cells or their neighboring ectodermal host cells does not prevent internalization, N-cad expression has been recently shown to be partly required for this process (Giger & David, 2017). Nonetheless, no obvious mesendoderm internalization defects are observed in *n-cad* loss-of-function mutant embryos (Lele et al., 2002; Warga & Kane, 2007), raising questions as to the role of N-cad as a key effector of Nodal signaling in controlling this process. More broadly, it remains to be tested whether cell-cell contact formation contributes for mesendoderm polarization as such or rather functions to coordinate internalization movements among neighboring cells.

In addition to Nodal signaling, Dickkopf 1 (Dkk1), a secreted inhibitor of Wnt signaling, was implicated in regulating mesendoderm cell internalization and migration. Dkk1 downregulation using morpholinos results in a mild acceleration of both mesendoderm internalization and convergence and extension movements, while Dkk1 overexpression slows down both morphogenetic movements. Since this phenotype is not observed upon modulation of either canonical or non-canonical Wnt signaling, it was proposed that Dkk1 may regulate cell motility by coordinating the activity of these two signaling pathways (Caneparo et al., 2007). Although it was shown that Dkk1 requires the Glypican 4/6 homolog Knypek (Kny), a component of the Wnt/Planar cell polarity pathway (Wnt/PCP), to regulate gastrulation movements (Caneparo et al., 2007), the molecular mechanisms by which Dkk1 mediates the switch between canonical and non-canonical Wnt signaling, and how this promotes cell motility during zebrafish gastrulation remain unclear.

2.1.3 Establishment of inside-out polarity in the zebrafish gastrula

The spatial cue(s) inducing mesendoderm cell polarization at the onset of gastrulation are largely unknown. Classical experiments showed that induced endodermal cells can internalize even when transplanted into the animal pole (David & Rosa, 2001), suggesting that mesendodermal cells can respond to inside-out polarity cue(s) present throughout the early gastrula. The observation that Nodal ligands initiate, but not directly guide, ingression of induced endoderm cells further argues against an instructive role of Nodal signaling in driving mesendoderm cell polarization and directed migration (Liu et al., 2018). Similarly, cell-matrix adhesion is unlikely to play a central role in directing mesendoderm polarization, as pronounced extracellular matrix accumulation only becomes obvious after the onset of the mesendoderm internalization movements (Latimer & Jessen, 2010; Smutny et al., 2017).

As both endogenous mesendoderm and induced endoderm cells form highly directed cell protrusions towards the YSL surface (Giger & David, 2017; Krens et al., 2017; Liu et al., 2018), signal(s) secreted from this extra-embryonic tissue might direct mesendoderm protrusion formation and migration. Yet, injecting RNAse into the YSL does not prevent internalization of induced endoderm cells in transplantation assays (Giger & David, 2017), suggesting that signal(s) secreted by the YSL are dispensable for mesendoderm polarization, or that such polarizing signal(s) may not be encoded by mRNAs. Alternatively, or in addition to a potential attractive signal from the YSL, the EVL on the surface of the blastoderm might produce a repulsive cue that helps to restrict the formation of mesendoderm protrusions towards the YSL surface. Since proper EVL differentiation is required for embryo integrity (Fukazawa et al., 2010; Sabel et al., 2009; Song et al., 2013), analyzing the specific role of the EVL in mesendoderm polarization will require the development of genetic tools for conditionally controlling EVL differentiation. Taken together, the spatial cues directing mesendoderm cell polarization and migration within the zebrafish gastrula are still unknown, and future experiments involving cell type-specific gene knock-outs and newly-generated compound mutants will be needed to dissect potentially redundant gene regulatory networks involved in this process.

2.1.4 Maintaining epiblast/hypoblast tissue boundary

Cell sorting entails both segregating different cell populations into distinct domains and maintaining their segregated state. Computational modeling of cell sorting during neural tube patterning in zebrafish embryos proposed that both chemotaxis and differential adhesion play important roles in mediating cell segregation in this process. While chemotaxis achieves correct, yet unstable, cell type segregation, differential adhesion promotes slow and error-prone, but very stable, cell sorting (Tan & Chiam, 2014). In analogy, stable and efficient segregation of germ layer progenitors during zebrafish gastrulation might require both differential adhesion and directed mesendoderm cell migration. Consistent with a role for differential cell-cell adhesion in germ layer segregation during gastrulation, E-cad down-regulation results in the formation of a less distinct, but still intact, border between epiblast and hypoblast (Montero et al., 2005; Smutny et al., 2017). This indicates that differential E-cad-mediated adhesion between epiblast and hypoblast promotes germ layer separation, but also suggests that additional molecules might be required for this process (Fig. 2C). N-cad-mediated cell-cell adhesion is unlikely to play a role here, since *n-cad* mutants only show defects in tissue boundary maintenance at later stages of gastrulation (Hong & Brewster, 2006; Lele et al., 2002; Warga & Kane, 2007). Ephrin signaling—a key pathway implicated in cell-cell repulsion—was shown to promote tissue separation in a variety of developmental processes (reviewed in Fagotto, 2014); yet, blocking Ephrin/Eph signaling in zebrafish embryos by several complementary methods does not prevent tissue separation at the epiblast/hypoblast boundary (Chan et al., 2001; Oates et al., 1999; Zhang, Jiang, Liu, & Meng, 2016), arguing against this pathway playing a critical role in germ layer segregation in zebrafish.

In addition to molecules mediating cell-cell attraction and/or repulsion, differential TST, set by the combined activities of cell-cell adhesion and actomyosin cortical tension (Glossary), was shown to play critical roles in mediating tissue separation in many developmental contexts (reviewed in Fagotto, 2014; Dahmann, Oates, & Brand, 2011). Consistent with classical observations in amphibians, differential TST can drive segregation of dissociated germ layer progenitor cells from gastrulating zebrafish embryos *in vitro* (Krieg et al., 2008; Maître et al., 2012; Schötz et al., 2008; Townes & Holtfreter, 1955). While more recent studies argue against an essential role for differential TST in driving mesendoderm cell internalization in zebrafish embryos (Krens et al., 2017; Krieg et al., 2008; Maître et al., 2012; Ninomiya et al., 2012; Schötz et al., 2008), it still might contribute to boundary maintenance at the epiblast/hypoblast interface. Consistent with this, upregulating Ndr2 signaling not only triggers mesendoderm specification, but also results in increased amounts of phosphorylated, and thus activated, Myosin II (Barone et al., 2017; Feldman et al., 1998; Gritsman et al., 1999; Rebagliati et al., 2002). Future work will be needed to determine whether this increase in cortical tension is sufficient to modulate TST and, ultimately, regulate tissue separation at the epiblast/hypoblast boundary.

While both directed cell migration and differential adhesion/TST are likely to be involved in germ layer segregation in zebrafish embryos, the nature and relative contribution of these mechanisms in initiating and maintaining tissue separation at the epiblast/hypoblast boundary is still unclear. Other potentially important molecular mechanisms that remain to be thoroughly tested in this context are the accumulation of extracellular matrix components (ECM) and interstitial fluid (IF) pockets at the epiblast/hypoblast boundary. Indeed, Fibronectin, Laminin and IF accumulate at this tissue boundary from mid-gastrulation onward (Fig. 2C) (Krens et al., 2017; Latimer & Jessen, 2010; Smutny et al., 2017), raising the intriguing possibility that ECM and/or IF accumulation at the epiblast/hypoblast interface can contribute for germ layer tissue separation, for instance by restricting cell movements and/or cell-cell adhesion over this boundary.

2.2 Animal pole-directed mesendoderm migration—A path that comes in many flavors

Following internalization, mesendodermal cells rapidly move away from the margin (Dumortier et al., 2012; Keller et al., 2008; Montero et al., 2003, 2005; Myers et al., 2002; Norris et al., 2017; Pauli et al., 2014; Pézeron et al., 2008; Sepich et al., 2005), and further differentiate into the mesoderm or endoderm lineages (Warga & Nusslein-Volhard, 1999). While in the dorsal domain of the embryo, internalized ppl cells continuously move towards the animal pole, cells on the ventral side of the germ ring change direction soon after internalization and instead move vegetally, where they will contribute to tail morphogenesis. In lateral regions of the germ ring, internalized cells show an intermediate behavior, initially migrating towards the animal pole like their dorsal counterparts, but eventually engaging in dorsal-directed convergence movements (Fig. 2A) (Arboleda-Estudillo et al., 2010; Dumortier et al., 2012; Keller et al., 2008; Montero et al., 2003, 2005; Myers et al., 2002; Norris et al., 2017; Pauli et al., 2014; Pézeron et al., 2008; Sepich et al., 2005). As reflected by their distinct migratory behaviors, the upstream signaling cues controlling mesendoderm migration vary along the DV axis of the zebrafish gastrula. In the next section, we will discuss the signaling cues and molecular pathways involved in determining the migratory properties of these distinct cell populations.

2.2.1 Dorsal mesendoderm migration: Integrating autonomous cell motility amid a collective

Dorsal anterior axial mesendoderm (ppl) migration is among the most conserved migratory processes observed during gastrulation (reviewed in Solnica-Krezel, 2005). In zebrafish, ppl cells migrate animally as a cohesive group of mesenchymal cells, thus contributing to anterior-posterior body axis extension and positioning of the neural anlage by exerting friction forces at the ppl-neurectoderm interface (Fig. 1A) (Dumortier et al., 2012; Montero et al., 2005; Smutny et al., 2017; Ulrich et al., 2003). All ppl cells display similar polarization and migration directionality (Dumortier et al., 2012; Smutny et al., 2017), suggesting that all cells actively migrate towards the animal pole. Yet, ppl movement is a collective process, since E-cad-mediated cell-cell contacts among ppl cells are required to both orient their actin-rich protrusions along the direction of movement and coordinate migration among neighboring cells (Dumortier et al., 2012; Montero et al., 2005; Smutny et al., 2017).

Several signaling pathways were implicated in regulating animal pole-directed migration of ppl cells, including Wnt/PCP, β-cat-Stat3 and Platelet-Derived-Growth-Factor (PDGF) signaling via the Phosphoinositide 3-kinase pathway (PI3K) (Čapek et al., 2019; Heisenberg et al., 2000; Miyagi et al., 2004; Montero et al., 2003; Ulrich et al., 2003; Yamashita et al., 2002, 2004). While the potential roles of β-cat-Stat3 and PDGF-PI3K in this process require further genetic confirmation, evidence for Wnt/PCP signaling playing a critical role in dorsal mesendoderm migration first arose upon characterization of mutant embryos for the non-canonical Wnt ligand Wnt11 (also called Silberblick) (Heisenberg et al., 2000; Liu, Sepich, & Solnica-Krezel, 2017; Miyagi et al., 2004; Montero et al., 2003; Yamashita et al., 2002, 2004). Indeed, loss-of-function mutations in Wnt11 and its receptor Frizzled 7 (Fz7) result, among other defects, in defective ppl migration (Čapek et al., 2019; Heisenberg et al., 2000; Ulrich et al., 2003). In zebrafish embryos, Wnt/PCP signaling likely plays a permissive rather than instructive role in ppl cell migration, since its uniform activation, either by Wnt11 overexpression or light-induced uniform activation of its receptor Fz7, is sufficient to rescue defective ppl migration in *wnt11* or *fz7* mutant embryos, respectively (Čapek et al., 2019; Heisenberg et al., 2000). Consistent with this, the ppl in *wnt11* and *fz7* mutants still moves towards the animal pole, albeit in a slower and less directed fashion (Čapek et al., 2019; Heisenberg et al., 2000; Ulrich et al., 2003), suggesting that other factors must also be involved in regulating this process.

Pharmacological inhibition of PDGF-PI3K signaling results in ppl migration defects similar to those described in *wnt11* and *fz7* mutants (Montero et al., 2003), suggesting that Wnt/PCP and PDGF-PI3K signaling may act together to regulate ppl cell migration, yet additional work is required to thoroughly test this possibility. Finally, although ECM components play an important role in guiding ppl migration in frogs, they only accumulate after the onset of ppl migration in zebrafish, and interfering with

Integrin α5 has little effect on ppl migration (Johnson, Darribere, Shi, Bache, & Delarue, 1990; Jülich et al., 2005; Latimer & Jessen, 2010; Smutny et al., 2017; Trinh & Stainier, 2004), arguing against a major role for cell-matrix adhesion in ppl migration in zebrafish.

While several signaling pathways with a function in ppl cell migration were identified, key questions as to their effector mechanisms remain to be answered. For instance, recent work showed that downregulation of Sin1, a component of the Torc2 complex, affects both ppl speed and migration directionality (Dumortier & David, 2015), raising the possibility that the Torc2 complex may be a key regulator of ppl cell migration. Furthermore, expression of Protocadherin-18a (Pcdh18a) in a previously unknown cell population positioned at the boundary between the ppl and the trailing notochord, termed notochord tip cells (NTCs), is required for proper migration of more posterior axial mesoderm cells (Bosze et al., 2018). This points at the intriguing possibility that not only ppl cells, but also NTCs, might be required for animal pole-directed migration of the dorsal axial mesendoderm during gastrulation.

2.2.2 Ventrolateral mesendoderm migration: The role of Apelin signaling

Following internalization at the germ ring margin, ventral and lateral mesendoderm migrate toward the animal pole during early-to-mid stages of gastrulation (Fig. 2A) (Keller et al., 2008; Myers et al., 2002; Sepich et al., 2005). Apela (also known as Toddler or Elabela), a secreted peptide able to elicit signaling via the APJ/Apelin G-protein-coupled-receptors (GPCR), was recently shown to regulate this process, independently of cell fate specification (Chng et al., 2013; Norris et al., 2017; Pauli et al., 2014). Consistent with its ubiquitous expression during late blastula and gastrula stages, Apela was proposed to promote ventrolateral mesendoderm cell motility, rather than directly regulating cell directionality (Chng et al., 2013; Pauli et al., 2014). Apela signaling is itself activated by Nodal signals, which again induces expression of the Apelin receptors (Pauli et al., 2014). Yet, how Apela regulates mesendodermal cell migration remains unclear, and RNA-sequencing efforts did not reveal major changes in gene expression in Apela mutant embryos during gastrulation (Norris et al., 2017). Notably, Apela mutants often show delayed epiboly movements (Chng et al., 2013; Pauli et al., 2014), raising the possibility that the observed defects in mesendodermal migration may be linked to reduced epiboly movements in the overlying ectoderm. Indeed, previous observations in *Xenopus* embryos suggest that pulling forces, sensed and transmitted by E-cad, can trigger mesendoderm cell polarization (Weber, Bjerke, & DeSimone, 2012). Subsequent work in the zebrafish embryo further showed that frictional forces generated at the interface between the animal-directed ppl and the vegetal-directed neurectoderm epiboly movements influence neurectoderm cell movements and final positioning along the AP axis (Smutny et al., 2017). It is thus conceivable that similar tissue interactions may occur at the ventrolateral side of the gastrula and be able to influence the relative movements of mesendoderm and ectoderm therein.

2.2.3 Endoderm migration—A 'guided' random walk?

Endodermal progenitors are among the first cells to internalize at the germ ring margin (Warga & Nusslein-Volhard, 1999). In contrast to internalized mesodermal cells, endoderm progenitors form a non-continuous monolayer located directly at the interface with the YSL (Warga & Nusslein-Volhard, 1999). Internalized endodermal cells were proposed to undergo non-oriented and non-coordinated movements, best fit by a random walk, which rapidly disperses this cell population over the yolk surface. Consistent with this, no correlation was found between endoderm cell collisions and changes in cell directionality (Pézeron et al., 2008). Using theoretical modeling, it was further shown that such undirected and uncoordinated movement is sufficient to explain the dispersal of endoderm cells over both the animal and vegetal portions of the embryo during the course of gastrulation (Pézeron et al., 2008).

From late gastrulation onwards, endodermal cells, along with their mesodermal counterparts, initiate convergence movements towards the dorsal side of the embryo, where they eventually re-adhere to line the gastrointestinal tract (Pézeron et al., 2008). Activation of the Rho GTPase Rac1 by Prex1, a transcriptional target of Nodal signaling, is required for both stages of endoderm migration, whereas cell-matrix adhesion and Cxcl12a chemokine signaling, via the Cxcr4a receptor, are specifically required for dorsal convergence (Latimer & Jessen, 2010; Mizoguchi, Verkade, Heath, Kuroiwa, & Kikuchi, 2008; Nair & Schilling, 2008; Norris et al., 2017; Woo et al., 2012). Endoderm convergence was proposed to rely on the convergence movements of the overlying mesoderm, with endoderm cells becoming tethered to mesoderm cells via Integrin-ECM adhesion (Norris et al., 2017). Consistent with this notion, blocking Integrin-Fibronectin adhesion or Cxcl12a/b signaling results in prolonged animal migration of endoderm cells (Latimer & Jessen, 2010; Mizoguchi et al., 2008; Nair & Schilling, 2008; Norris et al., 2017). Collectively, the ability of endoderm cells to switch their migratory behavior from random to persistent motion may be required to ensure their rapid and efficient cell dispersion, as well as their subsequent swift and coordinated dorsal translocation once mesoderm cells initiate convergence movements.

3. Outlook

Embryonic development entails both correct differentiation of cells into given lineages and their proper positioning along the future body axes. In zebrafish embryos, gastrulation starts with specification and synchronized ingression of mesendoderm progenitor cells at the blastoderm margin (Kimmel et al., 1995). Once internalized, mesendoderm cells commit into the mesoderm or the endoderm lineage and migrate away from the germ ring margin (Fig. 2A) (Chng et al., 2013; Dumortier et al., 2012; Keller et al., 2008; Montero et al., 2003, 2005; Myers et al., 2002; Norris et al., 2017; Pauli et al., 2014; Pézeron et al., 2008; Sepich et al., 2005; Warga & Nusslein-Volhard, 1999). While Nodal signaling was thought to be the prime signal inducing mesendoderm progenitors both along the DV and AV axes of the germ ring, recent findings are now changing this view and instead implicate an FGF-dependent relay mechanism in triggering the expression of key Nodal targets beyond the Nodal expression domain (Fig. 1C) (van Boxtel et al., 2017). These studies raise intriguing questions as to the mechanisms by which the FGF signaling domain is established and how the relative proportions of the Nodal and FGF signaling domains are set and maintained during gastrulation. Likewise, the long-held view that mesendoderm progenitors segregate from ectoderm progenitor cells by differences in their TST is changing, with directed cell migration playing a central role in triggering internalization via synchronized mesendoderm ingression and thus germ layer cell progenitor segregation (Giger & David, 2017; Krens et al., 2017). These paradigm-shifting findings support that germ cell segregation is a more active process than previously anticipated and highlight the necessity to understand the interplay between the onset of mesendoderm cell fate specification and acquisition of cell polarization and motility.

To understand the crosstalk between mesendoderm cell fate specification and segregation, it will be key to analyze how these fundamental processes are coordinated in space and time during gastrulation. While various highly multidisciplinary approaches, drawing from systems biology, physics and mathematics, have been developed over the last decade to bridge the scales from noisy signaling networks and gradient formation to large-scale tissue pattering and morphogenetic movements, methods to precisely monitor and modulate ongoing mesendoderm cell fate specification and movement are still lacking. The development of direct reporters of Nodal and FGF signaling, by e.g. tagging direct target proteins with destabilized GFP (Green Fluorescent Protein) (Soroldoni et al., 2014), is likely to provide a dynamic readout of ongoing mesendoderm specification and its relationship with the onset and dynamics of its internalization movements. Furthermore, recently generated optogenetic receptors (Sako et al., 2016) allowing tight spatiotemporal control over Nodal signaling strength at a cellular scale, will be useful tools to address how Nodal signaling translates into a distinct cell behavior. Finally, recently published high-throughput single cell RNA-sequencing datasets acquired throughout early zebrafish development provide unprecedented resolution into the transcriptional landscape and trajectory of germ cell progenitors during gastrulation and therefore offer a great opportunity to identify interesting candidate genes involved in coupling mesendoderm specification and internalization (Farrell et al., 2018; Wagner et al., 2018).

Collectively, gastrulation is a highly conserved process and the key signaling pathways involved therein (e.g. Nodal, Wnt, BMP) are implicated in many other developmental events. Thus, understanding how these pathways robustly couple cell fate specification, cellular mechanics, tissue patterning, and morphogenesis will be of general interest to cell, systems and developmental biologists alike.

Acknowledgments

We thank Alexandra Schauer, Nicoletta Petridou and Feyza Nur Arslan for comments on the manuscript. Research in the Heisenberg laboratory is supported by an ERC Advanced Grant (MECSPEC 742573), ANR/FWF (I03601) and FWF/DFG (I03196) International Cooperation Grants. D. Pinheiro acknowledges a fellowship from EMBO ALTF (850-2017) and is currently supported by HFSP LTF (LT000429/2018-L2).

Glossary

Organizer defined by Hans Spemann as a region, or a group of cells, able to induce and pattern adjacent embryonic cells (reviewed in Anderson & Stern, 2016). Using transplantation of different marginal regions into the animal pole, it was shown that, in zebrafish embryos, the entire blastula/gastrula margin is able to act as a continuous organizer (Fauny, Thisse, & Thisse, 2009).

Reaction-Diffusion Model based on the idea that a signaling pair composed of a slow diffusing activator and a fast diffusing inhibitor can self-organize and give rise to pattern formation during development. As initially formulated, the distribution of the activator/inhibitor pair is predicted to depend on their intrinsic diffusion coefficients and reaction rates (reviewed in Kondo & Miura, 2010).

Relay Model refers to a mechanism required to extend the signaling domain of an inducing signal beyond its expression domain. It can be mediated by a self-sufficient positive feedback loop, in which an inducing signal can activate its own expression in neighboring cells, effectively propagating the signaling domain at a distance from the initial signaling source. Alternatively, this may be achieved by the induction of a secondary signaling molecule with a longer signaling range. Finally, both mechanisms can also cooperate to establish and spread the initial signaling domain (reviewed in Rogers & Müller, 2019).

Morphogen term originally coined by Alan Turing, commonly used to describe molecular signals able to spread within tissues and control cell patterning and growth (reviewed in Rogers & Schier, 2011; Sagner & Briscoe, 2017).

French-flag Model based on the idea that cellular responses to morphogen signaling are critically determined by the morphogen concentration they experience. Since a morphogen is able to diffuse away from its localized secretion source, a signaling gradient is predictably formed within the tissue. Thus, cells located close to the source, which experience highest morphogen levels, adopt a certain response, whereas cells located further away from the source, which experience intermediate or low morphogen concentration, adopt a distinct response (reviewed in Rogers & Schier, 2011).

Differential Adhesion Hypothesis formulated in analogy with the physics of immiscible liquids and postulates that cells tend to maximize homotypic adhesion and minimize heterotypic interactions. The differential adhesive properties of distinct cell populations, if sufficiently strong, are therefore sufficient to drive their cell sorting (reviewed in Fagotto, 2014).

Differential Tension Hypothesis considers the contribution of other forces, namely interfacial tension, in regulating cell-cell adhesion and, consequently, cell sorting. This model therefore postulates that the sorting of distinct cell populations is determined by the combined effect of cell cortex contractility and cell-cell adhesion (reviewed in Fagotto, 2014).

References

Adams, R. J., & Kimmel, C. B. (2004). Morphogenetic cellular flows during zebrafish gastrulation. In C. D. Stern (Ed.), *Gastrulation: From cells to embryo* (pp. 305–316): Cold Spring Harbor Laboratory Press.

Alexandre, C., Baena-Lopez, A., & Vincent, J. P. (2014). Patterning and growth control by membrane-tethered wingless. *Nature, 505*, 180–185.

Almuedo Castillo, M., Bläßle, A., Mörsdorf, D., Marcon, L., Soh, G. H., Rogers, K. W., et al. (2018). Scale-invariant patterning by size-dependent inhibition of Nodal signalling. *Nature Cell Biology, 20*, 1032–1042.

Anderson, C., & Stern, C. D. (2016). Organizers in development. In *Current topics in developmental biology* (pp. 435–454): Academic Press.

Aoki, T. O., David, N. B., Minchiotti, G., Saint-Etienne, L., Dickmeis, T., Persico, G. M., et al. (2002). Molecular integration of casanova in the Nodal signalling pathway controlling endoderm formation. *Development, 129*, 275–286.

Arboleda-Estudillo, Y., Krieg, M., Stühmer, J., Licata, N. A., Muller, D. J., & Heisenberg, C.-P. (2010). Movement directionality in collective migration of germ layer progenitors. *Current Biology, 20*, 161–169.

Babb, S. G., & Marrs, J. A. (2004). E-cadherin regulates cell movements and tissue formation in early zebrafish embryos. *Developmental Dynamics*, *230*, 263–277.
Barone, V., Lang, M., Krens, S. F. G., Calin, C. G., & Heisenberg, C.-P. (2017). An effective feedback loop between cell-cell contact duration and morphogen signaling determines cell fate. *Developmental Cell*, *43*, 1–14.
Bennett, J. T., Joubin, K., Cheng, S., Aanstad, P., Herwig, R., Clark, M., et al. (2007). Nodal signaling activates differentiation genes during zebrafish gastrulation. *Developmental Biology*, *304*, 525–540.
Bisgrove, B. W., Su, Y. C., & Yost, H. J. (2017). Maternal Gdf3 is an obligatory cofactor in nodal signaling for embryonic axis formation in zebrafish. *eLife*, *6*, 1–19.
Bisgrove, B. W., Yost, H. J., & Essner, J. J. (1999). Regulation of midline development by antagonism of lefty and nodal signaling. *Development*, *126*, 3253–3262.
Bosze, B., Mattes, B., Sinner, C., Stricker, K., Gourain, V., Strähle, U., et al. (2018). Pcdh18a-positive tip cells instruct notochord formation in zebrafish. *bioRxiv*. https://doi.org/10.1101/257717.
Caneparo, L., Huang, Y. L., Staudt, N., Tada, M., Ahrendt, R., Kazanskaya, O., et al. (2007). Dickkopf-1 regulates gastrulation movements by coordinated modulation of Wnt/β-catenin and Wnt/PCP activities, through interaction with the Dally-like homolog Knypek. *Genes & Development*, *21*, 465–480.
Čapek, D., Smutny, M., Tichy, A.-M., Morri, M., Janovjak, H., & Heisenberg, C.-P. (2019). Light-activated Frizzled7 reveals a permissive role of non-canonical wnt signaling in mesendoderm cell migration. *eLife*, *8*, 1–24.
Carmany-Rampey, A., & Schier, A. F. (2001). Single-cell internalization during zebrafish gastrulation. *Current Biology*, *11*, 1261–1265.
Chan, J., Mably, J. D., Serluca, F. C., Chen, J. N., Goldstein, N. B., Thomas, M. C., et al. (2001). Morphogenesis of prechordal plate and notochord requires intact Eph/ephrin B signaling. *Developmental Biology*, *234*, 470–482.
Chen, Y., & Schier, A. F. (2001). The zebrafish nodal signal Squint functions as a morphogen. *Nature*, *411*, 607–610.
Chen, Y., & Schier, A. F. (2002). Lefty proteins are long-range inhibitors of squint-mediated nodal signaling. *Current Biology*, *12*, 2124–2128.
Chng, S. C., Ho, L., Tian, J., & Reversade, B. (2013). ELABELA: A hormone essential for heart development signals via the apelin receptor. *Developmental Cell*, *27*, 672–680.
D'Amico, L. A., & Cooper, M. S. (1997). Spatially distinct domains of cell behavior in the zebrafish organizer region. *Biochemistry and Cell Biology*, *75*, 563–577.
D'Amico, L. A., & Cooper, M. S. (2001). Morphogenetic domains in the yolk syncytial layer of axiating zebrafish embryos. *Developmental Dynamics*, *222*, 611–624.
Dahmann, C., Oates, A. C., & Brand, M. (2011). Boundary formation and maintenance in tissue development. *Nature Reviews. Genetics*, *12*, 43–55.
David, N. B., & Rosa, F. M. (2001). Cell autonomous commitment to an endodermal fate and behaviour by activation of Nodal signalling. *Development*, *128*, 3937–3947.
Dickmeis, T., Aanstad, P., Clark, M., Fischer, N., Herwig, R., Mourrain, P., et al. (2001). Identification of Nodal signaling targets by array analysis of induced complex probes. *Developmental Dynamics*, *222*, 571–580.
Diz-Muñoz, A., Krieg, M., Bergert, M., Ibarlucea-Benitez, I., Muller, D. J., Paluch, E., et al. (2010). Control of directed cell migration *in vivo* by membrane-to-cortex attachment. *PLoS Biology*, *8*, 1000544.
Dougan, S. T., Warga, R. M., Kane, D. A., Schier, A. F., & Talbot, W. S. (2003). The role of the zebrafish nodal-related genes squint and cyclops in patterning of mesendoderm. *Development*, *130*, 1837–1851.

Duboc, V., Röttinger, E., Besnardeau, L., & Lepage, T. (2004). Nodal and BMP2/4 signaling organizes the oral-aboral axis of the sea urchin embryo. *Developmental Cell, 6*, 397–410.

Dubrulle, J., Jordan, B. M., Akhmetova, L., Farrell, J. A., Kim, S.-H., Solnica-Krezel, L., et al. (2015). Response to nodal morphogen gradient is determined by the kinetics of target gene induction. *eLife, 4*, 1–27.

Dumortier, J. G., & David, N. B. (2015). The TORC2 component, sin1, controls migration of anterior mesendoderm during zebrafish gastrulation. *PLoS One, 10*(2), e0118474.

Dumortier, J. G., Martin, S., Meyer, D., Rosa, F. M., & David, N. B. (2012). Collective mesendoderm migration relies on an intrinsic directionality signal transmitted through cell contacts. *Proceedings of the National Academy of Sciences, 109*, 16945–16950.

Fagotto, F. (2014). The cellular basis of tissue separation. *Development, 141*, 3303–3318.

Farrell, J. A., Wang, Y., Riesenfeld, S. J., Shekhar, K., Regev, A., & Schier, A. F. (2018). Single-cell reconstruction of developmental trajectories during zebrafish embryogenesis. *Science, 360*, eaar3131.

Fauny, J.-D., Thisse, B., & Thisse, C. (2009). The entire zebrafish blastula-gastrula margin acts as an organizer dependent on the ratio of Nodal to BMP activity. *Development, 136*, 3811–3819.

Feldman, B., Concha, M. L., Saúde, L., Parsons, M. J., Adams, R. J., Wilson, S. W., et al. (2002). Lefty antagonism of Squint is essential for normal gastrulation. *Current Biology, 12*, 2129–2135.

Feldman, B., Dougan, S. T., Schier, A. F., & Talbot, W. S. (2000). Nodal-related signals establish mesendodermal fate and trunk neural identity in zebrafish. *Current Biology, 10*, 531–534.

Feldman, B., Gates, M. A., Egan, E. S., Dougan, S. T., Rennebecl, G., Sirotkin, H. I., et al. (1998). Zebrafish organizer development and germ-layer formation require nodal-related signals. *Nature, 395*, 181–185.

Flowers, V. L., Courteau, G. R., Poustka, A. J., Weng, W., & Venuti, J. M. (2004). Nodal/activin signaling establishes oral-aboral polarity in the early sea urchin embryo. *Developmental Dynamics, 231*, 727–740.

Fukazawa, C., Santiago, C., Park, K. M., Deery, W. J., Gomez De La, S., Canny, T., et al. (2010). poky/chuk/ikk1 is required for differentiation of the zebrafish embryonic epidermis. *Developmental Biology, 346*, 272–283.

Giger, F. A., & David, N. B. (2017). Endodermal germ-layer formation through active actin-driven migration triggered by N-cadherin. *Proceedings of the National Academy of Sciences, 114*(38), 10143–10148.

Glickman, N. S., Kimmel, C. B., Jones, M. A., & Adams, R. J. (2003). Shaping the zebrafish notochord. *Development, 130*, 873–887.

Gore, A. V., Maegawa, S., Cheong, A., Gilligan, P. C., Weinberg, E. S., & Sampath, K. (2005). The zebrafish dorsal axis is apparent at the four-cell stage. *Nature, 438*, 1030–1035.

Gritsman, K., Talbot, W. S., & Schier, A. F. (2000). Nodal signaling patterns the organizer. *Development, 127*, 921–932.

Gritsman, K., Zhang, J., Cheng, S., Heckscher, E., Talbot, W. S., & Schier, A. F. (1999). The EGF-cfc protein one-eyed pinhead is essential for nodal signaling. *Cell, 97*, 121–132.

Hagos, E. G., & Dougan, S. T. (2007). Time-dependent patterning of the mesoderm and endoderm by Nodal signals in zebrafish. *BMC Developmental Biology, 7*, 22.

Harvey, S. A., & Smith, J. C. (2009). Visualisation and quantification of morphogen gradient formation in the zebrafish. *PLoS Biology, 7*, e1000101.

Heisenberg, C.-P., Tada, M., Rauch, G.-J., Saúde, L., Concha, M., Geisler, R., et al. (2000). Silberblick/Wnt11 mediates convergent extension movements during zebrafish gastrulation. *Nature, 405*, 76–81.

Hill, C. S. (2018). Spatial and temporal control of NODAL signaling. *Current Opinion in Cell Biology*, *51*, 50–57.

Ho, R. K., & Kimmel, C. B. (1993). Commitment of cell fate in the early zebrafish embryo. *Science*, *261*, 109–111.

Hong, E., & Brewster, R. (2006). N-cadherin is required for the polarized cell behaviors that drive neurulation in the zebrafish. *Development*, *133*, 3895–3905.

Hong, S.-K., Jang, M. K., Brown, J. L., McBride, A. A., & Feldman, B. (2011). Embryonic mesoderm and endoderm induction requires the actions of non-embryonic Nodal-related ligands and Mxtx2. *Development*, *138*, 787–795.

Howell, M., Mohun, T. J., & Hill, C. S. (2001). Xenopus Smad3 is specifically expressed in the chordoneural hinge, notochord and in the endocardium of the developing heart. *Mechanisms of Development*, *104*, 147–150.

Jing, X., Zhou, S., Wang, W., & Chen, Y. (2006). Mechanisms underlying long- and short-range nodal signaling in Zebrafish. *Mechanisms of Development*, *123*, 388–394.

Johnson, K. E., Darribere, T., Shi, D., Bache, H. B., & Delarue, M. (1990). Fibronectin-rich fibrillar extracellular cell migration during amphibian matrix controls gastrulation. *The International Journal of Developmental Biology*, *147*, 139–147.

Jülich, D., Geisler, R., Holley, S. A., Van Bebber, F., Busch-Nentwich, E., Dahm, R., et al. (2005). Integrinα5 and delta/notch signaling have complementary spatiotemporal requirements during zebrafish somitogenesis. *Developmental Cell*, *8*, 575–586.

Keller, P. J., Schmidt, A. D., Wittbrodt, J., & Stelzer, E. H. K. (2008). Reconstruction of zebrafish early embryonic development by scanned light sheet microscopy. *Science*, *322*, 1065–1069.

Kiecker, C., Bates, T., & Bell, E. (2016). Molecular specification of germ layers in vertebrate embryos. *Cellular and Molecular Life Sciences*, *73*, 923–947.

Kimmel, C. B., Ballard, W. W., Kimmel, S. R., Ullman, B., & Schilling, T. F. (1995). Stages of embryonic development of the zebrafish. *Developmental Dynamics*, *203*, 255–310.

Kimmel, C. B., Warga, R. M., & Schilling, T. F. (1990). Origin and organization of the zebrafish fate map. *Development*, *108*, 581–594.

Kondo, S., & Miura, T. (2010). Reaction-diffusion model as a framework for understanding biological pattern formation. *Science*, *329*, 1616–1620.

Krens, S. F. G., Veldhuis, J. H., Barone, V., Capek, D., Maître, J.-L., Brodland, G. W., et al. (2017). Interstitial fluid osmolarity modulates the action of differential tissue surface tension in progenitor cell segregation during gastrulation. *Development*, *144*, 1798–1806.

Krieg, M., Arboleda-Estudillo, Y., Puech, P.-H., Käfer, J., Graner, F., Müller, D. J., et al. (2008). Tensile forces govern germ-layer organization in zebrafish. *Nature Cell Biology*, *10*, 429–436.

Langdon, Y. G., & Mullins, M. C. (2011). Maternal and zygotic control of zebrafish dorsoventral axial patterning. *Annual Review of Genetics*, *45*, 357–377.

Latimer, A., & Jessen, J. R. (2010). Extracellular matrix assembly and organization during zebrafish gastrulation. *Matrix Biology*, *29*, 89–96.

Lele, Z., Folchert, A., Concha, M., Rauch, G.-J., Geisler, R., Rosa, F., et al. (2002). Parachute/N-cadherin is required for morphogenesis and maintained integrity of the zebrafish neural tube. *Development*, *129*, 3281–3294.

Leung, T., Bischof, J., Soll, I., Niessing, D., Zhang, D., Ma, J., et al. (2003). bozozok directly represses bmp2b transcription and mediates the earliest dorsoventral asymmetry of bmp2b expression in zebrafish. *Development*, *130*, 3639–3649.

Link, V., Carvalho, L., Castanon, I., Stockinger, P., Shevchenko, A., & Heisenberg, C.-P. (2006). Identification of regulators of germ layer morphogenesis using proteomics in zebrafish. *Journal of Cell Science*, *119*, 2073–2083.

Liu, Z., Lin, X., Cai, Z., Zhang, Z., Han, C., Jia, S., et al. (2011). Global identification of SMAD2 target genes reveals a role for multiple co-regulatory factors in zebrafish early gastrulas. *The Journal of Biological Chemistry*, *286*, 28520–28532.

Liu, Y., Sepich, D. S., & Solnica-Krezel, L. (2017). Stat3/Cdc25a-dependent cell proliferation promotes embryonic axis extension during zebrafish gastrulation. *PLoS Genetics*, *13*(2), e1006564.

Liu, Z., Woo, S., & Weiner, O. D. (2018). Nodal signaling has dual roles in fate specification and directed migration during germ layer segregation in zebrafish. *Development*, *145*, dev163535.

Londin, E. R., Mentzer, L., & Sirotkin, H. I. (2007). Churchill regulates cell movement and mesoderm specification by repressing Nodal signaling. *BMC Developmental Biology*, *7*, 120.

Long, S., Ahmad, N., & Rebagliati, M. R. (2003). The zebrafish nodal-related gene southpaw is required for visceral and diencephalic left-right asymmetry. *Development*, *130*, 2303–2316.

Lu, F.-I., Thisse, C., & Thisse, B. (2011). Identification and mechanism of regulation of the zebrafish dorsal determinant. *Proceedings of the National Academy of Sciences of the United States of America*, *108*, 15876–15880.

Maître, J.-L., Berthoumieux, H., Krens, S. F. G., Salbreux, G., Julicher, F., Paluch, E. K., et al. (2012). Adhesion functions in cell sorting by mechanically coupling the cortices of adhering cells. *Science*, *338*, 253–256.

Massagué, J. (2012). TGFβ signalling in context. *Nature Reviews. Molecular Cell Biology*, *13*, 616–630.

Mathieu, J., Griffin, K., Herbomel, P., Dickmeis, T., Strahle, U., Kimelman, D., et al. (2004). Nodal and Fgf pathways interact through a positive regulatory loop and synergize to maintain mesodermal cell populations. *Development*, *131*, 629–641.

Melby, A. E., Beach, C., Mullins, M., & Kimelman, D. (2000). Patterning the early zebrafish by the opposing actions of bozozok and vox/vent. *Developmental Biology*, *224*, 275–285.

Meno, C., Gritsman, K., Ohishi, S., Ohfuji, Y., Heckscher, E., Mochida, K., et al. (1999). Mouse lefty2 and zebrafish antivin are feedback inhibitors of nodal signaling during vertebrate gastrulation. *Molecular Cell*, *4*, 287–298.

Miyagi, C., Yamashita, S., Ohba, Y., Yoshizaki, H., Matsuda, M., & Hirano, T. (2004). STATS noncell-autonomously controls planar cell polarity during zebrafish convergence and extension. *The Journal of Cell Biology*, *166*, 975–981.

Mizoguchi, T., Verkade, H., Heath, J. K., Kuroiwa, A., & Kikuchi, Y. (2008). Sdf1/Cxcr4 signaling controls the dorsal migration of endodermal cells during zebrafish gastrulation. *Development*, *135*, 2521–2529.

Montague, T. G., & Schier, A. F. (2017). Vg1-nodal heterodimers are the endogenous inducers of mesendoderm. *eLife*, *6*, 1–24.

Montero, J.-A., Carvalho, L., Wilsch-Brauninger, M., Kilian, B., Mustafa, C., & Heisenberg, C.-P. (2005). Shield formation at the onset of zebrafish gastrulation. *Development*, *123*, 1187–1198.

Montero, J.-A., Kilian, B., Chan, J., Bayliss, P. E., & Heisenberg, C.-P. (2003). Phosphoinositide 3-kinase is required for process outgrowth and cell polarization of gastrulating mesendodermal cells. *Current Biology*, *13*, 1279–1289.

Morokuma, J., Ueno, M., Kawanishi, H., Saiga, H., & Nishida, H. (2002). HrNodal, the ascidian nodal-related gene, is expressed in the left side of the epidermis, and lies upstream of HrPitx. *Development Genes and Evolution*, *212*, 439–446.

Müller, P., Rogers, K. W., Jordan, B. M., Lee, J. S., Robson, D., Ramanathan, S., et al. (2012). Differential diffusivity of nodal and lefty underlies a reaction-diffusion patterning system. *Science*, *336*, 721–724.

Myers, D. C., Sepich, D. S., & Solnica-Krezel, L. (2002). Bmp activity gradient regulates convergent extension during zebrafish gastrulation. *Developmental Biology*, *243*, 81–98.

Nair, S., & Schilling, T. F. (2008). Chemokine signaling controls endodermal migration during zebrafish gastrulation. *Science*, *322*, 89–92.

Ninomiya, H., David, R., Damm, E. W., Fagotto, F., Niessen, C. M., & Winklbauer, R. (2012). Cadherin-dependent differential cell adhesion in Xenopus causes cell sorting *in vitro* but not in the embryo. *Journal of Cell Science*, *125*, 1877–1883.

Norris, M. L., Pauli, A., Gagnon, J. A., Lord, N. D., Rogers, K. W., Mosimann, C., et al. (2017). Toddler signaling regulates mesodermal cell migration downstream of Nodal signaling. *eLife*, *6*, 1–18.

Nusse, R., & Clevers, H. (2017). Wnt/β-catenin signaling, disease, and emerging therapeutic modalities. *Cell*, *169*, 985–999.

Oates, A. C., Lackmann, M., Power, M. A., Brennan, C., Down, L. M., Do, C., et al. (1999). An early developmental role for Eph-ephrin interaction during vertebrate gastrulation. *Mechanisms of Development*, *83*, 77–94.

Ogata, S., Morokuma, J., Hayata, T., Kolle, G., Niehrs, C., Ueno, N., et al. (2007). TGF-β signaling-mediated morphogenesis: Modulation of cell adhesion via cadherin endocytosis. *Genes & Development*, *21*, 1817–1831.

Pauklin, S., & Vallier, L. (2015). Activin/nodal signalling in stem cells. *Development*, *142*, 607–619.

Pauli, A., Norris, M. L., Valen, E., Chew, G.-L., Gagnon, J. A., Zimmerman, S., et al. (2014). Toddler: An embryonic signal that promotes cell movement via apelin receptors. *Science*, *343*(6172), 1248636.

Pelliccia, J. L., Jindal, G. A., & Burdine, R. D. (2017). Gdf3 is required for robust nodal signaling during germ layer formation and left-right patterning. *eLife*, *6*, 1–23.

Pézeron, G., Mourrain, P., Courty, S., Ghislain, J., Becker, T. S., Rosa, F. M., et al. (2008). Live analysis of endodermal layer formation identifies random walk as a novel gastrulation movement. *Current Biology*, *18*, 276–281.

Pinheiro, D., & Bellaïche, Y. (2018). Mechanical force-driven adherens junction remodeling and epithelial dynamics. *Developmental Cell*, *47*, 3–19.

Pogoda, H. M., Solnica-Krezel, L., Driever, W., & Meyer, D. (2000). The zebrafish forkhead transcription factor FoxH1/Fast1 is a modulator of nodal signaling required for organizer formation. *Current Biology*, *10*, 1041–1049.

Rebagliati, M. R., Toyama, R., Haffter, P., & Dawid, I. B. (2002). cyclops encodes a nodal-related factor involved in midline signaling. *Proceedings of the National Academy of Sciences*, *95*, 9932–9937.

Robertson, E. J. (2014). Dose-dependent nodal/smad signals pattern the early mouse embryo. *Seminars in Cell & Developmental Biology*, *32*, 73–79.

Rodaway, A., Takeda, H., Koshida, S., Broadbent, J., Price, B., Smith, J. C., et al. (1999). Induction of the mesendoderm in the zebrafish germ ring by yolk cell-derived TGF-beta family signals and discrimination of mesoderm and endoderm by FGF. *Development*, *126*, 3067–3078.

Rogers, K. W., Lord, N. D., Gagnon, J. A., Pauli, A., Zimmerman, S., Aksel, D. C., et al. (2017). Nodal patterning without lefty inhibitory feedback is functional but fragile. *eLife*, *6*, 1–20.

Rogers, K. W., & Müller, P. (2019). Nodal and BMP dispersal during early zebrafish development. *Developmental Biology*, *447*, 14–23.

Rogers, K. W., & Schier, A. F. (2011). Morphogen gradients: From generation to interpretation. *Annual Review of Cell and Developmental Biology*, *27*, 377–407.

Ryu, S. L., Fujii, R., Yamanaka, Y., Shimizu, T., Yabe, T., Hirata, T., et al. (2001). Regulation of dharma/bozozok by the Wnt pathway. *Developmental Biology*, *231*, 397–409.

Sabel, J. L., D'Alençon, C., O'Brien, E. K., Van, O. E., Lutz, K., Cuykendall, T. N., et al. (2009). Maternal Interferon Regulatory Factor 6 is required for the differentiation of primary superficial epithelia in Danio and Xenopus embryos. *Developmental Biology, 325*, 249–262.

Sagner, A., & Briscoe, J. (2017). Morphogen interpretation: Concentration, time, competence, and signaling dynamics. *Wiley Interdisciplinary Reviews: Developmental Biology, 6*, e271.

Sako, K., Pradhan, S. J., Barone, V., Inglés-Prieto, A., Müller, P., Ruprecht, V., et al. (2016). Optogenetic control of nodal signaling reveals a temporal pattern of nodal signaling regulating cell fate specification during gastrulation. *Cell Reports, 16*, 866–877.

Schier, A. F., & Talbot, W. S. (2005). Molecular genetics of axis formation in zebrafish. *Annual Review of Genetics, 39*, 561–613.

Schneider, S., Steinbeisser, H., Warga, R. M., & Hausen, P. (1996). β-catenin translocation into nuclei demarcates the dorsalizing centers in frog and fish embryos. *Mechanisms of Development, 57*, 191–198.

Scholpp, S., & Brand, M. (2004). Endocytosis controls spreading and effective signaling range of Fgf8 protein. *Current Biology, 14*, 1834–1841.

Schötz, E. M., Burdine, R. D., Jülicher, F., Steinberg, M. S., Heisenberg, C.-P., & Foty, R. A. (2008). Quantitative differences in tissue surface tension influence zebrafish germ layer positioning. *Human Frontier Science Program, 2*, 42–56.

Schulte-Merker, S., Lee, K. J., McMahon, A. P., & Hammerschmidt, M. (1997). The zebrafish organizer requires chordino. *Nature, 387*, 862–863.

Sekine, R., Shibata, T., & Ebisuya, M. (2018). Synthetic mammalian pattern formation driven by differential diffusivity of Nodal and Lefty. *Nature Communications, 9*, 5456.

Sepich, D. S., Calmelet, C., Kiskowski, M., & Solnica-Krezel, L. (2005). Initiation of convergence and extension movements of lateral mesoderm during zebrafish gastrulation. *Developmental Dynamics, 234*, 279–292.

Shen, M. M. (2007). Nodal signaling: Developmental roles and regulation. *Development, 134*, 1023–1034.

Shih, J., & Fraser, S. E. (1995). Distribution of tissue progenitors within the shield region of the zebrafish gastrula. *Development, 121*, 2755–2765.

Shimizu, T., Yamanaka, Y., Nojima, H., Yabe, T., Hibi, M., & Hirano, T. (2002). A novel repressor-type homeobox gene, ved, is involved in dharma/bozozok-mediated dorsal organizer formation in zebrafish. *Mechanisms of Development, 118*, 125–138.

Shimizu, T., Yamanaka, Y., Ryu, S. L., Hashimoto, H., Yabe, T., Hirata, T., et al. (2000). Cooperative roles of Bozozok/Dharma and Nodal-related proteins in the formation of the dorsal organizer in zebrafish. *Mechanisms of Development, 91*, 293–303.

Smutny, M., Ákos, Z., Grigolon, S., Shamipour, S., Ruprecht, V., Čapek, D., et al. (2017). Friction forces position the neural anlage. *Nature Cell Biology, 19*, 306–317.

Solnica-Krezel, L. (2005). Conserved patterns of cell movements during vertebrate gastrulation. *Current Biology, 15*, 213–228.

Solnica-Krezel, L., & Sepich, D. S. (2012). Gastrulation: Making and shaping germ layers. *Annual Review of Cell and Developmental Biology, 28*, 687–717.

Solnica-Krezel, L., Stemple, D. L., Mountcastle-Shah, E., Rangini, Z., Neuhauss, S. C. F., Malicki, J., et al. (1996). Mutations affecting cell fates and cellular rearrangements during gastrulation in zebrafis. *Development, 123*, 67–80.

Song, S., Eckerle, S., Onichtchouk, D., Marrs, J. A., Nitschke, R., & Driever, W. (2013). Pou5f1-dependent EGF expression controls E-cadherin endocytosis, cell adhesion, and zebrafish epiboly movements. *Developmental Cell, 24*, 486–507.

Soroldoni, D., Jorg, D. J., Morelli, L. G., Richmond, D. L., Schindelin, J., Julicher, F., et al. (2014). A Doppler effect in embryonic pattern formation. *Science, 345*, 222–225.

Stachel, S. E., Grunwald, D. J., & Myers, P. Z. (1993). Lithium perturbation and goosecoid expression identify a dorsal specification pathway in the pregastrula zebrafish. *Development*, *117*, 1261–1274.

Steinberg, M. S. (1962). On the mechanism of tissue reconstruction by dissociated cells, I. Population kinetics, differential adhesiveness, and the absence of directed migration. *Proceedings of the National Academy of Sciences*, *48*, 1577–1582.

Tan, R. Z., & Chiam, K.-H. (2014). Computational modeling reveals that a combination of chemotaxis and differential adhesion leads to robust cell sorting during tissue patterning. *PLoS One*, *9*(10), e109286.

Thisse, B., Wright, C. V. E., & Thisse, C. (2000). Activin- and nodal-related factors control antero-posterior patterning of the zebrafish embryo. *Nature*, *403*, 425–428.

Tian, J., Andree, B., Jones, C. M., & Sampath, K. (2008). The pro-domain of the zebrafish Nodal-related protein Cyclops regulates its signaling activities. *Development*, *135*, 2649–2658.

Townes, P. L., & Holtfreter, J. (1955). Directed movements and selective adhesion of embryonic amphibian cells. *The Journal of Experimental Zoology*, *128*, 53–120.

Tran, L. D., Hino, H., Quach, H., Lim, S., Shindo, A., Mimori-Kiyosue, Y., et al. (2012). Dynamic microtubules at the vegetal cortex predict the embryonic axis in zebrafish. *Development*, *139*, 3644–3652.

Trinh, L. A., & Stainier, D. Y. R. (2004). Fibronectin regulates epithelial organization during myocardial migration in zebrafish. *Developmental Cell*, *6*, 371–382.

Ulrich, F., Concha, M. L., Heid, P. E., Voss, E., Witzel, S., Roehl, H., et al. (2003). Slb/Wnt11 controls hypoblast cell migration and morphogenesis at the onset of zebrafish gastrulation. *Development*, *130*, 5375–5384.

van Boxtel, A. L., Economou, A. D., Heliot, C., & Hill, C. S. (2017). Long-range signaling activation and local inhibition separate the mesoderm and endoderm lineages. *Developmental Cell*, *44*, 1–13.

Van, B. A. L., Chesebro, J. E., Heliot, C., Ramel, M.-C., Stone, R. K., & Hill, C. S. (2015). A temporal window for signal activation dictates the dimensions of a nodal signaling domain. *Developmental Cell*, *35*, 175–185.

Veil, M., Schaechtle, M. A., Gao, M., Kirner, V., Buryanova, L., Grethen, R., et al. (2017). Maternal Nanog is required for zebrafish embryo architecture and for cell viability during gastrulation. *Development*, *124*, dev155366.

Vopalensky, P., Pralow, S., & Vastenhouw, N. L. (2018). Reduced expression of the Nodal coreceptor Oep causes loss of mesendodermal competence in zebrafish. *Development*, *145*, dev158832.

Wagner, D. E., Weinreb, C., Collins, Z. M., Briggs, J. A., Megason, S. G., & Klein, A. M. (2018). Single-cell mapping of gene expression landscapes and lineage in the zebrafish embryo. *Science*, *360*, 981–987.

Wang, Y., Wang, X., Wohland, T., & Sampath, K. (2016). Extracellular interactions and ligand degradation shape the nodal morphogen gradient. *eLife*, *5*, 1–19.

Warga, R. M., & Kane, D. A. (2003). One-eyed pinhead regulates cell motility independent of Squint/Cyclops signaling. *Developmental Biology*, *261*, 392–411.

Warga, R. M., & Kane, D. A. (2007). A role for N-cadherin in mesodermal morphogenesis during gastrulation. *Developmental Biology*, *310*, 211–225.

Warga, R. M., & Kimmel, C. B. (1990). Cell movements during epiboly and gastrulation in zebrafish. *Development*, *108*, 569–580.

Warga, R. M., & Nusslein-Volhard, C. (1999). Origin and development of the zebrafish endoderm. *Development*, *126*, 827–838.

Watanabe, H., Schmidt, H. A., Kuhn, A., Höger, S. K., Kocagöz, Y., Laumann-Lipp, N., et al. (2014). Nodal signalling determines biradial asymmetry in Hydra. *Nature*, *515*, 112–115.

Weber, G. F., Bjerke, M. A., & DeSimone, D. W. (2012). A mechanoresponsive cadherin-keratin complex directs polarized protrusive behavior and collective cell migration. *Developmental Cell*, *22*, 104–115.

Woo, S., Housley, M. P., Weiner, O. D., & Stainier, D. Y. R. (2012). Nodal signaling regulates endodermal cell motility and actin dynamics via Rac1 and Prex1. *The Journal of Cell Biology*, *198*, 941–952.

Xiang, F., Hagos, E. G., Xu, B., Sias, C., Kawakami, K., Burdine, R. D., et al. (2007). Nodal signals mediate interactions between the extra-embryonic and embryonic tissues in zebrafish. *Developmental Biology*, *310*, 363–378.

Xu, C., Fan, Z. P., Müller, P., Fogley, R., DiBiase, A., Trompouki, E., et al. (2012). Nanog-like regulates endoderm formation through the Mxtx2-nodal pathway. *Developmental Cell*, *22*, 625–638.

Xu, P.-F., Houssin, N., Ferri-Lagneau, K. F., Thise, B., & Thisse, C. (2014). Construction of a vertebrate embryo from two opposing morphogen gradients. *Science*, *344*, 87–89.

Yamanaka, Y., Mizuno, T., Sasai, Y., Kishi, M., Takeda, H., Kim, C. H., et al. (1998). A novel homeobox gene, dharma, can induce the organizer in a non-cell- autonomous manner. *Genes & Development*, *12*, 2345–2353.

Yamashita, S., Miyagi, C., Carmany-Rampey, A., Shimizu, T., Fujii, R., Schier, A. F., et al. (2002). Stat3 controls cell movements during zebrafish gastrulation. *Developmental Cell*, *2*, 363–375.

Yamashita, S., Miyagi, C., Fukada, T., Kagara, N., Che, Y.-S., & Hirano, T. (2004). Zinc transporter LIVI controls epithelial-mesenchymal transition in zebrafish gastrula organizer. *Nature*, *429*, 1–5.

Yan, L., Chen, J., Zhu, X., Sun, J., Wu, X., Shen, W., et al. (2018). Maternal Huluwa dictates the embryonic body axis through b-catenin in vertebrates. *Science*, *362*, eaat1045.

Zhang, J., Jiang, Z., Liu, X., & Meng, A. (2016). Eph/ephrin signaling maintains the boundary of dorsal forerunner cell cluster during morphogenesis of the zebrafish embryonic left-right organizer. *Development*, *143*, 2603–2615.

Zhang, J., Talbot, W. S., & Schier, A. F. (1998). Positional cloning identifies zebrafish one-eyed pinhead as a permissive EGF-related ligand required during gastrulation. *Cell*, *92*, 241–251.

CHAPTER THIRTEEN

Cellular and molecular mechanisms of convergence and extension in zebrafish

Margot L.K. Williams, Lilianna Solnica-Krezel*
Department of Developmental Biology, Washington University in St. Louis, School of Medicine, St. Louis, MO, United States
*Corresponding author: e-mail address: solnical@wustl.edu

Contents

Abstract

Gastrulation is the period of development when the three germ layers, mesoderm, endoderm and ectoderm, are not only formed, but also shaped into a rudimentary body plan. An elongated anteroposterior (AP) axis is a key feature of all vertebrate body plans, and it forms during gastrulation through the highly conserved morphogenetic mechanism of convergence & extension (C&E). As the name suggests, this process requires that cells within each germ layer converge toward the dorsal midline to narrow the tissue in the mediolateral (ML) dimension and concomitantly extend it in the AP dimension. In a number of vertebrate species, C&E is driven primarily by mediolateral intercalation behavior (MIB), during which cells elongate, align, and extend protrusions in the ML direction and interdigitate between their neighbors. MIB is only one of many complex cellular mechanisms that contributes to C&E in zebrafish embryos, however,

Current Topics in Developmental Biology, Volume 136
ISSN 0070-2153
https://doi.org/10.1016/bs.ctdb.2019.08.001

where a combination of individual cell migration, collective migration, random walk, radial intercalation, epiboly movements, and MIB all act together to shape the nascent germ layers. Each of these diverse cell movements is driven by a distinct suite of dynamic cellular properties/activities, such as actin-rich protrusions, myosin contractility, and blebbing. Here, we discuss the spatiotemporal patterns of cellular behaviors underlying C&E gastrulation movements within each germ layer of zebrafish embryos. These behaviors must be coordinated with the embryonic axes, and we highlight the roles of Planar Cell Polarity (PCP) in orienting and BMP signaling in patterning C&E cell behaviors with respect to the AP and dorsoventral axes. Finally, we address the role of GPCR signaling, extracellular matrix, and mechanical signals in coordination of C&E movements between adjacent germ layers.

1. Overview of zebrafish C&E

Convergence & extension (C&E) is a highly conserved morphogenetic mechanism by which tissues become simultaneously longer and narrower. This type of tissue deformation is common during development and contributes to the shaping of a number of embryonic tissues and organ rudiments. The focus of this chapter will be on C&E of the zebrafish embryo during gastrulation, where it functions primarily to extend the primary body axis in the anteroposterior (AP) dimension while narrowing axial tissues in the mediolateral (ML) dimension. C&E is preceded by the specification of the embryonic axes and three primordial germ layers as described in the chapter "Setting up for gastrulation in zebrafish" by Marlow, and it occurs simultaneously with the other gastrulation movements of epiboly and internalization, discussed in the chapters "Zebrafish gastrulation: Putting fate in motion" by Pinheiro and Heisenberg and "Cellular and molecular mechanisms of convergence and extension in zebrafish" by Williams and Solnica-Krezel, respectively. During early gastrulation, epiboly thins and spreads the blastoderm over the syncytial yolk cell, driving the embryonic margin toward the vegetal pole. Once the blastoderm covers approximately 50% of the yolk surface (50% epiboly), the internalization of mesoderm and endoderm germ layers begins at the blastoderm margin first on the dorsal side, and then proceeds ventrally around the circumference of the margin. This mixed mesendoderm internalizes primarily as individual cells, which "turn the corner" to migrate animally (anteriorly) between the yolk cell and the overlying blastoderm. As gastrulation proceeds, epiboly continues to spread the germ layers vegetally (posteriorly) while the internalized mesendoderm continues to migrate animally, thereby contributing to the AP extension of the nascent primary body axis (Fig. 1A). Cells initiate convergence movements once the embryo reaches approximately 70% epiboly, contributing to axial extension together with continued epiboly

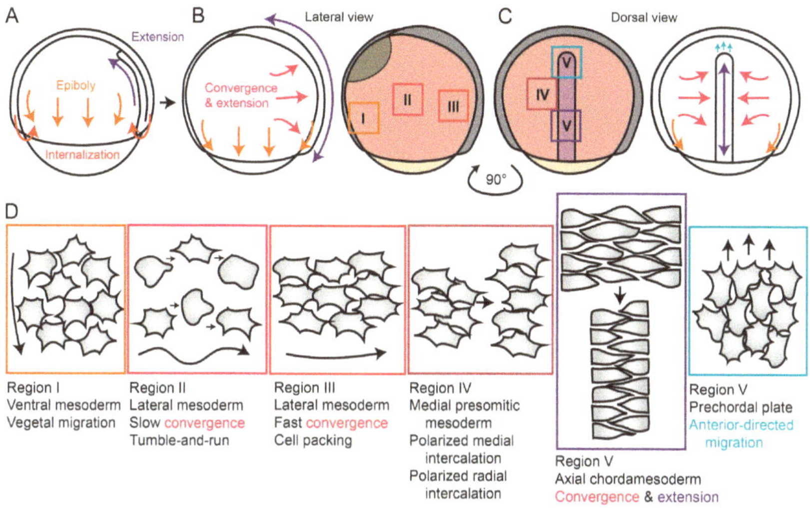

Fig. 1 Cell behaviors underlying convergence & extension (C&E) of zebrafish mesoderm. Gastrulation movements prior to (A) and after the onset of C&E in lateral (B) and dorsal (C) view. The purple stripe in C represents the axial mesoderm. (D) Cell behaviors underlying C&E movements of mesoderm in each of the five regions indicated in (B) and (C): vegetal migration in region I, dorsal migration in regions II and III, cell intercalations in regions IV and V, and anterior migration in region V. Dorsal is to the right in (A), (B), and (D); anterior is up in all panels.

and anterior migration (Fig. 1B and C). Blastoderm cells that remain on the surface of the embryo upon completion of internalization give rise to non-neural (ventrally) and neural (dorsally) ectoderm. As epiboly proceeds, these ectoderm cells continue to move vegetally while simultaneously undergoing C&E concomitant with similar movements in the underlying mesendoderm. At this point, the most ventral mesoderm, which does not engage in C&E, switches from animal- to vegetal-directed migration. When epiboly is complete, this population will undergo the subduction movements to contact the posterior dorsal mesoderm at the site of yolk plug closure and give rise to the tailbud and its derivatives. Each of these germ layers therefore exhibits a distinct, region-specific suite of cell behaviors which are coordinated to collectively drive C&E of the AP axis.

C&E is conserved throughout the animal kingdom and is achieved through a number of cellular mechanisms, including cell shape change, oriented cell division, directed cell migration, and cell intercalation (Keller, 2006). The best-described among these mechanisms is planar cell intercalation, by which a group of cells rearranges into a longer and narrower array through junctional exchange and/or interdigitation with neighboring cells termed "cell shuffling." Junctional exchange was identified as a mechanism for germ band extension during *Drosophila* gastrulation and involves remodeling of apical junctions between epithelial cells (Bertet, Sulak, & Lecuit, 2004; Zallen & Blankenship, 2008) (see chapter "Gastrulation in the sea urchin" by McClay et al.). In contrast, cell shuffling is driven by polarized cell protrusions thought to gain traction on neighboring cells and the underlying extracellular matrix (Pfister, Shook, Chang, Keller, & Skoglund, 2016). This is best exemplified by mediolateral intercalation behavior (MIB) of mesenchymal mesendodermal cells within *Xenopus* gastrulae (Keller et al., 2000). Recent evidence suggests that these two modes of cell intercalation are not mutually exclusive, and that the same cells can employ both junctional neighbor exchange and polarized protrusions to cooperatively drive C&E in both epithelial and mesenchymal tissues (Huebner & Wallingford, 2018; Sun et al., 2017; Williams, Yen, Lu, & Sutherland, 2014).

In these "classic" examples of C&E, convergence & extension movements are coupled by ML intercalation, which links narrowing in one dimension (ML) to a corresponding elongation in an orthogonal dimension (AP during C&E, dorsal-ventral during convergent thickening) (Keller et al., 2000; Shih & Keller, 1992a). For this reason, C&E in *Xenopus* is often referred to as "convergent extension." By contrast, convergence & extension movements within zebrafish gastrulae are relatively independent and often uncoupled from one another, and are driven by a number of cell behaviors in addition to ML intercalation (Glickman, Kimmel, Jones, & Adams, 2003; Sepich, Calmelet, Kiskowski, & Solnica-Krezel, 2005; Warga & Kimmel, 1990). Unlike in *Xenopus* and *Drosophila*, in zebrafish gastrulae, individual mesendodermal cells migrate over the surface of a large central yolk cell. Lateral mesendoderm cells in the equatorial region undergo directed dorsal migration to narrow the nascent primary axis, while the more animally- and vegetally-located cells each exhibit a bias toward their respective poles, causing their paths to "fan out" and contribute to extension as they converge dorsally (Fig. 1B) (Sepich et al., 2005). As these lateral mesendodermal cells migrate to drive C&E, they leave behind an "empty space" in the animal-ventral region termed the evacuation zone (Kimmel, Ballard, Kimmel, Ullmann, & Schilling, 1995). Concurrently, anterior migration of axial mesoderm cells extends the nascent axis without contributing to convergence (Concha & Adams, 1998; Ulrich et al., 2003; Warga & Kimmel, 1990). Such individual cell migration is facilitated not only by interactions with the yolk, but also by the reduced adhesion between zebrafish cells relative to the tightly-packed cellular sheets that comprise *Xenopus* and *Drosophila* gastrulae, which allows more freedom of movement.

Convergence & extension are coupled in the dorsal axial and paraxial mesoderm of zebrafish gastrulae as observed in "classic" mediolateral intercalation (Fig. 1D, region V purple box), but even these movements can be uncoupled in certain mutant conditions (Glickman et al., 2003). The distinct suite of cell behaviors underlying zebrafish C&E will be described in further detail below.

As in nearly all complex developmental processes, gastrulation morphogenesis is intricately linked with embryonic patterning. This is intuitive: for cells to move to the anterior or dorsal region of an embryo, they must first be provided with cues for where anterior and dorsal are. Indeed, the signals that specify cell fate often also dictate cell behavior within zebrafish gastrulae. For example, C&E movements are most prominent within the dorsal-most tissues, whereas ventral tissues on the other side of the gastrula exhibit essentially no C&E (Fig. 1). Mutant analyses demonstrated that signals that specify ventral fates (namely, Bone morphogenic protein (BMP)) actually inhibit C&E cell behaviors (Myers, Sepich, & Solnica-Krezel, 2002; von der Hardt et al., 2007), whereas widespread expansion of dorsalizing signals that inhibit BMP expands both dorsal cell fates (Mintzer et al., 2001) and C&E behaviors around the entire circumference of the embryo (Myers et al., 2002), squeezing the yolk cell into an abnormally elongated shape, sometimes until it ruptures (Mullins et al., 1996). BMP is thought to limit C&E movements through negative regulation of genes encoding PCP components and adhesion molecules (Myers et al., 2002; von der Hardt et al., 2007), acting in parallel rather than downstream of cell fate specification. This illustrates the close ties between patterning and morphogenesis, but with the exception of a few examples discussed further below, the molecular basis of this coordination is largely unknown.

In this chapter, we will describe C&E of each germ layer and tissue within the zebrafish gastrula, the underlying cellular mechanisms, and the known molecular regulators of each. Finally, we will discuss how interactions between germ layers influence C&E morphogenesis.

2. C&E of the mesoderm

Mesoderm within the zebrafish gastrula can be divided into five regions, each of which exhibit a distinct suite of cell behaviors (Fig. 1) (Myers et al., 2002; Roszko, Sawada, & Solnica-Krezel, 2009). Region I encompasses ventral mesoderm, regions II and III are lateral, region IV comprises medial presomitic mesoderm (PSM), and region V refers to the axial mesoderm.

2.1 Axial mesoderm: Chordamesoderm

Axial mesoderm within region V is comprised of two different tissues that exhibit distinct gastrulation cell behaviors: chordamesoderm and prechordal plate (Fig. 1C and D, purple and blue boxes). At the onset of gastrulation, both internalize at the dorsal margin and migrate toward the animal pole, thus extending the tissue prior to the onset of convergence. At approximately 70% epiboly, the chordamesoderm, which gives rise to the notochord, begins to converge and extend by mediolateral cell intercalation similar to that observed in *Xenopus* (Concha & Adams, 1998; Glickman et al., 2003). These cells elongate, align, and extend protrusions mediolaterally to facilitate intercalation between their anterior and posterior neighbors to produce a longer and narrower array (Fig. 1D, purple box). This continues until the chordamesoderm is only a single cell diameter across at early somite stages.

ML alignment of both cell shape and cellular protrusions is essential for efficient intercalation and is largely regulated by planar cell polarity (PCP) signaling. This highly conserved signaling system was first identified in *Drosophila*, where it functions at both local and global levels to promote polarization of cells within the plane of a tissue, ensuring that both sub- and supra-cellular structures (like wing trichomes and body hairs) all point in the same direction (Strutt & Strutt, 2005; Vinson & Adler, 1987; Wong & Adler, 1993). Vertebrate and *Drosophila* PCP share several so-called "core" components, including the four-pass transmembrane protein Van Gogh-like (Vangl), the Frizzled (Fz) seven-pass transmembrane receptor, the hybrid cadherin/G protein coupled receptor (GPCR) Celsr (also known as Flamingo), PDZ-domain-containing intracellular Disheveled (Dvl), and the LIM domain-containing Prickle (Pk) (Carreira-Barbosa et al., 2003, 2009; Gubb et al., 1999; Jessen et al., 2002; Lu, Usui, Uemura, Jan, & Jan, 1999; Park & Moon, 2002; Takeuchi et al., 2003; Taylor, Abramova, Charlton, & Adler, 1998; Usui et al., 1999; Veeman, Slusarski, Kaykas, Louie, & Moon, 2003). In addition to these "core" proteins, vertebrate PCP also shares a subset of components with the Wnt/β-catenin pathway, including Wnt ligands (Heisenberg et al., 2000; Kilian et al., 2003), Fz receptors (Čapek et al., 2019; Wallingford, Vogeli, & Harland, 2001), and intracellular Dvl effectors (Wallingford et al., 2000), and is therefore often referred to as non-canonical Wnt/PCP signaling. Zebrafish mutations that disrupt PCP signaling yield embryos with shorter AP and wider ML body axes, indicative of reduced C&E, but few patterning or cell fate specification defects. These include *vangl2 (trilobite), wnt5a (pipetail), wnt11 (silberblick)*, and *frizzled7a/b*; as well as PCP-related/interacting genes such as *glypican4* (*knypek*) and *ptk7* (Čapek et al., 2019; Hayes, Naito, Daulat, Angers, & Ciruna, 2013; Heisenberg et al., 2000; Jessen et al., 2002; Kilian et al., 2003; Sepich et al., 2000; Topczewski et al., 2001). Core PCP components like Prickle and Vangl2 become asymmetrically distributed or enriched at the anterior and Dvl to the posterior edge of intercalating mesoderm cells (Ciruna, Jenny, Lee, Mlodzik, & Schier, 2006; Roszko, Sepich, Jessen, Chandrasekhar, & Solnica-Krezel, 2015; Yin, Kiskowski, Pouille, Farge, & Solnica-Krezel, 2008) (Fig. 2A), leading to the hypothesis that PCP acts as a molecular compass that allows cells to orient with respect to the embryonic axes (Gray, Roszko, & Solnica-Krezel, 2011; Yin et al., 2008). Without this spatial "awareness," cells cannot acquire the ML polarity required for effective cell intercalation, leading to disrupted C&E morphogenesis.

Chordamesoderm cells within PCP mutant gastrulae exhibit inefficient intercalation, leading to wider and shorter notochords (Čapek et al., 2019; Heisenberg et al., 2000; Jessen et al., 2002; Kilian et al., 2003; Topczewski et al., 2001). This is due to reduced ML polarity of both cell shapes and cellular protrusions, which gain traction on neighboring cells to drive their intercalation. Wnt/PCP signaling is thought to regulate the polarity of these cell behaviors largely via an intracellular cascade that promotes phosphorylation of Myosin regulatory light chain (MRLC) to enhance Myosin contractility. Work in *Xenopus* found that upon binding of non-canonical Wnt ligands to Fz, Dvl is recruited to the membrane where it forms a complex with the Formin homolog protein Disheveled-associated activator of morphogenesis (Daam1) to activate RhoA GTPases (Habas, Kato, & He, 2001). RhoA in turn activates downstream effectors Diaphanous (Lai et al., 2008; Zhu, Liu, Korzh, Gong, & Low, 2006) and the Rho Kinase Rok2, which promotes C&E cell behaviors in a cell-autonomous and non-autonomous fashion (Marlow, Topczewski, Sepich, & Solnica-Krezel, 2002) likely by phosphorylating MRLC (Winter et al., 2001) (Fig. 2A). Additional work in *Xenopus* demonstrated that Septins restrict Myosin activation in a planar-polarized fashion to AP cell interfaces to ensure ML directional intercalation (Shindo & Wallingford, 2014), but whether similar mechanisms play a role in zebrafish C&E is unknown.

Wnts also regulate C&E cell behaviors via a variety of other non-canonical downstream signaling effectors, including Ca^{2+}, Protein Kinase C (PKC), and c-Jun amino terminal kinase (JNK) (Fig. 2A). In zebrafish

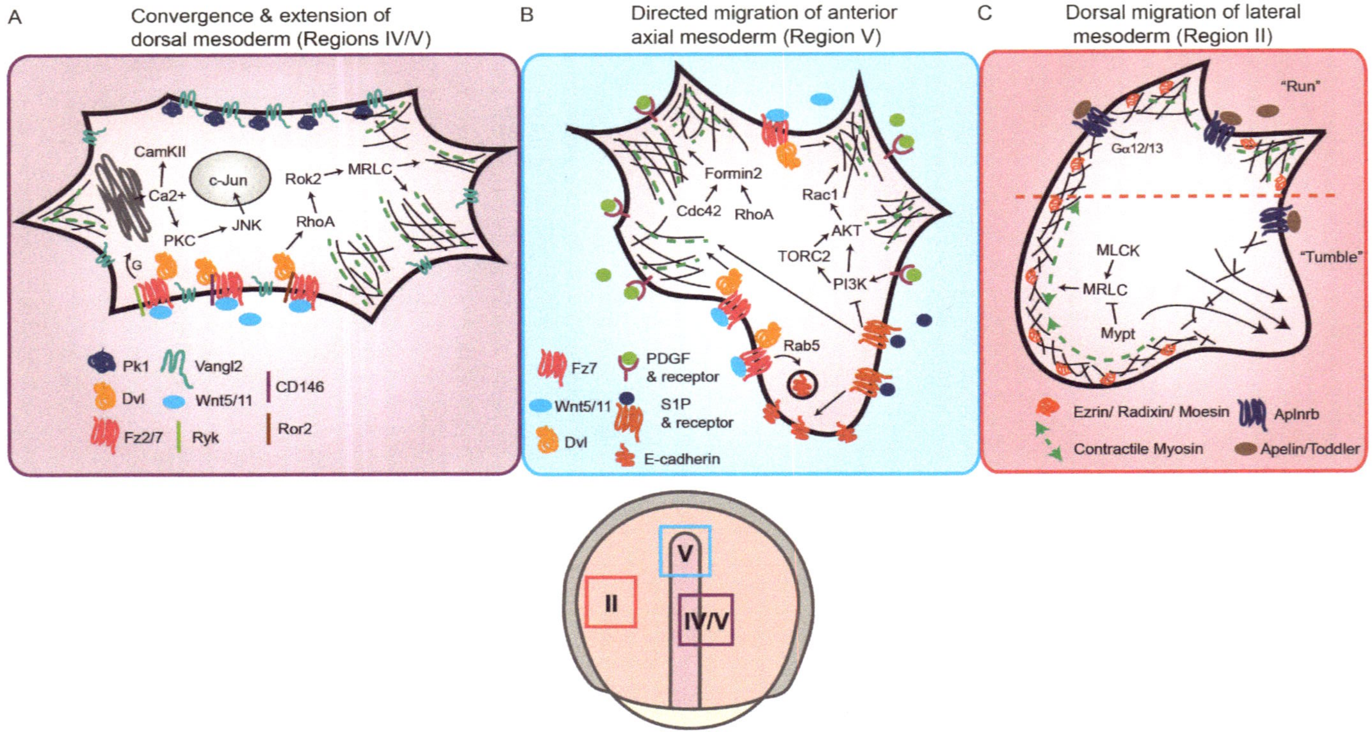

Fig. 2 Signaling pathways regulating C&E cell behaviors. (A) Within regions IV and V, non-canonical Wnt/PCP signaling regulates ML cell elongation and ML-biased cell protrusions via a number of downstream modules including JNK, Ca^{2+}, and RhoA. (B) Within region V, anteriorly-directed cell migration is regulated by sphingosine-1-phosphate (S1P), Wnt/PCP, and PDGF signaling that ultimately promote directed protrusions via localized Rac1 and RhoA activity. (C) Within region II, cells converge slowly by alternating phases of "tumble"—blebbing induced by loss of membrane-cortical attachment, and "run"—directed protrusion-driven migration requiring Apelin receptor signaling. Black hashes represent the actin cytoskeleton, green hashes represent non-muscle myosin.

embryos, signaling by Wnt5b—but not canonical Wnt ligands—was shown to promote release of calcium ions from intracellular stores in a G protein-dependent manner (Slusarski, Corces, & Moon, 1997; Slusarski, Yang-Snyder, Busa, & Moon, 1997). Work in a combination of zebrafish and *Xenopus* embryos found that upon Wnt binding to Fz7, Dvl is recruited to the membrane together with the Ca^{2+}-responsive PKC, which then stimulates JNK and its target c-Jun (Kühl, Sheldahl, Malbon, & Moon, 2000; Kühl, Sheldahl, Park, Miller, & Moon, 2000; Sheldahl, Park, Malbon, & Moon, 1999; Sheldahl et al., 2003). Wnt-induced Ca^{2+} release also activates Ca^{2+}/Calmodulin-dependent Kinase II (CamKII) in *Xenopus* (Sheldahl et al., 2003), and activated CamKII suppresses tail defects in *wnt5b/ppt* mutant zebrafish embryos (Westfall et al., 2003). The relationship between this Ca^{2+}-dependent non-canonical Wnt pathway and Wnt/PCP signaling is not fully understood.

The "choice" of signaling cascade downstream of Wnt signaling and its effect on C&E morphogenesis is dependent upon a number of alternative/co-receptors, including Ryk, Ror2, and CD146 (Fig. 2A). CD146 is a transmembrane cell adhesion molecule that interacts directly with both Wnt5 ligands and Dvl2 and is required for Dvl2 phosphorylation and downstream activation of JNK, which promotes cell motility and C&E movements upon Wnt ligand binding (Ye et al., 2013). Related to Tyrosine Kinase (Ryk), a kinase-dead transmembrane receptor tyrosine kinase (RTK), also acts downstream of Wnt5b, but functions independently of Dvl and Fz receptors to promote release of intracellular Ca^{2+}. Ryk-expressing cells extend protrusions away from sources of Wnt5b (but not Wnt11), implying that it mediates directional cell responses to non-canonical Wnt ligands through repulsive interactions (Lin, Baye, Westfall, & Slusarski, 2010). Another RTK, Ror2, was shown to act as an alternative receptor for Wnt5a in other vertebrate species (Ho et al., 2012; Schambony & Wedlich, 2007), but mediates Wnt11 signals during zebrafish gastrulation via downstream activation of Dvl and RhoA (Bai et al., 2014). The RTK Ptk7 also promotes planar polarization of both cell behaviors and intracellular distribution of PCP components (Hayes et al., 2013), and its loss of function, like all of the Wnt alternative/co-receptors discussed above, disrupts C&E gastrulation movements in zebrafish. While it is clear that such alternative/co-receptors are vital to the interpretation of morphogenetic cues, the nature of these signals—let alone how they are integrated by cells via multiple receptors to achieve specific behavioral outcomes—is not well understood. In particular, whereas it is clear that Wnt11 and Wnt5b are essential for ML cell polarity underlying C&E, it remains unclear whether they play an instructive or permissive role (see below).

G protein coupled receptors (GPCRs) have also been implicated in non-canonical Wnt/PCP signaling and C&E morphogenesis. Overexpression of the adhesion (ad)GPCR Gpr125/Adgra3, for example, disrupts polarization of PCP components and reduces ML cell polarity and C&E. While Gpr125 knockdown using antisense morpholino oligonucleotides does not cause any phenotypes in wild-type gastrulae, it exacerbates C&E defects in PCP mutants (Li et al., 2013). The intracellular domain of Gpr125 recruits PCP signaling components to cell membranes by clustering the PCP effector Disheveled into membrane sub-domains (Li et al., 2013), but how this activity is induced/regulated during normal C&E is unknown. One possibility is that it functions as a Wnt co-receptor, as is the case of the closely-related adGPCR GPR124/Adgra2, which acts together with Reck to confer receptor specificity for Wnt7 ligands in mammalian cells (Eubelen et al., 2018; Vallon et al., 2018).

The G proteins $G\alpha_{12/13}$ are also required cell-autonomously for proper C&E, as their loss- or gain-of function results in shorter and wider chordamesoderm and reduced ML cell polarity in zebrafish gastrulae without affecting embryonic patterning (Lin et al., 2005). Loss of $G\alpha_{12/13}$ exacerbates defects in C&E and underlying ML cell polarity in PCP mutant embryos, but it is unclear whether they function downstream of or in parallel to PCP signaling. G proteins have been implicated in non-canonical $Ca^{2+/}$PKC-mediated Wnt signaling (Slusarski, Corces, et al., 1997), and $G\alpha_{12/13}$ can function upstream of Rho GTPases to regulate shape and migration of cultured cells (Buhl, Johnson, Dhanasekaran, & Johnson, 1995; Sugimoto, Takuwa, Okamoto, Sakurada, & Takuwa, 2003). GPCRs like Gpr125 (Li et al., 2013), the core PCP component Celsr (Carreira-Barbosa et al., 2009), and chemokine receptor Cxcr4 (described further below) are all required for C&E of mesoderm and/or endoderm, allowing for the possibility that signaling via G proteins like $G\alpha_{12/13}$ is a critical component of Wnt-dependent signaling networks that drive C&E.

Beyond these complex, overlapping, and interacting signaling networks that regulate polarized cell behaviors and tissue morphogenesis, mechanical forces are also essential regulators of gastrulation. Surface tension, for example, contributes significantly to epiboly movements that occur contemporaneously with C&E (Morita et al., 2017). For more on the importance of additional physical properties during morphogenesis, such as tissue stiffness and tension, please see: Davidson (2017), Keller, Davidson, and Shook

(2003), Miller and Davidson (2013), and Zhou, Pal, Maiti, and Davidson (2015). One physical element that is critical for C&E of the chordamesoderm are its boundaries with the adjacent presomitic mesoderm (PSM). This tissue boundary between the axial and paraxial mesoderm was first shown to influence the behavior of adjacent chordamesoderm cells in the form of "boundary capture" in *Xenopus* gastrulae (Shih & Keller, 1992b), the phenomenon by which contact with the boundary quiets protrusions on one side of the cell, converting bipolar protrusive activity into monopolar activity that continues to narrow the axial mesoderm. The notochord boundary was found to similarly influence polarized cell behaviors within the chordamesoderm of zebrafish gastrulae. Boundary-associated cues, including cell adhesion and tissue tension, promote ML polarity of cells within approximately two cell diameters of the boundary and, importantly, function independently of Wnt/PCP signaling (Williams et al., 2018). Boundary cues and PCP were similarly shown to cooperate in C&E of Ascidian chordamesoderm (Veeman et al., 2008). Another physical attribute of the axial mesoderm that influences axis extension is cell number. Indeed, even a moderate reduction in proliferation rate significantly reduces extension of the chordamesoderm and AP axis length (Liu, Sepich, & Solnica-Krezel, 2017; Riley et al., 2010), likely due to fewer "building blocks" with which to extend this tissue. These studies demonstrate that an embryo's physical properties cooperate with PCP and other signaling to promote ML polarity and/or cell intercalation within the chordamesoderm.

2.2 Axial mesoderm: Prechordal plate

The anterior-most portion of the axial mesoderm, the prechordal plate (ppl), contributes to C&E of the AP axis via a different suite of cell behaviors than its neighbor to the posterior. Axial mesoderm cells are among the first to internalize at the dorsal margin and undergo highly directed migration toward the animal (future anterior) pole of the embryo. As described above, the chordamesoderm switches from anterior migration to ML intercalation at mid-gastrulation, while ppl cells continue to migrate as a cohesive population that produces polarized protrusions at its leading (anterior) edge while maintaining adhesions with fellow ppl cells and the chordamesoderm to its posterior (Ulrich et al., 2003) (Fig. 1C and D blue box). Migrating cells at the leading edge of the ppl exhibit a variety of protrusive behaviors, including filopodia, lamellipodia, and blebs. The balance between these modes of migration determines the speed and persistence of migration, and is achieved through precise regulation of attachment between the cell cortex and membrane (Diz-Muñoz et al., 2010). Notably, insufficient anterior displacement of the ppl (as is observed in many PCP mutants) can result from reduced C&E in the lateral mesoderm alone (Heisenberg et al., 2000). This suggests that full extension of the axial mesoderm requires both "pulling" by directed ppl migration at the anterior end and "pushing" by C&E of the chordamesoderm and lateral mesoderm populations.

Like C&E cell behaviors within the chordamesoderm, anterior migration of axial mesoderm cells is under the regulation of Wnt/PCP signaling (Fig. 2B). Wnt11, for example, is required for the formation of anterior-directed protrusions within newly-internalized mesendoderm populations (Ulrich et al., 2003). Loss of Wnt5b does not affect anterior migration in the same way, but injection of exogenous *wnt5b* RNA suppresses ppl defects in *wnt11/slb* mutants (Kilian et al., 2003), demonstrating that these ligands have redundant functions within the anterior axial mesoderm. It has been debated, however, whether PCP signaling plays a permissive or instructive role in C&E during zebrafish gastrulation. It was shown that Wnt5b is not only necessary for efficient C&E, but can also orient cell protrusions away from the source of the ligand (Lin et al., 2010), suggesting that Wnt5b signaling may instruct C&E cell behaviors. This is supported by reports that ubiquitous expression of *wnt5b* RNA does not rescue C&E defects in *wnt5b/ppt* mutant gastrulae (Kilian et al., 2003), consistent with an instructive rather than permissive role in these cell behaviors. This contrasts with another well-characterized Wnt/PCP mutant, *wnt11/slb*, whose C&E defects can be rescued by *wnt11* RNA injection (Ulrich et al., 2003). Similarly, the Fz7 receptor with which Wnt11 is thought to interact is essential for proper gastrulation, but uniform optogenetic-driven activation of Fz7 in *fz7* mutant gastrulae rescues anterior migration of the ppl (Čapek et al., 2019), again implying a permissive rather than instructive role for Wnt11 signaling in anterior migration of ppl cells. Wnt5b was shown to signal through a different receptor, Fz2 (Kilian et al., 2003), raising the possibility that each of these ligand/receptor pairs regulates C&E via a distinct mechanism: one instructive and one permissive. In *Xenopus* embryos, non-canonical ligands Wnt5a and Wnt11 were shown to be both necessary and instructive for planar-polarized distribution of core PCP components Pk and Vangl2 (Chu & Sokol, 2016; Ossipova, Kim, & Sokol, 2015), demonstrating similarities and potential differences with Wnt/PCP signaling in zebrafish.

Another well-characterized signaling pathway regulating anterior axial mesoderm migration in zebrafish is platelet-derived growth factor (PDGF) (Fig. 2B). PDGF receptor signaling activates PI3 Kinase, which signals through Target of Rapamycin 2 (TORC2) and/or AKT (Dumortier & David, 2015; Montero, Kilian, Chan, Bayliss, & Heisenberg, 2003) to locally activate Rac1 at the leading edge of ppl cells, possibly via the Rac Guanine nucleotide exchange factor (GEF) Prex1 (Campbell et al., 2013; Woo, Housley, Weiner, & Stainier, 2012). The sphingosine-1-phosphate GPCR S1pr2, encoded by the zebrafish *miles apart* gene, negatively regulates PI3K-dependent cell motility while promoting ppl cell cohesion to ensure cohesive, directed migration, but also regulates cell motility independent of PI3K (Kai, Heisenberg, & Tada, 2008). This may be mediated by $G\alpha_{12/13}$-dependent activation of Rho and inhibition of Rac GTPases, which has been observed in cultured cells (Sugimoto et al., 2003). The Diaphanous-related Formin2 is also critical for efficient ppl protrusive activity, but it appears to function in a Rac-independent manner and instead interacts with activated RhoA and/or Cdc42 (Lai et al., 2008).

Interestingly, although PDGF is essential for activation of this signaling cascade, it does not appear to instruct the direction of migration. This is supported by experiments in which isolated ppl cells are motile but fail to migrate anteriorly until they make contact with the intact ppl, at which point their migration becomes appropriately polarized. This instructive interaction requires not only Rac1 activity, but also E-cadherin (Dumortier, Martin, Meyer, Rosa, & David, 2012), indicating that cadherin-based contacts between ppl cells either directly provide directionality to this collective cell migration or are required for ppl cells to perceive additional directionality cues. This likely explains why cohesion of the ppl is so critical to proper morphogenesis, and why dispersal of these cells results in abnormal migration and reduced C&E (Kai et al., 2008). Cohesion of the ppl is regulated in part by Wnt11, which regulates endocytosis of E-cadherin through interactions with Rab5 to modulate cells' adhesive properties (Ulrich et al., 2005). This is consistent with evidence that Wnt11/Fz7 signaling plays a permissive rather than instructive role in ppl migration (Čapek et al., 2019; Ulrich et al., 2003). Cell-cell adhesion within the ppl can even influence cell fate specification, as prolonged cadherin-based contacts between cells enhance Nodal signaling and ppl cell fate as part of a positive feedback loop (Barone et al., 2017).

2.3 Paraxial mesoderm

Like the neighboring chordamesoderm, the paraxial mesoderm (region IV) also migrates anteriorly upon internalization at the embryonic margin, then switches to C&E movements at mid-gastrulation (Fig. 1C and D, maroon box). This population of cells directly abuts the chordamesoderm and ultimately gives rise to adaxial and somitic mesoderm that contributes to embryonic muscle, skeleton, and skin. Although the onset of C&E behaviors in both paraxial and the neighboring chordamesoderm is essentially simultaneous and utilizes many of the same cellular mechanisms, the rate of extension within the paraxial mesoderm is smaller compared to the axial compartment, resulting in shearing between these tissues (Glickman et al., 2003). Paraxial mesoderm cells are less densely-packed than chordamesoderm cells, and exhibit two distinct cell behaviors that cooperate to drive C&E: polarized medial (planar) intercalations and polarized radial intercalations. During planar medial intercalation, cells converge toward the dorsal midline and intercalate within the plane between their anteroposterior neighbors. Polarized radial intercalation, by contrast, entails the movement of cells vertically into a deeper/shallower cell layer to preferentially separate neighboring cells along the AP axis, and thus contributing to tissue extension (Yin et al., 2008).

As in the chordamesoderm, PCP signaling components become asymmetrically localized to the anterior or posterior edges of intercalating PSM cells at the onset of C&E (Roszko et al., 2015; Yin et al., 2008) and regulate both polarized planar and radial intercalation behaviors. Cell intercalations occur in PCP mutants, but their directional bias is lost. For example, in *tri/vangl2;kny/gpc4* double mutants, cells undergoing either planar or radial intercalations are significantly less likely to separate AP neighbors and significantly more likely to separate ML neighbors, resulting in embryonic axes that are severely shortened and widened (Yin et al., 2008). This intercalation behavior requires ML polarization of cell bodies and protrusions, both of which are reduced in paraxial mesoderm cells upon loss of the Wnt/PCP components Wnt5b, Vangl2, and Gpc4 (Kilian et al., 2003; Yin et al., 2008).

2.4 Lateral mesoderm

Regions II and III of C&E are contained within the lateral mesoderm, which gives rise to the heart, blood and kidney among other tissues. These cells are more loosely packed than those in regions IV and V, but still undergo dorsal convergence (Sepich et al., 2005). The speed of this convergence is the main distinction between region II, which converges slowly, and region III, which is defined by faster dorsal convergence (Myers et al., 2002) (Fig. 1B and D, pink boxes). Unlike the regions discussed above, slow dorsal convergence within region II is independent of PCP signaling (Jessen et al., 2002). These cells are not highly ML elongated, and instead migrate along complex trajectories (Jessen et al., 2002), exhibiting "front and back" polarity and a distinct "tumble-and-run" mode of movement during which individual cells alternate between intervals of protrusion-driven dorsally-directed migration (run) and less directed blebbing behavior (tumble) (Diz-Muñoz et al., 2016). Bleb-driven migration is promoted by reduced membrane-cortex attachment (MCA) and increased Myosin contractility, which is thought to increase intracellular pressure and push regions of cytoplasm outward into the space between the membrane and cortex (Diz-Muñoz et al., 2016; Ruprecht et al., 2015). As in the ppl, the ratio of run to tumble is regulated by levels of MCA, which is mediated by ERM (Ezrin/Radixin/Moesin) proteins (Fig. 2C), and tipping the balance toward either run or tumble disrupts migration speed and persistence during gastrulation (Diz-Muñoz et al., 2016). Indeed, excessive Myosin contractility induced by loss of the Myosin Phosphatase Mypt1 increases blebbing within lateral mesoderm, disrupting C&E morphogenesis (Weiser, Row, & Kimelman, 2009). Migratory behavior within the lateral mesoderm also requires signaling via the GPCR Agtrl1b/Aplnrb and its ligands Apelin and Toddler (also known as Apela and Elabela), the latter of which was shown to act as a "motogen" to promote cell motility without providing directionality (Pauli et al., 2014). Loss of Apelin/Toddler signaling disrupts mesendoderm internalization movements (Pauli et al., 2014) and later dorsal convergence of lateral mesoderm cells by impairing normal protrusive activity, resulting in wider and shorter body axes (Zeng, Wilm, Sepich, & Solnica-Krezel, 2007). $G\alpha_{12/13}$ are also required for convergence in this region, as cells with reduced $G\alpha_{12/13}$ are rounder in shape and migrate dorsally with reduced speed and persistence (Lin et al., 2005) (Fig. 2C). Mathematical modeling suggests that these convergence movements are regulated by a chemoattractant(s) emanating from dorsal midline (Sepich et al., 2005), but their identity remains elusive.

Fibronectin and laminin ECM components are laid down between nascent germ layers during zebrafish gastrulation concurrent with the onset of C&E (Latimer & Jessen, 2010). ECM assembly is abnormal in several different PCP loss of function mutants (Dohn, Mundell, Sawyer,

Dunlap, & Jessen, 2013), implicating cell-matrix adhesion as an important regulator of gastrulation morphogenesis in zebrafish. Indeed, abnormal ECM organization and increased expression of the ECM receptor Integrinβ1b are thought to contribute to C&E defects resulting from mutations in the transcription factor *pitx2c* (Collins et al., 2018). The matrix remodeling enzyme MMP14 is also required for proper cell polarity, migration speed, and axial extension, as its knockdown produces phenotypes similar to PCP mutants (Coyle, Latimer, & Jessen, 2008). PCP *vangl2/tri* mutant zebrafish embryos contain less Fibronectin than WT while *gpc4/kny* mutants exhibit a denser ECM network (Dohn et al., 2013; Williams et al., 2012), implying that both increased and decreased cell-matrix adhesion can adversely affect dorsal mesoderm migration.

Despite the relatively low density of lateral mesoderm cells, cell-cell adhesions are also critical for proper convergence movements. Although the most dramatic phenotype of N-cadherin (*cdh2*) loss-of-function mutants is within the neuroectoderm, close inspection revealed that lateral mesoderm cells are less ML elongated and exhibit slower and less persistent migration (Warga & Kane, 2007). Similar but even more severe mesoderm phenotypes are observed in *cdh2* gain-of-function mutants, which manifest shorter and wider mesoderm domains (Warga & Kane, 2007). Cadherin complexes are essential for cell-cell adhesion, as demonstrated by the catastrophic loss of embryonic integrity in *cdh1* (E-cadherin) and *ctnna1* (α-catenin) mutants (Kane et al., 1996; Kane, McFarland, & Warga, 2005), but they also have important signaling functions. Epithelial integrity in *ctnna1* mutants can be rescued by a *ctnna1* construct lacking the mechanosensitive Vinculin binding site, but these embryos still exhibit severe C&E defects (Han et al., 2016). This suggests that in addition to their adhesive function, adherens junctions regulate C&E cell movements (at least in part) through mechanosensitive signaling.

Unlike the slow convergence in region II, fast convergence within region III is PCP-dependent. PCP signaling is required for proper ML polarization and migratory speed/persistence of these lateral mesoderm cells, but notably, not their direction of migration (Jessen et al., 2002; Topczewski et al., 2001). This reinforces the notion that PCP is essential for cells to sense and/or efficiently respond to directional cues, but does not itself comprise the hypothesized dorsal convergence cue.

2.5 Ventral mesoderm

Like other mesodermal precursors, the ventral mesoderm (region I) initially moves toward the animal pole upon internalization, but unlike other regions of zebrafish mesoderm, it exhibits neither convergence nor extension and instead begins at mid-gastrulation to move vegetally as the blastoderm undergoes epiboly (Myers et al., 2002) (Fig. 1B and D, orange box). The ventral mesoderm contributes to the tailbud upon completion of epiboly, giving rise to posterior mesoderm in post-gastrulation stages. Evidence supports the notion that C&E is actively inhibited within this population by the BMP signals that specify ventral fates via at least two distinct mechanisms. First, BMP suppresses expression of PCP signaling components on the embryo's ventral side in parallel to fate specification (Myers et al., 2002). Second, BMP signaling modulates cell-cell adhesion within the ventrolateral mesoderm that generates an "adhesion gradient" that increases from ventral to dorsal (von der Hardt et al., 2007). Reduced adhesion between cells prevents neighbors from gaining traction on one another, thereby preventing dorsal convergence movements. This example illustrates how embryonic patterning cues regulate morphogenesis, and it is one of few in which underlying molecular mechanisms have been identified.

3. C&E of the endoderm

In zebrafish, mesoderm and endoderm lineages are intermingled as they internalize at the blastoderm margin during gastrulation (Warga & Nüsslein-Volhard, 1999). Despite their physical proximity, endoderm cells immediately diverge from the cell behaviors of their mesodermal neighbors. Unlike the dorsal- or anterior-directed migration exhibited by the mesoderm, upon internalization the sparser endoderm layer disperses evenly across the surface of the yolk by random walk of individual cells (Fig. 3A). This migration is characterized by high speed but low persistence and is poorly correlated between neighboring endoderm cells. Random walk behavior is cell-autonomous, as it is exhibited by isolated endoderm cells and requires neither other endoderm cells nor mesoderm to occur (Pézeron et al., 2008). This movement is achieved through the elaboration of abundant and randomly-oriented actin-rich protrusions which prefigure cell body translocation and are therefore thought to directly drive endoderm cell displacement (Pézeron et al., 2008; Woo et al., 2012).

Random walk behavior is intrinsic to endoderm cells and is induced by the same signaling cascade that specifies endoderm fate: high levels of Nodal signaling and its downstream transcriptional target, *sox32*. Cell-autonomous activation of Nodal signaling is sufficient to induce random walk behavior in isolated endoderm cells, but not when endoderm specification is blocked by simultaneous loss of *sox32* function (Pézeron et al., 2008). Nodal signaling

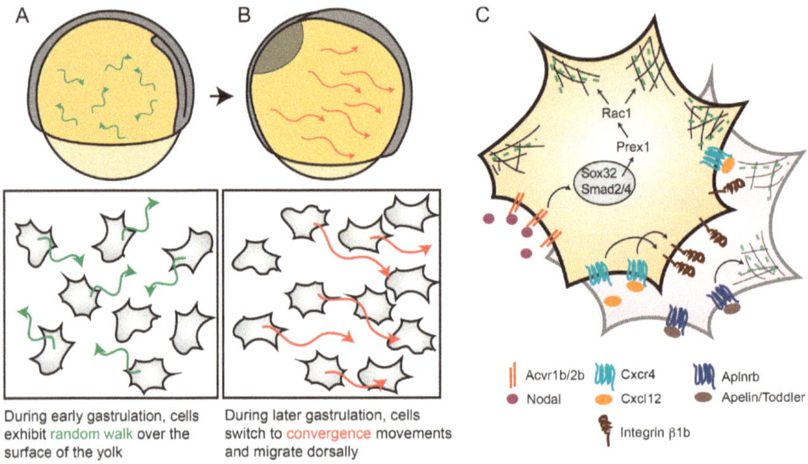

Fig. 3 Cell behaviors underlying convergence & extension of zebrafish endoderm. Endoderm cells move by random walk prior to the onset of C&E (A), then switch to dorsal migration at mid-gastrulation (B). (C) Endoderm migration is regulated by Nodal and Apelin/Toddler signaling and is tethered to the underlying mesoderm by increased Integrin-based adhesion downstream of Cxcl12/Cxcr4 chemokine signaling.

promotes actin dynamics underlying random protrusive behavior through activation of Rac1 via the Rac-specific guanine nucleotide exchange factor (GEF) Prex1 (Fig. 3C), which is also a transcriptional target of Nodal signaling (Woo et al., 2012). This cellular mechanism is notably similar to that regulating migration of lateral and anterior axial mesoderm (ppl, described above), another tissue induced by high levels of Nodal signaling. Given that directionality of migration is conferred to the ppl by interactions with other ppl cells (Dumortier et al., 2012), lack of cell-cell contacts may contribute, together with *sox32* expression, to the distinction between Nodal-induced random walk exhibited by endoderm and directed migration by ppl cells.

Although other cell types are not required for random walk behavior, interactions with the overlying mesoderm limits (or "tethers") endoderm migration toward the animal pole (Nair & Schilling, 2008). This tethering is mediated by signaling between Cxcl12 secreted by the mesoderm and its cognate chemokine Cxcr4 receptor in the endoderm. Rather than mediating chemotaxis, Cxcl12-Cxcr4 signaling modulates Integrin expression levels within endoderm cells to enhance their adhesion to the underlying ECM and consequently to limit their migration (Nair & Schilling, 2008) (Fig. 3C). Cxcl12-expressing mesoderm can recruit endogenous endoderm to ectopic sites, providing potential evidence for chemotaxis (Mizoguchi, Verkade, Heath, Kuroiwa, & Kikuchi, 2008), but this result could also be explained by enhanced endoderm adhesion to sites of increased Cxcl12-Cxcr4 signaling. Toddler/Apelin signaling via their GPCR is also critical for endoderm migration, as its loss prevents the dispersal of cells away from the embryonic margin during gastrulation (Pauli et al., 2014). However, the effect of Toddler on endoderm migration appears to be indirect via its requirement for mesoderm migration. Toddler only participates in migration of the endoderm when it is tethered to the underlying mesoderm, and this requirement is removed upon loss of Cxcr4 (Norris et al., 2017), indicating that the primary role for Toddler/Apelin signaling is within the mesoderm (Fig. 3C).

At mid-gastrulation, the mode of endoderm migration switches from random walk to dorsal convergence (Fig. 3B). The latter is driven by actin-rich protrusions, which reorient toward the embryo's dorsal side concurrent with the switch in migration behavior (Pézeron et al., 2008; Woo et al., 2012). Unlike random walk, endoderm convergence is not cell-autonomous, but rather requires mesoderm and the appearance of an environmental "convergence signal" at mid-gastrulation (Pézeron et al., 2008). Because convergence of the endoderm mirrors the movements of the overlying mesoderm and cannot occur in its absence, the switch from random walk is likely mediated by interactions with the mesoderm. In fact, the same Cxcl12-Cxcr4 signaling that tethers early endoderm is also required for convergence of late endoderm (Nair & Schilling, 2008) (Fig. 3C). Notably, mesoderm convergence is unaffected by loss of this interaction, indicating that Cxcl12 is not a "convergence signal," but rather promotes convergence of endoderm by tethering it to the converging mesoderm (Nair & Schilling, 2008). Indeed, live in toto imaging of endoderm gastrulation morphogenesis revealed cell movement patterns that are inconsistent with chemotaxis, instead supporting a tethering model (Schmid et al., 2013). Endoderm convergence is variably affected in PCP mutants with *vangl2/tri* mutants showing mild defects and *gpc4/kny* mutants showing strong defects (Hu et al., 2018; Miles, Mizoguchi, Kikuchi, & Verkade, 2017), and may be partially explained by defective C&E of mesoderm to which the endoderm is tethered. However, Glypican4 may play a more direct and cell-non-autonomous role in endoderm convergence by promoting expression of MMP14, a matrix metalloproteinase, which degrades ECM. Because enhanced cell-matrix adhesion limits endoderm migration (Nair & Schilling, 2008), a reduction in MMP14 expression and resulting increase in ECM may contribute to the reduced endoderm convergence observed in *gpc4/kny* mutant embryos (Hu et al., 2018).

4. C&E of the neuroectoderm

4.1 Neural plate

Once nascent endoderm and mesoderm cells have internalized during zebrafish gastrulation, those blastoderm cells that remain in the superficial layer of the embryo beneath the enveloping layer are fated to become non-neural (ventral) and neural (dorsal) ectoderm (Fig. 4A). Unlike in mouse, chick, and frog embryos, early zebrafish neuroectoderm does not develop as a columnar or pseudostratified epithelium, but instead consists of a closely packed, cohesive sheet of non-epithelial cells that comprise the neural plate (Fig. 4B). During gastrulation, cells within the neural plate converge toward the dorsal midline to extend the AP axis in much the same way as the underlying mesoderm, exhibiting ML alignment, elongation, and intercalation (Concha & Adams, 1998; Jessen et al., 2002; Kilian et al., 2003). Despite their ectodermal nature, these cells are highly motile and extend a variety of polarized protrusions that drive their dorsal migration (Kilian et al., 2003; Love, Prince, & Jessen, 2018). As in the underlying mesoderm, these cell behaviors are regulated by PCP signaling, ECM organization, and N-cadherin (Hong & Brewster, 2006; Jessen et al., 2002; Love et al., 2018).

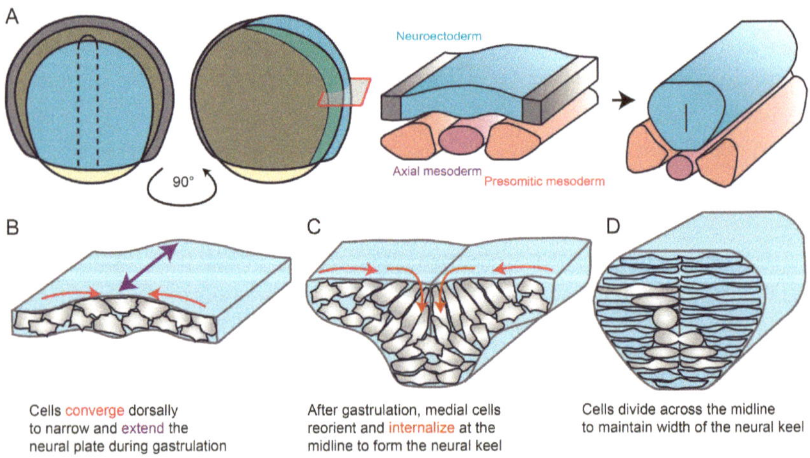

Fig. 4 Cell behaviors underlying convergence & extension of zebrafish neuroectoderm. (A) The zebrafish neuroectoderm transforms from a neural plate into the neural keel after completion of gastrulation. Convergence of neuroectoderm cells toward the dorsal midline first drives extension (B), then extension stops while convergence continues, instead driving internalization of cells at the midline to form the neural keel (C). Once the neural keel is established, mirror-image divisions across the midline maintain the width of the neural keel/neural rod (D).

PCP components Prickle1 and Vangl2 become asymmetrically localized to the anterior edges of neuroectoderm cells during gastrulation (Ciruna et al., 2006; Roszko et al., 2009; Yin et al., 2008), indicating planar polarization of the tissue. Disruption of PCP signaling by mutation or morpholino knockdown of PCP components *vangl2* or *wnt5b* reduces ML cell alignment and elongation, and loss of *wnt5b* reduces directed protrusive activity of neuroectoderm cells (Jessen et al., 2002; Kilian et al., 2003; Love et al., 2018), resulting in a shorter and wider neural plate. Although their ML elongation is reduced, *vangl2/tri* mutant ectoderm cells exhibit excessive protrusive behavior and are associated with reduced Fibronectin deposition between the ectoderm and mesoderm layers. Ectopic Fibronectin restores the number of protrusions in *vangl2/tri* mutant ectoderm to WT levels, but does not improve ML cell elongation or alignment, demonstrating a requirement for both PCP-dependent elongation and finely-tuned cell-matrix interactions for efficient migration (Love et al., 2018). Though less severe than the C&E defects in PCP mutants, both gain- and loss-of-function mutations in *cdh2* also produce wider neural plates by the end of gastrulation (Warga & Kane, 2007). These defects worsen during later development, as described in Section 4.2. C&E of the neural plate is also severely disrupted upon loss of Nodal signaling due to reduced cell migration speed, persistence, and directionality (Araya et al., 2014). However, complete loss of Nodal signaling also results in severe mesoderm and endoderm deficiencies, and evidence suggests that Nodal regulates neural C&E largely via the mesoderm it specifies (Aquilina-Beck, Ilagan, Liu, & Liang, 2007; Araya et al., 2014; Gonsar et al., 2016; Smutny et al., 2017). This and additional interactions between germ layers will be discussed in Section 5.

4.2 Neural keel

Unlike neurulation in other vertebrate embryos, in which the epithelial neural plate folds along the medial (and in some species, lateral) hinge point(s) to fuse along the embryo's dorsal side and form the neural tube, the zebrafish neural plate undergoes a series of complex shape changes to become the neural keel (Fig. 4), then a cylindrical solid neural rod, which subsequently opens a lumen to become the neural tube (Schmitz, Papan, & Campos-Ortega, 1993). As discussed above, tracking of neuroectoderm cells reveals robust C&E movements during gastrulation, but shortly thereafter, extension movements stop while convergence movements continue (Araya et al., 2019). Instead of contributing to extension, these converging cells begin to move deeper within the embryo as they "dive in" along the midline of the neural plate to thicken it in the dorsoventral (D-V) dimension (Fig. 4C) (Araya et al., 2019), after which this structure is known as the neural keel. This "diving in," or internalization, and the resulting convergent thickening are accompanied by dramatic changes in cell shape and orientation of medial cells (Araya et al., 2019; Hong & Brewster, 2006), including reorientation along the D-V axis and shrinking of their apical surfaces as they internalize at the midline. This shape change is driven by Myosin II contractility, which is enriched along the dorsal midline and whose localization is regulated by N-Cadherin (Araya et al., 2019; Hong & Brewster, 2006). Disruption of C&E (by loss of PCP function) or of internalization (by loss of Myosin contractility or N-cadherin) both result in an abnormally-shaped neural keel (Araya et al., 2019; Ciruna et al., 2006; Hong & Brewster, 2006; Lele et al., 2002). Although neural plate to neural keel morphogenesis entails cell behaviors typical of epithelia, a more complete suite of epithelial characteristics is expressed only during late segmentation (Buckley et al., 2013; Girdler, Araya, Ren, & Clarke, 2013; Tawk et al., 2007).

Cell divisions play an increasingly important role at subsequent stages of neuroectoderm morphogenesis. During gastrulation stages, neural precursors within the neural plate divide parallel to the plane of the tissue (Concha & Adams, 1998; Geldmacher-Voss, Reugels, Pauls, & Campos-Ortega, 2003). Within the neural keel (and also later in the neural rod stages), however, cells switch to "mirror-image" divisions perpendicular to the tissue plane and across the midline (Fig. 4D), producing one daughter that remains on the ipsilateral side and one that crosses contralaterally (Ciruna et al., 2006; Clarke, 2009; Geldmacher-Voss et al., 2003; Hong & Brewster, 2006; Kimmel, Warga, & Kane, 1994; Tawk et al., 2007). Proper midline crossing of daughter cells requires N-cadherin (Hong & Brewster, 2006) and polarization of ParD3 and Rab11 (Buckley et al., 2013; Tawk et al., 2007), and the location and orientation of these divisions is, like many other C&*E*-related events, also under the control of PCP. In *vangl2/tri* mutants, convergence of the neural plate is delayed, but cell divisions are not, causing them to occur in lateral regions of the abnormally wide neural keel rather than at the midline (Ciruna et al., 2006; Tawk et al., 2007). The result is a mass of cells in the medial region of the neural keel which is separated from each of the lateral regions by an ectopic "midline" (Tawk et al., 2007). Although reduced C&E is the primary defect within the neural plate of *vangl2/tri* mutants, the resulting ectopic divisions are responsible for their dramatic "double neural tube" phenotype, as inhibiting mitosis substantially improves neural keel morphology (Ciruna et al., 2006). The symmetry of cell divisions is similarly lost within the highly disorganized neural keel of Nodal signaling mutants, but these appear to be a symptom rather than the cause of this phenotype, as normal morphology is not restored to these mutants by blocking cell divisions (Araya et al., 2014).

5. Interactions between germ layers influence C&E

We have described above the cell behaviors that drive C&E morphogenesis of each of the germ layers within the zebrafish gastrula and many of the molecular mechanisms known to regulate them. However, none of these tissue movements occur in isolation, and disruption of C&E within one cell layer often influences the behavior of cells in neighboring compartments. Some of these interactions are chemical in nature, like the Cxcr4-Cxcl12 signaling that tethers migration of endoderm and mesoderm (described in Section 3), while others, like the influence of the notochord boundary on axial mesoderm (Section 2), are thought to be largely mechanical. Here we discuss additional examples of interactions between germ layers that influence C&E morphogenesis.

5.1 Mesoderm-neuroectoderm interactions

As discussed briefly in Section 4, C&E defects within the neuroectoderm of Nodal signaling mutants have been largely attributed to the lack of mesoderm within these embryos. In a phenotypic series of Nodal mutants or Nodal inhibitor-treated embryos, the degree of anterior mesoderm deficiency correlates well with the severity of neural tube closure defects (Aquilina-Beck et al., 2007; Gonsar et al., 2016). Furthermore, restoring mesoderm to Nodal signaling-deficient embryos is sufficient to restore neural tube morphology (Araya et al., 2014), indicating that neuroectoderm morphogenesis is tightly coupled to the mesoderm. Indeed, cell movements of the neuroectoderm and underlying mesoderm are highly correlated during normal gastrulation, but this correlation is severely reduced upon simultaneous loss of the ECM components Fibronectin and Laminin, resulting in neural tube defects similar to those seen in Nodal signaling mutants (Araya, Carmona-Fontaine, & Clarke, 2016). These results indicate that coupling of meso- and neuroectoderm by the ECM is essential for proper neural C&E. The anterior axial mesoderm was further found to promote C&E of the overlying neuroectoderm through E-cadherin-mediated friction forces between the two tissues (Smutny et al., 2017). Together, these results demonstrate the critical role of mesoderm interactions in C&E of the neuroectoderm.

5.2 Endoderm-mesoderm interactions

We described above how molecular tethering to the mesoderm is essential for proper endoderm migration during early gastrulation, but evidence suggests that endoderm can reciprocally influence morphogenesis of the overlying mesoderm at later stages. The *miles apart* mutation, which disrupts sphingosine-1-phosphate signaling, was so named for its cardia bifida phenotype resulting from the failure of mesodermal cardiac precursors to converge at the midline (Stainier et al., 1996). This effect on cardiomyocytes is cell-non-autonomous, and was instead found to be secondary to defective endoderm convergence (Xie, Ye, Sepich, & Lin, 2016; Ye & Lin, 2013; Ye, Xie, Hu, & Lin, 2015). Mutations in Toddler/Apelin signaling components also produce severe heart defects due to reduced migration of cardiac precursors (Zeng et al., 2007), and the Apelin receptor was similarly found to function cell-non-autonomously during heart development (Paskaradevan & Scott, 2012). Together these data indicate that convergence of at least the lateral plate mesoderm is coupled to that of the underlying endoderm. However, given that endoderm relies on mesoderm migration at earlier stages, it remains unclear precisely which cells are guiding which, and when.

6. Concluding remarks

In this chapter, we have described the cell behavior programs that drive convergence & extension of the zebrafish embryonic axis and, to the extent that they are understood, the signaling events that regulate them. While a subset of the cell behavior programs employed by zebrafish gastrulae, such as "tumble-and-run" migration, are not known to contribute significantly to axis extension in other model organisms, many other cellular and molecular mechanisms are highly conserved between multiple tissue types and vertebrate species, including PCP-dependent mediolateral intercalation, coupling of germ layers by ECM, and chemokine-directed migration (as described in other chapters of this volume). Due to their rapid external development, availability of transgenes and other genetic tools, and their optical clarity, zebrafish have proven an advantageous system in which to characterize both the detailed behavior of individual cells and their exquisite coordination that drives embryo-wide gastrulation movements. Because of the high degree of conservation of these behaviors, this wealth of zebrafish-derived knowledge is likely to inform the mechanisms at play in other vertebrate species.

References

Aquilina-Beck, A., Ilagan, K., Liu, Q., & Liang, J. O. (2007). Nodal signaling is required for closure of the anterior neural tube in zebrafish. *BMC Developmental Biology*, *7*, 126.

Araya, C., Carmona-Fontaine, C., & Clarke, J. D. (2016). Extracellular matrix couples the convergence movements of mesoderm and neural plate during the early stages of neurulation. *Developmental Dynamics*, *245*, 580–589.

Araya, C., Häkkinen, H. M., Carcamo, L., Cerda, M., Savy, T., Rookyard, C., et al. (2019). Cdh2 coordinates Myosin-II dependent internalisation of the zebrafish neural plate. *Scientific Reports*, *9*, 1835.

Araya, C., Tawk, M., Girdler, G. C., Costa, M., Carmona-Fontaine, C., & Clarke, J. D. (2014). Mesoderm is required for coordinated cell movements within zebrafish neural plate in vivo. *Neural Development*, *9*, 9.

Bai, Y., Tan, X., Zhang, H., Liu, C., Zhao, B., Li, Y., et al. (2014). Ror2 receptor mediates Wnt11 ligand signaling and affects convergence and extension movements in zebrafish. *The Journal of Biological Chemistry*, *289*, 20664–20676.

Barone, V., Lang, M., Krens, S. F. G., Pradhan, S. J., Shamipour, S., Sako, K., et al. (2017). An effective feedback loop between cell-cell contact duration and morphogen signaling determines cell fate. *Developmental Cell*, *43*. 198-211.e112.

Bertet, C., Sulak, L., & Lecuit, T. (2004). Myosin-dependent junction remodelling controls planar cell intercalation and axis elongation. *Nature*, *429*, 667–671.

Buckley, C. E., Ren, X., Ward, L. C., Girdler, G. C., Araya, C., Green, M. J., et al. (2013). Mirror-symmetric microtubule assembly and cell interactions drive lumen formation in the zebrafish neural rod. *The EMBO Journal*, *32*, 30–44.

Buhl, A. M., Johnson, N. L., Dhanasekaran, N., & Johnson, G. L. (1995). G alpha 12 and G alpha 13 stimulate Rho-dependent stress fiber formation and focal adhesion assembly. *The Journal of Biological Chemistry*, *270*, 24631–24634.

Campbell, A. D., Lawn, S., McGarry, L. C., Welch, H. C., Ozanne, B. W., & Norman, J. C. (2013). P-Rex1 cooperates with PDGFRβ to drive cellular migration in 3D microenvironments. *PLoS One*, *8*, e53982.

Čapek, D., Smutny, M., Tichy, A. M., Morri, M., Janovjak, H., & Heisenberg, C. P. (2019). Light-activated Frizzled7 reveals a permissive role of non-canonical wnt signaling in mesendoderm cell migration. *elife*, *8*, e42093.

Carreira-Barbosa, F., Concha, M. L., Takeuchi, M., Ueno, N., Wilson, S. W., & Tada, M. (2003). Prickle 1 regulates cell movements during gastrulation and neuronal migration in zebrafish. *Development*, *130*, 4037–4046.

Carreira-Barbosa, F., Kajita, M., Morel, V., Wada, H., Okamoto, H., Martinez Arias, A., et al. (2009). Flamingo regulates epiboly and convergence/extension movements through cell cohesive and signalling functions during zebrafish gastrulation. *Development*, *136*, 383–392.

Chu, C. W., & Sokol, S. Y. (2016). Wnt proteins can direct planar cell polarity in vertebrate ectoderm. *eLife*, *5*, e16463.

Ciruna, B., Jenny, A., Lee, D., Mlodzik, M., & Schier, A. F. (2006). Planar cell polarity signalling couples cell division and morphogenesis during neurulation. *Nature*, *439*, 220–224.

Clarke, J. (2009). Role of polarized cell divisions in zebrafish neural tube formation. *Current Opinion in Neurobiology*, *19*, 134–138.

Collins, M. M., Maischein, H. M., Dufourcq, P., Charpentier, M., Blader, P., & Stainier, D. Y. (2018). Pitx2c orchestrates embryonic axis extension via mesendodermal cell migration. *eLife*, *7*, e34880.

Concha, M. L., & Adams, R. J. (1998). Oriented cell divisions and cellular morphogenesis in the zebrafish gastrula and neurula: A time-lapse analysis. *Development*, *125*, 983–994.

Coyle, R. C., Latimer, A., & Jessen, J. R. (2008). Membrane-type 1 matrix metalloproteinase regulates cell migration during zebrafish gastrulation: Evidence for an interaction with non-canonical Wnt signaling. *Experimental Cell Research*, *314*, 2150–2162.

Davidson, L. A. (2017). Mechanical design in embryos: Mechanical signalling, robustness and developmental defects. *Philosophical Transactions of the Royal Society of London. Series B, Biological Sciences*, *372*, 20150516.
Diz-Muñoz, A., Krieg, M., Bergert, M., Ibarlucea-Benitez, I., Muller, D. J., Paluch, E., et al. (2010). Control of directed cell migration in vivo by membrane-to-cortex attachment. *PLoS Biology*, *8*, e1000544.
Diz-Muñoz, A., Romanczuk, P., Yu, W., Bergert, M., Ivanovitch, K., Salbreux, G., et al. (2016). Steering cell migration by alternating blebs and actin-rich protrusions. *BMC Biology*, *14*, 74.
Dohn, M. R., Mundell, N. A., Sawyer, L. M., Dunlap, J. A., & Jessen, J. R. (2013). Planar cell polarity proteins differentially regulate extracellular matrix organization and assembly during zebrafish gastrulation. *Developmental Biology*, *383*, 39–51.
Dumortier, J. G., & David, N. B. (2015). The TORC2 component, Sin1, controls migration of anterior mesendoderm during zebrafish gastrulation. *PLoS One*, *10*, e0118474.
Dumortier, J. G., Martin, S., Meyer, D., Rosa, F. M., & David, N. B. (2012). Collective mesendoderm migration relies on an intrinsic directionality signal transmitted through cell contacts. *Proceedings of the National Academy of Sciences of the United States of America*, *109*, 16945–16950.
Eubelen, M., Bostaille, N., Cabochette, P., Gauquier, A., Tebabi, P., Dumitru, A. C., et al. (2018). A molecular mechanism for Wnt ligand-specific signaling. *Science*, *361*, eaat1178.
Geldmacher-Voss, B., Reugels, A. M., Pauls, S., & Campos-Ortega, J. A. (2003). A 90-degree rotation of the mitotic spindle changes the orientation of mitoses of zebrafish neuroepithelial cells. *Development*, *130*, 3767–3780.
Girdler, G. C., Araya, C., Ren, X., & Clarke, J. D. (2013). Developmental time rather than local environment regulates the schedule of epithelial polarization in the zebrafish neural rod. *Neural Development*, *8*, 5.
Glickman, N. S., Kimmel, C. B., Jones, M. A., & Adams, R. J. (2003). Shaping the zebrafish notochord. *Development*, *130*, 873–887.
Gonsar, N., Coughlin, A., Clay-Wright, J. A., Borg, B. R., Kindt, L. M., & Liang, J. O. (2016). Temporal and spatial requirements for Nodal-induced anterior mesendoderm and mesoderm in anterior neurulation. *Genesis*, *54*, 3–18.
Gray, R. S., Roszko, I., & Solnica-Krezel, L. (2011). Planar cell polarity: Coordinating morphogenetic cell behaviors with embryonic polarity. *Developmental Cell*, *21*, 120–133.
Gubb, D., Green, C., Huen, D., Coulson, D., Johnson, G., Tree, D., et al. (1999). The balance between isoforms of the prickle LIM domain protein is critical for planar polarity in Drosophila imaginal discs. *Genes & Development*, *13*, 2315–2327.
Habas, R., Kato, Y., & He, X. (2001). Wnt/Frizzled activation of Rho regulates vertebrate gastrulation and requires a novel Formin homology protein Daam1. *Cell*, *107*, 843–854.
Han, M. K., Hoijman, E., Nöel, E., Garric, L., Bakkers, J., & de Rooij, J. (2016). αE-catenin-dependent mechanotransduction is essential for proper convergent extension in zebrafish. *Biology Open*, *5*, 1461–1472.
Hayes, M., Naito, M., Daulat, A., Angers, S., & Ciruna, B. (2013). Ptk7 promotes non-canonical Wnt/PCP-mediated morphogenesis and inhibits Wnt/β-catenin-dependent cell fate decisions during vertebrate development. *Development*, *140*, 1807–1818.
Heisenberg, C. P., Tada, M., Rauch, G. J., Saude, L., Concha, M. L., Geisler, R., et al. (2000). Silberblick/Wnt11 mediates convergent extension movements during zebrafish gastrulation. *Nature*, *405*, 76–81.
Ho, H. Y., Susman, M. W., Bikoff, J. B., Ryu, Y. K., Jonas, A. M., Hu, L., et al. (2012). Wnt5a-Ror-Dishevelled signaling constitutes a core developmental pathway that controls tissue morphogenesis. *Proceedings of the National Academy of Sciences of the United States of America*, *109*, 4044–4051.

Hong, E., & Brewster, R. (2006). N-cadherin is required for the polarized cell behaviors that drive neurulation in the zebrafish. *Development*, *133*, 3895–3905.

Hu, B., Gao, Y., Davies, L., Woo, S., Topczewski, J., Jessen, J. R., et al. (2018). Glypican 4 and Mmp14 interact in regulating the migration of anterior endodermal cells by limiting extracellular matrix deposition. *Development*, *145*, dev163303.

Huebner, R. J., & Wallingford, J. B. (2018). Coming to consensus: A unifying model emerges for convergent extension. *Developmental Cell*, *46*, 389–396.

Jessen, J. R., Topczewski, J., Bingham, S., Sepich, D. S., Marlow, F., Chandrasekhar, A., et al. (2002). Zebrafish trilobite identifies new roles for Strabismus in gastrulation and neuronal movements. *Nature Cell Biology*, *4*, 610–615.

Kai, M., Heisenberg, C. P., & Tada, M. (2008). Sphingosine-1-phosphate receptors regulate individual cell behaviours underlying the directed migration of prechordal plate progenitor cells during zebrafish gastrulation. *Development*, *135*, 3043–3051.

Kane, D. A., Hammerschmidt, M., Mullins, M. C., Maischein, H. M., Brand, M., van Eeden, F. J., et al. (1996). The zebrafish epiboly mutants. *Development*, *123*, 47–55.

Kane, D. A., McFarland, K. N., & Warga, R. M. (2005). Mutations in halfbaked/E-cadherin block cell behaviors that are necessary for teleost epiboly. *Development*, *132*, 1105–1116.

Keller, R. (2006). Mechanisms of elongation in embryogenesis. *Development*, *133*, 2291–2302.

Keller, R., Davidson, L., Edlund, A., Elul, T., Ezin, M., Shook, D., et al. (2000). Mechanisms of convergence and extension by cell intercalation. *Philosophical Transactions of the Royal Society of London. Series B, Biological Sciences*, *355*, 897–922.

Keller, R., Davidson, L. A., & Shook, D. R. (2003). How we are shaped: The biomechanics of gastrulation. *Differentiation*, *71*, 171–205.

Kilian, B., Mansukoski, H., Barbosa, F. C., Ulrich, F., Tada, M., & Heisenberg, C. P. (2003). The role of Ppt/Wnt5 in regulating cell shape and movement during zebrafish gastrulation. *Mechanisms of Development*, *120*, 467–476.

Kimmel, C. B., Ballard, W. W., Kimmel, S. R., Ullmann, B., & Schilling, T. F. (1995). Stages of embryonic development of the zebrafish. *Developmental Dynamics*, *203*, 253–310.

Kimmel, C. B., Warga, R. M., & Kane, D. A. (1994). Cell cycles and clonal strings during formation of the zebrafish central nervous system. *Development*, *120*, 265–276.

Kühl, M., Sheldahl, L. C., Malbon, C. C., & Moon, R. T. (2000). Ca(2+)/calmodulin-dependent protein kinase II is stimulated by Wnt and Frizzled homologs and promotes ventral cell fates in Xenopus. *The Journal of Biological Chemistry*, *275*, 12701–12711.

Kühl, M., Sheldahl, L. C., Park, M., Miller, J. R., & Moon, R. T. (2000). The Wnt/Ca2+ pathway: A new vertebrate Wnt signaling pathway takes shape. *Trends in Genetics*, *16*, 279–283.

Lai, S. L., Chan, T. H., Lin, M. J., Huang, W. P., Lou, S. W., & Lee, S. J. (2008). Diaphanous-related formin 2 and profilin I are required for gastrulation cell movements. *PLoS One*, *3*, e3439.

Latimer, A., & Jessen, J. R. (2010). Extracellular matrix assembly and organization during zebrafish gastrulation. *Matrix Biology*, *29*, 89–96.

Lele, Z., Folchert, A., Concha, M., Rauch, G. J., Geisler, R., Rosa, F., et al. (2002). Parachute/n-cadherin is required for morphogenesis and maintained integrity of the zebrafish neural tube. *Development*, *129*, 3281–3294.

Li, X., Roszko, I., Sepich, D. S., Ni, M., Hamm, H. E., Marlow, F. L., et al. (2013). Gpr125 modulates Dishevelled distribution and planar cell polarity signaling. *Development*, *140*, 3028–3039.

Lin, S., Baye, L. M., Westfall, T. A., & Slusarski, D. C. (2010). Wnt5b-Ryk pathway provides directional signals to regulate gastrulation movement. *The Journal of Cell Biology*, *190*, 263–278.

Lin, F., Sepich, D. S., Chen, S., Topczewski, J., Yin, C., Solnica-Krezel, L., et al. (2005). Essential roles of G{alpha}12/13 signaling in distinct cell behaviors driving zebrafish convergence and extension gastrulation movements. *The Journal of Cell Biology*, *169*, 777–787.
Liu, Y., Sepich, D. S., & Solnica-Krezel, L. (2017). Stat3/Cdc25a-dependent cell proliferation promotes embryonic axis extension during zebrafish gastrulation. *PLoS Genetics*, *13*, e1006564.
Love, A. M., Prince, D. J., & Jessen, J. R. (2018). Vangl2-dependent regulation of membrane protrusions and directed migration requires a fibronectin extracellular matrix. *Development*, *145*, dev165472.
Lu, B., Usui, T., Uemura, T., Jan, L., & Jan, Y. N. (1999). Flamingo controls the planar polarity of sensory bristles and asymmetric division of sensory organ precursors in Drosophila. *Current Biology*, *9*, 1247–1250.
Marlow, F., Topczewski, J., Sepich, D., & Solnica-Krezel, L. (2002). Zebrafish Rho kinase 2 acts downstream of Wnt11 to mediate cell polarity and effective convergence and extension movements. *Current Biology*, *12*, 876–884.
Miles, L. B., Mizoguchi, T., Kikuchi, Y., & Verkade, H. (2017). A role for planar cell polarity during early endoderm morphogenesis. *Biology Open*, *6*, 531–539.
Miller, C. J., & Davidson, L. A. (2013). The interplay between cell signalling and mechanics in developmental processes. *Nature Reviews. Genetics*, *14*, 733–744.
Mintzer, K. A., Lee, M. A., Runke, G., Trout, J., Whitman, M., & Mullins, M. C. (2001). Lost-a-fin encodes a type I BMP receptor, Alk8, acting maternally and zygotically in dorsoventral pattern formation. *Development*, *128*, 859–869.
Mizoguchi, T., Verkade, H., Heath, J. K., Kuroiwa, A., & Kikuchi, Y. (2008). Sdf1/Cxcr4 signaling controls the dorsal migration of endodermal cells during zebrafish gastrulation. *Development*, *135*, 2521–2529.
Montero, J. A., Kilian, B., Chan, J., Bayliss, P. E., & Heisenberg, C. P. (2003). Phosphoinositide 3-kinase is required for process outgrowth and cell polarization of gastrulating mesendodermal cells. *Current Biology*, *13*, 1279–1289.
Morita, H., Grigolon, S., Bock, M., Krens, S. F., Salbreux, G., & Heisenberg, C. P. (2017). The physical basis of coordinated tissue spreading in zebrafish gastrulation. *Developmental Cell*, *40*, 354-366.e354.
Mullins, M. C., Hammerschmidt, M., Kane, D. A., Odenthal, J., Brand, M., van Eeden, F. J., et al. (1996). Genes establishing dorsoventral pattern formation in the zebrafish embryo: The ventral specifying genes. *Development*, *123*, 81–93.
Myers, D. C., Sepich, D. S., & Solnica-Krezel, L. (2002). Bmp activity gradient regulates convergent extension during zebrafish gastrulation. *Developmental Biology*, *243*, 81–98.
Nair, S., & Schilling, T. F. (2008). Chemokine signaling controls endodermal migration during zebrafish gastrulation. *Science*, *322*, 89–92.
Norris, M. L., Pauli, A., Gagnon, J. A., Lord, N. D., Rogers, K. W., Mosimann, C., et al. (2017). Toddler signaling regulates mesodermal cell migration downstream of Nodal signaling. *eLife*, *6*, e22626.
Ossipova, O., Kim, K., & Sokol, S. Y. (2015). Planar polarization of Vangl2 in the vertebrate neural plate is controlled by Wnt and Myosin II signaling. *Biology Open*, *4*, 722–730.
Park, M., & Moon, R. T. (2002). The planar cell-polarity gene stbm regulates cell behaviour and cell fate in vertebrate embryos. *Nature Cell Biology*, *4*, 20–25.
Paskaradevan, S., & Scott, I. C. (2012). The Aplnr GPCR regulates myocardial progenitor development via a novel cell-non-autonomous, Gα(i/o) protein-independent pathway. *Biology Open*, *1*, 275–285.
Pauli, A., Norris, M. L., Valen, E., Chew, G. L., Gagnon, J. A., Zimmerman, S., et al. (2014). Toddler: An embryonic signal that promotes cell movement via Apelin receptors. *Science*, *343*, 1248636.

Pézeron, G., Mourrain, P., Courty, S., Ghislain, J., Becker, T. S., Rosa, F. M., et al. (2008). Live analysis of endodermal layer formation identifies random walk as a novel gastrulation movement. *Current Biology, 18*, 276–281.

Pfister, K., Shook, D. R., Chang, C., Keller, R., & Skoglund, P. (2016). Molecular model for force production and transmission during vertebrate gastrulation. *Development, 143*, 715–727.

Riley, B. B., Sweet, E. M., Heck, R., Evans, A., McFarland, K. N., Warga, R. M., et al. (2010). Characterization of harpy/Rca1/emi1 mutants: Patterning in the absence of cell division. *Developmental Dynamics, 239*, 828–843.

Roszko, I., Sawada, A., & Solnica-Krezel, L. (2009). Regulation of convergence and extension movements during vertebrate gastrulation by the Wnt/PCP pathway. *Seminars in Cell & Developmental Biology, 20*, 986–997.

Roszko, I., Sepich, S. D., Jessen, J. R., Chandrasekhar, A., & Solnica-Krezel, L. (2015). A dynamic intracellular distribution of Vangl2 accompanies cell polarization during zebrafish gastrulation. *Development, 142*, 2508–2520.

Ruprecht, V., Wieser, S., Callan-Jones, A., Smutny, M., Morita, H., Sako, K., et al. (2015). Cortical contractility triggers a stochastic switch to fast amoeboid cell motility. *Cell, 160*, 673–685.

Schambony, A., & Wedlich, D. (2007). Wnt-5A/Ror2 regulate expression of XPAPC through an alternative noncanonical signaling pathway. *Developmental Cell, 12*, 779–792.

Schmid, B., Shah, G., Scherf, N., Weber, M., Thierbach, K., Campos, C. P., et al. (2013). High-speed panoramic light-sheet microscopy reveals global endodermal cell dynamics. *Nature Communications, 4*, 2207.

Schmitz, B., Papan, C., & Campos-Ortega, J. A. (1993). Neurulation in the anterior trunk region of the zebrafish *Brachydanio rerio*. *Roux's Archives of Developmental Biology, 202*, 250–259.

Sepich, D. S., Calmelet, C., Kiskowski, M., & Solnica-Krezel, L. (2005). Initiation of convergence and extension movements of lateral mesoderm during zebrafish gastrulation. *Developmental Dynamics, 234*, 279–292.

Sepich, D. S., Myers, D. C., Short, R., Topczewski, J., Marlow, F., & Solnica-Krezel, L. (2000). Role of the zebrafish trilobite locus in gastrulation movements of convergence and extension. *Genesis, 27*, 159–173.

Sheldahl, L. C., Park, M., Malbon, C. C., & Moon, R. T. (1999). Protein kinase C is differentially stimulated by Wnt and Frizzled homologs in a G-protein-dependent manner. *Current Biology, 9*, 695–698.

Sheldahl, L. C., Slusarski, D. C., Pandur, P., Miller, J. R., Kühl, M., & Moon, R. T. (2003). Dishevelled activates Ca2+ flux, PKC, and CamKII in vertebrate embryos. *The Journal of Cell Biology, 161*, 769–777.

Shih, J., & Keller, R. (1992a). Cell motility driving mediolateral intercalation in explants of *Xenopus laevis*. *Development, 116*, 901–914.

Shih, J., & Keller, R. (1992b). Patterns of cell motility in the organizer and dorsal mesoderm of *Xenopus laevis*. *Development, 116*, 915–930.

Shindo, A., & Wallingford, J. B. (2014). PCP and septins compartmentalize cortical actomyosin to direct collective cell movement. *Science, 343*, 649–652.

Slusarski, D. C., Corces, V. G., & Moon, R. T. (1997). Interaction of Wnt and a Frizzled homologue triggers G-protein-linked phosphatidylinositol signalling. *Nature, 390*, 410–413.

Slusarski, D. C., Yang-Snyder, J., Busa, W. B., & Moon, R. T. (1997). Modulation of embryonic intracellular Ca2+ signaling by Wnt-5A. *Developmental Biology, 182*, 114–120.

Smutny, M., Ákos, Z., Grigolon, S., Shamipour, S., Ruprecht, V., Čapek, D., et al. (2017). Friction forces position the neural anlage. *Nature Cell Biology, 19*, 306–317.

Stainier, D. Y., Fouquet, B., Chen, J. N., Warren, K. S., Weinstein, B. M., Meiler, S. E., et al. (1996). Mutations affecting the formation and function of the cardiovascular system in the zebrafish embryo. *Development*, *123*, 285–292.

Strutt, H., & Strutt, D. (2005). Long-range coordination of planar polarity in Drosophila. *BioEssays*, *27*, 1218–1227.

Sugimoto, N., Takuwa, N., Okamoto, H., Sakurada, S., & Takuwa, Y. (2003). Inhibitory and stimulatory regulation of Rac and cell motility by the G12/13-Rho and Gi pathways integrated downstream of a single G protein-coupled sphingosine-1-phosphate receptor isoform. *Molecular and Cellular Biology*, *23*, 1534–1545.

Sun, Z., Amourda, C., Shagirov, M., Hara, Y., Saunders, T. E., & Toyama, Y. (2017). Basolateral protrusion and apical contraction cooperatively drive Drosophila germ-band extension. *Nature Cell Biology*, *19*, 375–383.

Takeuchi, M., Nakabayashi, J., Sakaguchi, T., Yamamoto, T. S., Takahashi, H., Takeda, H., et al. (2003). The prickle-related gene in vertebrates is essential for gastrulation cell movements. *Current Biology*, *13*, 674–679.

Tawk, M., Araya, C., Lyons, D. A., Reugels, A. M., Girdler, G. C., Bayley, P. R., et al. (2007). A mirror-symmetric cell division that orchestrates neuroepithelial morphogenesis. *Nature*, *446*, 797–800.

Taylor, J., Abramova, N., Charlton, J., & Adler, P. N. (1998). Van Gogh: A new Drosophila tissue polarity gene. *Genetics*, *150*, 199–210.

Topczewski, J., Sepich, D. S., Myers, D. C., Walker, C., Amores, A., Lele, Z., et al. (2001). The zebrafish glypican knypek controls cell polarity during gastrulation movements of convergent extension. *Developmental Cell*, *1*, 251–264.

Ulrich, F., Concha, M. L., Heid, P. J., Voss, E., Witzel, S., Roehl, H., et al. (2003). Slb/Wnt11 controls hypoblast cell migration and morphogenesis at the onset of zebrafish gastrulation. *Development*, *130*, 5375–5384.

Ulrich, F., Krieg, M., Schötz, E. M., Link, V., Castanon, I., Schnabel, V., et al. (2005). Wnt11 functions in gastrulation by controlling cell cohesion through Rab5c and E-cadherin. *Developmental Cell*, *9*, 555–564.

Usui, T., Shima, Y., Shimada, Y., Hirano, S., Burgess, R. W., Schwarz, T. L., et al. (1999). Flamingo, a seven-pass transmembrane cadherin, regulates planar cell polarity under the control of Frizzled. *Cell*, *98*, 585–595.

Vallon, M., Yuki, K., Nguyen, T. D., Chang, J., Yuan, J., Siepe, D., et al. (2018). A RECK-WNT7 receptor-ligand interaction enables isoform-specific regulation of Wnt bioavailability. *Cell Reports*, *25*, 339-349.e339.

Veeman, M. T., Nakatani, Y., Hendrickson, C., Ericson, V., Lin, C., & Smith, W. C. (2008). Chongmague reveals an essential role for laminin-mediated boundary formation in chordate convergence and extension movements. *Development*, *135*, 33–41.

Veeman, M. T., Slusarski, D. C., Kaykas, A., Louie, S. H., & Moon, R. T. (2003). Zebrafish prickle, a modulator of noncanonical Wnt/Fz signaling, regulates gastrulation movements. *Current Biology*, *13*, 680–685.

Vinson, C. R., & Adler, P. N. (1987). Directional non-cell autonomy and the transmission of polarity information by the frizzled gene of Drosophila. *Nature*, *329*, 549–551.

von der Hardt, S., Bakkers, J., Inbal, A., Carvalho, L., Solnica-Krezel, L., Heisenberg, C. P., et al. (2007). The Bmp gradient of the zebrafish gastrula guides migrating lateral cells by regulating cell-cell adhesion. *Current Biology*, *17*, 475–487.

Wallingford, J. B., Rowning, B. A., Vogeli, K. M., Rothbacher, U., Fraser, S. E., & Harland, R. M. (2000). Dishevelled controls cell polarity during Xenopus gastrulation. *Nature*, *405*, 81–85.

Wallingford, J. B., Vogeli, K. M., & Harland, R. M. (2001). Regulation of convergent extension in Xenopus by Wnt5a and Frizzled-8 is independent of the canonical Wnt pathway. *The International Journal of Developmental Biology*, *45*, 225–227.

Warga, R. M., & Kane, D. A. (2007). A role for N-cadherin in mesodermal morphogenesis during gastrulation. *Developmental Biology*, *310*, 211–225.

Warga, R. M., & Kimmel, C. B. (1990). Cell movements during epiboly and gastrulation in zebrafish. *Development*, *108*, 569–580.

Warga, R. M., & Nüsslein-Volhard, C. (1999). Origin and development of the zebrafish endoderm. *Development*, *126*, 827–838.

Weiser, D. C., Row, R. H., & Kimelman, D. (2009). Rho-regulated myosin phosphatase establishes the level of protrusive activity required for cell movements during zebrafish gastrulation. *Development*, *136*, 2375–2384.

Westfall, T. A., Brimeyer, R., Twedt, J., Gladon, J., Olberding, A., Furutani-Seiki, M., et al. (2003). Wnt-5/pipetail functions in vertebrate axis formation as a negative regulator of Wnt/beta-catenin activity. *The Journal of Cell Biology*, *162*, 889–898.

Williams, B. B., Cantrell, V. A., Mundell, N. A., Bennett, A. C., Quick, R. E., & Jessen, J. R. (2012). VANGL2 regulates membrane trafficking of MMP14 to control cell polarity and migration. *Journal of Cell Science*, *125*, 2141–2147.

Williams, M. L. K., Sawada, A., Budine, T., Yin, C., Gontarz, P., & Solnica-Krezel, L. (2018). Gon4l regulates notochord boundary formation and cell polarity underlying axis extension by repressing adhesion genes. *Nature Communications*, *9*, 1319.

Williams, M., Yen, W., Lu, X., & Sutherland, A. (2014). Distinct apical and basolateral mechanisms drive planar cell polarity-dependent convergent extension of the mouse neural plate. *Developmental Cell*, *29*, 34–46.

Winter, C. G., Wang, B., Ballew, A., Royou, A., Karess, R., Axelrod, J. D., et al. (2001). Drosophila Rho-associated kinase (Drok) links Frizzled-mediated planar cell polarity signaling to the actin cytoskeleton. *Cell*, *105*, 81–91.

Wong, L. L., & Adler, P. N. (1993). Tissue polarity genes of Drosophila regulate the subcellular location for prehair initiation in pupal wing cells. *The Journal of Cell Biology*, *123*, 209–221.

Woo, S., Housley, M. P., Weiner, O. D., & Stainier, D. Y. (2012). Nodal signaling regulates endodermal cell motility and actin dynamics via Rac1 and Prex1. *The Journal of Cell Biology*, *198*, 941–952.

Xie, H., Ye, D., Sepich, D., & Lin, F. (2016). S1pr2/Gα13 signaling regulates the migration of endocardial precursors by controlling endoderm convergence. *Developmental Biology*, *414*, 228–243.

Ye, D., & Lin, F. (2013). S1pr2/Gα13 signaling controls myocardial migration by regulating endoderm convergence. *Development*, *140*, 789–799.

Ye, D., Xie, H., Hu, B., & Lin, F. (2015). Endoderm convergence controls subduction of the myocardial precursors during heart-tube formation. *Development*, *142*, 2928–2940.

Ye, Z., Zhang, C., Tu, T., Sun, M., Liu, D., Lu, D., et al. (2013). Wnt5a uses CD146 as a receptor to regulate cell motility and convergent extension. *Nature Communications*, *4*, 2803.

Yin, C., Kiskowski, M., Pouille, P. A., Farge, E., & Solnica-Krezel, L. (2008). Cooperation of polarized cell intercalations drives convergence and extension of presomitic mesoderm during zebrafish gastrulation. *The Journal of Cell Biology*, *180*, 221–232.

Zallen, J. A., & Blankenship, J. T. (2008). Multicellular dynamics during epithelial elongation. *Seminars in Cell & Developmental Biology*, *19*, 263–270.

Zeng, X. X., Wilm, T. P., Sepich, D. S., & Solnica-Krezel, L. (2007). Apelin and its receptor control heart field formation during zebrafish gastrulation. *Developmental Cell*, *12*, 391–402.

Zhou, J., Pal, S., Maiti, S., & Davidson, L. A. (2015). Force production and mechanical accommodation during convergent extension. *Development*, *142*, 692–701.

Zhu, S., Liu, L., Korzh, V., Gong, Z., & Low, B. C. (2006). RhoA acts downstream of Wnt5 and Wnt11 to regulate convergence and extension movements by involving effectors Rho kinase and Diaphanous: Use of zebrafish as an in vivo model for GTPase signaling. *Cellular Signalling*, *18*, 359–372.

CHAPTER FOURTEEN

Movements of chick gastrulation

Octavian Voiculescu*
Department of Physiology, Development and Neuroscience, University of Cambridge, Cambridge, United Kingdom
*Corresponding author: e-mail address: ogv20@cam.ac.uk

Contents

Abstract

In birds as in all amniotes, the site of gastrulation is a midline structure, the primitive streak. This appears as cells in the one cell-thick epiblast undergo epithelial-to-mesenchymal transition to ingress and form definitive mesoderm and endoderm. Global movements involving tens of thousands of cells in the embryonic epiblast precede gastrulation. They position the primitive streak precursors from a marginal position (equivalent to the situation in anamniotes) along the future antero-posterior axis (typical for amniotes). These epithelial movements continue in modified form during gastrulation, when they are accompanied by collective movements of different class in the forming mesoderm and endoderm. Here I discuss the nature of these collective cell movements shaping the embryo, their interplay with signaling events controlling fate specification and significance in an evolutionary perspective.

Current Topics in Developmental Biology, Volume 136
ISSN 0070-2153
https://doi.org/10.1016/bs.ctdb.2019.11.015

1. Introduction: Morphology of the early embryo

At the time of egg laying, the embryo is a flat, round blastoderm 2–3 mm in diameter, sitting atop the yolk encased in the acellular vitelline membrane. The blastoderm already comprises more than 20,000 cells; most of these are organized as an epithelial monolayer, the epiblast, oriented with the apical side facing the vitelline membrane. The cells have prismatic morphology with the exception of the edge cells, which are flattened, attach the disk to the vitelline membrane, and keep it under tension. However, the epiblast is divided into three regions that can readily be distinguished upon dissection, because the tissues underneath differ in their cellular composition (see Fig. 1A and B).

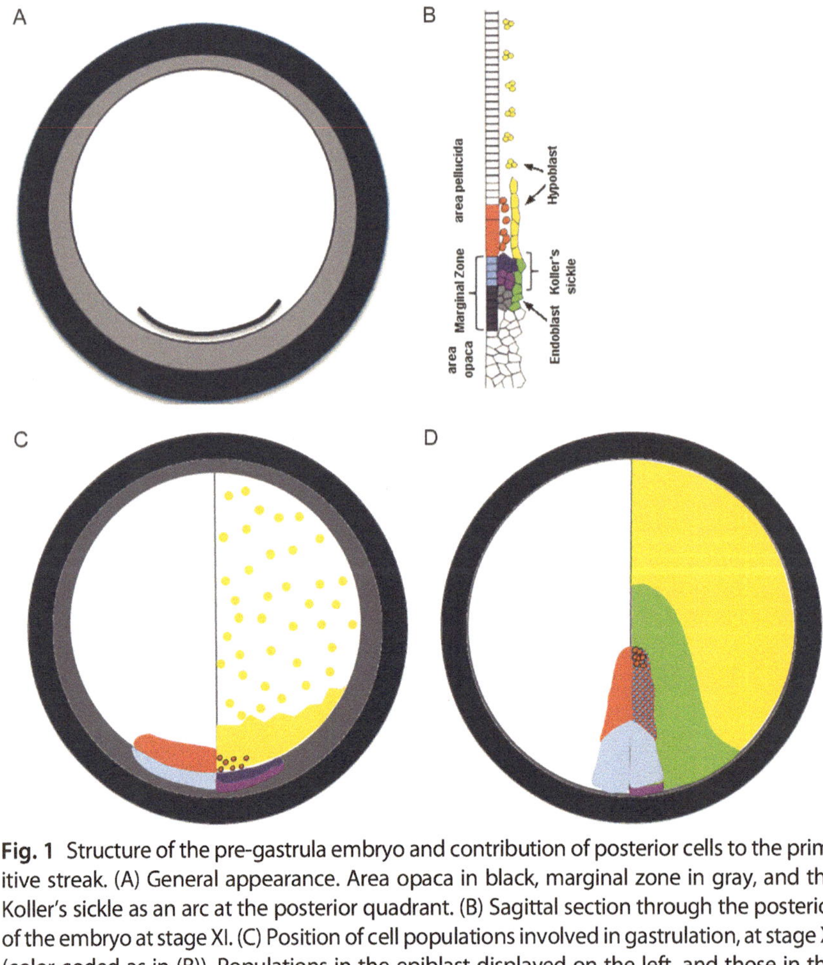

Fig. 1 Structure of the pre-gastrula embryo and contribution of posterior cells to the primitive streak. (A) General appearance. Area opaca in black, marginal zone in gray, and the Koller's sickle as an arc at the posterior quadrant. (B) Sagittal section through the posterior of the embryo at stage XI. (C) Position of cell populations involved in gastrulation, at stage XI (color coded as in (B)). Populations in the epiblast displayed on the left, and those in the lower layers on the right. (D) Position of these populations at the early primitive streak stage.

The rim of the blastoderm (area opaca, AO) is directly apposed to the yolk and appears darker due to several layers of very large, yolky cells (germ wall) that strongly adhere to its underside. This region is extra-embryonic in fate. The central part of the blastoderm (area pellucida) is separated from the yolk by a fluid-filled subgerminal cavity, making it transparent. The AP cells are short and columnar at first and (they columnarize more at subsequent stages), and constitute the embryo proper. The area pellucida is lined on its basal side by dozens of islands, each composed of 5–20 yolky cells that are larger than the epiblast cells but smaller than those of the germ wall. This loose collection of clumps constitutes the hypoblast layer, which will later form the extra-embryonic yolk sac.

The germ wall protrudes for a few cell diameters under the area pellucida; the cells of this lip (germ wall margin) are only loosely attached to the epiblast, unlike in the AO. The belt of the epiblast above the germ wall margin is called the marginal zone (MZ), and its cells are cuboidal. The MZ is somewhat wider on one side of the AO. The primitive streak will later start forming in the neighboring sector in the embryonic epiblast, which is therefore called 'posterior.' Between the posterior marginal zone (PMZ) and the AP, a crescent-shaped ridge of small cells (Koller's sickle, also termed Rauber's sickle) strongly adhere to the basal side of the epiblast. A few middle layer cells, between the epiblast and the hypoblast, are also observed around the Koller's sickle (see Fig. 1).

During the 12–14 h preceding the appearance of the primitive streak and the start of gastrulation, the hypoblast islands flatten and merge, forming a continuous sheet. This process proceeds from the posterior to the anterior, and defines the stages of development before gastrulation, denoted by Roman numerals (Eyal-Giladi & Kochav, 1976): from stage X at egg laying, when all hypoblast is insular, through stage XII when the sheet hypoblast covers the posterior half of the AP, to stage XIII when the process is completed.

2. Early movements in the epiblast position the gastrulation site before its formation

2.1 Initial positioning of gastrulation regulators

Precise fate-mapping experiments revealed that, at stages X–XI, the precursors of the early primitive streak are located in the Koller's sickle (both epiblast and deep layers) and a narrow band of the epiblast anterior to it (Bachvarova, Skromne, & Stern, 1998; Eyal-Giladi, Debby, & Harel, 1992; Wei & Mikawa, 2000) (see Fig. 1B and C). In contrast, the epiblast of the PMZ behind the Koller's sickle does not become incorporated into the primitive streak, instead remaining caudal to it at all stages.

It has been shown that the PMZ from stages X–XI can induce the formation a new axis upon ectopic transplantation, without contributing cells to it (Bachvarova et al., 1998; Eyal-Giladi & Khaner, 1989; Khaner & Eyal-Giladi, 1989; Skromne & Stern, 2001; Spratt & Haas, 1960). This inductive capacity of the PMZ declines sharply after these stages.

Several classes of signaling molecules are already distributed asymmetrically in both the embryonic and extra-embryonic tissues at the time of egg laying—the mechanisms of breaking the radial symmetry of the embryo at earlier stages are discussed in chapter "Signaling events regulating embryonic polarity and formation of the primitive streak in the chick embryo" by Raffaelli and Stern.

cVg1 is active in the posterior blastoderm at stage X, and its expression intensifies and concentrates in the epiblast of the PMZ at stages XI–XII (Shah et al., 1997). The action of cVg1 is potentiated by the products of the *Tsukushi* (*cTSK*) gene, expressed in the PMZ as well as in the area opaca and the Koller's sickle (Ohta et al., 2006). *Wnt8c* is expressed in all the marginal zone, in a gradient highest posteriorly (Hume & Dodd, 1993; Skromne & Stern, 2001). Vg1 can induce a new primitive streak if ectopically expressed in the marginal zone, where Wnt8C is present (Seleiro, Connolly, & Cooke, 1996; Shah et al., 1997), and even in the AP but only if combined with a similar Wnt activity (Skromne & Stern, 2001). It was also shown that Vg1 can induce components of the Wnt canonical pathway (Skromne & Stern, 2002). It is likely that, as in amphibians and fish, the cooperation between these pathways directly upregulates *goosecoid*: faintly in a few middle cells associated with the central Koller's sickle at stage X, then more strongly and in more cells in the Koller's sickle and some anterior to it, at stages XI and XII; these cells will later become part of the organizer (Izpisúa-Belmonte, De Robertis, Storey, & Stern, 1993). Another organizer gene, *Chordin,* is upregulated in a similar fashion. Chordin antagonizes the BMP4 produced by cells in the AO, and this is required for the axis to form. Conversely, Chordin can trigger the formation of the axis if expressed ectopically in the MZ (Streit et al., 1998).

FGF activity was also shown to be required early for a primitive streak to be initiated (Mitrani, Gruenbaum, Shohat, & Ziv, 1990). FGF8 is first seen expressed at stage X, in the middle layer associated with the Koller's sickle, and later also in the forming, posterior hypoblast (Streit, Berliner, Papanayotou, Sirulnik, & Stern, 2000). Exposure of the whole embryo to

FGF activity induces the expression of the pan-mesodermal marker *Bra* in the marginal zone, where Wnt8C is expressed; induction can be achieved in the central AP if both FGF and Wnt activities (Alev, Wu, Nakaya, & Sheng, 2013). Indeed, *Bra* is indeed expressed from very early on, in the Koller's sickle and the band of epiblast anterior to it (Knezevic, De, & Mackem, 1997). Other markers of the future primitive streak, such as *Tbx6* (Knezevic et al., 1997), *Snai2* (Acloque et al., 2011) and *Nodal* (Chapman, Schubert, Schoenwolf, & Lumsden, 2002) are upregulated in a similar pattern, at least from stages X–XI and gradually intensifying in the Koller's sickle and the epiblast surrounding it.

Importantly, the hypoblast also has a critical role as a repressor: it a source of Dkk1, Crescent and Cerberus (Foley, Skromne, & Stern, 2000), therefore being able to further restrict the fields of BMP, canonical Wnt and Nodal activities. The hypoblast was indeed shown to repress the formation of the primitive streak, and this is specifically due to the inhibition of Nodal activity (Bertocchini & Stern, 2002). This is maintained as long as the hypoblast underlies the entire AP, up to stage XIII. How Nodal activity triggers the formation of the primitive streak is discussed later (see Section 3). Of note here are the similarities between the roles of the chick hypoblast and the mouse visceral endoderm in controlling the timing of streak formation (Stern & Downs, 2012).

In sum, the chick PMZ can be seen to be equivalent to the amphibian Nieuwkoop center, both in its role of inducing an organizer and the signaling events underlying this function. Well before gastrulation, it induces the expression of the organizer genes along the central portion of the Koller's sickle and the epiblast ahead. At the same time, the signals from the MZ and the hypoblast position markers later associated with the entire primitive streak in a similar domain but extending more laterally; however, the hypoblast also blocks the formation of an actual primitive streak.

Thus, the expression of genes marking the organizer and the rest of the gastrulation site are initially (stages X–XII) positioned on the margin of the blastoderm, a situation reminiscent of that in anamniotes. The epiblast cells to participate in the formation of the initial primitive streak are also arranged in a similar manner: epiblast precursors around the central portion of the Koller's sickle, and cells later occupying the more caudal portions of the initial primitive streak more laterally along the Koller's sickle (Bachvarova et al., 1998; Wei & Mikawa, 2000). How, then, are these precursors repositioned along the midline?

2.2 Rearrangement of the primitive streak precursors and global movements in the epiblast

Remarkable cell movements start in the epiblast from at least stage XI (see Fig. 2A and Movie 1 in the online version at https://doi.org/10.1016/bs.ctdb.2019.11.015). These involve all the cells of the area pellucida (not only the precursors of the primitive streak), and are executed without the epiblast losing its epithelial character. They take the shape of counter-rotating flows that join at the posterior midline (Fig. 2 and Movie 1 in the online version at https://doi.org/10.1016/bs.ctdb.2019.11.015). They were initially observed as early as 1926 by (Gräper, 1929), using vital dye

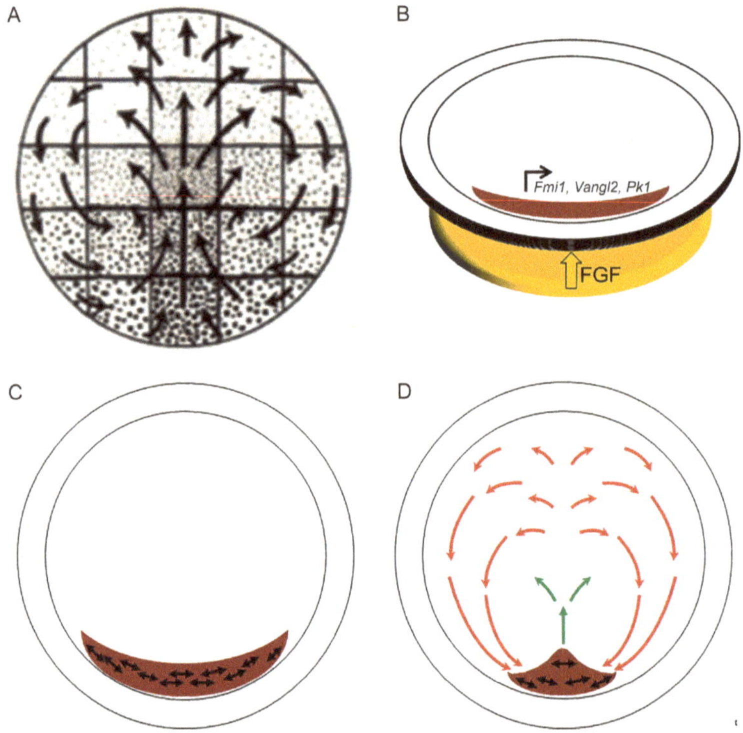

Fig. 2 Movements before gastrulation. (A) Diagrammatic display of the Polonaise movements. (B) FGF produced by the posterior hypoblast upregulates Wnt-PCP components in the epiblast around Koller's sickle. (C) Domain of epithelial intercalation at stage XII (brown). (D) Shape of intercalation domain at stage XIV. Lines of tension (red) and relaxation (green) conveyed in the rest of the epiblast by the narrowing and elongation of the intercalation domain. *Panel (A) from Wetzel, R. (1929). Untersuchungen am Hühnchen. Die Entwicklung des Keims Während der Ersten Beiden Bruttage. In R. Wetzel (Ed.),* Untersuchungen am Hühnchen: Die Entwicklung des Keims während der ersten beiden Bruttage *(pp. 188–321). Berlin, Heidelberg: Springer.*

markings and stereo videomicroscopy, and (Wetzel, 1929)—the former dubbing them 'Polonaise movements.' They were later confirmed by Spratt (1946), who carefully quantified them using carbon particles as tracers and (Vakaet, 1970). More recently, they were observed with the aid of methods permitting cellular resolution (Chuai et al., 2006; Cui, Yang, Chuai, Glazier, & Weijer, 2005; Rozbicki et al., 2015; Voiculescu, Bertocchini, Wolpert, Keller, & Stern, 2007; Voiculescu, Bodenstein, Lau, & Stern, 2014). These movements continue, but in modified form, during gastrulation stages.

Many theories have been put forward to account for these movements and how they relate to primitive streak formation.

- Differential or oriented cell divisions were proposed to take part in the process. First, it was hypothesized that excrescence in the anterior regions of the embryo pushes the cells at the posterior together, forcing them to create primitive streak as a crease (Duval, 1884). However, this theory has been abandoned, as it requires the blastoderm to be constrained, whereas the opposite is true: the epiblast is sensitive to tension generated by the area opaca outer cells attachment to the vitelline membrane (New, 1959), and proper tensioning of the epiblast is required for normal morphogenesis (Bellairs, Bromham, & Wylie, 1967). In addition, the rate of proliferation does not differ significantly across the epiblast at pre- and gastrulation stages (Cui et al., 2005; Ozato, 1969; Sanders, Varedi, & French, 1993; Stern, 1979).
- A large number of attraction and repulsion theories were also proposed, most recently based on chemotaxis. It had been first suggested FGF8 and FGF4 act as opposing chemotactic cues at primitive streak stages (Yang, Dormann, Münsterberg, & Weijer, 2002) and this notion was then extended at pre-streak stages (Chuai et al., 2006; Cui et al., 2005) because blocking FGF signaling to epiblast cells arrests the Polonaise movements. Many models based on putative chemotaxis can indeed simulate Polonaise-like movements (Sandersius, Chuai, Weijer, & Newman, 2011; Vasiev, Balter, Chaplain, Glazier, & Weijer, 2010). However, it is not clear how such paradigms could apply to the blastoderm, since the basement membrane under the epiblast cells is not attached to a fixed substrate—therefore, cells lack a support upon which to exert traction in response to either attractors or repellents (irrespective of their nature). In fact, the extracellular matrix was observed to move in concert with the epiblast cells, at least at gastrulation stages (Zamir, Rongish, & Little, 2008).

- More recently, epithelial medio-lateral intercalation was discovered to occur in a posterior domain, at stage XI located in a similar fashion to where the initial primitive streak precursors reside—the epiblast of the Koller's sickle and anterior to it (Voiculescu et al., 2007)—see Fig. 2B–D. This finding was later confirmed (Rozbicki et al., 2015) and extended by the observation of oriented apical acto-myosin cables in the intercalating cells, which are required for their oriented rearrangements. The intercalation domain is also distinguished molecularly by the co-expression of genes coding for components of the Wnt-PCP pathway (Flamingo1, Vangl2 and Prickle1), which are upregulated by FGF8 from the posterior hypoblast and the Koller's sickle. At subsequent stages, the oriented cell intercalations within this domain cause it to narrow laterally and elongate toward the center of the blastoderm, and correspond to the primitive streak epiblast once it forms.

The intercalation in the posterior domain is required for Polonaise movements, since they are completely arrested upon knocking down the three Wnt-PCP components or blocking the pathway downstream of them, specifically in the intercalation domain (Voiculescu et al., 2014). Is local intercalation in the posterior domain sufficient to drive the movements in the rest of the epiblast? An object-oriented computer model was built to represent the sheet of cohesive cells of the epiblast, kept under the tension created by the edge cells. When intercalation is explicitly implemented in a small posterior domain, the model fully reproduces the Polonaise movement (Voiculescu et al., 2014). Furthermore, the model also reproduces the effect of hypoblast rotation (Waddington, 1932), which creates a new set of Polonaise movements (Foley et al., 2000) following the ectopic activation of Wnt-PCP components (Voiculescu et al., 2007). It therefore appears that oriented intercalation in the posterior domain, directed by the underlying hypoblast, creates mechanical tension by the pulling of the lateral edges, and relaxes tension ahead of the medially advancing tip. These lines of tension are conveyed across the epiblast, whose epithelial cells are mechanically coupled, thus causing the cells passively to rearrange and move in the Polonaise pattern.

Together, these data reveal that the hypoblast, an evolutionary acquisition of amniotes, directs the rearrangement of the first mesendoderm precursors, from the edge of the chick blastoderm (where they are initially specified) to the midline of the embryo (where the primitive streak forms about 12 h later). A crucial observation is that in the embryos where Wnt-PCP-dependent intercalation is blocked, a genuine site of gastrulation does form with the appropriate timing, but it appears on the edge of the area opaca—reminiscent of the situation in anamniotes (Voiculescu et al., 2007). Thus, in avian embryos, the amniote-specific repositioning of precursors is independent of triggering gastrulation. What controls the formation of the primitive streak?

3. Cooperative EMT triggers the formation of the initial primitive streak

Morphologically, the primitive streak appears abruptly, within 20–30 min, in the form of a triangle at the posterior area pellucida. This defines stage 2 in the Hamburger and Hamilton system (Hamburger & Hamilton, 1951), which uses Arabic numerals and is adopted by the majority of investigators. The formation of the primitive streak is the result of many cells in that region of the epiblast initiating epithelial-to-mesenchymal transition (EMT), by adopting bottle-shape aspect (undergoing apical constriction and baso-lateral expansion).

Cells adopting bottle-shaped morphology are found scattered across the area pellucida epiblast during stages X–XIII, and in all regions (not only the prospective primitive streak region) some cells (about 1–2% per hour) do complete ingression and join the hypoblast underneath (Voiculescu et al., 2014) (see Fig. 3A). These 'pioneer' cells ingress as individuals. However, just before primitive streak formation (during the brief stage XIV), cells were seen attempting EMT as small groups, which causes the deformation of the apical surface of the epiblast and the appearance of 'pits' in the region where the streak will later form. This suggests that, in the streak-forming region, either the EMT is generally promoted in all cells, or that the early, 'pioneer' cells start to favor the EMT of their neighboring cells (Fig. 3B). The second hypothesis is strongly supported by earlier observations that a fraction of pre-streak epiblast cells carry the HNK-1 epitope (Canning & Stern, 1988) and have invasive character (Mogi, Toyoizumi, & Takeuchi, 2000); they normally contribute to the lower layers formed during gastrulation and, crucially, their ablation at stage XIII prevents a primitive streak to form (Stern & Canning, 1990). Together, these results indicate that the 'pioneer' cells are a subset of the HNK-1^+ population and absolutely required for the initiation of coordinated EMT at the start of primitive streak formation.

It seems, therefore, that the formation of the primitive streak is caused by the early cells engaging in EMT favoring the same behavior in their

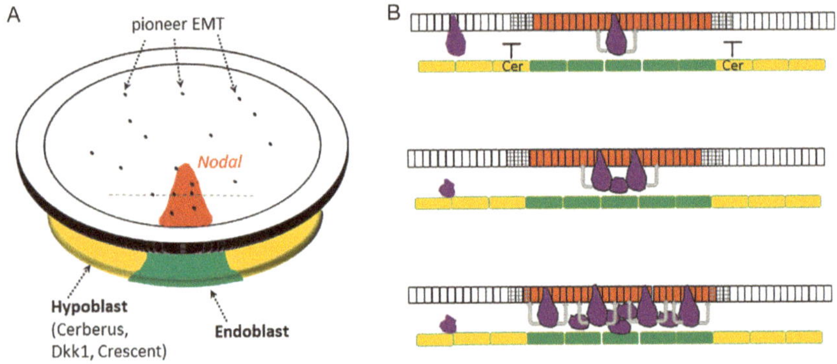

Fig. 3 The primitive streak forms and is maintained by a chain reaction of EMT events. (A) In the epiblast, pioneer cells ingressing singly are depicted as purple dots, and the domain of *Nodal* expression in brown. In the lower layer, the endoblast (green) pushes away the hypoblast that was repressing Nodal activity. (B) Sequence of EMT events triggering the formation of the primitive streak. Top: in the domain where Nodal is active (brown), ingressing cells (purple) induce EMT in their neighbors (gray curved arrows). Middle: ingression is amplified in the Nodal activity domain, which becomes a primitive streak. Bottom: as many cells ingress at the primitive streak, cells from the epiblast lateral to it are pulled toward the midline and are induced to express primitive streak genes, such as Nodal.

neighbors. This was demonstrated by grafting small groups of cells in the process of EMT in the early primitive streak into the margin of pre-streak embryos, away from the endogenous site of primitive streak formation. The early ingressing cells do indeed upregulate very quickly the expression of primitive streak markers, and trigger the formation of a new, ectopic primitive streak, mainly composed of host cells (Voiculescu et al., 2014).

How is this process triggered? Before gastrulation is initiated, the streak-inhibiting hypoblast (see Section 2.1) is invaded by cells starting to migrate anteriorly from the deep portions of the PMZ. These are more motile than the hypoblast cells and insinuate themselves between them, and form a new layer (endoblast or secondary hypoblast) that pushes the hypoblast anteriorly (Stern, 1990; Vakaet, 1970). Like the hypoblast, the endoblast is also extraembryonic in fate. The endoblast does not produce any of the inhibitors secreted by the hypoblast—Cerberus, Crescent and Dkk1 (Bertocchini & Stern, 2002; Foley et al., 2000). Consistent with the finding that the hypoblast represses the formation of a primitive streak via the Nodal inhibition of Cerberus, it was found that the early ingressing cells require Nodal activity (but not canonical Wnt or BMP) to induce EMT in the neighboring epiblast (Voiculescu et al., 2014).

Collectively, these findings indicate that, in the posterior area undergoing medio-lateral intercalation early and producing Nodal, early 'pioneer' cells (probably a subset of the HNK-1^+ cells) start inducing EMT in neighboring cells once the endoblast advances under it and displacing the hypoblast, the source of Nodal inhibition. This leads to a chain reaction of EMT events and the rapid accumulation of cells that form most of the initial primitive streak.

Concomitant with the endoblast invasion and the sudden increase in EMT events, very important movements occur in the Koller's sickle. The mechanisms are not understood, but the middle layer cells of the sickle undergo shearing, so that the anterior ones (expressing *gsc*) move anteriorly (Izpisúa-Belmonte et al., 1993; Streit et al., 2000). They thus come to be in contact again, at stages 2–3, with the epiblast *gsc*$^+$ cells which were already displaced anteriorly by intercalation (see Fig. 1C and D); this coincides with the tip of the primitive streak acquiring organizer properties.

4. Movements during gastrulation

4.1 Cell behaviors in the primitive streak

The short, triangular streak appearing at stage 2 narrows and elongates very quickly, becoming parallel-sided (stage 3); the apical side is flat and basement membrane under its cells is essentially intact. Soon afterward, a groove forms along the apical midline of the primitive streak, the basement membrane is broken down and mesenchymal cells start to emigrate away from the primitive streak—these features defining stage 3+. This stage lasts for about 8 h, during which the primitive streak slowly elongates to reach about 2/3 of the diameter of the AP. During elongation, the ingressing cells retain the same capacity to recruit cells from the epiblast to engage in EMT (Vakaet, 1973); as these new recruits narrow their apical surfaces and ingress, they pull on the epiblast cells near the primitive streak and draw them in to continue the process.

It is not entirely clear how signaling events control the EMT. At stage 3^+, it was shown that RhoA activity and microtubules dynamics in the basal aspect of epiblast cells are essential in maintaining the interaction of epiblast cells with the basement membrane. Net1, a RhoA GEF, is downregulated in cells in the primitive streak and would be a candidate for controlling EMT (Nakaya, Sukowati, Wu, & Sheng, 2008). It remains unknown how these activities are regulated in the first cells undergoing EMT through an intact basement (such as the 'pioneer cells' and those in the primitive streak at stages 2–3), and what is the causal sequence of events in cells entering the primitive streak at stage 3^+ and encountering an environment already devoid of basement membrane.

4.2 Tension created by ingression at the primitive streak drives movements in the entire epiblast

Once the primitive streak forms, global movements in the rest of the epiblast continue to occur in a modification of the pre-streak, Polonaise movements: now, cells in the lateral epiblast are seen to move into it at right angles, and the counter-rotating flows in the rest of the epiblast are shifted more anteriorly than in the pre-streak, Polonaise pattern (see Fig. 4 and Movie 2 in the online version at https://doi.org/10.1016/bs.ctdb.2019.11.015). The EMT chain reaction at the primitive streak is sufficient to simulate these observed movements when implemented in the object-oriented model of the pre-gastrula proliferative epiblast, without other modifications (Voiculescu et al., 2014). Recent observations showed that sister cells in the epiblast tend to separate immediately after division and create small rearrangements around them; it has been proposed that the neighbor exchanges fluidize the tissue and are required for the movements to occur, since the cell flows in the epiblast are indeed straighter if mitoses are blocked (Firmino, Rocancourt, Saadaoui, Moreau, & Gros, 2016). However, this

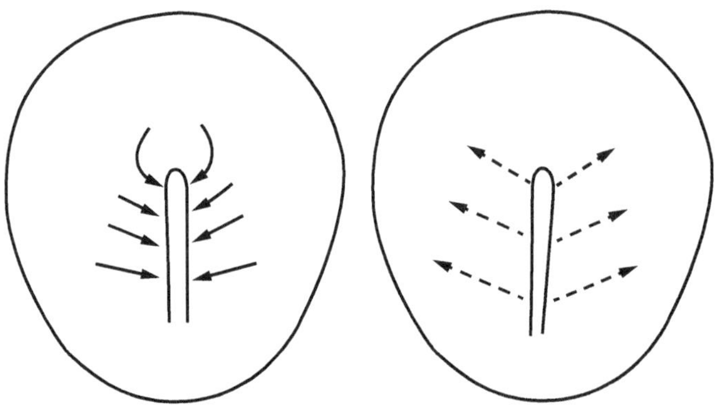

Fig. 4 Movements at full primitive streak stage. Left: pattern of movements in the epiblast. Right: movements of mesoderm away from the primitive streak. *From Gräper, L. (1929). Die Primitiventwicklung des Hühnchens nach stereokinematographischen Untersuchungen, kontrolliert durch vitale Farbmarkierung und verglichen mit der Entwicklung anderer Wirbeltiere.* Wilhelm Roux' Archiv für Entwicklungsmechanik, 116, *382–429.*

can be more simply explained by an increase in the tension of the epiblast in these experimental conditions, since cell proliferation partially counteracts the tension created in the epiblast by the margin of the blastoderm and the primitive streak.

4.3 Movements into the lower layers

From its formation, the primitive streak is molecularly patterned along its length. Its tip, where expression of *gsc* and *chordin* is positioned (see above), acquires organizer character. It appears bulbous during stage 3^+ (and is termed Hensen's node) and produces definitive endoderm (Kimura, Yasugi, Stern, & Fukuda, 2006; Lawson & Schoenwolf, 2003) through distinct routes: some cells directly insert into the lower layer, while others only after first traveling within the middle layer. All these cells displace the previous, extra-embryonic layers (endoblast and hypoblast) toward the periphery of the embryo. An excised node can be regenerated upon ablation (Joubin & Stern, 2001; Psychoyos & Stern, 1996a; Yuan & Schoenwolf, 1998), and it has been shown that the primitive streak is dynamically patterned by signaling centers along its length (Joubin & Stern, 2001). Indeed, cells along the primitive streak are emerging to have complex positional information (Alev et al., 2010; Chapman et al., 2002; Vermillion et al., 2018). However, transplantation experiments showed that the organizer cells are committed (will form notochord and somites when grafted into more posterior positions in the primitive streak), but cells emigrating from other regions of the primitive streak are flexible and can adjust their fate upon relocation into a different position in the primitive streak (Garcia-Martinez & Schoenwolf, 1992; Inagaki, Garcia-Martinez, & Schoenwolf, 1993; Wang et al., 2013).

The rest of the primitive streak generates mesoderm, migrating away from the streak (see Fig. 2 and Movie 2 in the online version at https://doi.org/10.1016/bs.ctdb.2019.11.015): initially at right angles from it, then curving anteriorly toward the regions devoid of middle layer cells or having a lower density of them. Cells migrating from the regions immediately behind the organizer migrate closer to the midline and form more dorsal types of mesoderm than those emerging from more posterior positions of the primitive streak (Lopez-Sanchez, Garcia-Martinez, & Schoenwolf, 2001; Psychoyos & Stern, 1996b; Sawada & Aoyama, 1999; Schoenwolf, Garcia-Martinez, & Dias, 1992). Thus, the rostro-caudal level of the primitive streak where cells ingress determines their eventual dorso-ventral position at later stages. For any given position of ingression along the streak, cells ingressing first will migrate more anteriorly than those following them and contributing to the same type of mesoderm.

Much more remains to be understood in terms of the mechanisms controlling the movements of mesodermal cells emerging from the primitive streak. On the one hand, it is in principle possible that local interactions between the cells (such as contact inhibition) is sufficient to drive directional and orderly migration of mesoderm, without their identity (established at the time of residence in the primitive streak) playing a specific role in migration itself. On the other hand, FGF8 (emanating from the primitive streak) was shown to act as a repellent for the mesoderm (Hardy, Yatskievych, Konieczka, Bobbs, & Antin, 2011; Yang et al., 2002), but it is not clear how far from the primitive streak this can act. Non-canonical Wnt signaling was also shown to affect emigration of cells from the streak (Hardy et al., 2008), through an unknown mechanism. Several pathways (Wnt/β-cat, BMP, PDGF) were implicated in guiding the migration of different subsets of mesoderm while away from the streak (see, for example, Song et al., 2014; Sweetman, Wagstaff, Cooper, Weijer, & Münsterberg, 2008; Yang, Chrisman, & Weijer, 2008), but it remains unclear whether they act at gastrulation stages or come into play later in development.

5. End of gastrulation stage

An abrupt transition marks the end of stage 3^+ and the transition to stage 4: the tip of the streak (Hensen's node and the region just behind it) stop recruiting cells from the epiblast lateral to it (Sheng, dos Reis, & Stern, 2003). This is accompanied by the acquisition of neural plate character by the epiblast lateral and anterior to the node. The neural plate, demarcated by the stable expression of *Sox2*, is composed of two distinct regions in terms of growth: anterior to the node, the cells are specified to form brain territories, while the regions close to the node behave as stem zones that will subsequently add more caudal material to the elongating neural plate (Brown & Storey, 2000; Mathis, Kulesa, & Fraser, 2001).

Perhaps surprisingly, it is only from stage 4 that the anterior streak starts laying down mesodermal types originating from the organizer in all vertebrates (notochord and presomitic mesoderm). It was shown that the anterior streak also starts behaving as a stem zone, containing resident cells that self-renew (Selleck & Stern, 1992) and also give off descendants into the forming prechordal mesendoderm and then notochord (from its median sector) and the medial presomitic mesoderm (from its lateral aspects) (Iimura, Yang, Weijer, & Pourquie, 2007; Selleck & Stern, 1991).

The primitive streak soon starts to shorten, as the anterior regions start forming the axis in a head-to-tail progression. Gastrulation movements (defined as movement of epiblast lateral to the streak into it and emergence of mesoderm away from it) continue in the regions of the primitive streak caudal to the node.

Almost nothing is known about the events setting up the late organizer as a stem zone and what controls the behavior of its resident cells. However, gastrulation as a stage can be said to end with its establishment.

6. Summary and conclusions

Global movements occur in the large epiblast of the early chick embryo well before gastrulation. They are driven by medio-lateral intercalation in a peripheral domain of the embryonic epiblast, and displace the precursors of the gastrulation site from the periphery of the blastoderm (as in anamniotes) to the midline (as in amniotes). The hypoblast, an evolutionary acquisition of amniotes, plays a dual role: it positions the intercalation domain and delays gastrulation until the precursors of the primitive streak assume their proper position, along the future antero-posterior axis of the embryo. A primitive streak then forms by a chain reaction of EMT in the intercalation domain, initiated by early, pioneer cells that ingress singly and infrequently before gastrulation. This mechanism continues to be at play subsequently, so that cells ingressing at the primitive streak (to form the definitive endoderm and mesoderm under the epiblast) recruit cells from the epiblast also to undergo EMT and continue gastrulation. The movements in the rest of the epiblast continue to be driven passively, by the tension created at the edge of the blastoderm and epithelial cells ingressing at the streak. Comparatively little is known about the mechanism of mesoderm migration away from the primitive streak, and even less about how endoderm cells insert themselves into the extra-embryonic lower layer. Gastrulation as a stage ends when the anterior primitive streak stops recruiting cells from the lateral epiblast and becomes a stem zone. At this stage, it also neuralizes the surrounding epiblast, and the regions neighboring the node become stem zones elongating the neural plate. As the embryo develops by gradual additions of material by these stem zones, gastrulation movements continue in the rest of the primitive streak. Very scarce information exists regarding the modes of cell division employed by the resident cells in this stem zone and how they lay down the axial and paraxial mesoderm throughout elongation. Nothing is known about the molecular mechanisms underpinning its function. Another important question regards the inferred existence of neuro-mesodermal precursors, which could contribute to both the mesodermal and neural stem zone. Currently, it is unknown where these precursors are, how much they contribute to axial elongation and how.

References

Acloque, H., Ocana, O. H., Matheu, A., Rizzoti, K., Wise, C., Lovell-Badge, R., et al. (2011). Reciprocal repression between Sox3 and snail transcription factors defines embryonic territories at gastrulation. *Developmental Cell*, *21*, 546–558.

Alev, C., Wu, Y., Kasukawa, T., Jakt, L. M., Ueda, H. R., & Sheng, G. (2010). Transcriptomic landscape of the primitive streak. *Development*, *137*, 2863–2874.

Alev, C., Wu, Y., Nakaya, Y., & Sheng, G. (2013). Decoupling of amniote gastrulation and streak formation reveals a morphogenetic unity in vertebrate mesoderm induction. *Development*, *140*, 2691–2696.

Bachvarova, R. F., Skromne, I., & Stern, C. D. (1998). Induction of primitive streak and Hensen's node by the posterior marginal zone in the early chick embryo. *Development*, *125*, 3521–3534.

Bellairs, R., Bromham, D. R., & Wylie, C. C. (1967). The influence of the area opaca on the development of the young chick embryo. *Development*, *17*, 195–212.

Bertocchini, F., & Stern, C. D. (2002). The hypoblast of the chick embryo positions the primitive streak by antagonizing nodal signaling. *Developmental Cell*, *3*, 735–744.

Brown, J. M., & Storey, K. G. (2000). A region of the vertebrate neural plate in which neighbouring cells can adopt neural or epidermal fates. *Current Biology*, *10*, 869–872.

Canning, D. R., & Stern, C. D. (1988). Changes in the expression of the carbohydrate epitope HNK-1 associated with mesoderm induction in the chick embryo. *Development*, *104*, 643–655.

Chapman, S. C., Schubert, F. R., Schoenwolf, G. C., & Lumsden, A. (2002). Analysis of spatial and temporal gene expression patterns in blastula and gastrula stage chick embryos. *Developmental Biology*, *245*, 187–199.

Chuai, M., Zeng, W., Yang, X., Boychenko, V., Glazier, J. A., & Weijer, C. J. (2006). Cell movement during chick primitive streak formation. *Developmental Biology*, *296*, 137–149.

Cui, C., Yang, X., Chuai, M., Glazier, J. A., & Weijer, C. J. (2005). Analysis of tissue flow patterns during primitive streak formation in the chick embryo. *Developmental Biology*, *284*, 37–47.

Duval, M. (1884). De la formation du blastoderme dans l'oeuf d'oiseau. *Annales Des Sciences Naturelles, Zoologie*, *18*, 1–208.

Eyal-Giladi, H., Debby, A., & Harel, N. (1992). The posterior section of the chick's area pellucida and its involvement in hypoblast and primitive streak formation. *Development*, *116*, 819–830.

Eyal-Giladi, H., & Khaner, O. (1989). The chick's marginal zone and primitive streak formation. II. Quantification of the marginal zone's potencies—Temporal and spatial aspects. *Developmental Biology*, *134*, 215–221.

Eyal-Giladi, H., & Kochav, S. (1976). From cleavage to primitive streak formation: A complementary normal table and a new look at the first stages of the development of the chick: I. General morphology. *Developmental Biology*, *49*, 321–337.

Firmino, J., Rocancourt, D., Saadaoui, M., Moreau, C., & Gros, J. (2016). Cell division drives epithelial cell rearrangements during gastrulation in chick. *Developmental Cell, 36*, 249–261.

Foley, A. C., Skromne, I., & Stern, C. D. (2000). Reconciling different models of forebrain induction and patterning: A dual role for the hypoblast. *Development, 127*, 3839–3854.

Garcia-Martinez, V., & Schoenwolf, G. C. (1992). Positional control of mesoderm movement and fate during avian gastrulation and neurulation. *Developmental Dynamics, 193*, 249–256.

Gräper, L. (1929). Die Primitiventwicklung des Hühnchens nach stereokinematographischen Untersuchungen, kontrolliert durch vitale Farbmarkierung und verglichen mit der Entwicklung anderer Wirbeltiere. *Wilhelm Roux' Archiv für Entwicklungsmechanik, 116*, 382–429.

Hamburger, V., & Hamilton, H. L. (1951). A series of normal stages in the development of the chick embryo. *Journal of Morphology, 88*, 49–92.

Hardy, K. M., Garriock, R. J., Yatskievych, T. A., D'Agostino, S. L., Antin, P. B., & Krieg, P. A. (2008). Non-canonical Wnt signaling through Wnt5a/b and a novel Wnt11 gene, Wnt11b, regulates cell migration during avian gastrulation. *Developmental Biology, 320*, 391–401.

Hardy, K. M., Yatskievych, T. A., Konieczka, J., Bobbs, A. S., & Antin, P. B. (2011). FGF signalling through RAS/MAPK and PI3K pathways regulates cell movement and gene expression in the chicken primitive streak without affecting E-cadherin expression. *BMC Developmental Biology, 11*, 20.

Hume, C. R., & Dodd, J. (1993). Cwnt-8C: A novel Wnt gene with a potential role in primitive streak formation and hindbrain organization. *Development, 119*, 1147–1160.

Iimura, T., Yang, X., Weijer, C. J., & Pourquie, O. (2007). Dual mode of paraxial mesoderm formation during chick gastrulation. *Proceedings of the National Academy of Sciences of the United States of America, 104*, 2744–2749.

Inagaki, T., Garcia-Martinez, V., & Schoenwolf, G. C. (1993). Regulative ability of the prospective cardiogenic and vasculogenic areas of the primitive streak during avian gastrulation. *Developmental Dynamics, 197*, 57–68.

Izpisúa-Belmonte, J. C., De Robertis, E. M., Storey, K. G., & Stern, C. D. (1993). The homeobox gene goosecoid and the origin of organizer cells in the early chick blastoderm. *Cell, 74*, 645–659.

Joubin, K., & Stern, C. D. (2001). Formation and maintenance of the organizer among the vertebrates. *The International Journal of Developmental Biology, 45*, 165–175.

Khaner, O., & Eyal-Giladi, H. (1989). The chick's marginal zone and primitive streak formation: I. Coordinative effect of induction and inhibition. *Developmental Biology, 134*, 206–214.

Kimura, W., Yasugi, S., Stern, C. D., & Fukuda, K. (2006). Fate and plasticity of the endoderm in the early chick embryo. *Developmental Biology, 289*, 283–295.

Knezevic, V., De, R. S., & Mackem, S. (1997). Two novel chick T-box genes related to mouse Brachyury are expressed in different, non-overlapping mesodermal domains during gastrulation. *Development, 124*, 411–419.

Lawson, A., & Schoenwolf, G. C. (2003). Epiblast and primitive-streak origins of the endoderm in the gastrulating chick embryo. *Development, 130*, 3491–3501.

Lopez-Sanchez, C., Garcia-Martinez, V., & Schoenwolf, G. C. (2001). Localization of cells of the prospective neural plate, heart and somites within the primitive streak and epiblast of avian embryos at intermediate primitive-streak stages. *Cells Tissues Organs, 169*, 334–346.

Mathis, L., Kulesa, P. M., & Fraser, S. E. (2001). FGF receptor signalling is required to maintain neural progenitors during Hensen's node progression. *Nature Cell Biology, 3*, 559–566.

Mitrani, E., Gruenbaum, Y., Shohat, H., & Ziv, T. (1990). Fibroblast growth factor during mesoderm induction in the early chick embryo. *Development, 109*, 387–393.

Mogi, K., Toyoizumi, R., & Takeuchi, S. (2000). Correlation between the expression of the HNK-1 epitope and cellular invasiveness in prestreak epiblast cells of chick embryos. *The International Journal of Developmental Biology, 44*, 811–814.

Nakaya, Y., Sukowati, E. W., Wu, Y., & Sheng, G. (2008). RhoA and microtubule dynamics control cell-basement membrane interaction in EMT during gastrulation. *Nature Cell Biology, 10*, 765–775.

New, D. a. T. (1959). The adhesive properties and expansion of the chick blastoderm. *Development, 7*, 146–164.

Ohta, K., Kuriyama, S., Okafuji, T., Gejima, R., Ohnuma, S., & Tanaka, H. (2006). Tsukushi cooperates with VG1 to induce primitive streak and Hensen's node formation in the chick embryo. *Development, 133*, 3777–3786.

Ozato, K. (1969). Cell cycle in the primitive streak and the notochord of early chick embryos. *Development, Growth & Differentiation, 10*, 297–311.

Psychoyos, D., & Stern, C. D. (1996a). Restoration of the organizer after radical ablation of Hensen's node and the anterior primitive streak in the chick embryo. *Development, 122*, 3263–3273.

Psychoyos, D., & Stern, C. D. (1996b). Fates and migratory routes of primitive streak cells in the chick embryo. *Development, 122*, 1523–1534.

Rozbicki, E., Chuai, M., Karjalainen, A. I., Song, F., Sang, H. M., Martin, R., et al. (2015). Myosin-II-mediated cell shape changes and cell intercalation contribute to primitive streak formation. *Nature Cell Biology, 17*, 397–408.

Sanders, E. J., Varedi, M., & French, A. S. (1993). Cell proliferation in the gastrulating chick embryo: A study using BrdU incorporation and PCNA localization. *Development, 118*, 389–399.

Sandersius, S. A., Chuai, M., Weijer, C. J., & Newman, T. J. (2011). A 'chemotactic dipole' mechanism for large-scale vortex motion during primitive streak formation in the chick embryo. *Physical Biology, 8*, 045008.

Sawada, K., & Aoyama, H. (1999). Fate maps of the primitive streak in chick and quail embryo: Ingression timing of progenitor cells of each rostro-caudal axial level of somites. *The International Journal of Developmental Biology, 43*, 809–815.

Schoenwolf, G. C., Garcia-Martinez, V., & Dias, M. S. (1992). Mesoderm movement and fate during avian gastrulation and neurulation. *Developmental Dynamics, 193*, 235–248.

Seleiro, E. A. P., Connolly, D. J., & Cooke, J. (1996). Early developmental expression and experimental axis determination by the chicken Vg1 gene. *Current Biology, 6*, 1476–1486.

Selleck, M. A., & Stern, C. D. (1991). Fate mapping and cell lineage analysis of Hensen's node in the chick embryo. *Development, 112*, 615–626.

Selleck, M. A. J., & Stern, C. D. (1992). Evidence for stem cells in the mesoderm of Hensen's node and their role in embryonic pattern formation. *NATO ASI Series*. Boston, MA: Springer (Series A: Life Sciences).

Shah, S. B., Skromne, I., Hume, C. R., Kessler, D. S., Lee, K. J., Stern, C. D., et al. (1997). Misexpression of chick Vg1 in the marginal zone induces primitive streak formation. *Development, 124*, 5127–5138.

Sheng, G., dos Reis, M., & Stern, C. D. (2003). Churchill, a zinc finger transcriptional activator, regulates the transition between gastrulation and neurulation. *Cell, 115*, 603–613.

Skromne, I., & Stern, C. D. (2001). Interactions between Wnt and Vg1 signalling pathways initiate primitive streak formation in the chick embryo. *Development, 128*, 2915–2927.

Skromne, I., & Stern, C. D. (2002). A hierarchy of gene expression accompanying induction of the primitive streak by Vg1 in the chick embryo. *Mechanisms of Development, 114*, 115–118.

Song, J., McColl, J., Camp, E., Kennerley, N., Mok, G. F., McCormick, D., et al. (2014). Smad1 transcription factor integrates BMP2 and Wnt3a signals in migrating cardiac progenitor cells. *Proceedings of the National Academy of Sciences of the United States of America*, *111*, 7337–7342.
Spratt, N. T. (1946). Formation of the primitive streak in the explanted chick blastoderm marked with carbon particles. *Journal of Experimental Zoology*, *103*, 259–304.
Spratt, N. T., & Haas, H. (1960). Integrative mechanisms in development of the early chick blastoderm. I. Regulative potentiality of separated parts. *Journal of Experimental Zoology*, *145*, 97–137.
Stern, C. D. (1979). A re-examination of mitotic activity in the early chick embryo. *Anatomy and Embryology*, *156*, 319–329.
Stern, C. D. (1990). The marginal zone and its contribution to the hypoblast and primitive streak of the chick embryo. *Development*, *109*, 667–682.
Stern, C. D., & Canning, D. R. (1990). Origin of cells giving rise to mesoderm and endoderm in chick embryo. *Nature*, *343*, 273–275.
Stern, C. D., & Downs, K. M. (2012). The hypoblast (visceral endoderm): An evo-devo perspective. *Development*, *139*, 1059–1069.
Streit, A., Berliner, A. J., Papanayotou, C., Sirulnik, A., & Stern, C. D. (2000). Initiation of neural induction by FGF signalling before gastrulation. *Nature*, *406*, 74–78.
Streit, A., Lee, K. J., Woo, I., Roberts, C., Jessell, T. M., & Stern, C. D. (1998). Chordin regulates primitive streak development and the stability of induced neural cells, but is not sufficient for neural induction in the chick embryo. *Development*, *125*, 507–519.
Sweetman, D., Wagstaff, L., Cooper, O., Weijer, C., & Münsterberg, A. (2008). The migration of paraxial and lateral plate mesoderm cells emerging from the late primitive streak is controlled by different Wnt signals. *BMC Developmental Biology*, *8*, 63.
Vakaet, L. (1970). Cinephotomicrographic investigations of gastrulation in the chick blastoderm. *Archives de Biologie*, *81*, 387–426.
Vakaet, L. (1973). Inductions by the nodus posterior of the primitive streak of birds. *Comptes Rendus des Seances de la Societe de Biologie et de Ses Filiales*, *167*, 1053–1055.
Vasiev, B., Balter, A., Chaplain, M., Glazier, J. A., & Weijer, C. J. (2010). Modeling gastrulation in the chick embryo: Formation of the primitive streak. *PLoS One*, *5*, e10571.
Vermillion, K. L., Bacher, R., Tannenbaum, A. P., Swanson, S., Jiang, P., Chu, L.-F., et al. (2018). Spatial patterns of gene expression are unveiled in the chick primitive streak by ordering single-cell transcriptomes. *Developmental Biology*, *439*, 30–41.
Voiculescu, O., Bertocchini, F., Wolpert, L., Keller, R. E., & Stern, C. D. (2007). The amniote primitive streak is defined by epithelial cell intercalation before gastrulation. *Nature*, *449*, 1049–1052.
Voiculescu, O., Bodenstein, L., Lau, I. J., & Stern, C. D. (2014). Local cell interactions and self-amplifying individual cell ingression drive amniote gastrulation. *eLife*, *3*, e01817.
Waddington, C. H. (1932). III. Experiments on the development of chick and duck embryos, cultivated in vitro. *Philosophical Transactions of the Royal Society of London. Series B*, *221*, 179–230. Containing Papers of a Biological Character.
Wang, X.-Y., Li, Y., Ma, Z.-L., Wang, L.-J., Chuai, M., Münsterberg, A., et al. (2013). Retention of stem cell plasticity in avian primitive streak cells and the effects of local microenvironment. *The Anatomical Record*, *296*, 533–543.
Wei, Y., & Mikawa, T. (2000). Formation of the avian primitive streak from spatially restricted blastoderm: Evidence for polarized cell division in the elongating streak. *Development*, *127*, 87–96.
Wetzel, R. (1929). Untersuchungen am Hühnchen. Die Entwicklung des Keims Während der Ersten Beiden Bruttage. In R. Wetzel (Ed.), *Untersuchungen am Hühnchen: Die Entwicklung des Keims während der ersten beiden Bruttage* (pp. 188–321). Berlin, Heidelberg: Springer.

Yang, X., Chrisman, H., & Weijer, C. J. (2008). PDGF signalling controls the migration of mesoderm cells during chick gastrulation by regulating N-cadherin expression. *Development, 135*, 3521–3530.

Yang, X., Dormann, D., Münsterberg, A. E., & Weijer, C. J. (2002). Cell movement patterns during gastrulation in the chick are controlled by positive and negative chemotaxis mediated by FGF4 and FGF8. *Developmental Cell, 3*, 425–437.

Yuan, S., & Schoenwolf, G. C. (1998). De novo induction of the organizer and formation of the primitive streak in an experimental model of notochord reconstitution in avian embryos. *Development, 125*, 201–213.

Zamir, E. A., Rongish, B. J., & Little, C. D. (2008). The ECM moves during primitive streak formation—Computation of ECM versus cellular motion. *PLoS Biology, 6*, e247.

CHAPTER FIFTEEN

Guts and gastrulation: Emergence and convergence of endoderm in the mouse embryo

Sonja Nowotschin*, Anna-Katerina Hadjantonakis*
Developmental Biology Program, Sloan Kettering Institute, Memorial Sloan Kettering Cancer Center, New York, NY, United States
*Corresponding authors: e-mail address: nowotscs@mskcc.org; hadj@mskcc.org

Contents

Abstract

Gastrulation is a central process in mammalian development in which a spatiotemporally coordinated series of events driven by cross-talk between adjacent embryonic and extra-embryonic tissues results in stereotypical morphogenetic cell behaviors, massive cell proliferation and the acquisition of distinct cell identities. Gastrulation provides the blueprint of the body plan of the embryo, as well as generating extra-embryonic cell types of the embryo to make a connection with its mother. Gastrulation involves the specification of mesoderm and definitive endoderm from pluripotent epiblast, concomitant with a highly ordered elongation of tissue along the anterior-posterior (AP) axis.

Current Topics in Developmental Biology, Volume 136
ISSN 0070-2153
https://doi.org/10.1016/bs.ctdb.2019.11.012

Interestingly, cells with an endoderm identity arise twice during mouse development. Cells with a primitive endoderm identity are specified in the preimplantation blastocyst, and which at gastrulation intercalate with the emergent definitive endoderm to form a mosaic tissue, referred to as the gut endoderm. The gut endoderm gives rise to the gut tube, which will subsequently become patterned along its AP axis into domains possessing unique visceral organ identities, such as thyroid, lung, liver and pancreas. In this way, proper endoderm development is essential for vital organismal functions, including the absorption of nutrients, gas exchange, detoxification and glucose homeostasis.

1. Preimplantation development, the blastocyst and emergence of the first endoderm population in the mouse

After fertilization, the totipotent zygote differentiates into the three lineages of the mammalian blastocyst that in mouse arise as two sequential, binary decisions (Chazaud & Yamanaka, 2016; Schrode et al., 2013; Simon, Hadjantonakis, & Schroter, 2018). The first lineage decision taking place in the mammalian embryo segregates the trophectoderm (TE, green, Fig. 1) from the inner cell mass (ICM, purple, Fig. 1). In the mouse embryo this decision takes place as the 8-cell morula stage embryo transitions to the 16-cell stage, around embryonic day (E)2.5. In the second lineage decision, which occurs in the mid-blastocyst stage around E3.5, cells of the ICM differentiate into pluripotent epiblast (EPI, red, Figs. 1 and 2) or primitive endoderm (PrE, blue, Fig. 1). EPI and PrE precursors emerge in a salt-and-pepper distribution within the ICM and eventually sort into two adjacent layers (Chazaud, Yamanaka, Pawson, & Rossant, 2006; Plusa, Piliszek, Frankenberg, Artus, & Hadjantonakis, 2008). The mechanisms driving lineage sorting are not well understood but have been proposed to involve active movement (Meilhac et al., 2009). Combined with the acquisition of apical-basal polarity by PrE cells (Gerbe, Cox, Rossant, & Chazaud, 2008; Rula et al., 2007), this results in the segregation of tissue layers and assembly of a basement membrane between them (Fig. 2). Interestingly, PrE cells share many cell behaviors (e.g., epithelialization) and molecular markers (e.g., transcription factors of the Gata, Sox and Fox families) with the definitive endoderm cells that emerge later at gastrulation (Artus, Piliszek, & Hadjantonakis, 2011; Blakeley et al., 2015; Burtscher & Lickert, 2009; Freyer et al., 2015; Plusa et al., 2008; Simon, Zhang, et al., 2018; Viotti, Nowotschin, & Hadjantonakis, 2014).

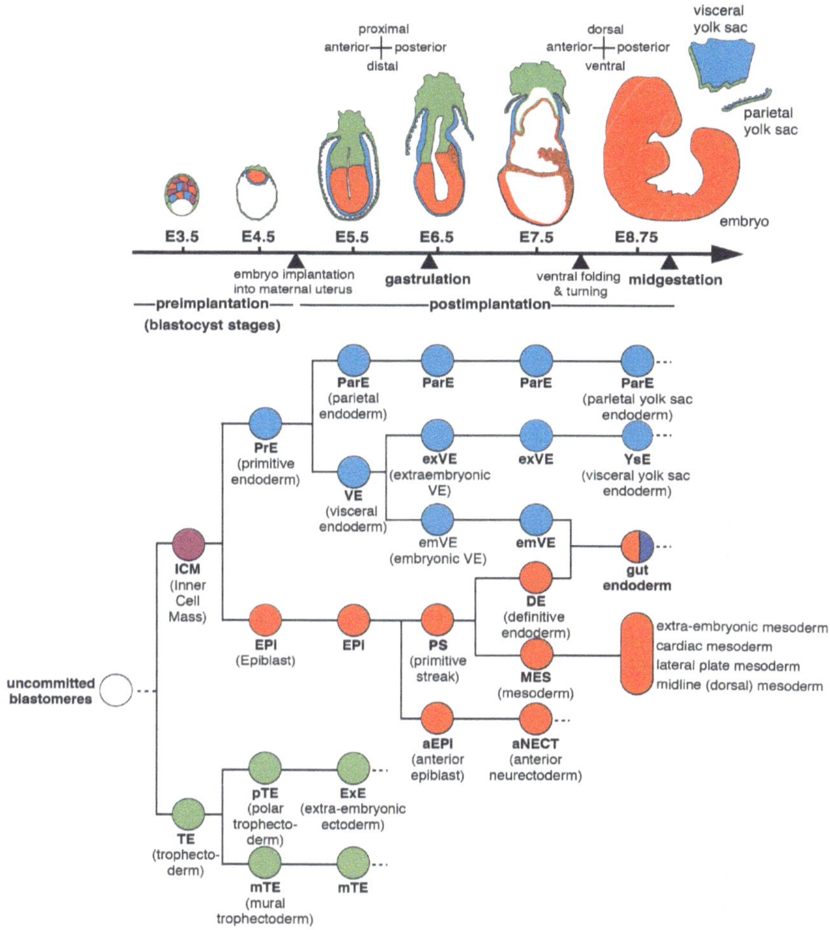

Fig. 1 The cell lineages of the early mouse embryo: from preimplantation to midgestation. The derivatives of the three lineages of the blastocyst and their descendants are color-coded: trophectoderm (TE, green), inner cell mass (ICM, purple), and its derivatives the primitive endoderm (PrE, blue), and the epiblast (EPI, red).

The TE can be further divided into mural and polar subtypes (based on their proximity to the ICM), which surround the blastocyst cavity and EPI, respectively. The polar TE will form the extra-embryonic ectoderm (ExE, at postimplantation), which will represent the fetal portion of the placenta, whereas the mural TE will form the outer layer of the parietal yolk sac (Fig. 1). As the first endodermal cell population arising in the developing mammalian embryo, the PrE differentiates into visceral endoderm (VE) and parietal endoderm (ParE) subtypes. These populations give rise to

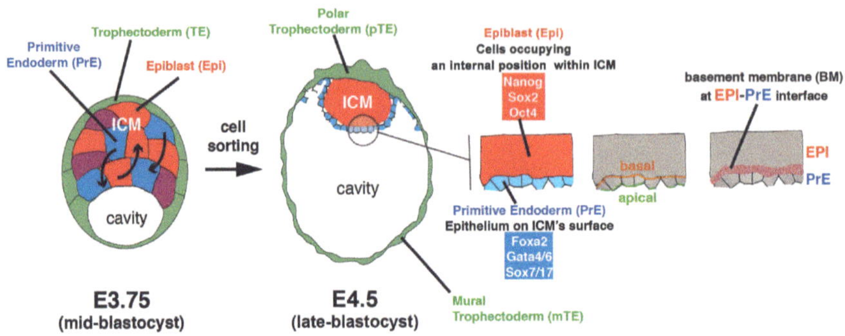

Fig. 2 Endoderm emerges twice during mouse development: the preimplantation blastocyst. The first endodermal population (the primitive endoderm, PrE) emerges at the blastocyst stage when inner cell mass (ICM) cells (purple) make a fate choice between pluripotent epiblast (EPI, red) and primitive endoderm (PrE, blue). PrE and EPI cells arise in a salt-and-pepper distribution (E3.75 embryo) within the ICM and subsequently sort into two layers (E4.5 embryo), whereby PrE emerges as an epithelium occupying the surface of the ICM. PrE cells exhibit many similarities with the definitive endoderm (DE) cells specified at gastrulation including expression of markers (including FOX, SOX, and GATA transcription factors) and morphogenetic cell behaviors which include the acquisition of apical-basal polarity, concomitant sorting to the tissue's surface. As in the gastrula, the emergence of (primitive) endoderm as an epithelium on the ICM's surface is accompanied by the assembly of a basement membrane separating it from the adjacent epiblast.

predominantly extra-embryonic tissues; the visceral and parietal yolk sacs, respectively. The VE epithelium is specialized in endocytosis and absorption making it the major tissue providing nutrients, gas and waste exchange for the early embryo before a maternal-embryonic circulation is established. Despite being predominantly extra-embryonic, a small number of VE cells also contribute descendants to the gut endoderm of the embryo later during development, as they intermingle with definitive endoderm, which emerges at gastrulation (Chan et al., 2019; Kwon, Viotti, & Hadjantonakis, 2008; Nowotschin et al., 2019; Pijuan-Sala et al., 2019; Viotti, Niu, Shi, & Hadjantonakis, 2012; Viotti, Nowotschin, & Hadjantonakis, 2011).

2. Anterior visceral endoderm (AVE), a derivative of the primitive endoderm (PrE) establishes the anterior-posterior axis of the mouse embryo

After the late blastocyst stage embryo has implanted into the maternal uterus, at the very earliest postimplantation stages (around E5.5), the proximal-distal axis of the embryo (relative to the implantation side) is reflected by the graded and convergent activities of the Nodal/Smad and Wnt/β-catenin signaling pathways (Arnold & Robertson, 2009). At this time the embryo takes on the shape of a radially symmetric "egg-cylinder" consisting of a bilaminar epithelium, which in turn is comprised of epiblast (EPI) and visceral endoderm (VE) distally, and EPI and extra-embryonic ectoderm (ExE) proximally (Fig. 1). Single-cell transcriptomic analyses have revealed a subdivision within the VE corresponding to the distally located embryonic (em)VE and the proximal extra-embryonic (ex)VE, surrounding the EPI and extra-embryonic ectoderm (ExE), respectively (Nowotschin et al., 2019). The basement membrane laid down at the PrE-EPI interface at the late blastocyst stage (Fig. 2) comes to separate the basolateral domains of the VE from the EPI in the embryonic, and VE from ExE in the extra-embryonic regions of the postimplantation embryo. Though exVE and emVE can be distinguished molecularly as early as E5.5, these contiguous epithelia exhibit a comparable cuboidal morphology at this stage, making them morphologically indistinguishable (Abe, Kutsuna, Kiyonari, Furuta, & Fujimori, 2018; Nowotschin et al., 2013; Trichas et al., 2011).

Signaling cross-talk between the pluripotent EPI and the adjacent VE and ExE tissue layers, is essential for setting up the AP axis of the embryo. Signals including Fgf, as well as Wnt, Bmp and Nodal emanating from these two predominantly extra-embryonic tissues are instructive to specify a unique subpopulation of the emVE at the distal tip of the E5.5 embryo, the distal visceral endoderm (DVE). The DVE expresses a distinct set of molecular markers which include transcription factors such as Hex and Hesx, as well as antagonists of the Nodal and Wnt signaling pathways (e.g., Lefty1, Dkk1, Cer1), which impact the signaling landscape of the adjacent EPI, restricting Nodal and Wnt signals to its posterior.

There is mounting evidence to suggest that DVE cells might be specified as early as the blastocyst stage. A small number of PrE cells express DVE markers such as Hex, Lefty1 and Cer1 (Takaoka, Yamamoto, & Hamada, 2011), and genetic fate mapping using a *Lefty1* reporter, as well as time-lapse imaging of *Cer1*-expressing cells from the blastocyst through to E5.5 has revealed their contribution to the DVE and associated anterior(A)VE (Takaoka et al., 2011, 2006; Takaoka, Nishimura, & Hamada, 2017; Torres-Padilla et al., 2007). Clonal analyses have demonstrated that at E5.75 AVE cells predominantly comprise DVE descendants expressing *Hex*, however by E6.25, the AVE comprises both *Hex*-expressing DVE descendants and *Hex*-expressing non-DVE descended cells (Rivera-Perez,

Mager, & Magnuson, 2003; Srinivas, Rodriguez, Clements, Smith, & Beddington, 2004; Takaoka et al., 2011). DVE/AVE cells collectively migrate within the VE epithelium in the proximal direction, and in doing so end up on one side of the embryo where they come to define the future anterior. At the contralateral posterior side of the embryo, another subpopulation of the VE, referred to as the posterior visceral endoderm (PVE) expresses *Wnt3* (Liu et al., 1999; Rivera-Perez & Magnuson, 2005) and *Mixl* (Pearce & Evans, 1999; Robb et al., 2000). Though a function has yet to be ascribed to the PVE, in addition to its unique molecular identity, it has been shown that unlike the rest of the emVE, PVE cells do not undergo major cell rearrangements but remain a posteriorly positioned cohesive cohort until gastrulation (Kwon et al., 2008; Lawson & Pedersen, 1987; Shioi et al., 2017).

The mechanism(s) initiating and driving proximal-ward DVE/AVE collective cell migration have been a matter of debate. DVE/AVE cells extend filopodial projections (Omelchenko et al., 2014), suggested as being oriented toward a chemoattractant (Srinivas et al., 2004) or repellant (Kimura-Yoshida et al., 2005). Alternatively, migration has been proposed in response to signals from the underlying basement membrane (Rodriguez, Srinivas, Clements, Smith, & Beddington, 2005; Salgueiro, Filipe, & Belo, 2006). In mouse embryos lacking NAP1, a component of the WAVE complex (Rakeman & Anderson, 2006), or its downstream effector Rac1 (Migeotte, Omelchenko, Hall, & Anderson, 2010) the migration of DVE/AVE cells is affected, suggesting that actin cytoskeletal rearrangements are critical for the collective migration of DVE/AVE cells. Live imaging studies using reporters of nuclei and adherens junctions have shown that DVE/AVE migration relies on global cell rearrangements taking place throughout the emVE epithelium (Abe et al., 2018; Shioi et al., 2017). Apical constriction of cells at the anterior of the DVE cohort initiate a global reorganization of emVE cells that in turn drives AVE formation. These studies further confirmed that by E6.5, most AVE cells are not descendants of the E5.5 DVE, but instead represented descendants of cells initially positioned behind the DVE. This raises the question of how cells might acquire a DVE/AVE identity; are they distinct or a derivative of the emVE as has been suggested by single-cell transcriptomic analyses (Nowotschin et al., 2019), perhaps being specified upon exposure to the appropriate environment? Thus is the future DVE specified early in development or alternatively is the entire emVE competent to form DVE/AVE when exposed to a particular signal?

3. Gastrulation: Convergence of signals driving exit from pluripotency and acquisition of distinct cell identities in time and space

Subsequent to the translocation of the DVE/AVE and establishment of the AP axis of the embryo, the primitive streak forms in the proximal-posterior epiblast adjacent to the ExE, and the process of gastrulation is initiated around E6.5. Over time the primitive streak elongates distally until it reaches the distal tip of the embryo. Gastrulation results in the formation of the three germ layers; the ectoderm (source of the skin and cells of the nervous system), the mesoderm (source of blood cells, muscles, kidney and heart) and the endoderm (source of the epithelial lining of the gut and associated organs).

Fate mapping studies using horseradish peroxidase labeling cells of the EPI and VE, as well as orthotopic tissue grafting experiments, complemented by mouse genetic studies and recent single-cell transcriptomics analyses have shown that cells ingressing through the primitive streak, depending on time and location, will adopt distinct mesodermal identities (Kinder et al., 1999; Lawson, Meneses, & Pedersen, 1991; Lawson & Pedersen, 1987; Peng et al., 2019; Scialdone et al., 2016). During gastrulation the precursors of both mesoderm and definitive endoderm (DE), exhibit transcriptional changes and become primed for differentiation. Single-cell transcriptomic analyses showed that there is increased transcriptional noise in uncommitted EPI cells and that this decreases in cells of the primitive streak leading to a decline in the expression of naïve pluripotency-associated factors and concomitant increase the expression of factors associated with primed pluripotency in preparation for lineage (mesoderm and DE) differentiation (Mohammed et al., 2017).

The acquisition of distinct cell fates takes place in an AP sequence and depends on a cell's exposure—concentration, duration and constellation—to BMP, WNT, NODAL and FGF signals converging at the primitive streak (Fig. 3, reviewed in Arnold & Robertson, 2009). Due to massive cellular proliferation and a concomitant increase in the embryo's size during mouse gastrulation (Snow, 1977), the signaling landscape in the vicinity of the primitive streak, and indeed across the embryo, is constantly changing. The first lineage to emerge from the primitive streak, the extra-embryonic mesoderm, arises at a proximal-posterior position in the embryo where BMP levels are highest. This is followed by cardiac mesoderm, lateral plate and intermediate mesoderm, and finally the paraxial mesoderm which exits at a more anterior position

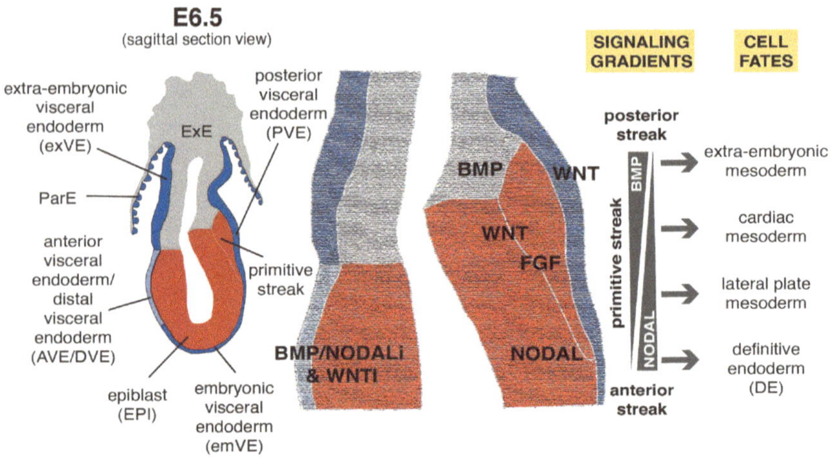

Fig. 3 A convergence of signals drive AP patterning and cell fate specification in time and space. Inhibitors of BMP/NODAL and WNT expressed by the anterior visceral endoderm (AVE) localize the primitive streak to the posterior specification of cell fates and associated morphogenetic cell behaviors. At the primitive streak cells exit pluripotency, acquire distinct lineage identities based on their position, and concomitantly adopt stereotypical morphogenetic programs, as they undergo an epithelial-to-mesenchymal transition (the gastrulation EMT, Fig. 4). Distinct cell identities can be mapped to different regions of cell emergence at the primitive streak and the superposition of opposing BMP-NODAL signaling gradients. Definitive endoderm (DE) cells emerge from the vicinity of the anterior primitive streak, where NODAL levels are high.

(Kinder et al., 1999; Scialdone et al., 2016; Tzouanacou, Wegener, Wymeersch, Wilson, & Nicolas, 2009). Some of the anterior-most populations to emerge from neighboring regions are the axial mesoderm and the definitive endoderm (DE, Fig. 3). A posterior domain of the T-box transcription factor BRACHYURY/T expression and an anterior domain of the forkhead transcription factor FOXA2 expression regionally demarcate the primitive streak, such that cells acquiring mesodermal fates are encompassed within the T domain, while cells fated to DE lie within the FOXA2 domain (Burtscher & Lickert, 2009), with the area of overlap presumably corresponding to axial mesoderm, which stereotypically expresses both T and FOXA2.

4. Morphogenetic cell behaviors at the primitive streak: The gastrulation EMT

As EPI cells ingress through the primitive streak, they undergo an epithelial-to-mesenchymal transition (EMT), which enables them to leave the epiblast epithelium and migrate some distance away within the so-called

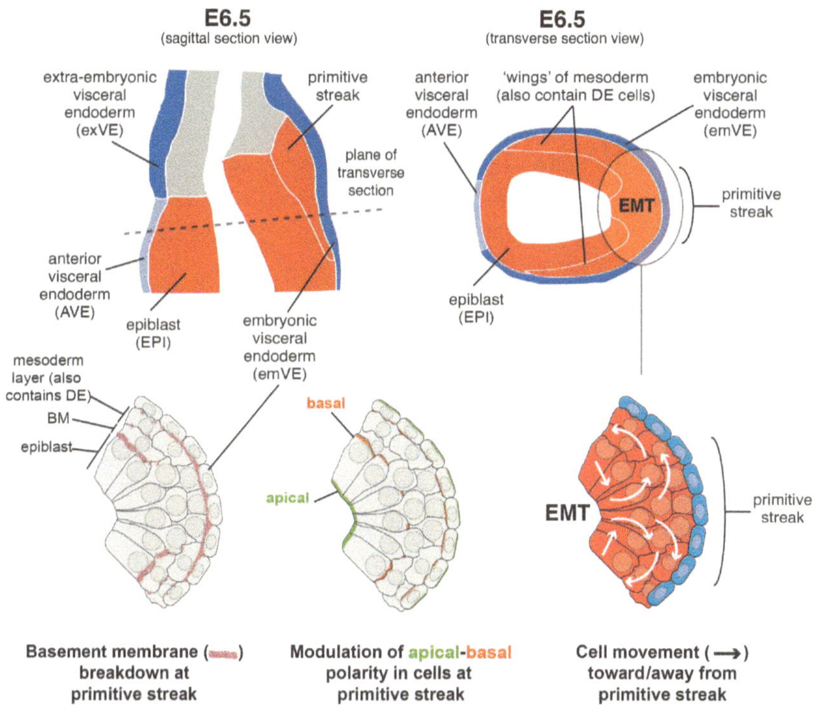

Fig. 4 The primitive streak: a hub of cellular dynamics, site of the gastrulation EMT and gateway for establishing the body plan. The establishment of a primitive streak heralds the start of gastrulation in the mouse. The primitive streak is a transient structure arising within the epiblast tissue layer at the posterior of the embryo at the onset of gastrulation (around E6.5). The basement membrane situated at the basolateral extremity of the epiblast cells (columnar epithelium) is broken down, consequently cells modulating their apical-basal polarity and undergoing epithelial-to-mesenchymal transition (EMT) can leave the epiblast epithelium. Epiblast cells move toward and enter the primitive streak, they then undergo an EMT thereby generating a migratory population which leaves its site-of-origin at the primitive streak to relocate to distant locations.

wings of mesoderm, an intervening layer between the epiblast (EPI) and visceral endoderm (VE) epithelia. During the gastrulation EMT the basement membrane at the EPI-VE interface breaks down, EPI cells become apically restricted and their nucleus shifts to the basal side of the cell (Ramkumar et al., 2016; Voiculescu, Bodenstein, Lau, & Stern, 2014). Finally, they lose apical-basal polarity by disassembling their tight and adherens junctions allowing them to break away from the epithelium (Fig. 4).

EMT at the primitive streak is dependent on the interplay of the key signaling pathways (Fig. 3), having several additional inputs ranging from transcription and cell polarity factors, to mechanical ones. Mouse mutants defective in any of these inputs display defects in the gastrulation EMT, with cells building up in the vicinity of the primitive streak. In mutants for FGF signaling pathway components, as well as mutants for the EMT-associated Zinc finger protein SNAIL, as well as the EPI lineage-specific loss of the T-box transcription factor EOMES cells ingress but cannot undergo an EMT, and consequently are incapable of migrating away from the primitive streak (Arnold, Hofmann, Bikoff, & Robertson, 2008; Carver, Jiang, Lan, Oram, & Gridley, 2001; Sun, Meyers, Lewandoski, & Martin, 1999; Yamaguchi, Harpal, Henkemeyer, & Rossant, 1994). These mutants thus fail to generate both mesoderm and definitive endoderm. Interestingly, embryos lacking the apical polarity protein CRUMBS2 do not display defects in cell polarity, but instead exhibit a failure in the delamination of EPI cells, which remain attached to the apical surface of the epiblast through E-cadherin, and hence cannot migrate away (Ramkumar et al., 2016). Defects in the behavior of cells migrating away from the primitive streak have also been reported in embryos lacking the Striatin-interacting protein STRIP1, a core component of the Striatin-interacting phosphatases and kinase (STRIPAK) complex, which is involved in signaling, cell cycle control, cell polarity, vesicular trafficking and cell migration. *Strip1* mutants show changes in the actin cytoskeleton and abnormal focal adhesions leading to a decrease in the velocity of migrating cells, consequently resulting in truncations of the body axis (Bazzi, Soroka, Alcorn, & Anderson, 2017). A live imaging study demonstrated that in addition to having distinct molecular identities and cell fates, embryonic mesoderm vs extra-embryonic cells exiting the primitive streak exhibit distinct migratory behaviors, with the former being meandrous but directional, and the latter showing little net displacement (Saykali et al., 2019). Further live imaging of large cohorts of cells within gastrulating mouse embryos revealed coherent directional movement of cells away from the primitive streak (McDole et al., 2018).

5. Morphogenetic cell behaviors driving gut endoderm formation

Cell labeling and orthotopic grafting studies reported that cells with a definitive endoderm (DE) fate can be identified at the anterior region of the primitive streak (APS) at E7.0. Subsequently, at E7.25–E8.0, DE cells can be detected on the surface of the embryo and thereafter in the forming gut tube. The first DE cells that emerge from the primitive streak give rise to the foregut region, followed by cells that will comprise the mid- and hindgut regions

(Franklin et al., 2008; Lawson, Meneses, & Pedersen, 1986; Lawson & Pedersen, 1987). DE cells are one of the last populations to exit the primitive streak, around the same time as the lateral plate, intermediate and paraxial mesoderm populations.

Recent live imaging and genetic fate mapping studies have provided insights into the cellular dynamics driving gut endoderm morphogenesis. By contrast to the long-standing (displacement) model, which posited that VE cells would be displaced proximally by DE cells exiting the primitive streak and intercalating into the VE epithelium, live imaging revealed that VE cells remain within the distal embryonic region becoming dispersed (dispersal model, Kwon et al., 2008) and interspersed with DE cells (Fig. 5). During dispersal, VE cells move apart, likely due to the massive tissue expansion resulting from the growing embryo, as well as the concomitant intercalation of DE cells. DE cells arising at the primitive streak, likely undergo a partial EMT, since they do not lose but redistribute E-cadherin (Viotti et al., 2014). Within a short period of time, DE cells move away from the primitive streak, possibly "piggy-backing" on actively migrating mesoderm, and subsequently re-epithelialize as they intercalate into the VE epithelium. In this way, the gut endoderm of the E7.5 embryo comprises a mix of DE and VE cells (Fig. 5), and the VE lineage persists in the embryo and contributes to the midgestational gut tube (Kwon et al., 2008; Nowotschin et al., 2019).

Interestingly VE descendants exhibit an anisotropic distribution along the gut tube, with a higher percentage contributing to the hindgut, than mid- and foregut (Kwon et al., 2008; Nowotschin et al., 2019; Pijuan-Sala et al., 2019). The elevated contribution of the VE lineage to the hindgut could be explained by the observation that PVE cells are not dispersed and remain as a cohort during gastrulation, and thereafter become incorporated into the hindgut (Balmer, Nowotschin, & Hadjantonakis, 2016; Kwon et al., 2008).

Studies on the mechanism of DE-VE cell intercalation are beginning to elucidate the complex sequence of stereotypical cell behaviors driving this process. At the pre-streak stage, the VE is a cuboidal epithelium surrounding the columnar epithelial epiblast (Fig. 5). Once the primitive streak has formed (E6.5), and EPI cells begin to ingress forming the mesodermal layer migrating between EPI and the VE epithelia. At this point the embryo is massively growing, and changes in the morphology of the VE can begin to be observed as it adopts a more squamous flatter epithelial like morphology. Around E6.75–E7.0, DE cells will ingress in the vicinity of the anterior streak, move laterally within the mesodermal layer and finally intercalate into the overlying VE epithelium. For intercalation to occur DE and VE

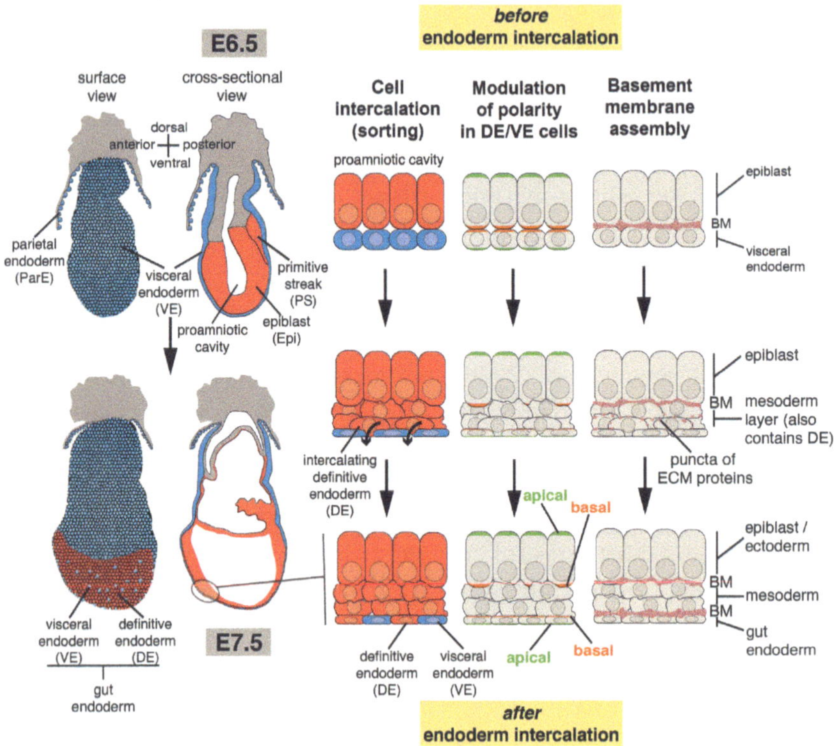

Fig. 5 Endoderm emerges twice during mouse development: the gastrula. The gut endoderm is one of the three layers generated at gastrulation. The gut endoderm is an epithelium on the surface of the mouse gastrula comprising cells of two distinct origins: primitive endoderm (PrE)-derived visceral endoderm (VE, blue) and move within the mesoderm layer and eventually emerge on the embryo's surface as they intercalate into the visceral endoderm (VE) epithelium thereby generating the gut endoderm. Morphogenesis of the gut endoderm involves modulation of apical-basal polarity in both DE and VE cell populations so as to facilitate intercalation, and the assembly of a basement membrane (BM) which segregates the (gut) endoderm tissue layer from the adjacent mesoderm. DE cells exhibit many similarities with the primitive endoderm (PrE) cells specified in the blastocyst including expression of molecular markers (e.g., FOX, SOX, and GATA transcription factors) and morphogenetic cell behaviors which include the acquisition of apical-basal polarity, and concomitant sorting to the tissue's surface.

cells must coordinate their behaviors. DE cells need to re-establish apical-basal polarity, and reassemble cell-cell junctions so as to undergo an MET and intercalate into the VE epithelium (Fig. 5). To facilitate this sequence, and in contrast to their mesoderm neighbors, DE cells only partially down-regulate the adherens junction apically-localized protein E-cadherin (Viotti et al., 2014). Given the time-frame between ingression at the primitive streak and intercalation into the VE epithelium on the embryo's surface (Fig. 5), it seems plausible that DE cells would undergo only a partial EMT enabling them to rapidly reassemble cell-cell junctions and re-establish polarity. Rapid and possibly partial EMT-MET cycles have been shown in tumor progression; from metastasis to intravasation (Ye & Weinberg, 2015), and merit further investigation in the context of gastrulation. This morphogenetic behavior also bears many similarities to epiboly movements that are well described in frogs and zebrafish.

Concomitantly, cells of the VE epithelium must transiently modulate their epithelial integrity; disassembling their cell-cell junctions, to allow DE cells to intercalate. The analysis of embryos lacking the SRY-box transcription factor SOX17 has provided some insight into the regulation of the de-epithelization and re-epithelialization process, as well as the association between gut endoderm morphogenesis and the assembly of a BM at the mesoderm-gut endoderm interface (see Section 6). *Sox17* mutants were initially reported to lack mid- and hindgut endoderm with cells of the gut tube possessing a morphology resembling VE rather than DE (Kanai-Azuma et al., 2002). Further analyses at earlier embryonic stages revealed a failure in VE dispersal (Viotti et al., 2012, 2014). Though EPI cells underwent a gastrulation EMT and were able to migrate away from the primitive streak, FOXA2-positive DE cells were unable to intercalate into the VE epithelium and consequently were retained within the mesodermal layer. The fate of these cells, which were unable to execute the DE morphogenetic program—whether they adopt a mesoderm identity or die—remains unknown. In the absence of SOX17 VE cells failed to relinquish their epithelial character, their cell-cell junctions remained intact, and their morphology remained cuboidal rather than becoming squamous (Viotti et al., 2014). These observations suggest cross-talk between DE and VE cells in these coordinate behaviors giving rise to gut endoderm.

Endocytosis has been shown to be important not just for the removal of activated receptors and ligands from a cell's surface, but also for relaying and modulating intercellular signals necessary for patterning of the embryo (reviewed in Bokel & Brand, 2014). The vitamin B12, albumin and Apolipoprotein A-I receptor CUBILIN is expressed throughout the VE. *Cubilin* mutants arrest at gastrulation, and have impaired endocytosis in VE cells, which fail to disperse. FOXA2-positive DE, as well as mesoderm cells accumulate at the primitive streak of *Cubilin* mutants, suggesting that cues from the adjacent VE, presumably the PVE, instruct nascent DE and mesoderm cells as they exit from the primitive streak (Perea-Gomez et al., 2019).

6. Gut endoderm morphogenesis coincides with germ layer segregation

With the formation of the three germ layers during gastrulation the embryo changes from comprising two tissue layers to three. Concomitantly, the embryo also transitions from having one basement membrane at the EPI-VE interface to one having two basement membranes separating the EPI-mesoderm and mesoderm-gut endoderm interfaces (Fig. 5). The basement membrane between the mesoderm-gut endoderm is formed when DE cells exit the mesoderm layer to intercalate into the overlying VE epithelium, most likely through de novo assembly of a BM, but not de novo transcription of BM proteins (Viotti et al., 2014). It is currently not known whether DE or VE cells, or both populations, lay down the new BM. Live imaging studies using fluorescently labeled extra-cellular matrix (ECM) components should help address this open question (Fig. 5).

7. Behavior of VE cells at the midline

VE cells at the embryonic midline exhibit a distinct behavior compared to their laterally positioned counterparts (reviewed in Balmer et al., 2016). Fate mapping studies using dye labeling revealed that VE cells overlying the midline move anteriorly not laterally (Thomas & Beddington, 1996), observations that were subsequently confirmed by imaging of embryos expressing a pan-VE (*Afp-GFP*) reporter (Kwon et al., 2006), which revealed the anterior displacement of VE by EPI-derived axial mesoderm, which gives rise to the node and notochord. Additional evidence for a distinct mechanism for midline morphogenesis has been provided by observations made in *Sox17* mutants where VE dispersal is affected, but midline displacement is not (Kanai-Azuma et al., 2002; Viotti et al., 2012).

Upon their displacement, VE cells align on either side of, and appear to decorate the midline (Kwon et al., 2008; Lee, Migeotte, & Anderson, 2010). It is tempting to speculate that this stereotypical arrangement may serve as structural scaffold for the invagination of the notochord or gut. The EPI-specific loss of the homeobox transcription factor LHX1 disrupts axial midline and node morphogenesis as well as resulting in anterior DE defects and delayed VE dispersal (Costello et al., 2015). Further studies will be needed to further investigate these midline behaviors and determine whether they are accomplished by active and coordinated behaviors of cell cohorts, as well as the underlying molecular mechanisms.

8. Single-cell transcriptomic studies support the dual origin of the gut endoderm

Several single-cell transcriptomic (scRNA-seq) studies of cells collected from mouse embryos have provided an unbiased computational investigation of the lineage contributions to the gut endoderm and confirm its dual origin (Chan et al., 2019; Nowotschin et al., 2019; Peng et al., 2019; Pijuan-Sala et al., 2019). These studies combined different approaches with scRNA-seq and pseudo-temporal analyses, including reporter-based lineage tracing (Nowotschin et al., 2019), barcoding-based recording of a cell's lineage history (Chan et al., 2019), or cataloging cells in defined spatial regions within the germ layers in a method termed GEO-seq (Peng et al., 2019). Notably these studies revealed that VE descendants acquired a transcriptional equivalence with their DE neighbors, and that VE and DE descendants become comparably patterned along the AP axis of the gut tube. Despite the overall transcriptional similarity of these two developmentally distinct populations, VE descendants still retain a memory of their lineage history through expression of a core set of genes including *Rhox5*, *Trap1a*, *Xlr3a*, *Cdkn2a* and *Ttr* (Chan et al., 2019; Nowotschin et al., 2019; Pijuan-Sala et al., 2019). An open question pertains the fate of VE cells. Future studies will need to address to what extent VE cells contribute to later gut tube derivatives beyond midgestation (Fig. 6), and whether they persist in the endoderm-derived visceral organs of the adult (Fig. 7).

9. The epigenetic landscape of embryonic (definitive) and extra-embryonic (visceral) endoderm

The epigenetic landscape during preimplantation development has been extensively described however, not much is known about the changes happening to the epigenome around the time of gastrulation (Lu et al., 2016; Messerschmidt, Knowles, & Solter, 2014; Smith & Meissner, 2013; Wu et al., 2016; Xu & Xie, 2018; Zhang et al., 2018). A single-cell nucleosome, methylome, transcriptome sequencing (scNMT-seq) study of transcriptome and epigenome dynamics in the mouse embryo between the late blastocyst through gastrulation stages has suggested a hierarchical epigenetic model of lineage decision making. Ectoderm is considered the default lineage, since

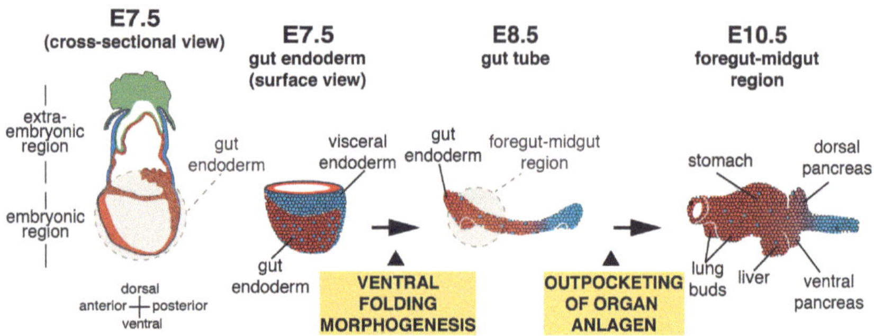

Fig. 6 The gut endoderm on the embryo's surface will go on to form the internal gut tube. After its formation on the surface of the embryo the gut endoderm will undergo a series of morphogenetic behaviors termed ventral folding, whereby it will transform into an internalized gut tube running the AP length of the embryo. Subsequent to its formation as an internalized tube, outpocketings emerge from the gut tube correlating with and representing organ-specific identities. These outpocketings reflect the initiation of organotypic morphogenetic programs.

pluripotent epiblast cells are primed at enhancer sites for ectoderm-specific genes. In this way, definitive endoderm and mesoderm identities need to be actively specified through exposure to a combination of signaling pathways at the primitive streak that induce demethylation and accessibility of chromatin at lineage-specific enhancers (Argelaguet et al., 2019).

An open question pertains the epigenetic status of VE descendants contributing to gut endoderm. In mice, the inactivation of the paternally inherited X-chromosome occurs at preimplantation in female embryos (cleavage stages ~E2.5) and is maintained in PrE and TE lineages (Fig. 1), while in EPI cells the inactivated X-chromosome is reactivated in a random fashion at E3.5–E5.5. Random X-inactivation of the either the paternally or maternally inherited X-chromosome is subsequently initiated in EPI cells (Mak et al., 2004; Okamoto, Otte, Allis, Reinberg, & Heard, 2004). Epiblast-derived DE cells will exhibit random X-inactivation, whereas in VE cells only the paternal X-chromosome will be inactivated. Since some VE cells contribute to gut endoderm of the embryo, it will be interesting to determine whether they retain their epigenetic status or whether they or their progeny undergo reprogramming/reversal of their X-chromosome inactivation status to become more like DE cells (by reactivation of both X-chromosomes, and consequent random inactivation).

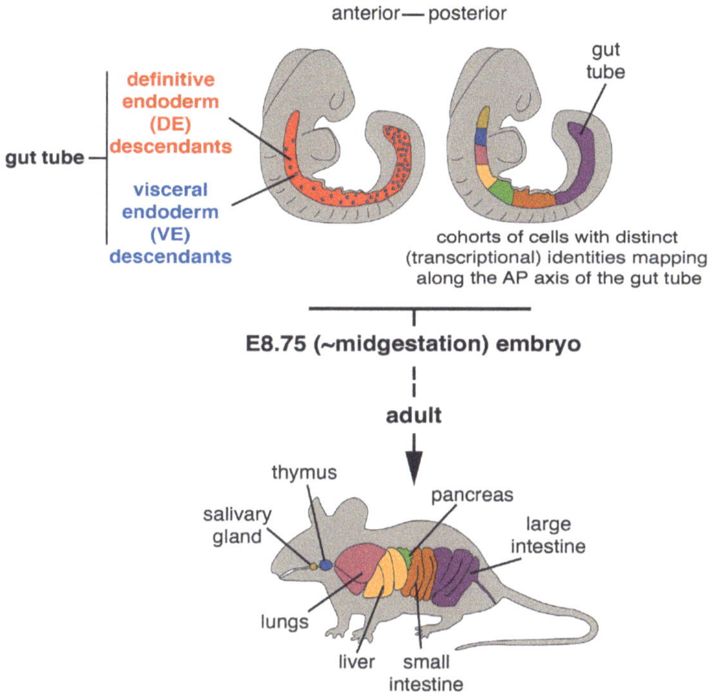

Fig. 7 The gut tube of the midgestation (~E8.75) mouse embryo is patterned into organ-specific territories. The gut tube comprises descendants of the pluripotent epiblast (EPI)-derived definitive endoderm (DE, red) and so-called extra-embryonic primitive endoderm (PrE)-derived visceral endoderm (VE, blue). Single-cell transcriptomic studies have revealed that at this stage the gut tube is already patterned along its AP axis into organ-specific territories mirroring the identity and stereotypical distribution of visceral organs along the body axis of adult animals. Cohorts of cells representing transcriptionally-distinct clusters positioned along the AP axis at E8.75 are color-coded and related to organotypic identities in the adult.

10. Mesendoderm and the mouse

A much-debated question pertains the existence of a bipotential cell population in the mouse that can give rise to both mesoderm and endoderm, as has been described for the fish (Warga & Nusslein-Volhard, 1999) and the chick (Hatada & Stern, 1994; Kimura, Yasugi, Stern, & Fukuda, 2006). Mesendoderm is generally considered to represent cells that have the potential to give rise to both mesoderm and definitive endoderm descendants (Tada et al., 2005). EPI cells, that are about to ingress at the anterior primitive streak already express FOXA2, a marker of DE and axial mesoderm, and the complementary expression of T and FOXA2, at the primitive streak suggests that these populations are segregated spatially as they acquire lineage identities (Burtscher & Lickert, 2009; Viotti et al., 2014). Clonal fate mapping studies found the majority (44/49) of endodermal clones did not exhibit other lineage contributions, suggesting the early segregation of definitive endoderm (Tzouanacou et al., 2009). However, support for a mesendoderm population has come from in vitro stem cell studies. Differentiation of human and mouse ES cells in the presence of ACTIVIN, a Tgf-β ligand used as a substitute for NODAL, gives rise to mesoderm and endoderm cells suggesting a common precursor (Kubo et al., 2004). However, these observations are difficult to disentangle as specification of mesoderm and definitive endoderm are intimately associated both temporally and spatially in the embryo. Mutations in various components of the Tgf-β signaling pathway, including *Nodal* or its downstream effectors, result in mesoderm as well as endoderm defects. Mutants in *Nodal* or its co-receptor *Cripto* result in a failure to form a primitive streak, and ensuing loss of both mesoderm and endoderm (Ding et al., 1998; Weinstein et al., 1998; Zhou, Sasaki, Lowe, Hogan, & Kuehn, 1993). Indeed, NODAL levels are critical for determining the fate of an EPI cell, with elevated levels biasing toward endoderm both in vivo at the anterior streak, and in vitro in stem cell differentiation paradigms. Thus, EPI cells at the primitive streak might indeed be bipotential, having the propensity to acquire both mesoderm and endoderm identities, but rather it is their position within the dynamic signaling landscape which dictates their fate (Fig. 3).

Wnt/β-catenin signaling has been implicated in directing the choice of EPI cells into either a mesoderm or endoderm identity. Endoderm-lineage-specific deletion of *β-catenin* leads to mesoderm-like cells within the endoderm layer suggesting a transfating of endoderm to mesoderm, specifically cardiac mesoderm (Lickert et al., 2002), or loss of endoderm when β-catenin is deleted in *Sox17*-expressing endoderm cells (Engert et al., 2013). In scenarios where cells are unable to commit to the endoderm lineage, perhaps they have the potential to become mesoderm? A *Foxa2* lineage tracing study identified descendants of *Foxa2*-expressing cells within the cardiac mesoderm (Bardot et al., 2017). It is therefore tempting to speculate that absence of β-catenin in cells fated to become definitive endoderm (DE), renders them unable intercalate into the VE epithelium, and that cells that express *Foxa2* have the propensity to adopt either a definitive endoderm or alternatively a mesoderm fate. Cells can only fully commit to an endoderm program upon activation of SOX17 which occurs just prior to their intercalation into the VE epithelium (Viotti et al., 2014). Failure in the activation of SOX17 or in intercalation might therefore result in cells being retained within the mesoderm layer (Fig. 5). Thus, might cells fated as DE, originating in the anterior primitive streak, have an expanded plasticity, and be able to modulate their fate if they unable to commit to a DE program?

11. Transforming the gut endoderm into the gut tube: Ventral folding

After its formation as an epithelial sheet on the surface of the embryo, the gut endoderm is internalized to form the gut tube in a process termed ventral folding morphogenesis (VFM) (Fig. 6). VFM includes the internalization of the gut endoderm, the closing of the body wall and the enclosing of the embryonic membranes around the embryo (reviewed by Gavrilov & Lacy, 2013). This wholescale morphogenetic event leads to the recognizable fetal position and the inversion of germ layers, whereby ectoderm is on the outside and endoderm is an internal tube.

The internalization of the gut tube initiates at the early somite stage with stereotypical cell movements leading to the formation of an anterior intestinal portal (AIP, also referred to as the foregut pocket) and subsequently a posterior (caudal) intestinal portal (CIP, also referred to as the hindgut pocket). After invagination of the AIP and CIP, the gut tube forms by extension of the anterior and posterior portals in a rostral-caudal and caudal-rostral manner (akin to a zippering toward the middle of the embryo), respectively. Fate mapping studies have been used to investigate the morphogenetic movements associated with gut tube formation and demonstrate that foregut and hindgut extensions are achieved by a similar repertoire of morphogenetic movements, achieved through folding of the lateral endoderm toward the ventral-medial side of the tube (Franklin et al., 2008).

The midgut region of the gut tube remains open at the site where the embryo is connected to the yolk sac until the completion of the internalization of the gut endoderm and the closure of the body wall, which happens in the mouse embryo through an axial rotation (turning) leading to the full enclosure of the embryo within its extra-embryonic membranes (visceral yolk sac and amnion, Fig. 1). Little is known of the mechanisms that drive the invagination of the gut endoderm. BMP2, a member of the TGF-β family, expressed within the emVE (Figs. 1 and 3) has been implicated in VFM.

VE-specific loss of BMP2 leads to a disorganized anterior morphology of the developing embryo (the heart is anterior and dorsal to the head, rather than posterior and ventral) leading to defects in the early steps of anterior foregut invagination (Madabhushi & Lacy, 2011). Since BMP2 has been suggested to regulate cell shape and cytoskeletal tension in human embryonic stem cell differentiation studies (Wang et al., 2012), it is plausible that it regulates the intrinsic cell behaviors driving VFM.

12. Patterning the gut tube and the emergence of organ identities

Fate mapping studies of gut endoderm cells using dye labeling at early somite stages showed regionalization of the gut endoderm along the AP axis. Specifically, anterior-axial gut endoderm contributes to dorsal foregut regions, whereas medial-axial endoderm cells around and lateral to the node populated dorsal midgut section, and posterior-axial endoderm, located posterior to the node, contributed to dorsal hindgut and posterior midgut regions. Cells from anterior-lateral regions, medial-lateral and posterior lateral regions contribute to ventral foregut, ventral foregut/midgut and ventral hindgut (Franklin et al., 2008).

Such fate mapping experiments did not however establish whether cells had acquired molecular identities corresponding to their location along the AP axis of the gut tube. A series of scRNA-seq studies have investigated emergent patterning within the gut endoderm. These studies demonstrated that as early as E7.5, when the gut endoderm epithelial sheet has just formed on the embryo's surface (Fig. 4), an anterior-posterior identity starts to be established well before a gut tube is generated (Nowotschin et al., 2019). By E8.25–E8.5 transcriptionally-distinct and AP regionalized cell populations were readily identified, with these regions corresponding to organ-specific territories (Nowotschin et al., 2019). Analysis of various signaling pathway components supported the predominantly posterior expression of key FGF, WNT and NOTCH signaling components, whereas BMP, HH, JAK/STAT and HIPPO pathways showed no regionalization in their activation and could be found at multiple locations along the gut tube. Besides the non-cell-autonomous signals, a small number of core transcription factors have been shown to be predictors of AP pattern as well suggesting that transcriptional regulation, likely in response to signals from the surrounding mesenchyme, plays an important role in gut tube patterning. Interestingly, AP patterning in the midgestational gut tube appears to be independent, or precedes, the establishment of a *Hox* code. Gut endoderm cells of distinct developmental origins are intermixed along the AP axis of the gut tube. By this time, DE and VE descendants have attained transcriptional equivalence and become similarly patterned.

13. Concluding remarks

Recent studies have shed new light on and revived interest in the formation of the gut endoderm at gastrulation by revealing novel insights into its dual developmental origin, and the novel morphogenetic behaviors which drive its formation, through a mechanism which exhibits many parallels with the epiboly movements characterized in frogs and fish. Signaling cross-talk between adjacent cell layers is a signature feature of the intrinsic programs that drive mammalian development and gastrulation mechanisms in particular. Cross-talk also likely operates across the emergent gut endoderm to potentiate a transcriptional equivalence attained by endodermal cells of distinct developmental origins. Future studies are likely to glean further details of the similarities and distinctions between gastrulation mechanisms operating in mice, vs other vertebrates. Moreover, with a heightened interest in the development of higher mammals, equipped with a foundational knowledge of events operating in the mouse, we can now move forward investigating the long-standing open question of how gastrulation is orchestrated—from inputs to cell behaviors to lineage relationships—in higher mammals, such as non-human primates and humans.

Acknowledgments

We thank Evan Bardot and Lila Solnica-Krezel for critical reading and insightful comments on this review. Grant sponsors: NIH R01-HD094868, R01-DK084391 and P30-CA008748.

References

Abe, T., Kutsuna, N., Kiyonari, H., Furuta, Y., & Fujimori, T. (2018). ROSA26 reporter mouse lines and image analyses reveal distinct region-specific cell behaviors in the visceral endoderm. *Development (Cambridge, England)*, *145*.

Argelaguet, R., Mohammed, H., Clark, S. J., Stapel, L. C., Krueger, C., Kapourani, C.-A., et al. (2019). Single cell multi-omics profiling reveals a hierarchical epigenetic landscape during mammalian germ layer specification. *bioRxiv*.

Arnold, S. J., Hofmann, U. K., Bikoff, E. K., & Robertson, E. J. (2008). Pivotal roles for eomesodermin during axis formation, epithelium-to-mesenchyme transition and endoderm specification in the mouse. *Development (Cambridge, England)*, *135*, 501–511.

Arnold, S. J., & Robertson, E. J. (2009). Making a commitment: Cell lineage allocation and axis patterning in the early mouse embryo. *Nature Reviews. Molecular Cell Biology*, *10*, 91–103.

Artus, J., Piliszek, A., & Hadjantonakis, A. K. (2011). The primitive endoderm lineage of the mouse blastocyst: Sequential transcription factor activation and regulation of differentiation by Sox17. *Developmental Biology*, *350*, 393–404.

Balmer, S., Nowotschin, S., & Hadjantonakis, A. K. (2016). Notochord morphogenesis in mice: Current understanding & open questions. *Developmental Dynamics*, *245*, 547–557.

Bardot, E., Calderon, D., Santoriello, F., Han, S., Cheung, K., Jadhav, B., et al. (2017). Foxa2 identifies a cardiac progenitor population with ventricular differentiation potential. *Nature Communications*, *8*, 14428.

Bazzi, H., Soroka, E., Alcorn, H. L., & Anderson, K. V. (2017). STRIP1, a core component of STRIPAK complexes, is essential for normal mesoderm migration in the mouse embryo. *Proceedings of the National Academy of Sciences of the United States of America*, *114*, E10928–E10936.

Blakeley, P., Fogarty, N. M., del Valle, I., Wamaitha, S. E., Hu, T. X., Elder, K., et al. (2015). Defining the three cell lineages of the human blastocyst by single-cell RNA-seq. *Development (Cambridge, England)*, *142*, 3151–3165.

Bokel, C., & Brand, M. (2014). Endocytosis and signaling during development. *Cold Spring Harbor Perspectives in Biology*, *6*.

Burtscher, I., & Lickert, H. (2009). Foxa2 regulates polarity and epithelialization in the endoderm germ layer of the mouse embryo. *Development (Cambridge, England)*, *136*, 1029–1038.

Carver, E. A., Jiang, R., Lan, Y., Oram, K. F., & Gridley, T. (2001). The mouse snail gene encodes a key regulator of the epithelial-mesenchymal transition. *Molecular and Cellular Biology*, *21*, 8184–8188.

Chan, M. M., Smith, Z. D., Grosswendt, S., Kretzmer, H., Norman, T. M., Adamson, B., et al. (2019). Molecular recording of mammalian embryogenesis. *Nature*, *570*, 77–82.

Chazaud, C., & Yamanaka, Y. (2016). Lineage specification in the mouse preimplantation embryo. *Development (Cambridge, England)*, *143*, 1063–1074.

Chazaud, C., Yamanaka, Y., Pawson, T., & Rossant, J. (2006). Early lineage segregation between epiblast and primitive endoderm in mouse blastocysts through the Grb2-MAPK pathway. *Developmental Cell*, *10*, 615–624.

Costello, I., Nowotschin, S., Sun, X., Mould, A. W., Hadjantonakis, A. K., Bikoff, E. K., et al. (2015). Lhx1 functions together with Otx2, Foxa2, and Ldb1 to govern anterior mesendoderm, node, and midline development. *Genes & Development*, *29*, 2108–2122.

Ding, J., Yang, L., Yan, Y. T., Chen, A., Desai, N., Wynshaw-Boris, A., et al. (1998). Cripto is required for correct orientation of the anterior-posterior axis in the mouse embryo. *Nature*, *395*, 702–707.

Engert, S., Burtscher, I., Liao, W. P., Dulev, S., Schotta, G., & Lickert, H. (2013). Wnt/beta-catenin signalling regulates Sox17 expression and is essential for organizer and endoderm formation in the mouse. *Development (Cambridge, England)*, *140*, 3128–3138.

Franklin, V., Khoo, P. L., Bildsoe, H., Wong, N., Lewis, S., & Tam, P. P. (2008). Regionalisation of the endoderm progenitors and morphogenesis of the gut portals of the mouse embryo. *Mechanisms of Development*, *125*, 587–600.

Freyer, L., Schroter, C., Saiz, N., Schrode, N., Nowotschin, S., Martinez-Arias, A., et al. (2015). A loss-of-function and H2B-Venus transcriptional reporter allele for Gata6 in mice. *BMC Developmental Biology*, *15*, 38.

Gavrilov, S., & Lacy, E. (2013). Genetic dissection of ventral folding morphogenesis in mouse: Embryonic visceral endoderm-supplied BMP2 positions head and heart. *Current Opinion in Genetics & Development*, *23*, 461–469.

Gerbe, F., Cox, B., Rossant, J., & Chazaud, C. (2008). Dynamic expression of Lrp2 pathway members reveals progressive epithelial differentiation of primitive endoderm in mouse blastocyst. *Developmental Biology, 313*, 594–602.

Hatada, Y., & Stern, C. D. (1994). A fate map of the epiblast of the early chick embryo. *Development (Cambridge, England), 120*, 2879–2889.

Kanai-Azuma, M., Kanai, Y., Gad, J. M., Tajima, Y., Taya, C., Kurohmaru, M., et al. (2002). Depletion of definitive gut endoderm in Sox17-null mutant mice. *Development (Cambridge, England), 129*, 2367–2379.

Kimura, W., Yasugi, S., Stern, C. D., & Fukuda, K. (2006). Fate and plasticity of the endoderm in the early chick embryo. *Developmental Biology, 289*, 283–295.

Kimura-Yoshida, C., Nakano, H., Okamura, D., Nakao, K., Yonemura, S., Belo, J. A., et al. (2005). Canonical Wnt signaling and its antagonist regulate anterior-posterior axis polarization by guiding cell migration in mouse visceral endoderm. *Developmental Cell, 9*, 639–650.

Kinder, S. J., Tsang, T. E., Quinlan, G. A., Hadjantonakis, A. K., Nagy, A., & Tam, P. P. (1999). The orderly allocation of mesodermal cells to the extraembryonic structures and the anteroposterior axis during gastrulation of the mouse embryo. *Development (Cambridge, England), 126*, 4691–4701.

Kubo, A., Shinozaki, K., Shannon, J. M., Kouskoff, V., Kennedy, M., Woo, S., et al. (2004). Development of definitive endoderm from embryonic stem cells in culture. *Development (Cambridge, England), 131*, 1651–1662.

Kwon, G. S., Fraser, S. T., Eakin, G. S., Mangano, M., Isern, J., Sahr, K. E., et al. (2006). Tg(Afp-GFP) expression marks primitive and definitive endoderm lineages during mouse development. *Developmental Dynamics, 235*, 2549–2558.

Kwon, G. S., Viotti, M., & Hadjantonakis, A. K. (2008). The endoderm of the mouse embryo arises by dynamic widespread intercalation of embryonic and extraembryonic lineages. *Developmental Cell, 15*, 509–520.

Lawson, K. A., Meneses, J. J., & Pedersen, R. A. (1986). Cell fate and cell lineage in the endoderm of the presomite mouse embryo, studied with an intracellular tracer. *Developmental Biology, 115*, 325–339.

Lawson, K. A., Meneses, J. J., & Pedersen, R. A. (1991). Clonal analysis of epiblast fate during germ layer formation in the mouse embryo. *Development (Cambridge, England), 113*, 891–911.

Lawson, K. A., & Pedersen, R. A. (1987). Cell fate, morphogenetic movement and population kinetics of embryonic endoderm at the time of germ layer formation in the mouse. *Development (Cambridge, England), 101*, 627–652.

Lee, J. D., Migeotte, I., & Anderson, K. V. (2010). Left-right patterning in the mouse requires Epb4.1l5-dependent morphogenesis of the node and midline. *Developmental Biology, 346*, 237–246.

Lickert, H., Kutsch, S., Kanzler, B., Tamai, Y., Taketo, M. M., & Kemler, R. (2002). Formation of multiple hearts in mice following deletion of beta-catenin in the embryonic endoderm. *Developmental Cell, 3*, 171–181.

Liu, P., Wakamiya, M., Shea, M. J., Albrecht, U., Behringer, R. R., & Bradley, A. (1999). Requirement for Wnt3 in vertebrate axis formation. *Nature Genetics, 22*, 361–365.

Lu, F., Liu, Y., Inoue, A., Suzuki, T., Zhao, K., & Zhang, Y. (2016). Establishing chromatin regulatory landscape during mouse preimplantation development. *Cell, 165*, 1375–1388.

Madabhushi, M., & Lacy, E. (2011). Anterior visceral endoderm directs ventral morphogenesis and placement of head and heart via BMP2 expression. *Developmental Cell, 21*, 907–919.

Mak, W., Nesterova, T. B., de Napoles, M., Appanah, R., Yamanaka, S., Otte, A. P., et al. (2004). Reactivation of the paternal X chromosome in early mouse embryos. *Science, 303*, 666–669.

McDole, K., Guignard, L., Amat, F., Berger, A., Malandain, G., Royer, L. A., et al. (2018). In toto imaging and reconstruction of post-implantation mouse development at the single-cell level. *Cell, 175*, 859–876.e33.

Meilhac, S. M., Adams, R. J., Morris, S. A., Danckaert, A., Le Garrec, J. F., & Zernicka-Goetz, M. (2009). Active cell movements coupled to positional induction are involved in lineage segregation in the mouse blastocyst. *Developmental Biology, 331*, 210–221.

Messerschmidt, D. M., Knowles, B. B., & Solter, D. (2014). DNA methylation dynamics during epigenetic reprogramming in the germline and preimplantation embryos. *Genes & Development, 28*, 812–828.

Migeotte, I., Omelchenko, T., Hall, A., & Anderson, K. V. (2010). Rac1-dependent collective cell migration is required for specification of the anterior-posterior body axis of the mouse. *PLoS Biology, 8*, e1000442.

Mohammed, H., Hernando-Herraez, I., Savino, A., Scialdone, A., Macaulay, I., Mulas, C., et al. (2017). Single-cell landscape of transcriptional heterogeneity and cell fate decisions during mouse early gastrulation. *Cell Reports, 20*, 1215–1228.

Nowotschin, S., Costello, I., Piliszek, A., Kwon, G. S., Mao, C. A., Klein, W. H., et al. (2013). The T-box transcription factor Eomesodermin is essential for AVE induction in the mouse embryo. *Genes & Development, 27*, 997–1002.

Nowotschin, S., Setty, M., Kuo, Y. Y., Liu, V., Garg, V., Sharma, R., et al. (2019). The emergent landscape of the mouse gut endoderm at single-cell resolution. *Nature, 569*, 361–367.

Okamoto, I., Otte, A. P., Allis, C. D., Reinberg, D., & Heard, E. (2004). Epigenetic dynamics of imprinted X inactivation during early mouse development. *Science, 303*, 644–649.

Omelchenko, T., Rabadan, M. A., Hernandez-Martinez, R., Grego-Bessa, J., Anderson, K. V., & Hall, A. (2014). beta-Pix directs collective migration of anterior visceral endoderm cells in the early mouse embryo. *Genes & Development, 28*, 2764–2777.

Pearce, J. J., & Evans, M. J. (1999). Mml, a mouse Mix-like gene expressed in the primitive streak. *Mechanisms of Development, 87*, 189–192.

Peng, G., Suo, S., Cui, G., Yu, F., Wang, R., Chen, J., et al. (2019). Molecular architecture of lineage allocation and tissue organization in early mouse embryo. *Nature, 572*, 528–532.

Perea-Gomez, A., Cases, O., Lelievre, V., Pulina, M. V., Collignon, J., Hadjantonakis, A. K., et al. (2019). Loss of Cubilin, the intrinsic factor-vitamin B12 receptor, impairs visceral endoderm endocytosis and endodermal patterning in the mouse. *Scientific Reports, 9*, 10168.

Pijuan-Sala, B., Griffiths, J. A., Guibentif, C., Hiscock, T. W., Jawaid, W., Calero-Nieto, F. J., et al. (2019). A single-cell molecular map of mouse gastrulation and early organogenesis. *Nature, 566*, 490–495.

Plusa, B., Piliszek, A., Frankenberg, S., Artus, J., & Hadjantonakis, A. K. (2008). Distinct sequential cell behaviours direct primitive endoderm formation in the mouse blastocyst. *Development (Cambridge, England), 135*, 3081–3091.

Rakeman, A. S., & Anderson, K. V. (2006). Axis specification and morphogenesis in the mouse embryo require Nap1, a regulator of WAVE-mediated actin branching. *Development (Cambridge, England), 133*, 3075–3083.

Ramkumar, N., Omelchenko, T., Silva-Gagliardi, N. F., McGlade, C. J., Wijnholds, J., & Anderson, K. V. (2016). Crumbs2 promotes cell ingression during the epithelial-to-mesenchymal transition at gastrulation. *Nature Cell Biology, 18*, 1281–1291.

Rivera-Perez, J. A., Mager, J., & Magnuson, T. (2003). Dynamic morphogenetic events characterize the mouse visceral endoderm. *Developmental Biology, 261*, 470–487.

Rivera-Perez, J. A., & Magnuson, T. (2005). Primitive streak formation in mice is preceded by localized activation of Brachyury and Wnt3. *Developmental Biology, 288*, 363–371.

Robb, L., Hartley, L., Begley, C. G., Brodnicki, T. C., Copeland, N. G., Gilbert, D. J., et al. (2000). Cloning, expression analysis, and chromosomal localization of murine and human homologues of a Xenopus mix gene. *Developmental Dynamics, 219*, 497–504.

Rodriguez, T. A., Srinivas, S., Clements, M. P., Smith, J. C., & Beddington, R. S. (2005). Induction and migration of the anterior visceral endoderm is regulated by the extra-embryonic ectoderm. *Development (Cambridge, England), 132*, 2513–2520.

Rula, M. E., Cai, K. Q., Moore, R., Yang, D. H., Staub, C. M., Capo-Chichi, C. D., et al. (2007). Cell autonomous sorting and surface positioning in the formation of primitive endoderm in embryoid bodies. *Genesis, 45*, 327–338.

Salgueiro, A. M., Filipe, M., & Belo, J. A. (2006). N-acetylgalactosamine 4-sulfate 6-O-sulfotransferase expression during early mouse embryonic development. *The International Journal of Developmental Biology, 50*, 705–708.

Saykali, B., Mathiah, N., Nahaboo, W., Racu, M. L., Hammou, L., Defrance, M., et al. (2019). Distinct mesoderm migration phenotypes in extra-embryonic and embryonic regions of the early mouse embryo. *eLife, 8*, e42434.

Schrode, N., Xenopoulos, P., Piliszek, A., Frankenberg, S., Plusa, B., & Hadjantonakis, A. K. (2013). Anatomy of a blastocyst: Cell behaviors driving cell fate choice and morphogenesis in the early mouse embryo. *Genesis, 51*, 219–233.

Scialdone, A., Tanaka, Y., Jawaid, W., Moignard, V., Wilson, N. K., Macaulay, I. C., et al. (2016). Resolving early mesoderm diversification through single-cell expression profiling. *Nature, 535*, 289–293.

Shioi, G., Hoshino, H., Abe, T., Kiyonari, H., Nakao, K., Meng, W., et al. (2017). Apical constriction in distal visceral endoderm cells initiates global, collective cell rearrangement in embryonic visceral endoderm to form anterior visceral endoderm. *Developmental Biology, 429*, 20–30.

Simon, C. S., Hadjantonakis, A. K., & Schroter, C. (2018). Making lineage decisions with biological noise: Lessons from the early mouse embryo. *Wiley Interdisciplinary Reviews: Developmental Biology, 7*, e319.

Simon, C. S., Zhang, L., Wu, T., Cai, W., Saiz, N., Nowotschin, S., et al. (2018). A Gata4 nuclear GFP transcriptional reporter to study endoderm and cardiac development in the mouse. *Biology Open, 7*.

Smith, Z. D., & Meissner, A. (2013). DNA methylation: Roles in mammalian development. *Nature Reviews. Genetics, 14*, 204–220.

Snow, M. H. L. (1977). Gastrulation in the mouse: Growth and regionalization of the epiblast. *Journal of Embryology and Experimental Morphology, 42*, 293–303.

Srinivas, S., Rodriguez, T., Clements, M., Smith, J. C., & Beddington, R. S. (2004). Active cell migration drives the unilateral movements of the anterior visceral endoderm. *Development (Cambridge, England), 131*, 1157–1164.

Sun, X., Meyers, E. N., Lewandoski, M., & Martin, G. R. (1999). Targeted disruption of Fgf8 causes failure of cell migration in the gastrulating mouse embryo. *Genes & Development, 13*, 1834–1846.

Tada, S., Era, T., Furusawa, C., Sakurai, H., Nishikawa, S., Kinoshita, M., et al. (2005). Characterization of mesendoderm: A diverging point of the definitive endoderm and mesoderm in embryonic stem cell differentiation culture. *Development (Cambridge, England), 132*, 4363–4374.

Takaoka, K., Nishimura, H., & Hamada, H. (2017). Both Nodal signalling and stochasticity select for prospective distal visceral endoderm in mouse embryos. *Nature Communications, 8*, 1492.

Takaoka, K., Yamamoto, M., & Hamada, H. (2011). Origin and role of distal visceral endoderm, a group of cells that determines anterior-posterior polarity of the mouse embryo. *Nature Cell Biology, 13*, 743–752.

Takaoka, K., Yamamoto, M., Shiratori, H., Meno, C., Rossant, J., Saijoh, Y., et al. (2006). The mouse embryo autonomously acquires anterior-posterior polarity at implantation. *Developmental Cell, 10*, 451–459.

Thomas, P., & Beddington, R. (1996). Anterior primitive endoderm may be responsible for patterning the anterior neural plate in the mouse embryo. *Current Biology, 6*, 1487–1496.

Torres-Padilla, M. E., Richardson, L., Kolasinska, P., Meilhac, S. M., Luetke-Eversloh, M. V., & Zernicka-Goetz, M. (2007). The anterior visceral endoderm of the mouse embryo is established from both preimplantation precursor cells and by de novo gene expression after implantation. *Developmental Biology, 309*, 97–112.

Trichas, G., Joyce, B., Crompton, L. A., Wilkins, V., Clements, M., Tada, M., et al. (2011). Nodal dependent differential localisation of dishevelled-2 demarcates regions of differing cell behaviour in the visceral endoderm. *PLoS Biology, 9*, e1001019.

Tzouanacou, E., Wegener, A., Wymeersch, F. J., Wilson, V., & Nicolas, J. F. (2009). Redefining the progression of lineage segregations during mammalian embryogenesis by clonal analysis. *Developmental Cell, 17*, 365–376.

Viotti, M., Niu, L., Shi, S. H., & Hadjantonakis, A. K. (2012). Role of the gut endoderm in relaying left-right patterning in mice. *PLoS Biology, 10*, e1001276.

Viotti, M., Nowotschin, S., & Hadjantonakis, A. K. (2011). Afp::mCherry, a red fluorescent transgenic reporter of the mouse visceral endoderm. *Genesis, 49*, 124–133.

Viotti, M., Nowotschin, S., & Hadjantonakis, A. K. (2014). SOX17 links gut endoderm morphogenesis and germ layer segregation. *Nature Cell Biology, 16*, 1146–1156.

Voiculescu, O., Bodenstein, L., Lau, I. J., & Stern, C. D. (2014). Local cell interactions and self-amplifying individual cell ingression drive amniote gastrulation. *eLife, 3*, e01817.

Wang, Y. K., Yu, X., Cohen, D. M., Wozniak, M. A., Yang, M. T., Gao, L., et al. (2012). Bone morphogenetic protein-2-induced signaling and osteogenesis is regulated by cell shape, RhoA/ROCK, and cytoskeletal tension. *Stem Cells and Development, 21*, 1176–1186.

Warga, R. M., & Nusslein-Volhard, C. (1999). Origin and development of the zebrafish endoderm. *Development (Cambridge, England), 126*, 827–838.

Weinstein, M., Yang, X., Li, C., Xu, X., Gotay, J., & Deng, C. X. (1998). Failure of egg cylinder elongation and mesoderm induction in mouse embryos lacking the tumor suppressor smad2. *Proceedings of the National Academy of Sciences of the United States of America, 95*, 9378–9383.

Wu, J., Huang, B., Chen, H., Yin, Q., Liu, Y., Xiang, Y., et al. (2016). The landscape of accessible chromatin in mammalian preimplantation embryos. *Nature, 534*, 652–657.

Xu, Q., & Xie, W. (2018). Epigenome in early mammalian development: Inheritance, reprogramming and establishment. *Trends in Cell Biology, 28*, 237–253.

Yamaguchi, T. P., Harpal, K., Henkemeyer, M., & Rossant, J. (1994). fgfr-1 is required for embryonic growth and mesodermal patterning during mouse gastrulation. *Genes & Development, 8*, 3032–3044.

Ye, X., & Weinberg, R. A. (2015). Epithelial-mesenchymal plasticity: A central regulator of cancer progression. *Trends in Cell Biology, 25*, 675–686.

Zhang, Y., Xiang, Y., Yin, Q., Du, Z., Peng, X., Wang, Q., et al. (2018). Dynamic epigenomic landscapes during early lineage specification in mouse embryos. *Nature Genetics, 50*, 96–105.

Zhou, X., Sasaki, H., Lowe, L., Hogan, B. L., & Kuehn, M. R. (1993). Nodal is a novel TGF-beta-like gene expressed in the mouse node during gastrulation. *Nature, 361*, 543–547.

FSC
www.fsc.org
MIX
Papier | Fördert gute Waldnutzung
FSC® C083411

Kolibri 360 GmbH
Ferdinand-Jühlke-Straße 7
99095 Erfurt, Deutschland
produktsicherheit@kolibri360.de